PEOPLE AND THE EARTH

As citizens of earth we have a vital relationship with our planet. While our numbers grow and our resource demands increase, we place greater stress on this planet's fragile environment. The earth's ability to provide essential products and absorb waste materials is a paramount concern for all of its inhabitants.

People and the Earth examines the numerous ways in which this planet enhances and limits our life-styles. The authors look at the geologic restrictions on our ability to withdraw resources – food, water, energy, and minerals – from the earth, the effect human activity has on the earth, and the lingering damage caused by natural disasters. *People and the Earth* examines the basic components of our interaction with this planet, provides a lucid, scientific discussion of each one, and speculates on what the future may hold. It provides the fundamental concepts that will enable us to make wise and conscientious choices on how to live our day-to-day lives.

Written with wit and remarkable insight, and illustrated with numerous case histories, this book provides a balanced view of the complex environmental issues facing our civilization. *People and the Earth* is an ideal introductory textbook and will also appeal to anyone concerned with our evolving relationship to the earth.

PEOPLE AND THE EARTH

BASIC ISSUES IN THE SUSTAINABILITY OF RESOURCES AND ENVIRONMENT

JOHN J. W. ROGERS
University of North Carolina,
Chapel Hill

P. GEOFFREY FEISS
College of William and Mary,
Williamsburg, Virginia

CAMBRIDGE
UNIVERSITY PRESS

PUBLISHED BY THE PRESS SYNDICATE OF THE UNIVERSITY OF CAMBRIDGE
The Pitt Building, Trumpington Street, Cambridge CB2 1RP, United Kingdom

CAMBRIDGE UNIVERSITY PRESS
The Edinburgh Building, Cambridge CB2 2RU, UK   www.cup.cam.ac.uk
40 West 20th Street, New York, NY 10011-4211, USA   www.cup.org
10 Stamford Road, Oakleigh, Melbourne 3166, Australia
Ruiz de Alarcón 13,28014 Madrid, Spain

First published 1998
Reprinted 1999
Printed in the United States of America

Typeset in Goudy and Futura

Library of Congress Cataloging-in-Publication Data
Rogers, John J. W. (John James William), 1930–
    People and the earth : basic issues in the sustainability of
resources and environment / John J.W. Rogers, P. Geoffrey Feiss.
        p.   cm.
    Includes bibliographical references.
    ISBN 0-521-56028-4 (hc). – ISBN 0-521-56872-2 (pbk.)
    1. Sustainable development.   2. Population – Environmental aspects.
3. Natural resources – Management.   I. Feiss, P. Geoffrey (Paul
Geoffrey), 1943– . II. Title.
HC79.E5R6313   1998
333.7 – dc21   97-25901
                                                                CIP

*A catalog record for this book is available from
the British Library*

ISBN 0 521 56028 4 hardback
ISBN 0 521 56872 2 paperback

CONTENTS

List of boxes		*page* xi
List of figures and tables		xiii
Preface		xix
Units		xxi
1	**PEOPLE AND LAND**	1
	1.0 Introduction	1
	1.1 Population densities and land use	1
	Variations in population densities	3
	Land use	4
	1.2 History of world population and growth rates	5
	1.3 Variation in population characteristics in the modern world	11
	Birth, fertility, and growth rates	11
	Life expectancy and age distributions	14
	Policy questions	15
	1.4 Population and wealth	15
	Policy questions	17
	1.5 Too many people? A comparison of lesser-developed countries with industrial countries	17
	Education and employment	17
	Income and wealth	20
	Women	20
	Policy questions	21
	1.6 Too many people? A discussion of the population of the United States	21
	Population in the 20th century	21
	Occupations and life-styles	22
	Income and race	25
	Policy questions	26
	1.7 Population control	26
	Carrying capacity	26

Quality of life	27
Population control	28
Policy questions	29
Problems	31
References	32

2 FOOD — 33

2.0 Introduction	33
2.1 Human dietary requirements	34
Water	34
Carbohydrates	34
Fats (lipids)	34
Proteins	37
Specific elements – minerals	37
Special chemicals	38
Policy questions	40
2.2 Crop production	40
Energy and plant food	40
Requirements for crop production	42
Grain production	45
Nongrain plants	49
Policy questions	52
2.3 Animal production	52
The commons	52
Animal raising	53
Policy questions	55
2.4 Fishing	55
The nurture of fish	55
The fishing industry	59
Animals and fish in human nutrition	61
Policy questions	63
2.5 Money and agriculture	63
Environmental issues	63
Money	65
The future	68
Policy questions	69
Problems	69
References	70

3 NATURAL HAZARDS — 71

3.0 Introduction	71
3.1 Earth's engines of change and their consequences	73
The consequences of change and the recognition of hazards	76
3.2 Thinking about risk	78
Natural hazards as a societal risk	79
Policy questions	81
3.3 Earthquakes	82
Causes and kinds of earthquakes	83
Destruction associated with earthquakes	83

Risk analysis and prevention of earthquake destruction 87
Policy questions 89
3.4 Volcanic eruptions 89
The nature of volcanic eruptions 89
Volcanic hazards 91
Risk and prediction 93
Policy questions 96
3.5 Tsunamis 96
Policy questions 97
3.6 River floods 98
Seasonal floods 98
Randomly occurring floods 99
Floods induced by human activity 104
Policy questions 104
3.7 Coastal flooding and erosion 104
Coastlines at risk 106
Human activity and the coastline 112
Policy questions 113
3.8 Landslides and related processes 113
Causes of rapid downslope movement of earth materials 116
Human influences on downslope movement 118
Policy questions 119
3.9 Who pays for natural disasters? 119
Policy questions 121
3.10 Conclusion 121
Problems 122
References 123

4 WATER 125
4.0 Introduction 125
4.1 Abundance of water 126
Global considerations 126
The (mostly) freshwater budget 127
Regional considerations of freshwater availability 130
4.2 Human uses of water 131
Where do we get water? 132
What do we do with water? 135
Choices versus necessities in water use 137
Policy questions 138
4.3 The surface water resource 138
Amount of surface runoff 138
Drainage basins and watersheds 139
Extraction of water for human use and the construction
of dams 141
The Colorado: A parable of a river 143
Policy questions 145
4.4 The groundwater resource 145
The geology of groundwater 147
Extraction of groundwater 150
A case study of groundwater use: The High Plains aquifer 151

           Policy questions   155
    4.5 Water quality   155
           Drinking water   155
           Other uses   159
           Policy questions   160
    4.6 What do societies do to control and apportion water?   160
           Who owns the water?   161
           Policy questions   165
    Problems   165
    References   167

5   **ENERGY**   168
    5.0 Introduction   168
    5.1 Temperature, energy, heat, and power   169
    5.2 Types of energy sources   171
           Direct energy   171
           Electricity   173
           Stored energy   177
           Policy questions   179
    5.3 Fossil fuels   180
           Products of decay of organic matter   180
           Combustion   180
           Geology of oil and natural gas   182
           Distribution and reserves of oil and gas   187
           Production, consumption, and transportation of oil   189
           Oil shale and tar sand   193
           Geology and distribution of coal   197
           Policy questions   199
    5.4 Nuclear power   200
           Radioactive decay   200
           Fission   201
           Fusion   202
           Nuclear reactors and bombs   202
           Fuel supplies for reactors   205
           Policy questions   206
    5.5 Present and future energy use   206
           Energy and the economy   206
           Present energy use   207
           Future energy use   209
           Policy questions   211
    Problems   211
    References   212

6   **MINERAL RESOURCES**   214
    6.0 Introduction   214
    6.1 Definition and characteristics of mineral resources   215
           Sustainable versus nonsustainable resources   215
           Scarcity of resources   217
           Resource and reserve inventories   218

Fixed location of mineral resources   221
Discovery risk   221
Policy questions   223
6.2 Classification and formation of mineral deposits   223
Classification   223
Geological ore-forming processes   225
6.3 Location of nonfuel mineral resources   233
Policy questions   235
6.4 Methods of mining ore and preparing it for market   235
Policy questions   239
6.5 Legal and property issues   240
Regalian versus accessory rights to minerals   240
The problem of the public lands in the United States   241
Policy questions   245
6.6 Thoughts on minerals and the future   245
Problems   245
References   246

7 WASTE AND POLLUTION   247
7.0 Introduction   247
7.1 Dispersal of waste in soil, surface water, groundwater, and air   248
7.2 Bulk waste   249
Policy questions   254
7.3 Hazardous chemicals   254
Industrial chemicals   255
Pesticides and herbicides   257
Disposal of hazardous waste   261
Policy questions   265
7.4 Radioactive waste   267
Dangers of radioactive waste   267
High-level nuclear waste   269
Low-level waste   273
Policy questions   274
7.5 Biologically active waste   274
Sewage   274
Excess nutrients (eutrophication)   277
Policy questions   279
7.6 Air pollution   279
Acid rain   280
Smog   280
Policy questions   284
7.7 Summary and conclusions   284
Bulk waste   284
Hazardous chemicals   284
Radioactive waste   284
Biologically active waste   284
Air pollution   285
Problems   285
References   286

8   GLOBAL CHANGE   287
   8.0 Introduction   287
   8.1 History and controls of climate and atmosphere   288
         Control of atmospheric composition   291
         Control of climate   293
   8.2 Human activity and the recent history of atmosphere and climate   297
         Recent history of the atmosphere and climate   297
         Possible effects of human activity   301
         Possible consequences of global warming   302
         Policy questions   304
   8.3 Extinctions and species diversity   305
         Extinction of the megafauna   305
         Modern species diversity   307
         Extinctions and endangered species   308
         Deforestation   313
         Policy questions   316
   8.4 Chemical modification of the earth's surface   317
         Chlorofluorocarbons (CFCs) and the ozone layer   318
         Lead   321
         Policy questions   323
   8.5 Summary and conclusions   323
   Problems   324
   References   325

9   A FINAL WORD   327
   People and land   327
   Food   327
   Natural hazards   328
   Water   328
   Energy   328
   Mineral resources   328
   Waste and pollution   329
   Global change   329
   A final word   329

   Author index   331
   Subject index   335

BOXES

Box 1.1   Squeezing the Afghan nomads   *page* 9
Box 1.2   Bushmen of the Kalahari Desert   12
Box 1.3   The children of Rwanda   19
Box 1.4   The southern border of California   24
Box 1.5   Birth control (Chinese style)   29
Box 1.6   Rice-roots family planning in Bangladesh   30
Box 2.1   How the great powers, with a lot of internal help, ruined
          the agriculture of Somalia   50
Box 2.2   Mad cows and caged hens   56
Box 2.3   The cod in the fog   58
Box 2.4   Golden apple snails   64
Box 2.5   Early agriculture around a Mexican lake   66
Box 3.1   The Tunguska event   77
Box 3.2   How to lose a continent   80
Box 3.3   Planning (?) in the aftermath of the 1906 San Francisco
          earthquake   81
Box 3.4   An eruption of Mt. Pelee   94
Box 3.5   Floods at Khartoum, Sudan   101
Box 3.6   The Johnstown Flood   105
Box 3.7   The Great Flood   107
Box 3.8   The lucky sheep of Huascaran   116
Box 4.1   Water budgets and personal budgets   127
Box 4.2   The Hidrovia Project   134
Box 4.3   The Libyan Arab People's Jamahiriya and the "Great
          Man-Made River"   136
Box 4.4   The Aswan High Dam of Egypt   142
Box 4.5   The Owens Valley water war   146
Box 4.6   Shrinking the Aral Sea   156
Box 4.7   Too many Floridians and the fate of the Everglades   158
Box 4.8   The Jordan River   162
Box 5.1   China's Three Gorges Dam   176
Box 5.2   Oil for Germany and Japan in World War II   186
Box 5.3   War in the Spratly Islands?   190
Box 5.4   Getting oil out of the Caspian Sea area   192

Box 5.5   The *Exxon Valdez*   194
Box 5.6   Core meltdown at Chernobyl   204
Box 6.1   Jason and the Argonauts   216
Box 6.2   Ok Tedi   222
Box 6.3   The Chile nitrate war   234
Box 6.4   Exporting bird droppings, stamps, and weight lifters   237
Box 6.5   The Butte underground war   242
Box 7.1   The great asbestos panic   251
Box 7.2   Fresh Kills Isn't so fresh anymore   252
Box 7.3   Coffee, tea, a little Russian gasoline?   256
Box 7.4   Poison gas at Bhopal, India   262
Box 7.5   Sinking the Brent Spar   264
Box 7.6   Cleaning up the Rhine River   266
Box 7.7   What's wrong with men?   275
Box 7.8   How to blacken a triangle   281
Box 8.1   Leaking Soviet pipelines   294
Box 8.2   The year without a summer   300
Box 8.3   A lake and its climate   303
Box 8.4   *Felis concolor coryi*   309
Box 8.5   Is the spotted owl a red herring?   314
Box 8.6   The trees of Haiti   317

FIGURES AND TABLES

FIGURES

Figure 1.1	Population density and the Nile valley	*page* 2
Figure 1.2	Types of land use	6–7
Figure 1.3	Nature of exponential curves	8
Figure 1.4	Nomad trails in the Afghanistan area	9
Figure 1.5	History of world population since 10,000 B.C.	11
Figure 1.6	Demographic transition	14
Figure 1.7	Distribution of ages in Finland and Panama	15
Figure 1.8	Per capita gross national product versus birthrate	16
Figure 1.9	Comparison of villages in poor and rich countries	18
Figure 1.10	U.S. population characteristics during the 1900s	22
Figure 2.1	Explanation of structures of organic compounds	35
Figure 2.2	Glucose and its polymers	36
Figure 2.3	Fatty acids	37
Figure 2.4	Amino acid	37
Figure 2.5	Steroid structures	39
Figure 2.6	Food pyramid	40
Figure 2.7	Distribution of rainfall	43
Figure 2.8	Soil profiles	44
Figure 2.9	European grain production	46
Figure 2.10	Tropical crops	52
Figure 2.11	Animal raising	54
Figure 2.12	Grand Banks, Newfoundland, fishing	58
Figure 2.13	Cores from Lake Patzcuaro, Mexico	66
Figure 3.1	Active volcanoes	72
Figure 3.2	Cross section of the earth	74
Figure 3.3	Inactive and active continental margins	75
Figure 3.4	Gulf of Thermakos, Greece	76
Figure 3.5	Strain energy and faulting	83
Figure 3.6	Earthquake damage at Kobe, Japan	84
Figure 3.7	Earthquake frequency vs. magnitude	85
Figure 3.8	Areas of seismic risk	86
Figure 3.9	Fault in Sierra Nevadas, California	87
Figure 3.10	New Madrid seismic zone	88

Figure 3.11   Volcanic materials   90
Figure 3.12   Volcanic phenomena   92–3
Figure 3.13   Destruction of trees near Mammoth Lake, California   96
Figure 3.14   Cross section of river valley   99
Figure 3.15   Flood areas of river valley   100
Figure 3.16   Hydrograph   100
Figure 3.17   Khartoum, Sudan, and Nile River   101
Figure 3.18   Flood frequencies for the Euphrates River   102
Figure 3.19   Flood risk in Raleigh, North Carolina   103
Figure 3.20   Area of biblical flood   108
Figure 3.21   Coasts on inactive and active margins   109
Figure 3.22   Features of shorelines   110
Figure 3.23   Consequence of shoreline construction   111
Figure 3.24   Effect of human activity along coasts   112
Figure 3.25   Types of earth movement   114–15
Figure 3.26   Slope angles on different types of rock   117
Figure 3.27   Angle of repose and consequence of oversteepening   118
Figure 3.28   Slippage on blue clay near Seattle, Washington   118
Figure 4.1   Hydrologic cycle   128
Figure 4.2   Water-vapor capacity of air as a function of temperature   129
Figure 4.3   Diurnal change in relative humidity and temperature   129
Figure 4.4   Latitudinal variations in relative humidity   130
Figure 4.5   Distribution of Hadley cells   131
Figure 4.6   Latitudinal variation in evaporation and precipitation   132
Figure 4.7   Major desert regions   133
Figure 4.8   Hidrovia Project   134
Figure 4.9   Colorado River drainage basin   140
Figure 4.10   Potential energy, fluid pressure, and hydraulic head   147
Figure 4.11   Effect of media on groundwater flow   148
Figure 4.12   Permeability variation   149
Figure 4.13   Depths to water tables   151
Figure 4.14   Perched water table and artesian system   152
Figure 4.15   Cone of depression and salt-water intrusion   153
Figure 4.16   Irrigation in High Plains region of West Texas   154
Figure 4.17   Map of High Plains reservoir   154
Figure 4.18   Changes in configuration of Aral Sea   156
Figure 4.19   Drainage in South Florida   158
Figure 4.20   Watershed of Jordan River   162
Figure 5.1   Life in an energy-poor country   169
Figure 5.2   Concepts of heat and temperature   170
Figure 5.3   Internal combustion engine   172
Figure 5.4   Geyser at Yellowstone   173
Figure 5.5   Electricity generator   174
Figure 5.6   Harnessing water power   175
Figure 5.7   Yangtze River and Three Gorges Dam, China   176
Figure 5.8   Parabolic mirror   177
Figure 5.9   Photovoltaic cell   178
Figure 5.10   Lead battery   178
Figure 5.11   Lake storage of energy   179
Figure 5.12   Hydrocarbons in crude oil and in gasoline   181

Figure 5.13 Preservation of organic material   182
Figure 5.14 Traps for oil and gas   183
Figure 5.15 Location of oil and gas basins   184
Figure 5.16 Oil deposits of Arabian Gulf   185
Figure 5.17 Distribution of thick sedimentary basins   188
Figure 5.18 Patterns of worldwide oil transfer   191
Figure 5.19 Pipelines from the Caspian Sea area   192
Figure 5.20 U.S. consumption and production of oil   195
Figure 5.21 Retorting oil shale   196
Figure 5.22 Accumulation of coal seams   197
Figure 5.23 Okefenokee swamp, Georgia   198
Figure 5.24 Patterns of worldwide coal transfer   199
Figure 5.25 Abundances of radioactive parent and daughter through time   201
Figure 5.26 Nuclear reactor   203
Figure 5.27 Containment vessels at nuclear power plant   205
Figure 5.28 Energy consumption versus gross domestic product   207
Figure 6.1   Worldwide reserves of copper, lead, and zinc   219
Figure 6.2   Types of reserves and resources   219
Figure 6.3   Levels of confidence in ore reserves   220
Figure 6.4   Prices of copper metal   221
Figure 6.5   Ore-forming environments   226
Figure 6.6   Deposition from hydrothermal solutions   227
Figure 6.7   Marble quarry near Carrara, Italy   228
Figure 6.8   Formation of alluvial gold deposits   229
Figure 6.9   Salt deposition in Death Valley, California   230
Figure 6.10 Formation of Missouri Valley Type (MVT) ores   231
Figure 6.11 Banded iron formation   232
Figure 6.12 Ore-exporting nations   235
Figure 6.13 Nations dependent on the export of one ore   236
Figure 6.14 Mining methods   238
Figure 6.15 Open-pit mine   239
Figure 6.16 Ownership of underground veins   242
Figure 7.1   Dispersal of pollutants   248
Figure 7.2   Diagram of sanitary landfill   253
Figure 7.3   Sanitary landfill, North Carolina   253
Figure 7.4   PCBs   257
Figure 7.5   Dioxins   257
Figure 7.6   DDT   258
Figure 7.7   Sarin   258
Figure 7.8   Parathion   259
Figure 7.9   Chlordane   259
Figure 7.10 Aldicarb   260
Figure 7.11 Pyrethrins   260
Figure 7.12 Carbaryl   261
Figure 7.13 Methods for dealing with toxic waste   263
Figure 7.14 Rhine River   266
Figure 7.15 Dose–response curve for radiation   268
Figure 7.16 Concept of threshold   270
Figure 7.17 Frequency versus atomic mass for fission products   271
Figure 7.18 Disposal of high-level nuclear waste   272

Figure 7.19 Sewage treatment plant   276
Figure 7.20 Eutrophication   278
Figure 7.21 Location of black triangle in central Europe   281
Figure 7.22 Vistula River at Cracow, Poland   282
Figure 8.1   Onsets of Antarctic and Arctic glaciation   289
Figure 8.2   Periodicity of glaciation in the past million years   289
Figure 8.3   Retreat of glaciers   290
Figure 8.4   Carbon reservoirs and transfers among them   291
Figure 8.5   Electromagnetic spectrum   295
Figure 8.6   Radiation balance at earth's surface   296
Figure 8.7   Path of North Atlantic Deep Water (NADW)   297
Figure 8.8   Ice core at Vostok, Antarctica   298
Figure 8.9   Climate and atmosphere during the past 1,000 years   299
Figure 8.10 Predicted temperatures in the Green River area   303
Figure 8.11 Extinction of large animals   306
Figure 8.12 Organisms endangered by human activity   312
Figure 8.13 Deforested areas   315
Figure 8.14 Antarctic ozone hole   319
Figure 8.15 Air conditioner   320
Figure 8.16 Lead in Greenland and Antarctic ice caps   322

TABLES

Table 1.1   Land use and population densities   3
Table 1.2   Estimated populations of megacities   4
Table 1.3   Population characteristics in selected countries   13
Table 1.4   Population characteristics of the United States   23
Table 2.1   Energy in 1 ounce of common foods   41
Table 2.2   Chemical analysis of corn plant   45
Table 2.3   Grain production and fertilizer use   47
Table 2.4   Grain and national wealth   48
Table 2.5   Grain trade and aid   49
Table 2.6   Meat production   57
Table 2.7   World fishing   60
Table 2.8   Nutrition energy for selected countries   62
Table 3.1   U.S. annual death rate   78
Table 3.2   Frequency of earthquakes   85
Table 3.3   Cost of recent geologic catastrophes   119
Table 4.1   Water on the earth's surface   126
Table 4.2   Global residence times of water   130
Table 4.3   Sources of freshwater in the United States   135
Table 4.4   Water-rich and water-poor nations   135
Table 4.5   Water consumption in the United States   137
Table 4.6   Hydraulic conductivities   150
Table 5.1   Units of energy and power and fuel equivalents   171
Table 5.2   Heats of combustion   182
Table 5.3   Oil and gas reserves and production   189
Table 5.4   Half-lives of nuclear fuels   202
Table 5.5   Energy consumption and production   208
Table 5.6   Costs of electricity production   210

Table 6.1   Eight ages of people   215
Table 6.2   Abundances and cutoff values of common metals   218
Table 6.3   Classification of nonfuel mineral resources   224
Table 7.1   Types of household plastic   254
Table 7.2   Natural background radiation   269
Table 7.3   Radioactive isotopes that must be stored   271
Table 7.4   Compositions of typical raw sewage and effluents   277
Table 7.5   Air pollution in 20 large cities   283
Table 8.1   Composition of the atmosphere   288
Table 8.2   Absorption of thermal infrared radiation   301
Table 8.3   Animal extinctions   310
Table 8.4   World forests   316

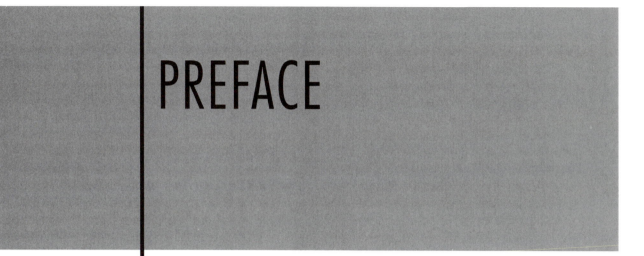

PREFACE

The geologist sat in a restaurant and listened to a conversation at the next table (the geologist knew that it is impolite to eavesdrop, but the conversation was conducted loudly, and besides, the geologist is not always polite).

The discussion was among four people employed by a client who wanted to open a steak restaurant. The first person was a financial adviser, who advised that the most important task was to make sure that adequate financing was arranged. The second person was a real-estate agent who felt that the most important issue was the location of the restaurant. The third person, a lawyer, was certain that the principal problem was to have all documents prepared properly. And the fourth person was an advertising specialist, who informed the others that first-rate publicity would be the major contributor to the future success of the restaurant.

"Quite wrong," thought the geologist to himself (the geologist was not so impolite as to interrupt the conversation). All of these activities (law, money, property, publicity) are secondary requirements. They *derive from* the primary necessity, without which there can be no restaurant. Very simply, if you wish to open a steak restaurant, first you need a cow.

This book is about cows, and corn, coal, copper, carbon dioxide, and other (not necessarily alliterative) subjects. Mostly it is about people. We start from the premise that people can live only because they are able to obtain from the earth the products necessary for their survival and are able to return to the earth the waste products of their lives. Consequently, as the population of the earth expands we must make decisions about our resources (food, energy, construction materials, etc.) and our waste (organic chemicals, carbon dioxide, heavy metals, radiation, etc.).

These topics are commonly regarded as lying within the special domains of environmentalists, developers, scientists, industrialists, economists, political scientists, sociologists, lawyers, and a number of other occupations. We do not address this book to any one of these groups, however, because we feel that, to some extent, all of us are in all of them. We are all interested in how we use our land, air, and water. We are all concerned about our economic, social, and political condition. We have written this book because we feel that the decisions that all of us must make on these issues can be made effectively only by people who are fully informed about the ways in which the earth affects – and, in many situations, controls – the decisions.

In short, this is a book about being aware of the earth as we go about the daily business of living. Although this is a serious subject, there is no reason to be ponderous, and we have adopted a somewhat lighthearted style that we hope will make the book both readable and provocative.

Although we are responsible for the contents of this book, we could not have written it unless numerous people had given us information, advice, and encouragement and had been willing to review and edit various parts of the book. We are particularly grateful to Trileigh Stroh, who reviewed and edited the entire manuscript, and Laurel Kaczor for help with the illustrations. We would also like to acknowledge the help of Mark Alperin, Steve Bradt, Cathy Cash, Meredith Clason, Catherine Flack, Philip Hirsch, Tom Holzer, Holly Johnson, Mary Lee Kerr, Suzanne Kjentrup, Caroline Martens, Barbara Rogers, Miriam Sheaves, and Nancy West.

John J. W. Rogers
P. Geoffrey Feiss

UNITS

This book is written for people who use many different terms – units – when they make measurements. Most of the world's people use metric measurements (liters, grams, etc.), but Americans primarily use "English" units (gallons, pounds, etc.), and much of the English-speaking world is also familiar with English units. To compound the problem, different scientific disciplines use slightly different terms for the same type of measurement – for example, the use of acre-feet instead of gallons or liters in studies of water supply. In this book, we attempt to use the terminology most common to the subject being discussed in order to make the transition from this book to other books as easy as possible for readers who wish to pursue a subject in more depth than we present here. This means that somewhat different units are used in different sections, and we maintain clarity by frequent reference to alternative units and also by supplying the table of equivalents and abbreviations for various units below. (Also see Table 5.1 for various units of energy and power.)

LENGTH

1 meter (m) = 1.094 yards = 3.281 feet
1 centimeter (cm) = 10 millimeters (mm) = 0.394 inch
1 kilometer (km) = 0.621 mile
1 micron (μ) = 0.01 (10^{-3}) mm = 1,000 (10^3) nannometers (nm) =
$\qquad$ 10^4 Angstroms

AREA

1 square kilometer (sq km; km^2) = 0.386 square mile
1 square mile = 640 acres
1 hectare = 10,000 (10^4) square meters = 2.471 acres

VOLUME

1 liter (l) = 0.264 gallon = 0.035 cubic foot
$\qquad$ (Note – we use U.S. gallons instead of Imperial gallons.)
1 cubic kilometer (cu km; km^3) = 0.235 cubic mile
1 acre-foot = 21 million (2.1×10^7) gallons

WEIGHT (MASS)

1 kilogram (kg) = 2.204 pounds

1 metric ton = 100 kilograms

1 ton = 2,000 pounds

CONCENTRATION

percent (%) = one part in 100

part per thousand (ppt) = one part in 1000 = 0.1%

part per million (ppm) = 10^{-4}% = milligrams per liter (mg/l)

part per billion (ppb) = 10^{-7}%

(Concentrations in liquids are given by weight and in gases by volume.)

1 | PEOPLE AND LAND

1.0 INTRODUCTION

Be fruitful, and multiply. – *Genesis 1:22*

People who count people count a lot of us, and more all the time. The world now holds about 5.5 billion people, adding approximately 100 million each year. Are there too many of us, using up the earth's resources and polluting its surface? Are there not enough, leaving work undone that could make the world's people more prosperous? Are there too many in some places and not enough in others, or too many that we don't like and not enough that we do? We address these issues in this chapter and throughout this book.

This chapter discusses the size of the world's population, its distribution, and the types of land available to people (1.1); the history of population growth (1.2); variation in such characteristics as age distributions and growth rates in different parts of the world (1.3); the relationship between population and wealth (1.4); and three sections that explore the question of whether the world has too many people – a comparison of lesser-developed countries with industrial ones (1.5), a discussion of the U.S. population (1.6), and finally an exploration of the thorny issue of population control (1.7).

FURTHER READING: This chapter depends on an abundance of statistics from the following sources: Population Reference Bureau (1994); World Resources Institute (1992, 1994); United Nations (annual a, annual b); and United States (annual).

1.1 POPULATION DENSITIES AND LAND USE

Most national governments pay lip service to the idea of conservation of natural resources, but in practice soil conservation is not a vote-winning issue with the electorate. – *N. Hudson,* Land Husbandry, *p. 32*

The earth currently (mid 1990s) holds about 5.5 billion people. If they were spread evenly over the entire inhabitable land area of the earth, we could calculate a population density as follows:

- The surface area of the earth is 197 million (197 × 10^6) square miles.
- Assume that 75% of the earth is uninhabitable. This figure includes the 71% that is ocean and 4% in Antarctica and the Greenland icecap, but it leaves large areas that are only marginally inhabitable, such as the Sahara Desert, in the 25% of the earth that we can live on. Thus, the inhabitable land surface is 0.25 × 197 × 10^6 square miles.
- The population density is the population divided by the area, or 5.5 × 10^9/0.25 × 197 × 10^6 = 111.7 people per square mile.

Instead of 111.7, we cite the population density as 110 people per square mile. The reason for the approximation is that we have made two prior approximations in the data. One is the population of the earth as 5.5 billion, and the other is the 25% of the earth that is inhabitable. Both of these figures are meaningful only to two "significant

(a)

(b)

Figure 1.1 Population along the valley of the Nile River in Egypt: (a) river meandering through a valley that is three to four miles wide and ranges from a few 10s to more than 100 feet below the surrounding desert. It contains farms, date orchards (large dark areas), small villages, and a few areas of high ground that are white because they are not irrigated by the river. Only a few irrigated areas (dark areas toward the left of the photograph) have been established in the surrounding desert. Because the desert is nearly uninhabitable, virtually all of Egypt's 60 million people, increasing by nearly 1 million each year, are crowded into the valley and delta formed where the river enters the Mediterranean. These areas contain only 4% of Egypt's total land area and have a population density of nearly 5,000 people per square mile; (b) date palms in the valley as seen from the surrounding desert. It is possible to stand with one foot in irrigated land of the valley and one in the desert that supports no vegetation.

figures," so we report the density as 110 people per square mile, showing only two significant figures (1 and 1). To put a population density of 110 per square mile in perspective, each person would have about 6 acres of space as a private preserve – a square mile contains 640 acres, and an acre is about the size of a football field (American foot-ball field, which is about 20% smaller than a soccer field, known as a football field in the rest of the world).

Calculation of the population density of the entire earth masks important variations in densities from place to place (Fig. 1.1). Many of these variations are controlled by variability in the types of land available for hu-

Table 1.1 *Land use and population densities in selected countries*

Countries	Percentage of area in				Density in people per square mile as population divided by	
	Cropland	Pasture	Forest	Other	Total area	Inhabitable area
World	11	26	31	32	110	310
United States	21	26	32	21	74	157
Canada	5	4	39	52	8	89
Netherlands	27	32	9	32	1174	1190
United Kingdom	29	46	10	16	626	835
Germany	35	16	30	19	602	1180
France	35	21	27	17	273	569
Russia	8	5	44	44	22	169
Malta	41	0	0	59	2954	7205
Nigeria	34	44	14	8	279	358
Saudi Arabia	0.5	40	0.5	59	22	48
India	57	4	22	17	794	1302
China	10	34	14	42	331	762
Japan	12	2	67	19	586	4186
Brazil	9	20	66	5	48	166
Trinidad and Tobago	23	2	43	31	650	2600
Australia	6	56	14	24	6	10

Notes: Because of lack of information, the areas of forest and other in Russia are arbitrarily assigned as one-half of the area not used for crops or pasture.
Sources: Population Reference Bureau (1994); World Resources Institute (1994).

man use. We discuss population densities first and then their relationship to land use.

Variations in population densities

A herder on the steppes of Mongolia, where the population density averages 4 people per square mile, may look for miles across the countryside without seeing anyone. Conversely, inhabitants of major cities can live only by building apartments on top of each other and thus achieving a level of crowding up to the population density of Tokyo (the world's most crowded city), with a population of nearly 8 million squeezed into slightly more than 200 square miles for a population density of 34,000 per square mile. We show two different measures of the population densities of selected countries in Table 1.1. One measure, which we used for the population density of the whole earth, is the population divided by the total land area. Another, perhaps more meaningful, number is the population divided by the land area on which people can be reasonably expected to live. Because wasteland

generally cannot be inhabited and forests should not be cut down for habitation, we define this inhabitable area as the sum of the areas of cropland and pasture.

The countries with the largest populations are China (1.2 billion) and India (900 million). China is still growing, although stringent birth control programs have reduced the rate (Section 1.5), and India is predicted to surpass China as the world's most populous country sometime during the next century. Large populations raise images of great crowding, well past the burden borne by the rest of the world. These images are false. The population densities (total area) are 300 people per square mile in China, well below the average in many wealthy European countries, and 800 in India, not much more than the European average. The perception of crowding in a country such as India comes from the cities, deluged with an influx of people from the countryside at a rate faster than services can expand. Bombay, for example, has grown from a city of 1 million in 1948 to more than 10 million today largely by absorbing young people pushed out of rural India by a birthrate that overwhelmed the supply of sufficient land or

Table 1.2 *Estimated populations of megacities in early 21st century*

City	Population (in millions)
Bangkok	11
Beijing	12
Bombay	16
Buenos Aires	13
Cairo	13
Calcutta	17
Delhi	13
Jakarta	15
Karachi	9
Los Angeles	11
Manila	12
Mexico City	25
Moscow	10
New York	17
Rio de Janeiro	14
São Paulo	23
Seoul	14
Shanghai	15
Tokyo	22

Notes: Some estimates include populations of metropolitan areas, but some cities would be much larger if their metropolitan areas were included.
Sources: Data are from various sources, mostly publications of the United Nations.

jobs to make a living. This influx is predicted to continue well into the 21st century, creating megacities with populations greater than 25 million people (Table 1.2).

Europe and Japan demonstrate that high population densities do not necessarily cause poverty. The Netherlands is one of the world's most crowded countries, and all of Europe exceeds the average world population density by a factor of 3 to 4. Europe is rich, however, because it has a climate and geography that permit about half of its land to be used for crop and animal production and a level of political and social organization that enables Europeans to use these resources. The overall population density of Japan is about twice that of Europe, but most of Japan is a land of steep-sided mountains, and only about 15% of the area can be farmed or even used for cities and towns. Thus, several thousand Japanese must fit into each square mile of arable and inhabitable land (see the difference between the two measures of population density in Table 1.1). Paradoxically, some of the world's most crowded countries are also its smallest and

some of its major vacation spots. The islands of Barbados, Maldives, and Malta must pack their populations of a few hundred thousand people into very small areas circumscribed by the ocean.

Land use

Land can be classified in various ways. We use four categories here – cropland, pasture, forest, and other land. As well as discussing the four categories, we list their distribution in selected countries in Table 1.1 and illustrate them in Figure 1.2.

Cropland is land used for standard production of grains, vegetables, and other plants consumed by people or fed to animals for human consumption. Land that can be used naturally must have adequate soil and rainfall, and most food crops have a limited range of temperatures to which they can be exposed (Section 2.2). Cropland also includes land that is not naturally fertile but can be used agriculturally through application of fertilizer and/or irrigation water (Sections 2.2 and 4.3).

The area covered by cropland can either expand or contract from year to year. Expansion is caused by conversion of other land to agriculture by application of fertilizer and water. It occurs when governments or private individuals attempt to increase agricultural productivity, either to provide more subsistence for the local population or to produce wealth through trade (Section 2.5). Most of the converted land was formerly used for pasture, and because animals can generally be raised more efficiently on hay grown domestically than on natural pasture (Section 2.3), this conversion temporarily increases total productivity. Some expansion of cropland is also accomplished by clearcutting of forests, generally accompanied by burning. Because most forested land, particularly in the tropics, does not have adequate soil for crops, this expansion usually requires addition of massive amounts of fertilizer; we discuss the disastrous consequences of clearing tropical forests in Section 8.3.

Area classified as cropland can shrink for two reasons. One is the deliberate decision, now being made in many European countries, to reduce overproduction of food and allow some land to return to a natural state. A second, and very unfortunate, reason is environmental degradation caused by soil erosion, leaching of nutrients out of the soil, or contamination of soil and water by human activity. This reduction is particularly serious in lesser-developed countries, but the industrialization of agriculture may have long-term effects in all countries, as we discuss in Section 2.5.

Pasture comprises areas that domestic animals graze on. Much of it is natural grassland, which is particularly useful for cattle and sheep, but goats can also browse on bushes and other vegetation. Some grasslands, especially in northern temperate climates, are very productive and support large populations of animals. In semiarid regions such as southern North America, much of Africa, and Australia, however, pasture may offer so little edible vegetation that animals can survive only if they can roam over large areas (more than 1 square mile per cow in some regions). Although some pasture is being converted to cropland, the principal change in the area devoted to pasture is a reduction caused by excessive animal populations. We discuss this problem further in Section 2.3

Forest and woodland include area covered by trees. This category includes vast stretches of conifers (pines and related varieties) in subarctic areas, mixed hardwood and conifer forests that cover the temperate regions of much of North America and some parts of Europe, and the environmentally fragile rainforests of the tropics. Tree-covered areas may expand because of abandonment of cropland or may contract by conversion to cropland (see earlier discussion and Section 8.3).

Other land constitutes the final category. Just when you thought the classification was going to be simple, we must now introduce this nebulous term. It results from the difficulty of further subdivision and generally means places without usable vegetation. Most of this other land could be called wasteland. It includes Arctic tundra, treeless and covered by small plants; steep mountain sides with large exposures of bare rock; and deserts so arid that they contain virtually no vegetation. It also includes areas covered by human habitations, industry, malls and their vast parking lots, roads, and airports – in short, places where nothing grows because people have covered them with asphalt, concrete, bricks, wood, and other construction materials.

Although the area that people have covered seems to be large, it actually is a very small percentage of total land area. It appears large because, when we are in the center of a city, we see only streets, sidewalks, and buildings and think that the entire city and its surrounding suburbs have demolished every living thing. But cities contain parks, and houses have gardens and lawns. A large city (megacity) like London, for example, seems to sprawl interminably over the countryside, but examination of a map showing a 30-mile radius around London reveals that more than half of one of the most urbanized places on earth consists of forests, parks, and agricultural land. A few cities, like Cairo, are compressed and lack

these amenities, but generally they cover only a few tens of square miles. The only country in which people have built on most of the land is Singapore, which consists essentially of the city of Singapore and a small outlying region. In short, an exact calculation of the amount of "other" land covered by constructed areas is difficult to make, but it is less than 1% in most countries and not more than a few percent even in densely populated Europe.

FURTHER READING: Hudson (1992).

1.2 HISTORY OF WORLD POPULATION AND GROWTH RATES

If the population of the world were to continue to grow at the present rate for six hundred years, there would then be only one square yard per person. It is inconceivable that this should happen, but the important question is: Why will it not happen? – *P. Appleman, The Silent Explosion, p. 137*

How has the earth reached its population of 5.5 billion people? The consequences of having children are illustrated in Figure 1.3, which shows a constantly increasing population that follows an "exponential" growth curve (the mathematics of exponential curves is explained in the caption for Fig. 1.3). This exponential curve demonstrates an excess of births over deaths that causes the population to increase at a "constant rate" through time. For example, if 1,000 people show an increase of 0.5% per year, to 1,005, then an 0.5% increase of 10,000 people will add 50 to population to increase to 10,050, and so on to larger populations. A constancy of growth rate signifies a constant period of time in which a population doubles its size. We refer to this period as the "doubling time" (t_{doub}). Starting from any arbitrary time, the population is twice as large after 1 t_{doub}, four times as large after 2t_{doub}, 8 times after 3t_{doub}, and so on. Inevitably, as the absolute number of people increases, the number of people added each year will increase also, yielding the concave upward curve of Figure 1.3, and populations with short doubling times become very large very quickly.

Having already used the figure of 5.5 billion for the present population of the world, we now investigate the population at different stages of earth history. We must start by recognizing that the world's population is not easy to count. A modern census suffers from the problems that births and deaths change the population while the counting is in progress; that people move, particularly nomads who may cross national borders; that some peo-

(a)

(b)

Figure 1.2 Examples of the four types of land use discussed in the text: (a) cropland in central Sweden; (b) pasture in the Dolomite Mountains of northern Italy; (c) forest in the Karelian region of western Russia; (d) wasteland along the border between Saudi Arabia and Yemen.

(c)

(d)

ple avoid being counted, especially if they are living in a country illegally, or aren't paying their taxes. Some people just do not like government officials who ask questions and refuse to cooperate. Census figures become shakier as we go further into the past. We know the population of England in 1085 because in that year William the Conqueror ordered the preparation of the Domesday Book in order to control his new kingdom more effectively. Few other areas of the world were that organized in the 11th century or kept any records at all. Furthermore, the idea that every square inch of the world belongs to one country or another is a modern invention (see discussion of Afghan nomads in Box 1.1). Most of human history was a time of massive migrations, ephemeral kingdoms, borderless open lands, and no written records. Using the best estimates that we can find for past

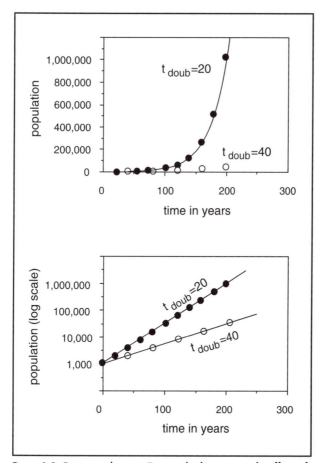

Figure 1.3 Exponential curves. Two graphs demonstrate the effects of growth at "constant" rates and different "doubling times" (see text for definition). The graphs show the sizes of two different populations, each starting with 1,000 people, after the doubling period for each population (40 years for one and 20 years for the other).

The top graph shows the growth of both populations with time plotted on the horizontal axis and population on the vertical axis. On this scale, the growth of the population with a 20-year doubling time is so fast that the population with a 40-year doubling time seems barely to change.

A curve drawn through plotted points for the population with the

20-year doubling time shows the typical shape for a "constant" rate of growth (see text) that leads to larger and larger increments as time passes. This curve is described by the equation

$$P_t = P_0 e^{kt},$$

where P_t = the population at any time t; P_0 = the population at any starting time designated as $t = 0$; t = the time of growth from the arbitrary starting time; k = a constant characteristic of the population; e = the base of natural logarithms (2.718). The value of k (the rate constant) varies among different populations. Where k is high, the population grows rapidly; a low k signifies slow growth. The position of kt as a "power" of e makes the equation and curve "exponential."

The constancy of growth shown by the exponential relationship can be demonstrated further using calculus. If we say

rate of increase = constant X population at any time,

the calculus expression is

$$dP = kPdt$$

or

$$dP/P = kdt$$

which, on integration from a starting time $t_0 = 0$ to a finishing time of t, yields

$$P_t = P_0 e^{kt}$$

The curvature of the exponential graph in the top graph shows the effect of constant growth but makes the graph more difficult to interpret than a straight line. Linearity is accomplished by expressing the relationship logarithmically as

$$lnP_t/P_0 = kt,$$

where ln is a natural logarithm (to the base e). The logarithmic relationship enables us to plot the logarithm of P_t/P_0 on an arithmetic axis versus t, yielding the straight line shown in part B.

Rates of growth of various populations can be described by the rate constant k. As discussed in the text, however, they are more easily understood by the "doubling time" (t_{doub}). The value of t_{doub} can be determined both graphically and from the logarithmic form of the growth equation. When $P_t/P_0 = 2$, the t in the growth equation is t_{doub}. Then

$$ln\ 2 = 0.693 = kt_{doub},\ or$$

$$t_{doub} = 0.693/k.$$

populations, we obtain the curve shown in Figure 1.5 for the change in world population through time. This figure shows three basic stages in human history.

The period prior to about 10,000 B.C. was a time of hunter–gatherer societies. People lived on the animals they could kill and the fruit and other edible vegetation they could find. Although the diet was varied and probably nutritious, these societies were limited by the primary productivity of the land (see Box 1.2 for the life of a modern hunter–gatherer society). Thus, the family/tribal

groups were small and widely scattered and had to move frequently as they exhausted the resources of one area. Several factors kept growth rates low. Life expectancy was short, perhaps about 20 years, with most children dying in infancy. The food gathered would not have been edible by infants, causing mothers to breast-feed their children for 2 to 3 years, thus reducing their periods of fertility. Also, very young children are hard to travel with, a fact confirmed by any modern parents who have packed a car to take a baby on a family outing. The total

BOX 1.1 SQUEEZING THE AFGHAN NOMADS

The collision of the Indian subcontinent with Asia that began about 40 million years ago, and still continues, lifted the Himalayas and adjacent lands to the highest elevations on the earth. Where the western edge of the Indian plate pushed against Asia, the crushing was so intense that it created what geologists call a "syntaxis," or in rare moments of clarity, a "knot" in the Pamir region of central Asia. Various mountain ranges radiate from this knot, and the southwesterly one occupies most of central and northern Afghanistan (Fig. 1.4).

For many centuries, the mountains and valleys of the Pamir and neighboring areas have been the home of nomads. They lived in tents and pastured their herds, mostly sheep and goats,

in sparse mountain meadows up to 10,000 feet high during the summer. When winter closed in on them they moved to the valleys, where they continued to tend the herds and where some of the wealthier families had permanent houses and agricultural plots. The spring and autumn migrations commonly lasted 2 months and covered several hundred miles.

The migrations did much more than provide feed for the herds. The nomads needed grains and vegetables to eat, clothing that could not be produced solely from wool, and some very limited consumer items. For these goods, they needed money, and they made it in several ways. One was transporting the merchandise (grain, etc.) for a fee. More common was the pur-

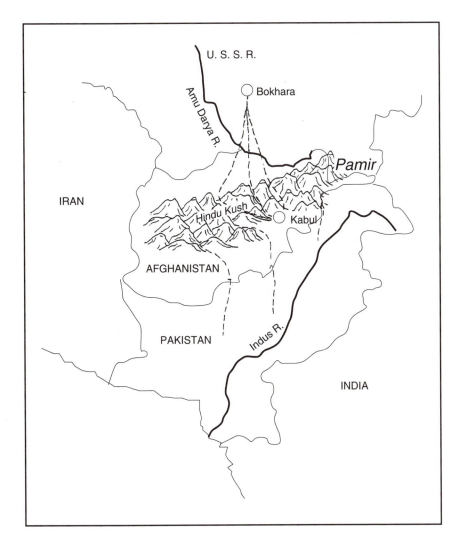

Figure 1.4 Nomad trails in the Afghanistan area. Modern Afghanistan extends from the northern border along the Amu Darya River, closed by Russia in 1920, to the southeastern border that Pakistan closed in 1960.

BOX 1.1 SQUEEZING THE AFGHAN NOMADS (continued)

chase of grain, clothing, and other goods in one area and its sale at a profit elsewhere. The nomads also sold the produce of their flocks as milk, wool, and occasionally as meat.

Many of the major trading routes crossed the present Afghanistan, although individual groups generally traveled over only part of the distance (Fig. 1.4). The low valleys of the Indus plain were fertile regions in which to spend the winter. The high reaches of the Hindu Kush Range provided summer pasture. Trading was good in Bokhara and other cities of central Asia made rich by the east–west traffic between Europe and China (the Silk Road). This migratory life suited the terrain of Afghanistan. Rain falls only in the mountains, and agriculture in the valleys depends on irrigation from streams or wells. Even with irrigation, only 12% of the country is arable, and much of the rest is too rocky or too dry even for grazing.

Wealthy countries have now eliminated most of this nomadic life. Three events are most important in this elimination. The first was in the 1800s, when the British colonial government in India, which included Pakistan, reached an agreement with the tsarist government of Russia on territorial borders between them. Russia would have control over areas north of the Amu Darya River, including Bokhara and other major cities. British India would have control of areas south and east of the present western border of Pakistan. Between Russia and India would be a buffer state, known as Afghanistan, where the local inhabitants would be free to live their own lives just as long as they didn't annoy the two major powers on either side.

The subdivision of the 1800s had little effect on the nomads, who were still able to cross borders with little trouble. The Bolshevik Revolution in Russia, however, led the new Soviet empire to close its border along the Amu Darya in 1920, effectively sealing off access to the northern part of the nomadic routes. Independence of India from Britain in 1947 was accomplished only by splitting mostly Hindu India and mostly Moslem Pakistan into two separate countries. Then in 1960, Pakistan closed its border to nomads and virtually all other Afghans. Not only did this closing keep the nomads away from the fertile Indus valley, but it also prevented visits between family groups, who had lived on both sides of the original British border and who had not been affected as long as cross-border movement was permitted.

Even with these restrictions, some nomadic movement still occurred within Afghanistan. Further modernization, however, terminated most of what remained. Instead of camels and ox-carts to haul goods, the Afghans bought trucks. The trucks were much more efficient, covered longer distance, even on poor Afghanistan roads, and permitted the opening of markets in settled locations instead of the traveling bazaars operated by the nomads. Herders who could afford to buy trucks hauled their sheep and goats between pastures in the back.

So civilization "won," and the Afghanistan nomadic life-style became extinct.

FURTHER READING: Pedersen (1994).

---

world population in 10,000 B.C. is estimated at approximately 6 million.

The date of 10,000 B.C. approximately marks the beginning of agriculture, the deliberate growing of grain and vegetables and raising of animals for meat (see Sections 2.2 and 2.3). Although some societies remained, and still remain, mixed hunting–gathering and agricultural, other areas favorable to crop growing became completely reliant on domestic plant and animal production. The benefits of agriculture over hunting and gathering are an increase in the amount of food produced and the ability to remain in one place, at least until the soil is depleted. Thus, population densities could increase, particularly in those parts of the world with rich soils (largely areas of abundant rainfall or river water). This growth averaged approximately 0.1% per year and brought esti-

mated world populations from 250 million in A.D. 0 to 750 million at the end of the dominantly agrarian period in 1750 (t_{doub} varying from ~2,000 years early in the agrarian period to ~1,000 years later). The growth resulted both from an increase in life expectancy, to a worldwide average of about 25 years, and an increase in the number of children produced per woman. Population growth would have been higher except for the bubonic plague ("black death") that scourged most of Europe and parts of Africa and Asia toward the latter part of the agricultural period. In 1750, the population density of Europe, the world's highest, was about 25 people per square mile.

The industrial revolution in 1750 ushered in an age of rapidly increasing populations in those parts of the world that it affected. In 100 years, the combined populations

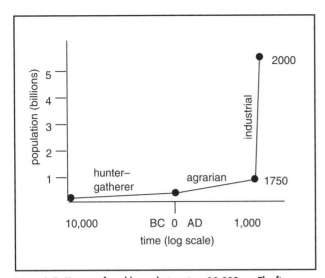

Figure 1.5 History of world population since 10,000 B.C. The figure shows three periods of growth delineated by two major boundaries (further explanation in the text). Slow growth of hunter–gatherer societies lasted until about 10,000 B.C., the beginning of the "Neolithic" age. At this time, introduction of agriculture enabled the world's population to begin growing at a faster rate. The next major boundary is shown at 1750 A.D., the beginning of the industrial age, when growth rates also accelerated. For further discussion, see Livi-Bacci (1992).

of Europe and North America more than doubled (to ~400 million by 1850). People could expect to live for 50 years or more. Growth in other parts of the world was somewhat slower, with an average t_{doub} of approximately 125 years between 1750 and 1950. After World War II, the rate of population growth increased even faster (see the quotation at the start of this section), with people in most of the world expecting to live considerably longer than 50 years and the world's population increasing from 2.5 billion to more than 5 billion (t_{doub} = 40 years).

The progressive increase in population density can be explained by increase in per capita available energy. We define the energy unit "calorie" in Section 5.1, and in most of this book we refer to kilocalories (kcal – i.e., 1,000 calories). For reference, a piece of toast provides about 50 kcal of nourishment (although diet books refer to it as 50 calories; see Section 2.2). Active adults need some 2,000 to 3,000 kcal of food per day and must find about 7,000 to 8,000 kcal more for cooking and heating. Thus, each person living a rudimentary life requires approximately 10,000 kcal of energy daily. In nonindustrial societies, cooking and heating are accomplished mostly by burning trees and brush. These energy requirements restrict the population density of hunter–gatherer societies to only a few people per square mile (Box 1.2). In-

creasing efficiency of food production in agricultural societies permits densities up to a maximum of about 100 per square mile, as is common in much of sub-Saharan Africa. These societies, however, are highly constrained by the available wood supply (Section 8.3).

In industrial societies where energy is obtained largely by burning coal and oil (Chapter 5), wood is less necessary and more land can be devoted to crop production. Access to extra energy also improves the transportation of goods to markets and permits increase in food production by providing fertilizers and agricultural machines (such as mechanized plows) that are more efficient than people. The availability of energy permits comfortable life-styles even with population densities of 1,000 per square mile, as in parts of modern Europe (Section 2.2).

FURTHER READING: Appleman (1965); Livi-Bacci (1992).

1.3 VARIATION IN POPULATION CHARACTERISTICS IN THE MODERN WORLD

Trains will arrive and leave within one minute of their scheduled times. – Instructions to foreign geologists traveling to various parts of Japan to meet field excursions at the 1992 International Geological Congress

Some of us feel crowded and dream longingly of riding a horse across the empty spaces of the Old West of the United States. Others strive to have many children so that there will be someone to take care of us in our old age. Children in affluent countries will probably be old some day; those in impoverished countries are likely to die young. Differences in attitude, plus the natural variation in land and climate, cause enormous variability in population characteristics around the modern world. We discussed population density in Section 1.1, and in this section we describe birth and growth rates, fertility rates, life expectancies, and age distributions.

Birth, fertility, and growth rates

The 140 million babies born to the world each year must be cared for until they can care for themselves. Generally, we nurture children more effectively in places where the birthrates are low, where mothers need not care for large numbers of babies. The birthrates vary enormously from country to country and even between groups within a country (Table 1.3). We use two closely related terms to describe births:

BOX 1.2 BUSHMEN OF THE KALAHARI DESERT

A book by Wannenburgh, Johnson, and Bannister and accompanying photographers provides vivid account of the life of a hunter–gatherer society in the Kalahari Desert of southern Africa during the 1970s. The Bushmen of this part of Botswana and Namibia are racially neither negroid nor Caucasian and had very little interaction with the sparse populations around them. Thus, at the time they were studied, they were virtually unaffected by modern society and lived as their ancestors had lived for centuries, perhaps millennia.

The Kalahari Desert ranges from arid to semiarid. Some parts have a cover of thorn trees and, during the short rainy season, a fairly lush crop of grasses. Other parts have a rock or dirt cover through most of the year. In most of the desert, water is available in pools at the surface only during the rainy season, but a few permanent water holes are fed by springs. The climate is generally warm to hot, although winter nights can be cold when the wind blows from the south. This harsh climate limits the vegetation and animal life. Some plants produce nuts and berries that are highly nutritious, and native vegetables grow sparsely. An important plant is the melon, whose fruit is a major source of water. Small antelopes of different varieties provide some animal food, but their population is too small to allow them to be a major source of nourishment.

Bushman society is organized around informal bands of approximately 50 people, who live, travel, hunt, and gather together — they do not separate into recognizable tribes. Grouping into bands is necessary for survival because many of their activities require more than one pair of hands. Men are the hunters. They use poisoned arrows, with the poison applied to the wooden shaft just behind the tip in order to prevent the men from dying by accidental cuts. Arrows are now tipped with iron, although before limited contact with outside society, the arrows were completely wood with a sharpened end. Group hunting permits the men to surround the swift antelope and enables one of them to get a good shot as the animal runs away. The owner of the arrow that killed the animal is the owner of the meat, although custom requires that it be shared not only with his family but also with his relatives and, in lesser proportion, with the rest of the band. Women are the gatherers, taking their small children with them as they forage. Because of the scarcity of meat, the livelihood of the band rests in their hands. They carry their finds in homemade baskets and distribute their bounty by the same customs as the men do their meat.

The size of bands is limited not only by the availability of food but, just as importantly, by the availability of water. Water is a problem except during the rainy season. Some bands camp near semipermanent springs, but there they run the risk of exhausting the local food supply or of being attacked by snakes or predators attracted to the water. Most bands migrate in search of new food. For these bands, water is provided mostly by melons or, in limited amounts, by various succulent tubers. In order to keep hydrated, adults eat around 15 pounds of melons per day during the dry season. Thus, the Bushmen do not have homes in the traditional sense. Huts are made of grass over a framework of small branches, but they are used only temporarily, and they soon collapse and decay into the ground. Eating, sleeping, and virtually all other activities are simply done in the outdoors. Family life centers around the fire, which is kept going for most of the time because of the difficulty of starting a new one.

For the Bushmen, education is the learning of survival skills. They must grasp the basic fact that people need at least as much energy input as they expend in order to survive. Thus, men intuitively determine whether chasing an antelope will use more energy than the meat will provide and continue or discontinue their hunts on that basis. Women learn to recognize and gather edible plants. All adults know that they must expend little or no energy or water during the middle of hot days.

As we write this description in the 1990s, a few Bushmen still lead this style of life. They can do so because the Kalahari Desert is large enough that they do not have to leave it and bleak enough that no one else wants to use it. Their numbers are decreasing, however, and sooner or later civilization will probably encroach upon them, and their way of life will become as extinct as that of the Afghan nomads that we discussed in Box 1.1.

FURTHER READING: Wannenburgh et al. (1979).

Table 1.3 *Population characteristics in selected countries*

Country	Birthrate (per 1,000)	Fertility rate (total)	Life expectancy (in years)
United States	16	2.1	76
United Kingdom	13	1.8	76
Italy	10	1.3	77
Hungary	11	1.7	69
Kenya	44	6.3	59
Turkey	29	3.5	67
Pakistan	39	5.5	51
Kuwait	35	4.9	76
Japan	10	1.5	79
Thailand	20	2.2	69
Argentina	21	2.9	61
El Salvador	33	3.9	66
Marshall Islands	49	7.2	61

Sources: Population Reference Bureau (1994); World Resources Institute (1994).

- *Birthrate* is expressed as the number of children born per 1,000 people per year. The world's birthrate in the mid 1990s is 25, resulting from the fact that approximately 10% of the world's women of child-bearing age are pregnant at any one time.
- *Fertility rate* is the average number of children that women in a society have during their lifetimes. In a stable population, women have an average of 2.1 children apiece, and the average fertility rate for the world's women in the mid 1990s is 3.1.

Wealthy societies generally enjoy, and in some cases actively promote, low birth and fertility rates – fewer children place fewer responsibilities of time and money on their parents (Table 1.3). Both birth and fertility rates are particularly low in Europe (12 and 1.6, respectively) and in Japan (10 and 1.5). The lowest rates are in Italy, leading to a population doubling time estimated as greater than 2,000 years. Similar low rates are shared by the white (non-Hispanic) people of the United States, but rates are slightly higher for African Americans and higher still for Hispanic Americans (Section 1.6). Birth and fertility rates are very high throughout Africa, where birthrates average 42 per 1,000 and the typical woman has 6 children (Table 1.3). The other area of particularly high birth and fertility rates is the dominantly Moslem region of the Middle East (we discuss the issue of religion and population planning in Section 1.7).

The growth rate of the world's population is its birth minus its death rate. In the mid 1990s, the growth rate is 100 million, the difference between 140 million births and 40 million deaths per year. (As discussed at the beginning of this section, the world population has a doubling time, t_{doub}, of 40 years.) For countries or any area smaller than the entire earth, the equation for growth rate must account for movements of people and becomes slightly more complicated: growth rate = births − deaths + immigration − emigration. For most countries, immigration and emigration are too small to have a significant effect on t_{doub}, and high growth rates result from high birthrates. Thus, Africa, most of Asia, and Central America have t_{doub} of 15 to 35 years. Conversely, t_{doub} for the United States and Canada is slightly less than 100 years and in the hundreds of years for Europe and Japan. The only major country in the mid 1990s with a shrinking population is Russia, where the death rate has increased since the collapse of the Soviet government and the birthrate has fallen because families appear to be reluctant to have children.

Lesser-developed countries have not always had the high population growth rate that they have experienced in the past half century. This exceptional growth results from a reduction in the death rate, caused largely by an improvement in worldwide health conditions and expansion of medical care following World War II. The process, referred to as the "demographic transition," is illus-

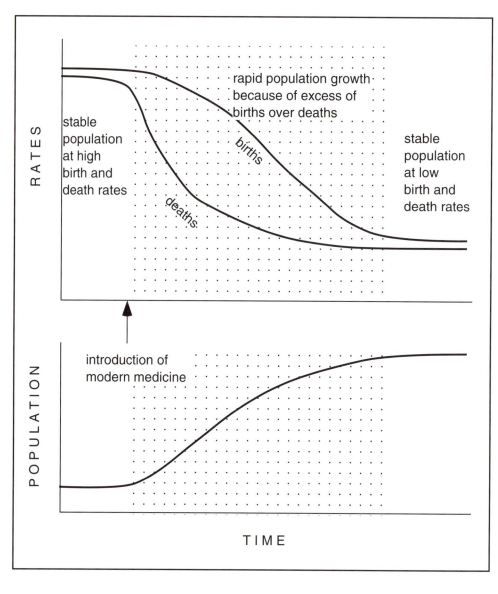

Figure 1.6 The "demographic transition." Prior to about the 1950s, most countries had populations in which both birth and death rates were high (left side of the graph). Generally, they were growing slowly because of a slight excess of births over deaths. Improved health policies in these countries during the later 1900s reduced the death rate, partly because of decline in infant mortality. Without reduction in the birthrate, the excess of births over deaths became very high, and the populations expanded rapidly (shaded part of the graph). Population stability can be achieved only by reduction of the birthrate to near the death rate (right side of the graph).

trated in Figure 1.6. Populations with high birthrates can be relatively stable if the death rate (particularly including infant deaths) is high. If the death rate is reduced by better health but the birthrate remains high, then the population explodes. Countries that succeed in reducing their birthrates (Section 1.7) can have stable populations even with low death rates. Most of the lesser-developed world has not achieved such stability.

Life expectancy and age distributions

Women in the United States today can expect to live 79 or more years, and men at least 72 years. These figures represent the anticipated ages by which half of the group has died; that is, life expectancy is the median age of death (Table 1.3). In the mid 1990s life expectancies are high and increasing in industrial countries, leading to predictions that nearly one-third of their populations will be over 65 by some time early in the 21st century. Life expectancies are lower in much of Asia and Latin America, and especially low in sub-Saharan Africa (commonly in the 40s). They continue to increase in almost all areas of the world except Russia, where life expectancy for men has decreased about 5 years in the 5 years since the end of the Soviet regime (also see the preceding discussion of growth rates).

The relationship between life expectancy and birthrate can be illustrated with the concept of "zero popula-

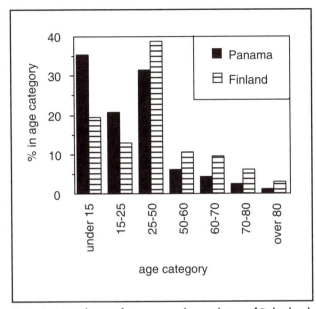

Figure 1.7 Distribution of ages among the populations of Finland and Panama. (Finland has a total population of about 5 million and Panama about 2.5 million.) The low birthrate and long life expectancy in Finland yield a population with a relatively small percentage of children and a large number of older people. Conversely, the high birthrates and short life expectancies in Panama yield a population "skewed" toward a concentration of people in younger ages and few older ones. Data are from United Nations, *Demographic Yearbook* (1992).

tion growth," which has particular significance for population planning. This concept, signifying no change in the population over long periods of time, requires an equality of births and deaths. This equality also requires an inverse proportionality of birthrate and median lifetime. For example, consider a population with 20 births per 1,000 people per year. If the population is not to grow, this requires 20 deaths per year among each 1,000 people (1 in every 50). An annual death rate of 1 in 50 could be accomplished if everyone lived to age 50 and then died. No population would ever organize itself so that everyone died at the same age, but the median age of death would have to be 50 for a birthrate of 20 per 1000 per year and no population growth. Thus, our desire for a long life within a stable population can only be satisfied by reducing birth rates. Otherwise, long lifetimes lead to a rapidly expanding population.

Because people are born young (a fact we should all remember), high birth rates naturally cause high proportions of the population to be children. Conversely, high life ex-

pectancies increase the percentage of elderly and decrease the proportion of children. The age distributions shown in Figure 1.7 demonstrate the different problems facing industrial and lesser-developed countries. In most industrial countries, about 20% of the population is under the age of 15 and 10% to 15% is over 65. The increase in the over-65 population because of increasing life expectancy is causing profound debate over the question of how a dwindling young population can care for an increasing number of old people who cannot care for themselves.

Conversely, in lesser-developed countries, high birthrates and low life expectancies cause the age distribution to be skewed (tilted) toward young ages. In much of Africa and western Asia, for example, children under 15 constitute 40% to 50% of the population, and people over 65 less than 5%. The proportion of children will probably increase in the next few years as today's children move into childbearing years and start to have even more children of their own. High proportions of children place a large, and thus far not resolved, need for nurture and education on their societies (Section 1.4).

Policy questions

We have suggested that the major population problem facing industrial countries is too many old people, and the major problem facing lesser-developed countries is too many young people. Do you agree with this assessment and, if so, do you think that you can do anything about these problems?

FURTHER READING: Jones (1990).

1.4 POPULATION AND WEALTH

A diminished power of supporting children is an absolutely unavoidable consequence of the progress of a country towards the utmost limits of its population. — T. R. Malthus, An Essay on the Principle of Population, 2:219

Prior to any further consideration of the world's population, we must describe a basic relationship that influences all discussions. A principal argument that world population is a serious problem is shown in the graph in Figure 1.8, which plots birthrate versus per capita gross national product (GNP); an approximately equivalent gross domestic product (GDP) has recently replaced the GNP in most tabulations. Shown on the graph are:

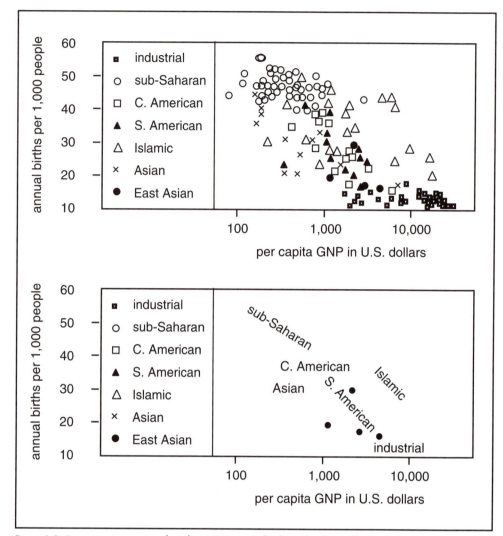

Figure 1.8 Per capita gross national product (GNP) versus birthrate for the world's countries. GNP is measured by a complicated economic formula and is generally slightly higher than the per capita income. Countries are classified into seven groups with different symbols:

1   Industrial (Europe, United States, Canada, Japan, Australia, New Zealand)
2   Sub-Saharan Africa
3   Central American and the Caribbean
4   South American
5   Islamic countries of North Africa, the Middle East, and western Asia
6   Asian (but not with Islamic governments)
7   Rapidly industrializing countries of eastern Asia (South Korea, Thailand, Taiwan, Malaysia)

The upper part of the diagram shows individual points for 130 countries. The lower part of the diagram shows the general area of the graph occupied by each of the various groups (except for the three individual points for group 7). With some exceptions, the graph shows a strong decrease in GNP with increase in birthrate (negative correlation). Countries that plot toward the upper right of the diagram show relatively high GNP for their high birthrate. Countries toward the lower left have low GNP despite low birthrates. The differences among the seven sets of countries are discussed in the text. Data are from World Resources Institute (1994).

- the world's most industrialized nations, all members of a group known as the Organization for Economic Co-operation and Development (OECD), whose wealth and low population growth rates place them at one end of the range;
- sub-Saharan Africa, at the other end of the range from the wealthy countries;
- Central America, with growth rates almost as high as those of sub-Saharan Africa;
- South America, somewhat more toward the wealthy end of the range in both wealth and growth rate;
- Islamic countries of the Middle East and North Africa, which have high population birthrates but include many countries that are wealthy because of oil production;
- rapidly developing countries of eastern Asia (principally South Korea, Thailand, Taiwan, Malaysia) whose birthrates have declined as their economic growth rates have increased.

Figure 1.8 clearly illustrates the tendency of slow-growing countries to be more wealthy – statistically, we could say that birthrate and per capita GNP are inversely correlated. Furthermore, because wealth is essentially the same as per capita consumption, this relationship means that people who live in countries with low birthrates consume more than people in countries with high birthrates. Before interpreting the figure too broadly, however, we should provide two warnings. First, the graph plots population birthrate, *not* population density. A plot of income versus density shows little relationship. For example, as mentioned in Section 1.1, most of the world's richest countries are in Europe, where population densities exceed those of many impoverished nations. Conversely, many of the world's poorest countries (e.g., Bangladesh, Nigeria) are also extremely crowded. The second warning about too-eager interpretation of Figure 1.8 is the statisticians' standard cry – a correlation does not prove a cause-and-effect relationship between the variables plotted. Either variable may be the cause or the effect, or both of them may simply be caused by some other factor not considered on the graph. We discuss this issue further in Section 1.7.

Policy questions

Do you think high birthrates cause poverty or that poverty causes high birthrates?

FURTHER READING: Sarre and Blunden (1995).

1.5 TOO MANY PEOPLE? A COMPARISON OF LESSER-DEVELOPED COUNTRIES WITH INDUSTRIAL COUNTRIES

We gave women freedom when we gave them jobs. — Michael Manley, former prime minister of Jamaica

The twin problems of the lesser-developed world are rapid population growth and poverty. To bring these issues into focus, we begin our discussion with a comparison of very poor and very wealthy countries. A visual indication of some of the differences is shown in Figure 1.9. The purely demographic differences have been discussed in previous sections and need only to be summarized here (also see Box 1.3). The lesser-developed countries have:

- birthrates of 30 to 50 per 1,000, compared with 10 to 15 in industrial countries;
- fertilities that average 6 to 8 children per woman, compared with approximately 2 in industrial countries;
- life expectancies of 40 to 60 years, compared with greater than 70 in industrial countries;
- age distributions in which 40% to 50% of the population is under 15, compared with approximately 20% in industrial countries.

These population characteristics are related to profound social and cultural differences between lesser-developed and developed countries. The differences cause misunderstanding – perhaps "friction" is a better word – in discussions of world population problems.

Education and employment

Compared with an almost universal ability to read and write in the industrialized world, literacy rates in many underdeveloped countries are less than 50%, and are marginally higher for men than women. The path to improvement is not clear in a country in which half of the population is under 15. Many of the women must remain home to care for the very young, and if the society tries to use a high proportion of the remaining adult population as teachers, who is left to produce the food and other goods that are necessary for survival? Unfortunately, most countries have necessarily solved this problem by limiting the number of years of required schooling and providing higher education to only a small percentage of the population.

The terms "labor force" and "unemployment" have somewhat different meanings in industrial and lesser-

(a)

(b)

Figure 1.9 Comparison of villages in poor and rich countries: (a) farm huts in Swaziland; (b) village on the northern side of Swiss Alps.

developed countries. The developed world regards the labor force as consisting almost exclusively of adults who have not reached the age of retirement. Employment of children is permitted only after they have reached some agreed-upon age, commonly around 16. Those adults who do not have jobs are unemployed, although we should note that the counting of unemployed people follows different rules in different countries; for example, in the United States a person who does not have a job is not unemployed unless he or she is actively looking for one. Using these various procedures, industrial countries commonly show unemployment rates of 5% to 10%.

BOX 1.3 THE CHILDREN OF RWANDA

In 1994 the world watched with horror the breakdown of Rwandan society. The killing of the Rwandan president (by shooting down his airplane) ignited a period of bloodshed that caused the deaths of perhaps a half million people in a population of about 8 million (exact figures are impossible to obtain) and sent about one-third of the Rwandan people fleeing into neighboring countries. Many of the dead were small children hacked to pieces with machetes, but murder was not the only form of enforced death – disease and starvation were also effective. The complete disruption of a society destroys whatever sanitation existed before, with the resultant spreading of disease. Cholera is almost impossible to treat in the absence of clean water and was responsible for many deaths. The displacement of farmers prevents planting and harvesting crops, which affects following years. Relief supplies cannot be delivered down roads clogged with refugees.

The responsibility for the savagery has been hotly debated. The presence of two very different tribal groups in Rwanda has contributed to tension since before the country gained its independence in 1962 from a United Nations Trusteeship administered by Belgium. During most of the brief history of the country, however, the groups had maintained an uneasy, occasionally bloody, peace. Some analysts attribute much of the tension in Rwanda to patterns of economic oppression continued since colonial times. For example, tea is a major export crop. It is produced by large organizations owned partly by the government, partly by individuals, and partly by foreign investors. Most profits in the tea industry go to a small number of Rwandans and to foreigners. Ordinary Rwandans employed on the tea plantations or in other, largely government, industries earn only a small wage that supplements the subsistence provided by crops that can be grown on a family farm. The per capita income in Rwanda is the equivalent of a few hundred U.S. dollars per year.

This book cannot discuss the political and economic issues behind the Rwanda problem. It is appropriate, however, to suggest that the underlying cause of the problems was population growth that outstripped resources. The population of Rwanda quadrupled between 1950 and 1994, a growth rate matched only by three other countries in the world (all in Africa and the Middle East). In 1994, 50% of Rwandans were under the age of 15, and the average woman had more than six children by the time she was in her 30s. The burden of caring for all of these children was placed on a society that is almost completely rural, with an economy dependent on agriculture.

Only about one-fourth of Rwanda is fully arable, and some of that land is allocated to tea and coffee for export. By 1994, each acre available for farming had to supply the needs of five Rwandans. The principal crop is maize (corn), supplemented by sweet potatoes, various lentils, and other vegetables. It is possible to feed five people from an acre of production (Section 2.2), but only with a highly efficient agricultural system. That system requires the use of expensive fertilizers and a marketing program that enables farmers to grow one crop and trade for their other needs rather than requiring each farm to produce food for its own subsistence. Some of that efficiency had been provided by coffee production by a large number of small landowners, but the worldwide 50% reduction in coffee prices in 1989 seriously reduced the ability of many Rwandans to buy the necessities of life.

By 1994, the concentration of power and wealth and the reduction in export earnings for ordinary citizens made it impossible to feed five people from an acre of subsistence farming. Domestic food production plus imports and aid provided Rwandans with only about 80% of the nourishment that they needed to sustain life, and much of the population was perpetually undernourished. Consequently, the country exploded into anarchy when an event triggered the exposure of underlying frustrations and stronger suppressed emotions, which many people would like to believe, are not contained in the human spirit. Regardless, the result was that people began to kill each other's children.

FURTHER READING: Newbury (1988); von Braun, de Haen, and Blanken (1991).

A further characteristic of employment in industrial countries is the small percentage of the labor force involved in agriculture, commonly less than 10%. The high agricultural productivity (Section 2.2) in industrial countries is the result of mechanization and other uses of energy. The nonagricultural labor force in the industrial world is about one-third in industry and two-thirds in services (a catchall term that includes a lot of people who attend to other people or spend their waking lives sitting in a chair).

The lesser-developed world provides a strong contrast to the industrial world in two ways. One is the use of children, some of them very young. Most of the children work with their parents on farms, but some have migrated to cities, where their plight is commonly terrible. The second contrast is that agriculture occupies half or more of the labor force in most poor countries, and production commonly requires the muscle power of both people and animals. People who do not have agricultural employment commonly migrate to cities (as will be discussed), where they support themselves by such sporadic jobs as street vending or manual labor. All of these conditions make it difficult to count either the labor force or the number of people officially employed in poor countries.

Income and wealth

Wage scales reflect the differences between industrial and lesser-developed countries. In terms of U.S. currency as a standard of comparison, the average person in the world's poorest countries earns a few hundred dollars per year. Because price and wage scales in a country tend to be roughly comparable, people can live for a year on a few hundred dollars by buying inexpensive food and clothes. Those fortunate enough to own land can grow their own food on small farm plots. A consequence of low wages, however, is that people in lesser-developed countries spend one half or more of their income on food, compared with less than 20% for people in industrialized countries.

Differences in per-person dwelling space starkly illustrate the contrast between the industrial and lesser-developed world. Americans have a lot of land to expand into and build the largest houses, most of them providing a few hundred square feet of living space per person – a typical bedroom in a medium-price home is about 150 square feet. These large houses also require a lot of energy for heating in the winter and cooling in the summer (see the discussion of energy use in Section 5.5). Conversely, housing for most people in lesser-developed countries consists of small rooms or apartments in cities or, in the countryside, huts made from local materials (mud, thatch, poor-quality plaster). Many of the dwellings provide less than 50 square feet per person and lack both electricity and running water. Crowding results both from the large size of families and the tradition that extended families live together.

In lesser-developed countries, the discrepancy of incomes between rich and poor is somewhat larger than in industrial countries (we discuss wealth discrepancies shortly). In terms of household (not individual) income, the wealthiest 20% of households receive some 40% to 60% of total national earnings in lesser-developed countries, compared with 30% to 40% of total earnings in industrial countries. Members of professions and the most successful merchants commonly have the same low birthrates and may live as well as similarly placed people in the industrial world, although they are hampered by the high cost of imported goods and of foreign travel. In some countries, they also spend a high proportion of their incomes on their children's education. By contrast, the lowest 20% of households receive considerably less than 5% of total income in lesser-developed countries and 5% to 10% in industrial countries. The discrepancy between rich and poor in lesser-developed countries would be larger than shown by these figures except for the tendency of poor households to receive some wage contribution from many family members, including children.

A low average wage scale provides an opportunity to industrialize that many countries have seized. The low wages and a plentiful supply of unskilled workers combine to make it possible for labor-intensive industries to produce goods more cheaply than is possible in wealthy countries. For example, most of the toys sold in the United States are imported from China, and the U.S. domestic textile industry has been severely reduced because of import of inexpensive textiles made abroad, primarily in Asia. Countries with low wages but relatively high educational levels have become major producers of electronic equipment, computers, automobiles, and other complex equipment. The growth of industries in low-wages countries and export of the products to the wealthy part of the world raise numerous ethical problems. Are wealthy countries "exporting jobs"? Are international corporations "taking advantage of" poor people in poor countries? Are these foreign-owned factories merely an effort of the industrial world to "export pollution"? We return to these issues throughout this book.

Women

Differences in employment and birth patterns between industrial and lesser-developed countries correlate closely with differences in the status of women. As with other issues, the cause-and-effect relationships are obscure. Here are two contrasting arguments.

1. The economic condition of lesser-developed countries controls the status of women. In a society that is poor and cannot afford an extensive educational system, children must spend most of their time close to home, and women must also remain there in order to care for them. These duties limit women's participation in education or in jobs that must be done away from home. Their occupations, therefore, involve agriculture chores on the family farm, making clothes for the family, possibly doing piecework at home (e.g., sewing or making small items such as toys, largely for export). Because most jobs in the society require rather heavy physical effort, the only industrial jobs available to women are tedious (e.g., textile making) and pay low wages. The only path to a better life for women lies in economic improvement that will enable their husbands to earn more money and relieve them of many of their present burdens.

2. Social attitudes toward women cause many countries to be poor. The cultures of many lesser-developed countries believe that men and women are so physically different that they have different roles in life. If women are unemployed and do not have an independent income, men become responsible for family support and thus assume total control. Men are respected for their ability to father many children and women for the size of their families. These attitudes lead to high birthrates because they prevent women from taking jobs that keep them away from their families, and the only path to economic improvement lies through the release of women from these arbitrary cultural restrictions.

Adherents of one or the other of the two philosophies find little or no common meeting ground, except when they come together to fight. The issue is clearly at the heart of the debate over "population control," and it will affect our discussion in Section 1.7.

Policy questions

Do you think that the best way to improve the condition of women in lesser-developed countries is to improve their general economies or that the economies will not improve until the status of women improves? Do you think that improving the educational level of people in lesser-developed countries will lead to greater wealth? For each question, do you think that people in industrial countries can, or should, do anything about the status of women and children in lesser-developed countries?

FURTHER READING: Ornstein and Ehrlich (1989).

1.6 TOO MANY PEOPLE? A DISCUSSION OF THE POPULATION OF THE UNITED STATES

Essentially three categories of work are emerging. . . . Call them *routine production services, in-person services,* and *symbolic-analytic services.* – Reich, The Work of Nations, p. 174

The United States holds 5% of the world's people in 7% of its habitable land area, yielding a U.S. population density of 70 people per square mile, some one-third less than the world average. In addition to this lower average density, the United States is also fortunate that 20% of its land can be used for crop production, compared with only 10% worldwide (Chapter 2). As we discuss in Chapters 5 and 6 about various commodities, the 5% of the world's people that are Americans consume approximately 30% of the world's resources (this rate of consumption does not please all non-Americans). A first glance at the statistics suggests that the United States should not have any population problems. Yet it seems to. Why? Let us start with a brief history of population changes during the past century.

Population in the 20th century

The population of the United States has descended almost exclusively from people who came here from other countries (some willingly and some unwillingly) within the past few centuries. Thus, the United States has an ethnic and cultural diversity and a range of social and economic conditions much larger than those of any other industrial country. Much of the discussion of population in the United States in the 1990s centers around this diversity. Figure 1.10 summarizes information about the U.S. population in the census years of 1910, 1950, and 1990 (also see Table 1.4). In interpreting this information, we must realize that the data were collected somewhat differently and summarized into different categories in these years. Thus, comparisons cannot be rigorously precise, and we keep the following discussion very general.

Total population growth from about 90 million in 1910, to 150 million in 1950, and 250 million in 1990 indicates a roughly constant doubling time of 60 years throughout the century. This doubling time is consistent with annual birthrates of 20 to 25 per 1,000 plus immigration averaging approximately 500,000 people per year. The total fertility (lifetime births per woman) has decreased from 2.5 to 3 in the early part of the century to 2.1 now. Approximately two-thirds of U.S. population

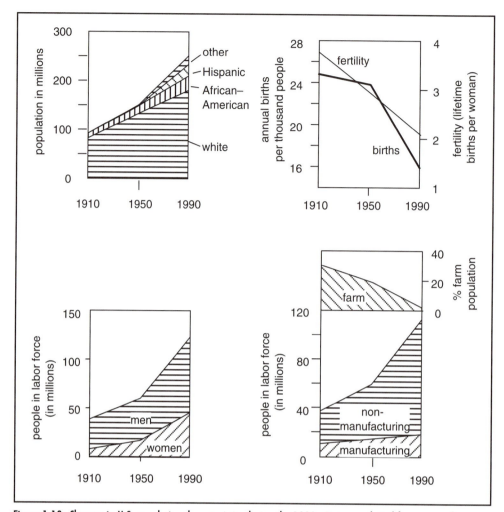

Figure 1.10 Changes in U.S. population characteristics during the 1900s. Data are plotted from census figures for 1990 and discussed more completely in the text. Some uncertainties and inadequacies result from the fact that the Bureau of Census did not use the same classifications of data or collect the same information in each census. In racial classifications, the term "white" refers to people not of Hispanic ancestry. Data are from United States, *Statistical Abstract of the United States* that provide census data for the years 1910, 1950, and 1990.

growth is the result of an excess of births over deaths, and one-third represents new immigrants. The nature of the immigrant population has undergone major changes during the 1900s. In the early part of the century, it was almost exclusively European, but the proportions have shifted to a vast predominance of Hispanics, about two-thirds from Mexico (Box 1.4), and a significant proportion from eastern Asia.

Occupations and life-styles

Not surprisingly, a century has brought numerous changes in the way Americans live (we do not refer only to the current hypnotic effect of television or the number

of people wired into cyberspace). At the beginning of the century, the United States was primarily an agricultural country, with approximately one-third of the population living on farms. That percentage has dropped to the present 2%, with three-fourths of the population living in urban areas. Consequently, many Americans have a limited knowledge of the methods of production of the food that they eat. This is the result of ease of buying a vast array of produce and manufactured meals and the ability to spend less time producing meals from scratch in kitchens across the country. For example, bread is something that is found inside plastic bags in markets, and fruit pies come from the freezer cabinet in supermarkets. Time once spent in the garden and kitchen is now devoted to shop-

Table 1.4 *Selected population characteristics of the United States in 1990*

Race	Population (%)	Birthrate (per 1,000)	Fertility rate (total)	Immigration rate (in 1,000s)	Family income (in 1,000 $)
Total	100[a]	16	2.1	1,500	30
White	72	15	1.9	100	37
African American	12	23	2.2	20	21
Hispanic	9	29	2.8	950	23
Other	7			450	

Notes: White does not include Hispanic. Other races include mostly Asiatic plus some Native American. Per capita income for women is 71% of the income for men.
[a]100% = 249 million.
Sources: United States, *Statistical Abstract of the United States,* census data for 1990.

ping and to "leisure" time. Despite the reduction in agricultural workers, consolidation of farms and improvements and intensification of food production have increased the total amount of food available and made the United States the world's largest exporter of agricultural products (Section 2.2). Food processing and marketing now employ more people than farming and ranching.

An additional major change in occupations during the past century is the decrease in percentage of the work force employed in factories or other mechanical and manufacturing jobs. In the first half of the 1900s, one-third to one-quarter of employed people held jobs in manufacturing or repair that required mechanical labor ranging from semiskilled to highly skilled. By 1990, that percentage was down to almost 10%. The decrease in the need for unskilled labor was even more dramatic. The reason for this decline in jobs involving physical (manual) work is hotly debated, but certainly part of the explanation is the continual improvement in technology, including automated processes that replaced repetitive operations by people. Another part of the explanation, stressed by labor interests, is that manufacturing jobs have been exported to low-wage countries (Section 1.5).

The people who used to be employed in agriculture and other physical labor now must look elsewhere for work. Of necessity, the new opportunities that have appeared are in a broad, and ill-defined, category referred to as the "service" industry, which according to some definitions includes restaurant waiters, medical workers, teachers, salespeople, lawyers, secretaries, financial analysts, even people who write books. Most of the new service jobs require a higher level of education than the labor of the past, and to the extent that educational achievement is measured by number of years in school, the United

States clearly has improved. About one-third of the high-school-age population was in school in 1910, but almost all of this population attended school in 1990. Similarly, 3% of the appropriate age group was in college in 1910 and about 40% in 1990. It is unclear, however, whether this extra time in school provides sufficient relevant education for the tasks of U.S. society in the 1990s. In international comparisons, the United States generally ranks low in mathematics and science skills relative to other industrialized nations.

The simple educational and employment data gloss over the profound adjustments that many American families and individuals have had to make. In the last half of the 20th century, families who had made their way for many years by "good honest labor" found that they and their offspring would either have to learn to do more complex jobs or support themselves with very low paying positions (commonly referred to as "hamburger-flipping") that were generally less satisfying and often less secure. People who were able to make the adjustment to more complex work have enjoyed rising incomes, while people who have not developed this ability remained at the low end of the economic scale and experienced stagnant or declining incomes. In addition to the increasing difference in incomes, upper-income people were better able to move from one job to another and were less likely to be put out of work as their employers reduced their number of employees in order to remain competitive with foreign companies operating at lower costs. All of these disparities have led to increasing contention and animosity among segments of the U.S. work force.

Another major facet of employment change in the United States is the participation of women in the labor force. In 1910, farm wives worked domestically and took

BOX 1.4 THE SOUTHERN BORDER OF CALIFORNIA

Few areas of the world offer such an exciting blend of cultures as the 2,000-mile-long border between Mexico and the United States. This cultural mix also causes intense furor.

For many years, Mexicans have crossed the border looking for work. At one time, the United States encouraged this process, permitting farm laborers to enter the country on temporary visas and then to return home with most of their earnings intact. In the 1970s and 1980s, Mexican immigration to the United States began to accelerate and now has reached what some refer to as "crisis" proportions. Some of the border-crossers arrive with official permission from the United States. In the past few years, however, illegal immigration has reached an uncertain number (it is hard to count illegal people) that is thought to be high, perhaps as much as 1 million per year. Most of them do not go home but stay in the United States to work. Many send a large portion of their earnings back to their families in Mexico (an aggregate of about $3 billion per year). In a sense, the United States has acted as a safety valve for the excess population caused by Mexico's high birthrate (43 per 1,000 people per year in the 1970s, "reduced" to 27 in the 1990s).

The problem is centered in, although not restricted to, southern California. In an area with more than 10% unemployment in the 1990s and a high demand on public welfare, schools, and medicine, many citizens are incensed. They point to these facts: that some illegal immigrants receive public money that could be spent on legal Californians; that many unemployed Californians could have jobs if they were not occupied by illegal aliens; that the increased number of people makes life in a crowded city even more difficult. They want the government to send the illegal Mexicans home and to prevent any more from entering the United States.

Sending them home is more easily said than done. The illegals work for low pay, and many industries, services, and agricultural businesses depend on them — by one estimate, more than 90% of the employees in the garment industry of Los Angeles are illegally residing in the United States. The employers will not send them back to Mexico without government pressure, and policing the approximately 7 million commercial operations of southern California requires more resources than are available to all of the governments involved at the federal, state, and local level. Keeping illegal immigrants out by force is also virtually impossible. Can the United States build an impregnable fence along its entire 2,000-mile border? How many police and military personnel would be needed to patrol it?

Solutions to the problem involve a combination of persuasion and coercion. Here are some that have been suggested:

- The United States should enforce minimum-wage and workplace safety laws in all of its businesses. If illegal workers could not be hired for a lower wage than U.S. citizens, then there would be no economic incentive to employ them and there would be little reason to cross the border illegally.
- California should eliminate all public assistance to illegal aliens. This elimination might include welfare support for food and housing, public education, and medicine except for emergencies.
- Mexico should make provision to absorb its burgeoning population. Because most of the population is from rural areas, land reform that ensures farmers title to enough land to support their families would retain many displaced people. Creation of labor-intensive industries in the countryside would have a similar effect.

The effect of the North American Free Trade Agreement, inaugurated in 1993, on immigration is uncertain. One U.S. argument in favor of NAFTA was that expansion of industry in Mexico would reduce the need for emigration, although a strong counterargument was that any jobs created in Mexico would simply reduce the number of jobs in U.S. industries. Conversely, import of cheap food produced on the mechanized farms of the United States may put Mexican farmers at a disadvantage and cause further displacement.

We will not go further into the details of the various issues surrounding illegal immigration. We can, however, point out that it is closely related to a birthrate that is higher than the rate of resource development.

FURTHER READING: Andreas (1994).

care of the house and children, while heavy industry was staffed mostly by men, and only 20% of adult women were counted as "employed." Official employment of women rose to nearly 50% in the 1990s, although most women are in lower-paid service jobs and their average income is about two-thirds that of men. Another contentious issue – we have so many – is whether the difference in men's and women's salaries is the result of women having jobs mostly in lower-paying occupations or whether men receive higher salaries when they and women have identical positions.

Income and race

The issues of wealth and race are so inextricably tangled in modern America that we treat them together. National debates over the distribution of income and wealth have grown more intense as the United States has become more racially diverse during the 20th century. They have also become more acrimonious. We attempt to clarify a few of the problems here.

As the U.S. population has increased throughout this century, there has been a major shift in its ethnic makeup. Throughout the first half of the 20th century, the U.S. population consisted of approximately 90% of whites descended almost entirely from European immigrants and 10% of African Americans descended from people brought to the country during the slave trade in the 16th through 18th century. Starting about 1950, the proportion of African Americans has risen to about 12%, the result of slightly higher birthrates and total fertility. During the same period of time, the percentage of European (non-Hispanic) whites declined to approximately 80%, largely as a result of immigration of Hispanic whites from Latin America and partly because of Asian immigrants. About two-thirds of this Hispanic influx was from Mexico, partly through normal immigration procedures and partly by illegal crossing of the border from Mexico to the United States (estimated to be more than 500,000 per year in the 1990s; Box 1.4). The high birth and fertility rates of Hispanic immigrants have contributed to rapid growth of their percentage of the population from less than 1% in 1900 to about 9% in the 1990s.

The change in racial proportions during the 1900s has not significantly affected the distribution of U.S. wealth. Most assets remain in the hands of non-Hispanic whites, and their average household income of more than $30,000 per year is considerably higher than those of Hispanics ($22,000) and African Americans ($19,000). The average

current income of African Americans at 60% that of non-Hispanic whites is only slightly higher than the 50% rate in 1950, when discrimination against African Americans was both legal and official. Many people attribute the small improvement in African American income to government programs begun in the 1960s to recruit African Americans to better-paying jobs. In the mid 1990s, however, unemployment among African Americans runs about 15% and reaches more than 50% in the desperately poor parts of the nation's large cities. Unemployment rates for Hispanics are slightly less than rates for African Americans and less than 5% for the white population.

Not only are wealth and income concentrated in the hands of non-Hispanic whites, but they are also highly concentrated in a small percentage of this population. The richest one-fifth (upper quintile) of Americans earns about 45% of all the money earned in the country; the lowest 20% earns 4%. These proportions do not appear to have changed significantly from 1950 to 1990. Disparity in family incomes, however, is accentuated by recent trends to two-income families, in which both spouses have professional positions and resulting paychecks. Wealth measured as total assets, not just annual income, shifted greatly into the hands of a small number of Americans during the 1980s and 1990s, and the richest 1% of Americans now owns more than one-third of the nation's assets.

The relationship between wealth and birthrate within the United States is similar to that between wealthy (industrial) and poor (lesser-developed) countries (Section 1.3). Higher-income families have fewer children, and in the 1990s the fertility rate for non-Hispanic whites is below the 2.1 children per woman that is necessary to maintain a stable population. Different segments of the society attribute this low rate to (choose one or more depending on your attitude): greater opportunities for women to have interesting jobs outside of the home; economic conditions that require a family to have two incomes (husband and wife) in order to live well or even survive; greater use of birth control, including abortion; a high rate of divorce that results in fewer child-bearing years within a relationship and leaves parents, mostly women, raising children by themselves.

Among African Americans, the relatively high fertility rate of almost four children per woman in the early 1900s has now declined to 2.2, just marginally above replacement. This reduction, however, has been primarily in higher-income families. Fertility rates are particularly high among unemployed African Americans, many of whom are supported by some form of federal and local welfare and commonly are not living as two-parent families. Approximately half of the African American children born in the 1990s are born to single women, mostly younger than 18. This burgeoning group of children leads to an intense debate over who, if anyone, will support them.

Fertility of Hispanic women (approximately 3) is the same in the U.S. as in Mexico, their principal country of origin. This fertility gives Mexico a population doubling time of 31 years and should yield the same rate of increase for the Hispanic population of America. Thus, assuming no further immigration and growth of the non-Hispanic white or African American groups, at current rates the U.S. population will be almost one-third Hispanic by the year 2050. Most Hispanic children live in two-parent families, and the high birthrates are primarily related to the dominantly Catholic faith and traditions of the group.

Policy questions

Here are a few questions about your attitude toward population issues in the United States.

- Do you think that the United States has done enough to improve the condition of African Americans? Are more programs and more opportunities the only way to solve the problems of family disintegration and the elevated birthrates among that segment of the African American population whose children are most likely to require government assistance?
- Do you think that programs that give African Americans preference for jobs and education have reduced the competitiveness of the African American population and the employment opportunities for the white population and that, consequently, these programs must be discontinued?
- Do you think that immigration from Mexico, particularly illegal border crossing, must be stopped immediately, even if the United States must erect an impenetrable fence along the entire border? Or do you think that, because the United States owes its existence to immigration, people from all countries should be welcomed without restriction?
- Do you think that income distribution must be equalized, possibly by imposing higher taxes on the wealthiest segment of the population, or do you think that there must be a reduction of government involvement in the lives of individual Americans?

FURTHER READING: Lutz (1991); Reich (1991); United States (annual).

1.7 POPULATION CONTROL

The only sensible course for our species is to do everything possible to stop its population growth rapidly and humanely and begin a slow population decline. – R. Ornstein and P. Ehrlich, New World, New Mind, p. 245

Earlier in this chapter (Section 1.4), we asked whether the world or any part of it is growing too fast. Now we ask the related, but quite different, question: Should we attempt to do anything about population growth? Are we out of control, threatening our future existence, or are we following a natural and self-correcting path to a world in which a larger population does not pose a problem to survival and happiness? Before proceeding, we must realize that different people ask the questions in very different ways. Thus, we distinguish between "carrying capacity" and "quality of life," neither of which can be defined easily.

Carrying capacity

Carrying capacity refers to the number of people that the earth or some part of it can support – keep alive – in a minimal fashion. As we discussed in Section 1.2, populations need an absolute minimum of ~1,500 kcal of food per person per day plus shelter and some form of energy, which is used for heating and cooking. The energy needed for heating is small in the tropics, but unless we try the almost impossible task of living on raw fruits and vegetables, all of us must have energy to cook grain and meat. Thus in the simplest sense, carrying capacity is based on the amount of food that can be grown by subsistence agriculture and the amount of fuel that can be harvested. More realistically, carrying capacity depends on the ability of a society to obtain energy, by purchase abroad if necessary.

In a nontechnological, subsistence, society, energy is mostly supplied by wood and brush (Section 5.2). The use of wood both for shelter and as an energy source places a double stress on resources in many countries. Many people, commonly young children, spend their days scavenging scrubby trees or brush from deforested countrysides, sometimes walking 10 to 20 miles a day to return with enough fuel for the evening meal. This scavenging either strips semiarid land with little initial vegetation or prevents regrowth of forests that have been logged over for commercial use, and, as we discuss in Section 8.3, deforestation of trees and brush seriously affects about one-third of the world's countries. Some countries are losing forest land at a rate of 4% per year, which obviously cannot be maintained for more than 25 years without almost complete elimination of the resource. Deforestation not only reduces wood supplies but also enhances erosion of agricultural lands by removal of one of nature's principal methods of dispersing flood waters through reduction of runoff.

Areas of high population density require more energy than less-crowded ones. The additional energy permits the operation of an industrial agriculture and distribution system (Section 2.2). Because this energy costs money and must be imported as oil or coal by most of the lesser-developed world, countries with high population densities cannot rely solely on subsistence agriculture but must allocate some of their production to export crops that earn a foreign income. This allocation is done by the free market in some countries and by government planning in others.

Based on these principles, we ask whether the carrying capacity of the earth, or any part of it, has been exceeded. For the world as a whole, the answer is no. In the mid 1990s, only 10% of the world's 193 countries are unable to produce at least 90% of the calories that their populations need, and other countries, particularly industrial ones, produce a considerable excess of food (Sections 2.2 and 2.3). The problem is not quantity of world food production but its physical distribution to people who need it. Almost all of the people in the world who are hungry simply do not have enough resources to grow food or the money to buy it, and extensive starvation commonly occurs only in countries racked by civil wars or other internal conflicts (e.g., Afghanistan and Angola). Extreme examples include Somalia (see Box 2.1) and Cambodia, where the breakdown or overthrow of governments allowed armed groups to carry out a policy of deliberate mass extermination by starvation. Some governments also use starvation as a method of controlling rebel or other minority groups. For example, attempts by the international community to deliver food aid to the black population of southern Sudan and the Arab population of northern Niger were blocked by the Arab government of Sudan and the black government of Niger respectively.

Although entire countries usually do not exceed their carrying capacity, some areas within them may do so, at least temporarily. A prime example is the flight of nomadic people southward from the savannah lands on the southern edge of the Sahara Desert – the Sahel – when a combination of drought and overgrazing in the 1970s destroyed their ability to feed themselves. Whether these areas can again become habitable with more rain and less population is unanswered at this time. Another temporary example is the migration of Americans out of the "dust bowl" of the Midwest during the 1930s. The problem there was a combination of drought and improper agricultural practices, but the area has now been restored to a region of extreme productivity.

We come, once more, to the issues of cause and effect. Are local carrying capacities exceeded, if only temporarily, because of improper use of the land, or is there some natural limit? Does starvation occur in countries because of civil war, or do wars start because the population is outstripping resources? Careful study of the variety of complex situations that have evolved in a range of countries may shed light on this issue, but we can only offer a platitude – we must always be aware of the earth's limitations but remain confident that proper use of it will enable us to live well.

Quality of life

The preceding discussion brings us to the second potential problem – the quality of life. Perhaps the earth can support 5.5 billion people and many more at a certain standard of living, but do we want it to do so at this standard? If we were all to have a high standard of living, the pressure on the earth's resources from a high-consuming society would be far higher.

Quality of life is a nebulous concept. In a general sense, we all know what high quality means: plenty of good, tasty, nourishing food; open spaces to roam around in; a comfortable house that pleases the eye and provides not only shelter but a place for recreation; the ability to spend much of our time in enjoyable and rewarding jobs and in recreation – that is, a minimum of drudgery; inexpensive transportation so that we can go to the store, visit friends, take a vacation. The problem is that people's views of the "good life" vary enormously. Some people are unhappy with a house with fewer than four bedrooms and a summer that does not include at least one trip abroad to an expensive resort. Others are content with a small apartment and a summer visit to their family.

The level that we use as an "acceptable" quality of life has changed in recent decades as technology has provided us with more goods. Both in industrial countries and lesser-developed ones we tend to expect more from life, at least in the form of material goods. As we do so, we put more pressure on the earth's ability to provide the goods that we want. We discuss this issue more in Chapters 5 (energy) and 6 (mineral resources), but should mention here that the rate of consumption by people living in industrial countries is far higher than by people in poor countries. If poor people attempt to increase their consumption to anywhere near the level of rich ones, then the world may well come close to its carrying capacity.

All industrial countries provide resources, either domestic or imported, well above the minimum that defines a country's carrying capacity. (In that case, we might wonder why we meet so many grumpy, dissatisfied people, but that is a philosophical problem far beyond this book.) This availability of resources does not mean that the developed world has no population problem, merely that the problem is locally less severe than in other areas. Consider these miscellaneous observations:

- The average commuter in Los Angeles spends 3 to 4 hours per day in a car (the traffic and commute in Tokyo are worse).
- The North Atlantic, which borders only industrial countries, has populations of commercial fish reduced to 10% of their size 50 years ago (Section 2.4).
- The effort to maintain high food production has caused depletion of groundwater and contamination of water runoff with pesticides in many areas of Europe and North America (Sections 4.4 and 7.3).
- Unemployment, mostly among ethnic minorities, is 50% or higher in some inner cities in the United States.

These problems might not exist if the populations were smaller. Does that mean that we must reduce birthrates and thus reduce the number of people, or do the problems result from inadequate planning that is easily remedied? Thus, we ask the question, Although we can support ourselves in a more crowded world, will we enjoy it? How many wilderness areas must be destroyed to build residential communities? How many more people can live in Mexico City, or Cairo, or New York, without reaching unacceptable levels of poverty, violence, and smog? This brings us to the great debate of population control.

Population control

The world is not clearly divided into two sides in the debate over population control. As a first approximation, however, we can distinguish the attitudes of the rich industrial countries and the attitudes of the lesser-developed ones.

The world's industrial countries tend to want population control in the poor ones. Rich countries argue that creation of prosperity is impossible in populations that are expanding too rapidly. They also point to the fact that some rich countries now send almost 5% of their grain production to lesser-developed countries. Because this grain generally cannot be paid for, it is simply donated free, and the donors ask the recipients to keep their populations from growing to a level at which even more aid must be provided. Some rich countries have considered discontinuing any form of aid to countries that do not promote birth control.

The attitude that population reduction must be the first priority faces several challenges. One is religious opposition by Islamic and Catholic groups in both rich and poor countries. The effect of religious attitudes varies from country to country. Among predominantly Catholic populations, birthrates are high in Central America, in Mexico, and among the Hispanic people of the United States. Catholics in industrial countries, however, show little difference in birthrates and fertility from other segments of the population; for example, Italy has one of the lowest birth and population growth rates in the world (Section 1.1). Birthrates among Moslems are very high in the Middle East and North Africa but lower among Islamic people in other countries such as Malaysia and Indonesia.

The concept of population control meets significant philosophical resistance among many people in poor countries. They can be divided into three groups: (1) those who are concerned about increasing populations; (2) those who view population increase as normal and healthy; and (3) the vast majority who do not have enough education to understand the issue. Regardless of their viewpoint, however, all of them resent being dictated to by rich folks, and particularly they dislike the idea of being forced to adopt birth-control or other policies as a condition of receiving food or other aid. Lesser-developed countries also have one other important argument against external coercion. As we shall discuss throughout this book, a person in a lesser-developed country consumes far less of the world's productivity and causes less of its environmental pollution than a person in a rich country. Thus, the birth of many people in poor nations may be less of a problem for the world than a few births in rich nations. Thus population control, if possible at all, can only be effective if it follows policies that do not interfere with the deeply held cultural beliefs of the people who practice it (Boxes 1.5 and 1.6).

BOX 1.5 BIRTH CONTROL (CHINESE STYLE)

The Communist forces that established the People's Republic of China (mainland China) in 1949 inherited a country that was already the world's most populous, with more than 500 million people. It was also relatively poor, and the new rulers decided that population growth must be restrained or even stopped. The methods used to reduce population growth have been harsh even by Chinese standards and draconian by the standards of most of the rest of the world. In a purely technical, numerical, sense, the control has worked. Between 1950 and the early 1990s, the population rose to approximately 1.1 billion, a doubling time of 40 years. During that same period, many of the world's lesser-developed countries tripled or almost quadrupled their populations. By about 1980 fertility rates were lowered to about 2.3 children per woman, and in 1991 the government announced that they had dropped below the replacement rate of 2.1.

Does the world applaud? Some people have — they see China as keeping its house in order. Before doing so, we should realize that the Chinese birth control program has not been voluntary. The methods used center around cadres appointed by local governments to oversee "reproduction" in some local area (village, rural district,etc.). The milder methods of control include public humiliation of families who have more than one (perhaps two) children, and the harsher ones range up to denial of access to food and water as punishment. Government programs gradually have become more formalized, and in 1979 China adopted a one-child policy, as follows:

- After the birth of the first child, the wife was required to have an intrauterine birth-control device inserted.
- A second child was permitted without penalty if the first one died.
- Birth of a second child without permission led to sterilization, particularly of the wife.
- Children could not be born without permission from local au-

thorities; penalties could include forced abortion, imprisonment of the husband, and confiscation of property; infanticide appears to have been rare but did occur.

The enormous size of China and the resistance of many people to these stringent laws prevented them from working effectively all of the time. Periods of several years witnessed little or no coercion. One reason for resistance was that few Chinese had any public income security and, thus, were wholly dependent on their family for support during illness or old age.

The effort to control births is only part of the Chinese government's effort to control the entire country. A demonstration of this desire was provided in the summer of 1995, when the United Nations asked world governments to send delegates to a conference in Beijing on the status of women (see Section 1.5 and Box 1.6 for further discussion of the relationship between birthrates and the status of women). A parallel conference of nongovernmental organizations was held at the same time in a town outside of the capital. The Chinese government realized, too late, that extending an invitation to the conferences had been a mistake. Fearful that an improvement in the status of women would undermine government authority, both in family planning and other aspects of society, the government resorted to such tactics as not granting visas, harassing and arresting delegates on false pretexts, and preventing the dissemination of news or interviewing of delegates.

Chinese-style birth control is presumably impossible in any country with an elected government. It may become both unnecessary and impossible as China industrializes and its economy improves. Not only will the more powerful economy lead to the type of fertility reduction found in other industrial countries, but the increased wealth of the people may give them greater freedom from government control.

FURTHER READING: Aird (1994).

The consequence of these arguments is that lesser-developed countries take an exactly opposite approach to the interaction of population and wealth than industrial countries. That is, the poor and the rich reverse the cause-and-effect relationship of prosperity and birthrate. Instead of following the suggestions (demands?) of rich countries to reduce the birthrate as a way to gain prosperity, poor countries commonly believe that increasing prosperity will reduce the birthrate. Their evidence is that developed countries have low birthrates.

Policy questions

We have raised two related, but slightly different, issues and ask you questions about each one.

BOX 1.6 RICE-ROOTS FAMILY PLANNING IN BANGLADESH

The Brahmaputra and Ganges Rivers gather rain and meltwater from the Himalayas, flow together, and then discharge through a meshwork of channels into the Bay of Bengal in the northern Indian Ocean. During the past 20 million years, the Himalayas have risen as India has driven northward and compressed the collision zone with Asia into slabs of upthrusted blocks. The rapidly growing mountains supplied so much sediment to the rivers that they built the Asiatic shoreline southward as a complex series of deposits much like the Mississippi delta.

Bangladesh occupies the new land created by the Brahmaputra and Ganges Rivers. It was part of British India until it became East Pakistan when India and Pakistan gained independence in 1947. Bangladesh began its existence as a country when the eastern and western parts of Pakistan broke into separate nations in 1971 following a bloody civil war. The country is low and flat and rainy. It is prone to flooding when the summer monsoon sweeps northward out of the Indian Ocean, sometimes bringing rainfall of more than 10 inches per day. The southern coast is dotted with low islands that suffer almost complete devastation when typhoons cross them.

At the time of independence in 1971, the predominantly Moslem population of Bangladesh was 70 million. With a doubling time of 25 to 30 years and a fertility rate of more than 5 children per woman, by the 1990s the population grew to about 120 million, 45% of whom were under the age of 15. The high birthrate resulted from a combination of factors: the respect accorded to both men and women who have many children; the necessity of maintaining large extended families to provide support for the elderly and for other family members who have no land and cannot find employment; the importance the Islamic religion attaches to home life and women's roles in it; the absence of opportunities for women to support themselves or engage in activities outside the home. The country has one of the world's lowest per capita incomes — about $200 (U.S.) per year.

The population density in Bangladesh is now about 2,300 per square mile, one of the world's highest. We can offer the following comparison. The state of Louisiana lies mostly on the delta and landward deposits of the Mississippi River. Thus, it has a geology and geography similar to that of Bangladesh, although Louisiana has some (slightly) higher elevations on rocks older than the recent river deposits. Louisiana has a population of about 5 million and an area 10% smaller than that of Bangladesh, giving it a population density of about 100 per

square mile. If one-half of the entire population of the United States were to move to Louisiana, it would have the population density of Bangladesh.

The fertile soils deposited by the rivers combine with the rainfall to give Bangladesh a high agricultural productivity. The major, but not only, crop is rice, which is planted and harvested almost totally by hand, plus some use of water buffalo and oxen for heavy pulling. Bangladesh produces about 250 pounds of rice per person per year in addition to large amounts of other crops. Despite this high productivity, the population growth of Bangladesh has overwhelmed the country's economy. By the 1990s, less than 90% of the calories needed for adequate nourishment of the population were available from all sources (domestic production, imports, foreign aid), thus giving Bangladesh one of the world's most undernourished populations.

Private and governmental groups of Bangladeshis saw the problem looming and began to introduce family planning in the 1980s. They had a fundamental set of principles. The people must accept the importance of birth control, not have it forced upon them. The message should be communicated both to women and men, both of whom must accept the fact that having numerous children was no longer the sole mark of importance in the community. The best instructors would be women who knew the communities in which they worked. Simple methods of birth control, primarily by inoculation of women, would be far more effective than complicated sequences of pills or other techniques.

None of the messages about birth control would have been effective without efforts to change the economic and social situation of women. One program grouped women into economic cooperatives and provided loans that let them plan their own moneymaking activities, giving them some financial independence and comradeship and support within the community. This last factor is particularly important in a society in which women are commonly isolated in the home and have few friends outside of it.

By these efforts, Bangladesh has gained some control over its population growth. Instead of a fertility rate of 7 and a doubling time of 25 years, the fertility rate has now dropped to about 3 and a doubling time probably greater than 30 years. These rates are still high by the standards of industrial countries, and, at this time, the future of Bangladesh is unclear. It is, however, less murky than it was 10 years ago.

FURTHER READING: Amin et al. (1994).

- Do you think that any form of birth control is morally correct, and do you think that governments should be involved in any way, either mandating, supporting, or preventing the practice of birth control by individuals?
- Do you think that industrial countries should provide aid to lesser-developed countries if those countries do not make efforts to control their population growth?

FURTHER READING: Ornstein and Ehrlich (1989); Hardin (1993); Sarre and Blunden (1995).

QUANTITATIVE QUESTIONS

This book contains three types of quantitative questions. One type provides information for a precise numerical answer, which can be obtained by arithmetic for most problems but requires a small amount of algebra and trigonometry for other problems. A second type of question does not provide enough information to obtain a precise numerical answer but can yield a numerical estimate. The third type deals with scientific concepts and can be answered without numerical calculation or estimation.

PROBLEMS

1 According to a theory proposed by T. R. Malthus in 1798, population should increase exponentially, while food supply should increase linearly. Using this assumption, start with a population that produces just enough food for adequate nourishment. After 10 years, the food supply has doubled and continues to increase linearly with time. The population, however, grows exponentially, with a doubling time of 20 years. Does this population ever reach a time when it does not produce enough food for adequate nourishment? This question can be solved using both diagrams and calculations.

2 The equation for exponential population growth can be expressed as

$$\ln(P_t/P_0) = kt$$

where

P_t = population at time t
P_0 = starting population
k = rate constant
t = time

Using the very approximate table of natural logarithms shown below:

a. For a starting population of 1 million and a rate constant of $k = 0.02$, how large will the population be after 80 years?

b. Calculate the rate constant for a population that quadruples (four times original) in 35 years, and calculate the P_t/P_0 after 55 years.

ln1 = 0	ln6 = 1.79
ln2 = 0.69	ln7 = 1.95
ln3 = 1.10	ln8 = 2.08
ln4 = 1.39	ln9 = 2.20
ln5 = 1.61	ln10 = 2.30

3 The life expectancy in a country is 60 years, and the birth rate is 25 children per 1,000 people per year. Is the population increasing, decreasing, or remaining stable?

4 Data for two countries (A and B) are shown.

	Country A	Country B
population	20 million	30 million
total area	50,000 square miles	100,000 square miles
% inhabitable land	80%	50%
% arable land	40%	30%

From the standpoint of population density and potential per capita food production, which country do you think would have the highest standard of living?

5 Use the data in Table 1.3 to make an estimated rank order of the various countries from fastest to slowest growing.

6 The country of New Zealand occupies two islands (North and South Islands). North Island has approximately two-thirds of the New Zealand population

and 40% of the total land surface. Quantitatively compare the population densities of North Island and South Island.

7 A country with a population of 10 million has 90,000 births per year and a life expectancy of 70 years. What do you think will happen to the country's population over time?

8 In this chapter, we estimated that ~10% of the world's women of childbearing age are pregnant at any one time. Make your own estimate, using the following information: the world population; the percentage of women in the population; the percentage of women that are not too young or old to bear children; the birthrate for the world. Estimates of these numbers can be made from data available in the text and tables.

9 Calculate the doubling time for a population that shows $P_t/P_0 = 8$ at $t = 60$ years.

10 The table below gives three types of data for each of three countries: population density in numbers of people per square mile; birthrate in numbers of children per 1,000 people per year; per capita gross national product calculated in U.S. dollars. What conclusions do you draw from these data?

	Population density	Birthrate	Per capita GNP
Netherlands	1,100	13	16,000
Rwanda	760	50	310
Saudi Arabia	17	42	6,230

REFERENCES

Aird, J. W. (1994). China's family planning terror. *Human Life Review* **20**(3): 83–204.

Amin, R., Ahmed, A. U., Chowdhury, J., and Ahmed, M. (1994). Poor women's participation in income-generating projects and their fertility regulation in rural Bangladesh: Evidence from a recent survey. *World Development* **22**(4): 555–65.

Andreas, P. (1994). The making of Amerexico: (Mis)handling illegal immigration. *World Policy Journal* **11**(2): 45–56.

Appleman, P. (1965). *The Silent Explosion.* Boston: Beacon Press.

Hardin, G. J. (1993). *Living within Limits: Ecology, Economics, and Population Taboos.* New York: Oxford University Press.

Hudson, N. (1992). *Land Husbandry.* Ithaca, N.Y.: Cornell University Press.

Jones, H. R. (1990). *Population Geography.* New York: Guilford Press.

Livi-Bacci, M. (1992). *A Concise History of World Population.* Translated by C. Ipsen. Cambridge, Mass.: Blackwell Publishers.

Lutz, W., ed. (1991). *Future Demographic Trends in Europe and North America – What Can We Assume Today?* London: Academic Press.

Malthus, T. R. (1826). *An Essay on the Principle of Population: A View of Its Past and Present Effects on Human Happiness, with an Inquiry into Our Prospects Respecting the Future Removal or Mitigation of the Evils Which It Occasions.* 3rd ed. 2 vols. London: Printed by T. Bensley for J. Johnson.

Newbury, C. (1988). *The Cohesion of Oppression: Clientship and Ethnicity in Rwanda.* New York: Columbia University Press.

Ornstein, R., and Ehrlich, P. (1989). *New World, New Mind: Moving toward Conscious Evolution.* New York: Doubleday.

Pedersen, G. (1994). *Afghan Nomads in Transition.* Copenhagen: Rhodos International.

Population Reference Bureau (1994). *World Population Data Sheet.* Washington, D.C.

Reich, R. B. (1991). *The Work of Nations: Preparing Ourselves for 21st-Century Capitalism.* New York: A. A. Knopf.

Sarre, P., and Blunden, J., eds. (1995). *An Overcrowded World? Population Resources and the Environment.* Oxford: Oxford University Press.

United Nations (annual a). *Demographic Yearbook.* New York.

United Nations (annual b). *Statistical Yearbook.* New York.

United States (annual). *Statistical Abstract of the United States.* Washington, D.C.: Government Printing Office.

von Braun, J., de Haen, H., and Blanken, J. (1991). *Commercialization of Agriculture under Population Pressure: Effects on Production, Consumption, and Nutrition in Rwanda.* International Food Policy Research Institute Report 85. Washington, D.C.

Wannenburgh, A., Johnson, P., and Bannister, A. (1979). *The Bushmen.* New York: Mayflower Books.

World Resources Institute (1992, 1994). *World Resources.* Oxford: Oxford University Press.

2 FOOD

2.0 INTRODUCTION

There is plenty for everyone's need, but not enough for everyone's greed.— Mahatma Gandhi

"Open wide – seven minutes until the school bus arrives." Gulp. Down goes breakfast. "Two minutes to brush your teeth and comb your hair." Out they run, hands grasping for lunch boxes and brown paper bags. "Now class, we have a forty-minute lunch break." Good, the kids think – four minutes for lunch and thirty-six to run around in the school yard. "Welcome home. Your snack's on the counter." Later comes dinner, and possibly milk and cookies before bed. Another day of eating ends.

What goes down all of those throats? For that matter, what do any of us eat, whether we live in rich industrial countries or scratch out a living for our families in subsistence economies? What food do we require, and what do we eat just because we like it? (Hint – 5% of food energy consumed by people in industrial countries is alcohol.) Where does the food come from, and what does its production do to the earth? Would soil erosion be one result, along with pollution from animal wastes and fertilizers? Do we have the capacity to feed the earth's burgeoning population without introducing unacceptable changes in our environment? What about right now – is the earth's population being fed adequately?

To begin, we compare plants and animals (including ourselves). The only necessities of life that animals can obtain without eating are water and oxygen. Everything else comes by consuming plants if the animal is a herbivore, other animals if the animal is a carnivore, or both if the animal is an omnivore (which includes us). Relative to animals, plants have an easy life because they have very simple requirements. They consist almost entirely of carbon (C), hydrogen (H), and oxygen (O) and need to take in water, carbon dioxide (CO_2), oxygen, and a few other chemical elements, principally potassium (K), nitrogen (N), and phosphorus (P). Given the uptake of those ingredients, plants do not need to "eat" but can synthesize all of their requirements within the plant itself.

A plant's life-style is made possible by chlorophyll. One of the forms of chlorophyll, which all "true" plants must contain, permits photosynthesis. In the presence of chlorophyll, sunlight causes carbon dioxide (CO_2) and water (H_2O) to combine to form organic compounds according to the reaction

$$CO_2 + H_2O + \text{light energy} \rightarrow \left(\begin{array}{l} \text{a series of complex compounds} \\ \text{containing C, H, and O} \end{array} \right) + O_2$$

The complex compounds are referred to as carbohydrates and form the basic building materials of the plant (wood, leaves, flowers, etc.). Some light energy is used for the plant's metabolism, and some is stored in the plant and then released when the plant dies and decomposes or when we use it, for example, by burning wood and coal or by eating it (Section 5.2). The release of oxygen by photosynthesis replenishes oxygen in the atmosphere that is used by animals and thus makes animal life possible on the earth. Plants also "breathe" oxygen, but the net effect of a plant's life is to release more than it consumes.

Some organisms that most people regard as plants are not true plants because they do not contain chlorophyll and, therefore, must live by consuming other organisms. They include bacteria and other microorganisms that are abundant in soil, and fungi, including mushrooms and different varieties of yeast. Although these organisms are not a major part of our diet (not many of us live on mushrooms), we will see in Section 2.1 that a few required elements of human nutrition can be supplied by these "non-plants."

But before discussing plant and animal production, we consider human nutrition.

FURTHER READING: Publications of the United Nations Food and Agriculture Organization (FAO) and World Resources Institute (1992, 1994). See also Toussaint-Samat (1992).

2.1 HUMAN DIETARY REQUIREMENTS

Everyone has an opinion about food, because everyone has to eat! – F. M. Lappe, Diet for a Small Planet, p. 185

People must eat or drink an enormous variety of chemical compounds in order to survive. We summarize them here, starting with the simplest – water.

Water

Our body cells are mostly water, and we require body fluids, such as blood, for all of our life processes. Thus, although a person who is in good health at the start of a fast can survive without food for about a month, no one can live more than a few days without water.

Carbohydrates

As we already discussed, compounds of carbon (C), hydrogen (H), and oxygen (O), commonly with phosphorus (P), are initially formed by plant photosynthesis. The simplest product is glucose, one of many types of sugar, and plant metabolic processes proceed from there to construct large carbohydrate molecules (polysaccharides) consisting of linked glucose and similar "building blocks" (Fig. 2.1 explains the structural diagrams of organic compounds in general, and glucose and polysaccharides are shown in Fig. 2.2). These large molecules fall generally into two categories, starch and cellulose, both of which have similar formulas but differ in the number and arrangements of the building blocks. Cellulose is particularly important to plants because it is the "stiffener" that enables them to stand upright.

Although people eat a lot of carbohydrate (bread, vegetables, etc.), the human body contains very little. The reason is that people use carbohydrates almost exclusively as an energy source rather than as building blocks.

The energy-forming process is essentially the reverse of the photosynthetic reaction (Section 2.0). The light energy that is absorbed and stored by a plant when CO_2 and H_2O combine to form carbohydrates is released when carbohydrates are "burned" with oxygen to form CO_2 and H_2O in an animal. This "burning" in the body requires that people (and all other animals) inhale O_2 and exhale the released CO_2.

Only glucose and other very simple sugars are actually burned in the body. Thus, complex starch molecules, which form most of our intake of energy, must be broken down to glucose before they can be utilized. This process occurs in two steps. The first is generally cooking. People can obtain very little energy from raw potatoes and virtually none from uncooked wheat, but cooking breaks down the complex molecules into simpler ones. The second step in making starch useful takes place in the body, where complex molecules that remain after cooking or in raw vegetables are acted on by enzymes (as discussed later) in the saliva of the mouth and the digestive system.

Cellulose is completely useless as a human energy source – for example, people cannot digest grass. Even after cooking, the human body cannot break cellulose down into glucose or other simple molecules. Some cellulose, however, is needed in human diets. We refer to it as fiber, and it serves the important function of cleaning out the intestines.

The human digestive system is somewhat less efficient than those of other animals. Mice, for example, can eat raw potatoes, raw wheat, and apparently just about everything else – there's no need to cook dinner for the mouse in your attic. Cows, goats, sheep, and most other grazing animals are ruminants. They possess a "rumen," essentially the first part of their stomachs, where microbes break cellulose down into simpler molecules that can be passed into the rest of the stomach and further processed into useful sugars.

Fats (lipids)

Fats are generally organic acids, consisting of a carbon dioxide and hydrogen group (carboxyl; COOH) attached to a very large molecule constructed almost entirely of carbon and hydrogen atoms (Fig. 2.3). Fats can be either saturated or unsaturated (see Fig. 2.1). In a saturated fat, no more hydrogen can be added to the molecule because all of the carbon atoms are linked by single bonds, and the substance is generally solid. Unsaturated fats are molecules to which more hydrogen can be added because they contain some carbon-carbon bonds that are double,

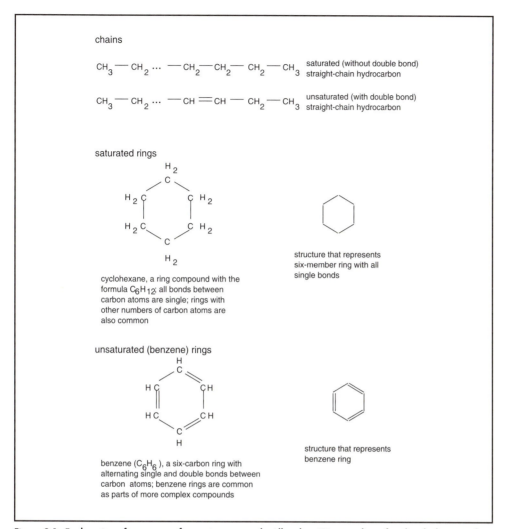

Figure 2.1 Explanation of structures of organic compounds. All carbon (C) atoms have four bonds that must be filled by connection to other atoms. Single bonds are represented by single lines and double bonds by double lines. Bonds to hydrogen atoms (H) are all single and are conventionally not shown but implied by the presence of hydrogen in the formula. All molecules are intricate 3-dimensional forms, and are shown here as two dimensional only for convenience. The diagrams show three basic types of organic structures:

(1) Chains. Carbon atoms are linked to two other carbon atoms, and the two remaining bonds of each carbon are filled by linkage to other elements. In the diagrams, all linkages are to hydrogen atoms. In the top chain, carbon atoms are linked to each other by single bonds and to two hydrogen atoms, filling the four bonds of each carbon with two to other carbon atoms and two to hydrogen atoms. This linkage is referred to as saturated. In the lower chain, the linkage between the central two carbon atoms is a double bond, and each atom is linked only to one hydrogen. The four bonds of these two carbon atoms are filled by the double bond, one bond to the next carbon in the chain, and one bond to hydrogen. This bonding forms an unsaturated compound.

(2) Rings in which all carbon atoms are linked by single bonds. The four bonds consist of two to adjoining carbon atoms in the ring and two to hydrogen atoms.

(3) Benzene ring, with three double bonds alternating with three single bonds. Each carbon atom has two of its bonds filled by the double bond, one by the single bond to an adjoining carbon, and one to hydrogen.

In other diagrams in this book, we commonly show only the simple ring or benzene ring without drawing in the individual carbon and hydrogen atoms.

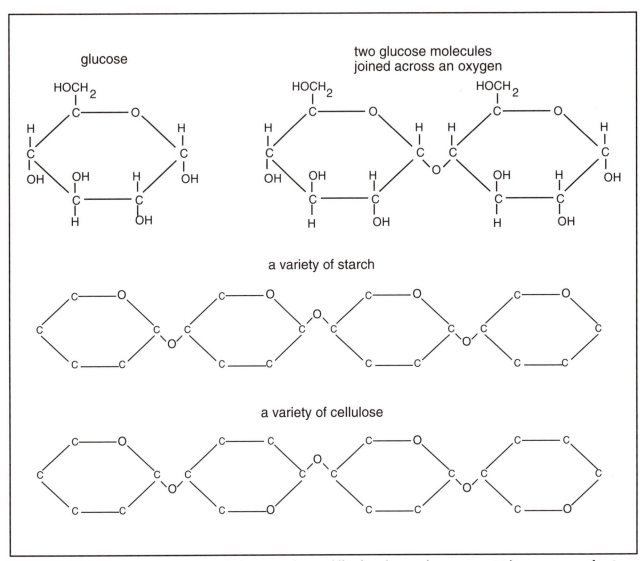

Figure 2.2 Structure of glucose and its polymers. The basic 6-member ring differs from the rings shown in Figure 2.1 because it contains five C atoms and one O. Bonds to other groups and atoms are shown to H, OH, and HOCH$_2$. Two molecules of glucose can bond together by removing one OH group apiece and linking across a common O. These links can be between various orientations of glucose molecules, producing either different types of starches or different types of cellulose as long chains of glucose molecules (known as "polysaccharides" or "polymers"). The atoms and groups bound to the carbon atoms are not shown in the lower two diagrams. Further explanation of organic structures is in Figure 2.1.

several if the fat is polyunsaturated; some of these compounds are liquid and referred to as "oils." Unsaturated fats can be converted into saturated ones by adding a hydrogen atom to each carbon in the double bond. This process of hydrogenation is used, for example, to convert some of the double bonds in vegetable oils to single ones and make margarine solid enough to be used on the dinner table. Individual fat molecules can be combined in various ways, commonly including phosphate-bearing groups, to make a large variety of complex fats. Plant fats and oils tend to be more unsaturated than animal fats.

Animals use fats in much the same ways as plants do, and about 20% of the average human body is fat. Fats are necessary building blocks of cells, and their presence in body fluids permits solution and transport of materials that are not soluble in water (e.g., vitamin A). The human body can synthesize most, but not all, of the fatty acids that it requires from its intake of carbohydrates. The fatty acids that cannot be synthesized can be obtained from plants or from meat, where they are more abundant. Fats can be broken down into simple sugars and used as an energy source, but they are not utilized

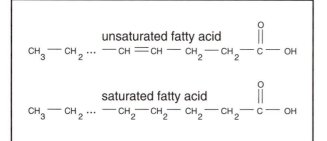

Figure 2.3 Fatty acids. Saturated fatty acids with no (or few) double bonds between C atoms occur primarily in animals. Unsaturated acids with one or more double bonds are found primarily in plants (used, e.g., in margarine). The acids form long and linked chains, and we show only representative small portions here. Further explanation of organic structures is in Figure 2.1.

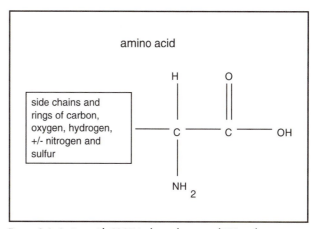

Figure 2.4 Amino acid. COOH is the acid part, and NH₂ is the amine (amino) part. Different attachments yield the 20 different varieties of amino acids that we need as food. Proteins are clusters of many, commonly thousands, of amino acids linked together. Further explanation of organic structures is in Figure 2.1.

unless the daily carbohydrate intake drops below the level needed to maintain the body's functions, and the body burns stored fat to compensate, which occurs during dieting and starvation. Fat-soluble compounds also include steroids, which are special chemicals that perform various vital functions in the body, including regulating the reproductive systems (see the section on special chemicals). Steroids are not present in plants, but the human body is generally capable of synthesizing all of the varieties that it needs, including cholesterol and the sex hormones from carbohydrates and other food.

Proteins

Proteins are complex molecules constructed from building blocks known as amino acids (Fig. 2.4). Amino acids are organic acids (Fig. 2.3) that contain an amine (NH_2) group, and 20 different molecular species occur in human proteins. The acids can be linked together to form proteins that differ in the proportions of the various acids and the number and method of linking of the individual molecules. Many proteins contain hundreds, to hundreds of thousands, of acids in a commonly globular cluster. Proteins are necessary functional and structural components of cells (14% of average human body weight is protein). They also serve as hormones and enzymes that regulate all metabolic processes (see the section on special chemicals). They can be partly broken down into sugars and serve as a "last resort" energy source when carbohydrates are not available and stored fat has been consumed, which occurs in cases of severe undernourishment to starvation.

The human body needs all 20 of the amino acids in order to synthesize all of its required proteins. It gets these acids by eating proteins (about 2 ounces per day for the average person), and it commonly breaks down the plant and animal proteins into their component building blocks and rearranges them into the required human proteins. During this reorganization, the body can synthesize 12 of the amino acids, but it cannot make 8 of them and must obtain them specifically from food. Animal products (meat, eggs, milk, etc.) are a rich source of proteins, including all 20 of the amino acids. Plants also contain all 20 acids, but their proportion varies from one plant species to another, and people who do not eat animal products must carefully combine different plants in order to obtain the necessary mix of amino acids.

Specific elements – minerals

The list of elements that people need encompasses a large part of the periodic chart. Some are very minor, such as cobalt (Co), which is an essential component of vitamin B_{12}, obtained from animals (vitamins are discussed with special chemicals). In the interests of brevity, we restrict the discussion to calcium (Ca), phosphorus (P), and iron (Fe).

PHOSPHORUS AND CALCIUM The predominant use of phosphorus and calcium is in bones and teeth. They consist of the mineral apatite, with the general formula $Ca_5(PO_4)_3(F,Cl,OH)$. A higher fluorine (F) content in apatite causes bones and teeth to be harder, which is the reason that some toothpaste contains fluorine and some public water supplies are fluoridated. A principal requirement for healthy bones is a continuing supply of calcium and phosphorus, and an ordinary diet generally provides all that people need. Milk is rich in Ca, and some people take Ca in pills, generally in the form of the mineral calcite ($CaCO_3$). Phosphorus is also needed, as small amounts are essential components of other parts of the body, for example in DNA. Maintenance of strong bones requires a delicate balance of the various hormones that regulate the relationship between bones and body fluids. One of these hormones is vitamin D, and if it is deficient, bone begins to dissolve even in the presence of adequate supplies of calcium and phosphorus.

IRON   The principal use of iron in the human body is in hemoglobin. Hemoglobin is an Fe–organic compound that circulates as the red cells in blood and is responsible for transportation of oxygen throughout the body. An inadequate supply of iron results in anemia and can be fatal.

Special chemicals

The proper functioning of the human body requires an extraordinary number of specific chemicals ranging from simple molecules to macromolecules (polymers) of great complexity and weight. Because this is not a book on human nutrition, we can only discuss a few in the context of human dietary requirements.

HORMONES   Known as the body's regulators, they are, with one exception, produced in endocrine glands (glands that secrete their products within the body), then move to another part of the body (the movement is an essential part of the definition of hormone), and then attach themselves within a cell or on its surface and control the cell's processes. Many hormones are proteins or other molecules bearing an amino (NH_2) group. Examples of some of the many essential hormones include the various hormones that regulate the relationships between bone and fluids mentioned earlier and the thyroid hormone, which influences the body's metabolic rate and, thus, the rate at which energy must be supplied by food.

One important group of hormones, steroids, does not contain nitrogen. They are all various modifications of the basic steroid formula shown in Figure 2.5, and the different ones have different functions in the operation of the body. Two of the most important steroid groups are the male hormones, principally testosterone, and various female hormones of the estrogen group. The male and fe-male hormones are comparatively simple steroids, and their functioning can be interrupted if other steroids are taken into the body. For example, pregnant women do not use the steroid cortisone on rashes, and production of steroidlike compounds in plastics, herbicides, and other chemicals may be having a major effect on reproduction rates in industrial societies (see Box 7.7).

One hormone is not produced in an endocrine gland. It results from the action of sunlight on a steroid in the skin (cholesterol). It was originally referred to as vitamin D, but because it is synthesized by the body, it is more properly referred to as a hormone. In the skin, sunlight breaks down the steroid core of the cholesterol molecule and then builds upon some of the fragments to synthesize vitamin D. A deficiency of vitamin D causes release of calcium and weakening of bones. Healthy people can obtain adequate supplies of vitamin D by exposure of their arms and faces to only 15 minutes of sunlight per day. Vitamin D is also abundant in animal products, but not in plants, and people in polar regions must obtain it from food during the winter – fortunately for the Eskimos, fish oil is rich in vitamin D.

Normal diets provide all of the ingredients that the body needs for hormone synthesis. One exception is a deficiency of iodine (I) in some soils and a consequent deficiency in plants and animals produced on them. Because iodine is a necessary component of the thyroid hormone, its deficiency results in severe disturbance of the metabolic processes regulated by thyroid hormone, leading in some cases to a slower metabolic rate, sluggishness, and weight gain, and in other cases to faster metabolism, irritability, and weight loss. As a precaution, industrial societies commonly add iodine (as potassium iodide) to table salt. Societies without access to iodized salt can supplement their diets by chewing on seaweed.

ENZYMES AND VITAMINS   These compounds are grouped together for this discussion because of their similar physiological activities. Enzymes promote chemical processes in the body; that is, they are "catalysts," a term that we will encounter in many other places in this book. Enzymes include an enormous number of chemicals ranging from complex proteins to simple small molecules. Examples, mentioned earlier, are the enzymes in saliva and the stomach that break complex carbohydrates down into usable sugars. All enzymes are synthesized in the human body and require only normal diets for their preparation.

Vitamins are, by definition, chemicals that the human body cannot synthesize and that must, consequently, be eaten (this is why vitamin D, discussed earlier, is not properly a vitamin). Vitamins fall into two classes, water

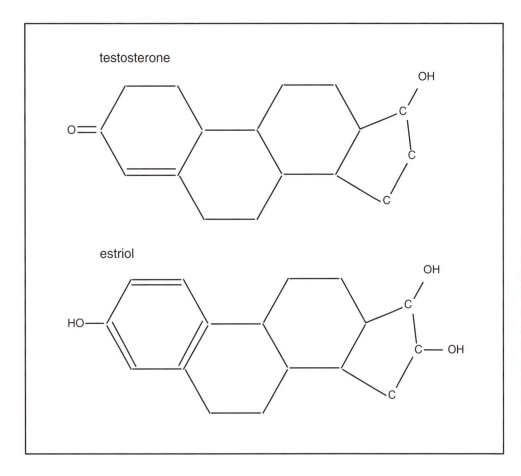

testosterone

estriol

Figure 2.5 The basic steroid structure is a 17-member carbon group of the type shown here. Different attachments and arrangements of double and single bonds yield different steroids. The two shown are the principal male hormone testosterone and estriol, one of the family of female hormones, generally known as estrogens. Further explanation of organic structures is in Figure 2.1.

soluble and fat soluble, and perform two general functions. Water-soluble vitamins are protein or amino based and include the numerous varieties of B and vitamin C. The B complex acts together with various enzymes as cofactors for a variety of reactions involving carbohydrates and fats, and C is necessary for the formation of collagen, a set of proteins that occur in bones and connective tissues in tendons and the skin. Fat-soluble vitamins include vitamin A, which helps maintain the retina of the eye, and vitamin E, which serves primarily as an antioxidant, to prevent oxidation of human tissue. Because vitamin E is more easily oxidized (combined with oxygen) than cell tissue, it removes oxygen that would otherwise damage tissue. Because their activities are vital to the success of the organism, the vitamins are regarded as "essential" foods.

Because all vitamins must be eaten, healthy diets must be designed so that these substances are present. With the exception of B_{12} and D, all vitamins can be obtained from plants, although some are more abundant in animal products. Cereals, such as wheat, corn, and rice, which make up a large part of many people's diet, are not rich in vitamins. Furthermore, industrial societies commonly remove the more vitamin-rich parts of the grains before they are cooked. Prime examples include the "polishing" of brown rice to form the more appealing white rice by removing the outer husk, which is rich in Vitamin B, and removal of the more fibrous (more vitamin-rich) outer husks of wheat used in packaged breakfast cereals, which then are artificially "revitaminized" before packaging. Many of the vitamin deficiencies that result from a diet that consists largely of grains can be made up from other plants. Vitamin A is abundant in yellow vegetables such as carrots and vitamin C in citrus fruits and other fresh vegetables such as tomatoes. The old mariner's disease, scurvy, which resulted from a lack of fresh vegetables on long voyages, was reduced in the middle 1700s when ships began to carry sauerkraut and, later, barrels of lime juice.

Animal products are the richest source of vitamins A, B, and D, and fish oils are particularly rich in A and D. Yeast contains most of the B complex, and bread that is "leavened" (rises) contains more B than unleavened bread made with baking flour. Vitamin B_{12}, however, occurs only in animal products. Its deficiency in humans leads to a variety of anemia that may ultimately result in neurological damage. People who do not eat meat can overcome the potential deficiency in B_{12} by drinking milk or eating cheese and eggs.

Industrial societies commonly avoid all forms of vitamin deficiency by giving people synthetic vitamin pills; they do not taste as good as a glass of milk or an orange, but they are a convenient method of ensuring adequate levels within a busy life-style. Poor countries, however, commonly cannot afford such supplements, and deficiencies of various types are common.

Now that we have discussed the food that people need, which is not particularly controversial (Fig. 2.6), we must move on to issues concerning how food is produced, the costs, and environmental consequences of its production. We will also deal with the very contentious issue of how food is distributed around the world and why many people are constantly hungry.

Policy questions

After reading this section, are you going to change your eating habits?

FURTHER READING: Gershoff (1996).

2.2 CROP PRODUCTION

If they have no bread, let them eat cake. — Attributed (perhaps falsely) to Marie Antoinette, queen of France, during riots over the scarcity of food before the French Revolution (in this sense, cake means coarse, fried grain that is less desirable than baked bread)

People get their nourishment from both plants and animals. This section is concerned with the production of plants as food. We discuss the use of plants as animal food here, but animals and fish as direct food are discussed in the next section. We start by describing the relationships between the energy in plants and the energy that people can obtain from them.

Energy and plant food

Viable societies must obtain more energy by eating the grain than is expended in growing it. We investigated a similar issue in Section 1.2, showing how population densities are related to the availability of energy that

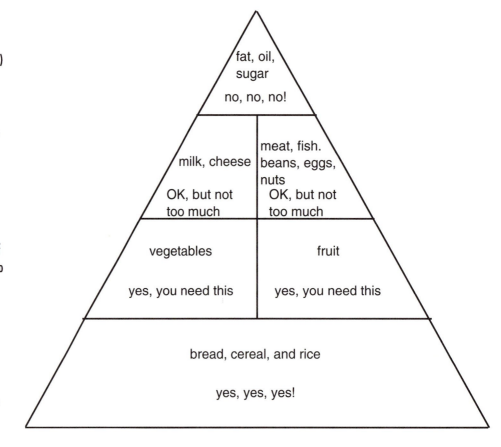

Figure 2.6 The "food pyramid" proposed by the U.S. Department of Agriculture (USDA) in 1992. The general purpose is to indicate that grain foods at the bottom of the pyramid should form the base of a diet, and foods farther toward the top of the pyramid are progressively less nutritious. The USDA originally specified the number of "servings" of each food type that a healthy diet should contain, but the sizes of these servings were not clear to many people. Furthermore, such groupings as beans and nuts with meat and also with fish are not accepted by many nutritionists, who regard red meat as far less healthy than other food. For these reasons, we have greatly "generalized" the advice.

people need for food, heating, and other activities in different styles of life from hunter–gatherer societies to modern industrial ones. Energy is measured in kilocalories and more completely defined in Section 5.1. Calories rather than kilocalories are used in books solely about food, but for consistency in our discussion of fossil fuels and other energy sources we measure food energy in kilocalories.

As discussed in Section 2.1, the food energy of a plant is the energy that was absorbed during photosynthesis. The energy released from food by the body's metabolism is equal to the energy that would be released if we set the edible part of the plant on fire with oxygen (O_2) and formed CO_2 and H_2O. Because the most energy-consumptive part of plant growth is conversion of CO_2 and H_2O to the simple sugar glucose, from which starch and cellulose are made, the food energy in most plants is the "heat of combustion" of glucose (3.7 kcal per gram or ~100 kcal per ounce). The heat of combustion is the difference between the heat of formation (energy used in formation) of glucose and the heats of formation of CO_2 and H_2O. (We return to this concept in Section 5.3.)

Table 2.1 shows the amount of energy supplied by 1-ounce (28-g) portions of various plant foods (we include meat products for comparison). One ounce of dry corn kernels or wheat grains yields about 100 kcal of energy, the heat of combustion of glucose. Energy values less than 100 kcal per ounce result from the fact that many plant foods are sold with a high water content, in contrast to the sale of wheat as dry flour. For example, rice contains some water and consequently yields about 75 kcal per ounce, and potatoes are nearly 80% water and 20% starch. For the same reason, leafy and other vegetables provide very little energy, and they are needed largely because of their content of proteins, vitamins, and fiber. Nuts and other plant products that yield more than 100 kcal per ounce are rich in fats, which have a heat of combustion approximately twice that of glucose.

We now use these concepts of food energy to compare two different methods of crop production: (1) subsistence farming in a poor country, based solely on manual labor and the productivity of the soil; and (2) mechanized farming in an industrial country, using fuel and artificial fertilizer.

An acre of corn in a subtropical country produces about 1 ton of kernels if it is farmed purely by manual labor and without fertilizers. This corn requires a farm couple to invest about 1,200 hours of labor by one person, roughly one-half of a year's work. That labor could be allocated as full 8-hour days for one person during a 6-month growing season, half-time labor for one person

Table 2.1 *Energy in 1 ounce (28 g) of common foods*

Food	Energy (in kcal)
Wheat, corn	100–25
Rice	75
Beans	100
Potato	30
Apple	25
Peanuts	150
Tree nuts	200
Steak (with fat)	100
Chicken breast	50
Fish	40
Cheese	100
Egg	40
Sugar	60
Chocolate	125

Notes: Beans includes brown, white, black, etc., but not green beans. Tree nuts includes almonds, pecans, walnuts, etc.
Sources: Data are generalized from Bowes (1980).

year round, or one quarter-time labor for two people year round (commonly with men doing the plowing and both men and women doing the harvesting). Regardless of the allocation of labor, a farm couple cannot allocate more than one half of a year's work between them to the mechanics of growing corn, simply because of the time that they must spend on all of their other duties – such as raising vegetables to provide nourishment that cannot be obtained from corn, repairing or building their hut, digging a well, caring for children, and cooking.

The corn raised by the subsistence farm couple has a food value of about 3 million (3×10^6) kcal (2000 lb × 16 oz × 100 kcal/oz). A person doing manual labor needs at least 2,500 kcal per day for a reasonably healthy life – roughly 1 million kcal per year. Thus, the farm couple working *full time* on the farm consumes about two-thirds of the energy that they produce, leaving only about one third to be allocated to other family members, including children too young and parents too old to work. Clearly the family has no unused food to sell. In fact, if the family is large, its members may have to live on less than 2,500 kcal per day and thus be undernourished and prone to diseases caused by an inadequate diet. Various survival strategies for large families include having family members earn money by working at least part time off the family farm and putting children to work so larger areas can be farmed, if available, or farming the allocated plot more intensely. School is an unaffordable luxury (Section 1.5). Most consumer goods are also unaffordable, even those that might be considered necessities in industrial societies, such as plumbing.

An acre of wheat in western Europe that is farmed with machinery, is thoroughly fertilized, and is heavily treated with pesticides and herbicides, produces about 2 tons of wheat grains, a food equivalent of 6 million kcal. Instead of the 1,200 man-hours of labor needed by the subsistence family to feed themselves minimally during the year, these 2 tons of wheat require about 10 man-hours of labor (just 10!). This means that one farmer working an 8-hour day 250 days per year (time off for weekends and birthdays of all of the family) could farm 200 acres, producing about 1 billion (1×10^9) kcal of food energy that would support more than 1,000 people for a year. All of these other people are freed from farm labor so that they can do something besides growing wheat – other types of agriculture, education, industry, the arts, television.

The high productivity of an industrial society results from the use of fertilizer and machinery. Both topics will be discussed more completely, but we mention here the energy input for the 2 tons of wheat. Just to produce the fertilizer and run the plows, harvesters, and other farm machinery requires about 1.5 million kcal of energy from gasoline and other fossil fuel. That is roughly the amount of energy in a barrel of crude oil, which could be purchased for about $50 after refining to gasoline (Section 5.3). This $50 and other costs of mechanization are trivial in comparison to saving several hundred man-hours of labor that a subsistence farmer would have to expend in producing the same amount of wheat solely by human labor.

Let us bear these comparisons in mind as we investigate worldwide crop production.

Requirements for crop production

Plants need light, water, and nutrients. Different parts of the world provide these necessities in different amounts.

LIGHT   All true plants require light as their source of photosynthetic energy. The leaves of plants absorb solar radiation mostly in the visual wavelengths, primarily in the red (long-wavelength) and blue (short-wavelength) ends of the spectrum (Section 8.1). The leaves are green in color because they do not absorb the middle, green, parts of the spectrum very well. The decrease in light intensity from the equator to the poles requires adaptation of different plants to different latitudes. For example,

high-latitude species must endure long dark winters, and tropical species must be adapted to periods of high temperature.

One of the principal adaptations to light intensity is the type of photosynthesis conducted by the plant. Most plants, including most commercial crops, are classified as C3 plants. The nomenclature results from the fact that the first product of photosynthesis is a three-carbon organic compound that includes phosphate (referred to as "triosephosphate"). The basic building blocks of carbohydrates, including glucose, are produced later. The principal plants of temperature climates follow C3 photosynthesis, which is ineffective during hot periods of the day when plants absorb very little CO_2 because they have closed their pores to prevent water loss. Some plants, known as C4 plants, manufacture a four-carbon phosphate as their first photosynthetic product, and they can survive in tropical areas because they photosynthesize during hot periods with limited CO_2 absorption. C4 plants include corn and sugarcane, and they grow plant matter more rapidly than most C3 plants.

The total amount of solar radiation that arrives at each acre of the earth's tropical and temperate land surface is about 20 billion (2×10^{10}) kcal per year. This figure represents about 40% of the radiation arriving at the top of the atmosphere, with the remainder of the energy reflected back into space or absorbed in the atmosphere (Section 8.1). About 1% to 2% of the radiation reaching the surface is absorbed by plants (including trees), with the remainder reflected from the ground or absorbed into it and reradiated. A field of wheat that produces 2 tons of wheat grains per acre is using approximately 0.1% of the incident radiation.

Calculations of the amount of radiation absorbed provides one insight into efforts to improve agricultural productivity by genetic breeding of plants. Plants with larger leaves absorb more radiation and thus produce more grain. Furthermore, plants with shorter stalks use more of the radiation to produce grain and less for stalks and leaves. The ideal plant, not yet achieved, would be an enormous grain of wheat that sits directly on the ground and has just enough leaf cover to run its photosynthesis.

WATER   A harvest of 1 ton of wheat grains has used about 500 tons (approximately 250 metric tons) of water for the growth of the wheat plants. Plants need this water for cells and fluids in much the same ways that animals do, and they "transpire" water back into the atmosphere (Section 4.1). Ideally, the water needed by plants is supplied by rain, and the very high agricultural productivity of northern Europe results partly from its abundant rain-

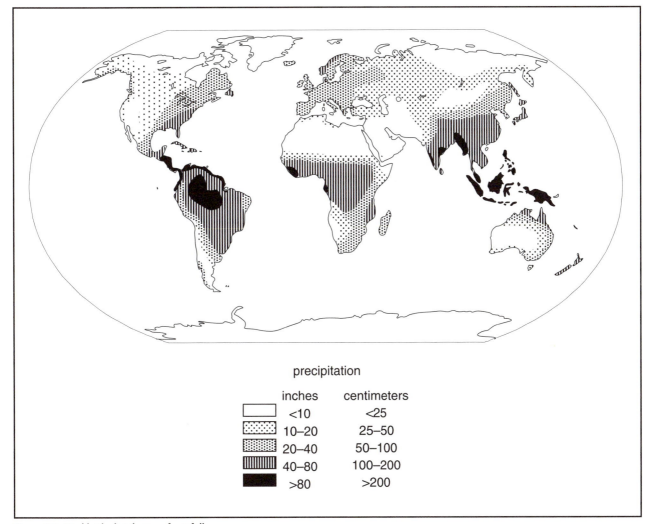

Figure 2.7 Worldwide distribution of rainfall.

fall (Fig. 2.7). Water quantities are commonly measured in acre-feet, a volume of water 1 foot deep over an acre, which weighs approximately 1,500 tons (Section 4.3). If we use the 500 : 1 ratio just mentioned, this amount of water on 1 acre should produce 2 to 3 tons of wheat.

The preceding calculation shows that high productivity requires 1 foot of water per year. One foot of rain, however, is far less than adequate. Good agricultural areas require rainfall of about 40 inches (100 cm) per year, approximately three times the amount of water used. The extra rain is needed because most of the water that falls either evaporates directly from the soil, runs off into rivers, or percolates into the ground and moves away (Section 4.1). Also, rainfall is seasonal in most areas, and periods of very heavy rain cause most of the water to be lost to runoff. Thus, where rainfall is less than about 40 inches per year, either crop productivity is low or the area

must be irrigated. As with other ways of enhancing productivity, industrial countries have a much better irrigation system than most underdeveloped countries. In the United States, approximately 80% of all water use is for irrigation, and more than half of farmland receives some water from stream or groundwater sources (Section 4.2).

NUTRIENTS Plants obtain their carbon from the air and water from the ground, and some plants obtain at least part of their nitrogen from the air, but everything else that the plant needs depends on the fertility of the soil. That is why we must take care of our soil if we want to continue eating.

Soil forms when rocks at the earth's surface interact with air and water (Fig. 2.8). The first part of the soil-forming process is weathering and results from the fact that most rocks now exposed to the surface were formed at some depth in the earth. At these depths, tempera-

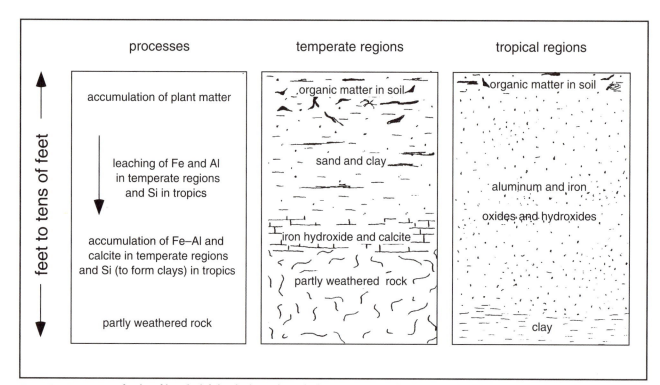

Figure 2.8 Formation of soil profiles. The left-hand column shows the basic processes in development of soil from rock that has undergone partial weathering to break down its original minerals. The uppermost part of the soil commonly contains organic matter formed by decay of plants, such as dead leaves. The upper horizon (the A horizon) is a zone of downward percolating rainwater that leaches some of the elements from the weathered rock. In temperate regions, plant matter commonly decays to organic acids that dissolve in the rainwater, whereas tropical climates cause decay to dissolved organic bases. The B horizon below the A horizon is a zone of accumulation of elements leached from the A horizon. The B horizon grades downward into partly weathered rock. Soil profiles generally range from a few feet to a few tens of feet thick.

The center column shows a typical soil profile in humid temperate regions, such as most of the industrial countries of Europe and North America. The organic zone is commonly thick. Leaching by acids carries Fe and Al downward, leaving a soil rich in sand and clay. Al and Fe form a "hardpan" of oxides and hydroxides in many areas. Where soils are thin in areas of relatively low rainfall, calcite may be precipitated throughout much of the soil profile.

The right-hand column shows the result of leaching of rocks by organic bases in tropical areas. Organic matter decays thoroughly under tropical conditions and leaves only a thin layer of accumulation. Leaching in the A horizon removes Si, leaving a thick residue enriched in iron (Fe) and aluminum (Al) hydroxides (see Al ores in Section 6.2). The B horizon is generally a thin zone of clays. Because of the extensive leaching, tropical soils are commonly poor in nutrients.

tures and pressures are higher than at the surface, whereas concentrations of water, CO_2, and O_2 (the active ingredients of weathering at the surface) are vanishingly small. Thus, the original rocks from depth are "out of equilibrium" at the earth's surface. As water, CO_2, and O_2 act on the rocks they chemically decompose to minerals that are stable and break into small grains as the decomposition disrupts the original hard rock. Interaction of iron (Fe) in the rocks with O_2 and water generates Fe oxides and hydroxides – this could be called "rusting" of the rock – and gives the earth's land surface a reddish color except where it is covered by vegetation or ice.

Interaction and mixing of weathered rock with plant and animal organic matter, plus groundwater transport of some elements, generates the soil in which the plants grow (Fig. 2.8). The soil-forming process makes the original elements of the rock more mobile and, thus, available to plants as nutrients. The ability to provide nutrients varies greatly among different soils, depending largely on the composition of the original rock, but also on the average temperature and rainfall of the region. The richest soils occur in temperate regions, where the world's richest people live – this is not an accidental relationship and does not imply greater moral virtue for people who live in temperate climates. Because of high rainfall and the nature of tropical vegetation, most nutrients in tropical soils have been leached out by migrating groundwater. Desert soils have high concentrations of nutrient elements but are generally so poorly weathered, as the lack of water limits chemical breakdown, that the nutrients are immobile and unavailable.

The six major nutrient elements are nitrogen (N), phosphorus (P), potassium (K), calcium (Ca), magnesium (Mg), sulfur (S) (Table 2.2 lists the elements found in a typical corn plant). P is a component of many carbohydrates and fats, and N is an ingredient of all proteins. K is not part of the structure of the carbohydrate and fat molecules that constitute most of a plant, but it is soluble in plant fluids and is required for a plant's metabolic process. Various niches and roles are occupied by calcium (Ca), magnesium (Mg), sulfur (S), and a host of minor elements. A plant cannot function effectively if it is deficient in any one of its nutrients. That is, having most of its food is not enough – it needs everything.

Native vegetation that is not harvested dies and thus recycles its nutrients into the soil for the next generation of plants. When animals eat grass and hay, and people harvest grain and vegetables, the nutrients in the edible parts of the plant are removed from the cycle and so the soil becomes depleted. The nutrients must be replaced in order for the soil to remain productive. The replacement can occur naturally – for example, animals leave manure, and sometimes their carcasses, to recycle into the soil. Natural harvesting, perhaps a squirrel eating acorns, may proceed at the same rate as soil formation releases additional nutrients from rock into the soil. The productivity needs of agricultural societies, however, require faster nutrient replacement than can occur naturally. Several strategies are employed, depending on the intensity of land use:

- In areas of low population density, people can move from exhausted fields to new ones. Thus, nomads drive their animals to better pastures, and planters "slash and burn" a new area of forest, taking advantage of nutrients accumulated over many years in the soil. In both situations, old areas must be left fallow for many years in order to rebuild soil nutrients.
- Agricultural societies can use "natural" methods to enhance nutrients. One is to ensure that animal manure is recycled into the soil, and the other is to use legumes to increase soil nitrogen. Purely "organic" gardening generally requires that vegetative mulch be brought from other areas to the site of crop produc-

Table 2.2 *Chemical analysis of typical corn plant*

Element	Abundance in weight (%)
C	44
O	44
H	6
N	1.5
K	1
Si	1
P	0.2
Ca	0.2
Mg	0.2
S	0.2

tion, which requires energy for transportation and removes nutrients from the site where the mulch was collected.

- Industrial societies use fertilizer, and we discuss its mining and preparation in Section 6.4. The principal nutrients added by fertilizer are nitrogen (N), phosphorus (P), and potassium (K). Large amounts of nitrogen are needed because nitrogen fertilizer is soluble and is leached from the soil rapidly. Phosphorus can be added either as natural "rock" phosphate or as the more soluble "superphosphate" made by treating phosphate rock with sulfuric acid. In addition to adding nutrient chemicals, industrial societies commonly use herbicides to kill weeds that might compete with crop plants for light and food, and they use pesticides to prevent crop damage by insects (in Section 7.3 we discuss the environmental consequences of the distribution of these artificial chemicals into soil, groundwater, and surface water).

Supplied with this information about required inputs for agricultural production, we now turn to the issue of the outputs, starting with grains.

Grain production

About two-thirds of the calories that sustain people around the world are obtained by eating some variety of grain (cereal). The major crops are corn (maize), wheat, and rice. Very roughly, wheat is grown only in temperate climates, primarily in Europe, North America, and Australia. Corn originated in the Americas, then was taken to Europe by colonizers, and now is grown all over the world. Tropical countries depend highly on it (remember, corn is a C4 plant that grows particularly well in equatorial regions). Rice is the main crop of the well-watered lands of eastern Asia. Before discussing the present status of grains around the world, we briefly trace the history of improvements in productivity, starting with that agricultural machine known as "western Europe" (Fig. 2.9).

In the Middle Ages, until about 1500, western Europe used the same form of subsistence agriculture as the subsistence farmers discussed earlier. Best estimates, which are very rough, suggest that a person working full-time for a year could produce about 1 ton of wheat grain. For the same reason that poor countries today have most of their population engaged in agriculture, more than half of the people of western Europe were then farmers. Improvement from the Middle Ages to the present time involved two principal steps (Fig. 2.9). Up to approximately 1900, farmers gradually learned the value of fertilizing, using a combination of animal manure and the less-potent "green manure" composed of vegetative mulch (plant stalks, etc.). This development increased production to about 1 ton of grains per acre by the late 1800s. The industrial revolution brought the second phase of improvement in productivity. The use of machinery for plowing and harvesting not only increased yields but freed grain farmers for other activities. Along with machinery, industrial societies developed the use of inorganic fertilizers, both natural and artificial. The increase in yields accelerated after World War II, partly by the use of new varieties of genetically improved plants.

European grain yields can now reach more than 3 tons per acre.

Increases in grain production since the 1940s are even greater in lesser-developed countries than in the industrial world. Largely because of this improvement, worldwide food supply increased from approximately 2,300 kcal per person in 1960 to approximately 2,700 kcal per person in 1990, a period of time in which the world population increased by 2 billion people (an approximate doubling). As just one example, during the 50 years since India became independent in 1947, its population increased 400% and the food supply increased 500%. The increase in grain production resulted largely from technologies imported from industrial nations, including increased use of fertilizer and the introduction of more efficient varieties of plants (the "green revolution"). Although these improvements have been largely responsible for maintaining adequate nutrition in countries with rapidly growing populations, they have caused both environmental and social problems, which we discuss in Section 2.5.

The total world production of grain in the mid 1990s is about 2 billion tons per year. If it were all consumed directly by people, it would supply approximately 7 quadrillion (7×10^{15}) kcal of energy. For a world population of approximately 5.5 billion, this grain energy provides approximately 4,000 kcal of energy per person per day. Thus, the energy supplied only by grain, not counting other agricultural products, is well in excess of the approximately 2,500 kcal per person recommended for

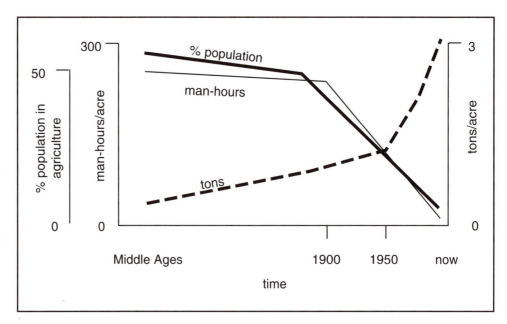

Figure 2.9 History of grain production in Europe from the Middle Ages to the present. The diagram shows productivity in terms of tons of grain per acre and the number of man-hours customarily devoted to farming 1 acre. As productivity increases, the percentage of the population engaged in agriculture declines.

Table 2.3 *Selected data on grain production and fertilizer use*

Countries	Production (in 10^6 tons/year)	Land used for crops (%)	Cropland used for grain (%)	Fertilizer used (in pounds/acre)	Production (in tons/acre/year)
United States	290	20	35	90	2.1
United Kingdom	21	27	35	150	2.8
Netherlands	1.4	27	20	540	3.1
Ukraine	44	58	35	110	1.4
Japan	15.4	12	50	350	2.5
Israel	0.3	21	25	210	1.1
Egypt	15	3	100	320	2.3
Tanzania	4.1	4	100	10	0.6
India	220	57	65	60	0.9
Vietnam	23	20	90	80	1.4
Peru	1.9	3	100	30	1.1
Guatemala	1.5	17	50	60	0.8

Sources: United Nations, Food and Agriculture Organization, *Production* Yearbook (1993), with calculations by the present writers.

good health. Why, then, does the world contain an estimated 1 billion undernourished people? We will not arrive at a simple answer, but we start our search for one by discussing variation in grain production around the world and then pursue the issue more completely in Section 2.5.

Table 2.3 presents data on grain production and land use in various countries that have been chosen as examples of the very large variations found around the world. Here are some of the major observations:

- The efficiency of production varies from more than 3 tons per acre to less than 1. Some of that variation can be attributed to differences in rainfall – high in the temperate countries of northern Europe and North America, and low in the dry areas of Africa and Asia, including countries such as Tanzania and India. Low rainfall can be supplanted by irrigation water. A prime example is Egypt (you rarely need an umbrella in Egypt), where the waters of the Nile River are now regulated by a dam and provide year-round irrigation (see Box 4.4). Mostly, however, efficiency is determined by the amount of fertilizer applied to the ground, which usually correlates with the application of pesticides and herbicides (Table 2.3).
- The percentage of a country's land that is available for crops is determined by factors out of, and within, the country's control. Japan and Peru, for example, have little choice; both can use less than 25% of their land

for crops because the rest of Japan is mountainous and the rest of Peru is either mountainous or covered by tropical forest. Almost 97% of Egypt is desert, and the only fertile part is along the Nile River and its delta. More than half of Tanzania is occupied only by sparse grass and brush. India and Ukraine are largely arable, although a better rainfall occurs in Ukraine, which was once referred to as the "breadbasket" of the Soviet Union.

- Although many countries have no choice in their selection of agricultural land, some countries with a favorable topography deliberately choose to restrict the production of grain. The Netherlands, for example, devotes more land to animal pasture than to harvested crops of any kind. The United States also has a large area of pasture; some of the cattle pastures in the eastern part of the country have sufficient rainfall to permit a conversion to crops, and this change could be matched by converting cropland in the semiarid west to grassland pasture. These choices are based on the desire of people in both countries for a meat-based diet, and, obviously, both countries are wealthy enough to afford the choice.

The total productivity of a country is not as significant as the productivity per person. Table 2.4 shows per capita production of some countries in terms of kilocalories of energy produced (at 100 kcal per ounce of grain) divided by the population of the country. The United States is

Table 2.4 *Selected data on grain and national wealth*

Countries	Population density (people/sq mile)	Grain fed to cattle (%)	Production (in 10^6 kcal/person/year)
United States	70	69	4.6
United Kingdom	600	59	1.2
Netherlands	1,140	42	0.31
Ukraine	220	n.a.	2.9
Japan	840	48	0.41
Israel	680	60	0.16
Egypt	2,200	33	0.85
Tanzania	80	3	0.46
India	770	3	0.75
Vietnam	550	0	10
Peru	50	35	0.25
Guatemala	230	27	0.5

Notes: Production in kcal per person per year is kcal available in grain produced before feeding to animals divided by population and does not represent total nourishment available to the people of the countries. For this calculation, 1 ounce of grain is assumed to yield 100 kcal of energy. n.a. = not available.
Sources: United Nations Food and Agriculture Organization, *Production* Yearbook (1993), with calculations by the present writers.

second only to China (not shown in Table 2.4) in total production but clearly leads all countries in the world in per capita production. Most countries produce grain that supplies about 0.25 million to 0.75 million kcal per person per year, which is approximately 25% to 75% of the 1 million kilocalories that a person needs annually for adequate nourishment (Section 2.1). The remaining energy needs are supplied by other foods, both plant and animal.

In discussions of worldwide nutrition, few subjects are as controversial as the use of grain as animal feed. Table 2.4 shows this feed as a percentage of grain consumed in the country (consumed grain is produced grain plus imports and minus exports). Almost all societies devote at least some of their grain to feed, and many countries import grain largely for that purpose. The United States uses about two-thirds of its consumed grain for animals, and Canada and most of western Europe show comparable rates. Poor countries, such as Tanzania and Guatemala, cannot afford to use grain for animals. India does not use it because the dominantly Hindu population regards cattle (bullocks) as sacred, employs them only for transportation and plowing, and lets them feed only on natural pasture. Vietnam feeds virtually no grain to cattle, partly because the Vietnamese eat very little meat and partly because cattle don't like rice. We discuss the amount of grain needed for animal raising in more detail in Section 2.3.

Imports and exports are controlled not only by grain production but also by the economics of the modern agricultural system (Table 2.5). Thus, Japan, the Netherlands, and Israel all import large quantities of grain, which is possible only in rich economies. As mentioned above, the Netherlands uses much of its imported grain for animals, which is simply one indication of the importance of meat in European diets. Some wealthy countries are fortunate enough to export a large part of their grain. They include Canada, Australia, and Argentina, all of whom sell about half of their grain abroad; the United States exports about one-third of its production and western Europe about 10%.

Poor countries try not to expend money on grain imports (Table 2.5). Some must import, however, in order to survive, and they pay for imports in currency generated by selling other products abroad. Some countries that need grain but cannot afford to pay for it receive foreign aid, either in the form of money to buy grain or free shipments of the grain itself. Of the total world production of 2 billion tons of grain, slightly less than 10 million tons (< 0.5%) are exchanged as aid. Principal recipients are sub-Saharan Africa, Egypt, and several countries in Asia (about half of Asian aid is to Bangladesh). Both the value and ethics of food aid are controversial, and we illustrate the problems in Box 2.1. Major donors are the

Table 2.5 *Selected data on grain trade and aid*

Countries	Import (+) and export (−) (in 10⁶ tons/year)	Aid received (+) and given (−) (in 10⁶ tons/year)	Production (in 10⁶ tons/year)
United States	−100	−7	290
United Kingdom		−1	23
Netherlands	+3.5		1.4
Ukraine	n.a.	n.a.	44
Japan	+30	−0.5	15.4
Israel	+1.9		0.3
Egypt	+9	+1.5	15
Tanzania			4.1
India	−0.4	+0.4	220
Vietnam	−13		23
Peru	+1.5	+0.3	1.9
Guatemala	+0.3	+0.3	1.5

Notes: Figures for total production are repeated from Table 2.3. n.a. = not available.
Sources: United Nations Food and Agriculture Organization, *Trade* Yearbook (1993), with calculations by the present writers.

United States (approximately 2% of its grain production), Canada, and western Europe; Japan donates approximately 3% of its rice crop. Please note that the amount of grain shipped to India as aid is equal to the amount that it exports. We pursue these issues more completely in Section 2.5.

Nongrain plants

We do not live by bread alone. Bread, made primarily from wheat, and other grains such as corn and rice provide more than half of the calories that most people in the world live on. Grains, however, are poor in many proteins, vitamins, and other necessary human nutrients. Consequently we supplement grains both with animal products (Section 2.3) and an extraordinary variety of plant products. The quantities of these nongrain plant products are considerably smaller than those of grains, and we can only briefly survey them here. Despite the low volume, the product can bring a good profit per acre to their growers. We also mention inedible fibers, principally cotton, but leave the production of wood and other biomass for energy to Chapter 5.

TUBERS Most plants grow from seeds (sexual reproduction), but tubers grow from runners that spread out along the ground from the plants. The runners leaf out at various intervals and send a new set of roots down into the ground, a process referred to as budding (asexual reproduction). Each newly formed plant stores energy as starch in a large bulbous mass in the soil below the plant. We can dig up these tubers and eat them or cut them into pieces and plant them back in the soil, where each piece can bud again to a new plant. The two major tubers are the potato, grown mostly in cool climates, and the cassava, which is restricted to the tropics. Potatoes originated in the Andes, where they were discovered by early Spanish explorers, brought to Europe, and then spread throughout most of the rest of the world.

With sufficient nutrients, the weight of tubers produced in an area of farmland is greater than that of any other edible plant, but three-fourths of that weight is water and only one-fourth starch. Because only the starch yields energy, the total nourishment from an acre of tubers is about the same as from an acre of wheat grown under similar conditions despite the greater total weight of the tubers. People who depend almost solely on potatoes for their nourishment (see the subsequent discussion on monocultures) must eat from 5 to 10 pounds per day, for an energy equivalent of 1,500 to 3,000 kcal. Current world production of potatoes is about 300 million tons, and because of their weight, potatoes are not traded or transported very far and are an important part of diets only where they are grown.

LEGUMES The major legume in world trade is the soybean, native to China and now grown extensively in North America (the United States produces half of the

BOX 2.1 HOW THE GREAT POWERS, WITH A LOT OF INTERNAL HELP, RUINED THE AGRICULTURE OF SOMALIA

The strategically important "Horn of Africa" is occupied by Somalia, bordered on the west by Ethiopia and on the south by Kenya. Until 1974, Ethiopia was ruled by the legendary Emperor Haile Selassie, who maintained cooperative ties with the West during the cold war, and the United States and western Europe armed his country. Somalia, being next door, was naturally a Soviet ally, and the Soviets armed the Somali army and built a naval base on the Gulf of Aden.

In 1969, a military government under strongman Siad Barre "elected" itself in Somalia by shooting anyone who objected. Then in 1974, a revolutionary group overthrew Haile Selassie and allied Ethiopia with the Soviet Union. At nearly the same time, the Somalis threw the Soviets out of their country and became new firm friends with the West. Mutual attacks and invasions ensued for several years, with temporary movement of the border one way or another but virtually no permanent change (except for the deaths of numerous combatants and noncombatants on both sides).

The West set about to support its new ally in Somalia, "President" Siad Barre. One way to strengthen the Somali regime was by providing aid in the form of free food. This aid was not needed as a humanitarian gesture because Somalia, in the early 1970s, was not only self-sufficient but an exporter of some types of food, mostly goats. The food aid did, however, strengthen the government by allowing it to sell food to urban residents below the price that they would have had to pay for food raised by Somali farmers. Because the food did not cost the Somali government anything, the profits from low-cost sales could be distributed among government supporters, which naturally included the army.

Gradually, the price of food in urban stores dropped below the cost of raising it on domestic farms. As Somali farmers were

increasingly driven out of business, it became necessary for donor countries to increase the amount of aid in order to prevent starvation. By the late 1980s, aid accounted for nearly half of the food consumption in Somalia. By 1991, rival factions in Somalia exploded, ousting Siad Barre and fighting each other so vigorously that the entire society was brought to a near standstill. Relief agencies established "feeding centers," and at one point in time 70% of the population was being kept marginally alive by relief feeding. In December 1992, the United Nations sent in troops to protect the relief workers. The troops withdrew in March 1993. By the end of 1993, 350,000 people were dead, mostly of starvation, including one-fourth of all children under 5.

The lessons from the Somali aid program are succinctly stated by Dr. Y. H. Farzin, and we summarize them as follows:

- Governments of countries receiving food aid must maintain adequate and stable prices for domestic farmers. Food prices to urban consumers should not be reduced below the market price of domestic production.
- Governments should aid farmers by such means as promoting conservation of rangelands and providing low-cost loans to farmers for fertilizer and land purchase.
- Governments should regard food aid as temporary and not as an "increasing free gift."
- Foreign governments providing aid should encourage (preferably insist on) proper behavior from recipient countries.

We hope that the tragedy of the starvation of the Somali children will underscore these conclusions.

FURTHER READING: Farzin (1991); Samatar (1994).

---

world's crop). Soybeans are raised largely for the oil that can be pressed out of them, but the bean is also edible (just barely). Legumes include a variety of other beans plus peas, peanuts, and alfalfa (used as a cattle feed). Their most important food quality is that their proteins and vitamins complement those of grains. Consequently, a diet of beans and rice or beans and corn (a tamale) is very nutritious. The diet of typical rural Mexicans contains about 1 pound of corn and 1 quarter pound of beans

plus other foods per day. Peanuts are rich in fats as well as starch, and consequently their food value is about twice that of ordinary grains (Table 2.1).

Legumes have a variety of uses in addition to providing food for people. Soybean meal, alfalfa, and the inedible (to humans) parts of other legumes are a rich animal feed. Also, because of their ability to capture atmospheric nitrogen and add it to the soil, legumes can replace artificial nitrogen fertilizers. This can be done either by alternating crops (legumes one year, grains another) in the same field or by planting rows of legumes, which are not harvested, between rows of grain crops grown at the same time.

OILS The term "oils" refers to mostly unsaturated fats made from vegetables. The most common oil in world trade is made from soybeans, but a host of other plants (corn, olives, etc.) are also used. Vegetable oils are used for baking and cooking and, in general, have taken over many of the uses for which butter, with its saturated fat, was once the principal source. Some plant oils have also replaced petroleum products as industrial lubricants, and the yellow-flowering rape is grown extensively for that purpose in India, China, and in vast areas of central and eastern Europe.

LEAF VEGETABLES Vegetables provide very little of the energy that people live on, but they are essential sources of vitamins and other nutrients, plus fiber. These needs are supplied by hundreds of different kinds of plants around the world, with total world production of about 400 million tons. People's opinions, particularly those of small boys, diverge widely about the relative merits and tastes of different "veggies" (a U.S. president was quite specific about his dislike for broccoli). Vegetables require very little farmland in order to supply the needs of a large population, and many of them are grown only in a few places. For example:

- The 70 million tons of tomatoes grown in the world each year require only a few thousand square miles of farmland.
- Three-fourths of all the artichokes produced in the United States are grown within 15 miles of the town of Castroville, on the central California coast; this region contains one area of 6 square miles that produces half of all U.S. artichokes.
- The 1.5 million people who live in the city of Vienna grow half of their vegetables within the city limits.
- More than three-fourths of the world's olives are grown in the Mediterranean area, mostly Italy and Spain.

FRUITS AND BERRIES Fruits and berries serve the same nutrient purpose as leafy vegetables, although with different proportions of amino acids, vitamins, fibers, and the like. Most fruit and berry production is climatically controlled. For example, citrus trees are limited to tropical or semitropical regions, and many berries are restricted to higher latitudes (virtually all of the world's strawberries are grown in Europe). Many types of fruit

and berries require the same type of limited space as vegetables. One index of productivity is the announcement of the European Common Market in the mid 1980s that the *excess* (unsold) production of wine would fill 10,000 Olympic-size swimming pools. Some fruits, however, occupy large portions of the arable land in some lesser-developed countries. They include principally bananas and pineapples, grown mostly in tropical lesser-developed countries, and we discuss the financial and ethical issues of their production in Section 2.5.

SUGAR Common table sugar is sucrose, a "disaccharide" consisting of one molecule of glucose (Fig. 2.2) and one molecule of another simple sugar known as fructose. As we will see in Section 2.5, sucrose and other sugars provide about 15% of all of the food energy ingested by people in industrial countries. Sucrose is extracted from sugar beets and sugarcane. One-third of world consumption is from beets, most of which are grown in Europe, and two-thirds from cane grown in tropical to subtropical areas (cane is a C4 plant: see Section 2.1). Cane was once the dominant source of sugar, much of it from plantations established during European colonization, but its use has been greatly reduced by the growth of the sugar beet industry, especially in Europe, and to an even greater extent by the manufacture of fructose from corn. This displacement has caused great hardship for many small countries, particularly in the Caribbean, which depended on cane as a major – in some cases, the only – export and the principal source of employment.

GOODIES Lacking an inclusive technical term, we use the term "goodies" to refer to such diverse products as coffee (Fig. 2.10), tea, chocolate, pepper, mustard, vanilla, and various spices and herbs. They have little or no nutritive value and are used almost wholly to make eating and drinking more enjoyable. Spices and herbs are produced in extremely small amounts that require very little agricultural land but a great deal of hand labor. The world's most labor-intensive agricultural product is the vanilla bean, formed by a climbing orchid native to Mexico and the circum-Caribbean region but transported to other colonial areas such as Reunion and Mauritius, two small islands in the Indian Ocean. Coffee and tea, along with bananas, pineapples, and sugarcane, occupy land that could be used for subsistence crops in tropical countries (Section 2.5).

FIBERS (MOSTLY COTTON) World production of cotton is approximately 50 million tons, one-fourth in China. This production is enough to make roughly 20 shirts for everyone on the earth. Cotton occupies large growing areas in a number of countries, and it is so affected by in-

Figure 2.10 Crops in tropical areas. Coffee in Costa Rica, showing areas of bare dirt subjected to erosion.

sect and animal pests that most growers use heavy doses of pesticide. For example, cotton is grown on 5% of the arable land of India, but more than 50% of all the pesticide used in India is applied to that 5% area. We discuss environmental consequences of this use in Sections 2.5 and 7.3.

Policy questions

Do you think that people have any responsibility for nourishment of other people except for friends and members of their own family? If you do, which (if any) of the following actions should be taken? Reduce population growth? Distribute fertilizer and industrial agricultural practices throughout the world? Restrict the use of grain for animal feed?

FURTHER READING: Pimentel and Hall (1984); Chadwick and Marsh (1993); Wild (1993); Chrispeels and Sadava (1994).

2.3 ANIMAL PRODUCTION

I was unprepared for a reader who was puzzled by a recipe she'd read; it asked her to skin chicken breasts. Did this mean, she asked, that she was to peel the plastic film off the styrofoam tray? – R. Mather, A Garden of Unearthly Delights, p. 1

Before the invention of planes, cars, and buses, the preferred way to travel was a "hay burner." The term comes from the fact that a horse has all the enzymes necessary to "burn" the carbohydrates in grasses. The hay does not burn rapidly, however, and a horse must continually graze in order to keep moving. For that reason, a person can always win a race with a horse provided the distance is long enough, perhaps 25 miles or more. Over a few miles, the speed of the horse is too much for a person, but after a few miles the horse must slow down, or stop, to graze. The human does not have to stop and finally overtakes the horse. We use both hay burners (land-based herbivores such as cows and sheep) and meat eaters (fish) for food. Although fish are animals, we follow conventional practice in reporting on food production by using the term "animal" for land-dwellers and "fish" for aquatic animals. In order to understand the various issues connected with animal production, we start with the concept of "commons."

The commons

A commons is a place that is accessible to everyone. A simple example is a village in which all residents may pasture their cows in a field owned by the village as public land. Thus, the villagers own the cows, and their meat and milk, but not the land on which they are raised. Ideally, where there is plenty of fertile land, there is no fee for use of the commons. The commons, however, may be supported by taxing everyone in the village, by charging a fee for each cow pastured in the commons regardless of how much meat and milk the cow produces, or by assigning royalties (a percentage of the profits) to the village when the meat and milk are sold.

The use of land as a commons encounters two kinds of problems. One is simply running out of space because of population growth. For example, population pressure on sparse grazing lands (and a tradition of warfare) sent waves of Mongols and related people out of central Asia during the Middle Ages. The second, but related, problem with a commons is any attempt to make it overproductive. If village farmers sell milk, they make more money by having more cows and goats. Thus, each farmer tries to add more cows and goats to the commons, and if the land is unregulated, the commons may eventually be overgrazed and become virtually useless (see the discussion of Sahel in Section 2.5).

A commons need not be in a settled position. For example, nomads try to "expand" their commons by moving from area to area (see Box 1.1). Migrations of animals, not exactly nomadism, also characterized the American West in the 1800s, when cattle and sheep could be driven across unhindered countryside, referred to as an "open range." Increasing immigration gradually reduced the open range by fencing in private land – the architect Frank Lloyd Wright stated that "barbed wire is a sign of advancing civilization only west of the Mississippi." Some open range existed through the 1950s, a menace to automobile drivers who not only had to repair their cars if they hit a cow on the highway but also had to pay for the cow.

Animal raising

Most of the animals that we use for food are not simply hay burners but ruminants (Section 2.1) The ability of ruminants to rework their feed and continually redigest it enables them to gain more nourishment from the feed but at a slower pace than a horse would (cows are seldom in a hurry). It also means that cows can eat almost any plant material, and goats can eat even more diverse material.

People obtain animal products in five different ways. The first method is hunting, which clearly cannot supply the world's population with meat. In fact, as we discuss in Section 8.3, hunting depletes its source even in sparsely populated areas.

The second model for animal use is pastoralism, in which animals browse (on bushes) or graze on a grassland commons, either in a settled location or in the herds of nomads (Fig. 2.11). Owners of these animals use the milk as food and the fur (wool and goathair) for clothes and shelter. Unless the pastureland is very rich, meat is generally eaten only at special feasts. The animal diet is supplemented by grain grown by the pastoralists or purchased with money earned from the sale of milk, hides, and meat or by "renting" animals for transportation of the goods of other farmers. Areas that raise animals in this fashion also use them extensively for labor, particularly in Africa, southern Asia, and parts of Latin America where, for example, oxen are used to haul carts and bullocks to pull plows and turn mill wheels.

A third method of raising and using animals is rotation of grazing animals among fenced pastures and feeding of pigs and chickens in separate enclosures. Animals can be driven from a pasture where they have been grazing to a fresh one with uncropped grass. This procedure lets the old pasture grow a new crop and also allows some pasture to be set aside to grow hay or other plants as a diet supplement and as feed for the winter. This model is particularly useful for the dairy industry and works well in areas of high productivity, such as in Europe and the eastern United States, or in low-productivity areas where there is a lot of space, as on the ranches of central Australia, which support fewer than one cow per square mile. Chickens and pigs can be fed with grain and other vegetable matter (swill) unfit for human consumption.

The fourth model is a combination of grassland and grain feeding, and applies particularly to raising of cows for meat. Young cattle graze in pastures and mature cattle, just before marketing, are "fattened" by feeding with corn or other grains. This fattening produces beef that has more fat (it is "marbled") and is much less tough (chewy) than meat from range-fed cattle. Worldwide, approximately one-third of all grain is used for cattle feed, and Table 2.4 shows percentages in selected countries. Cows use approximately 10 pounds of grain to produce 1 pound of beef, which contains approximately the same energy (in kilocalories) as 1 pound of grain. Thus, the energy ultimately derived by eating the meat is about 10% of the energy in the grain used as feed.

A fifth method of animal raising has been developed recently by industrial countries. It consists of keeping animals in pens for most of their lives, where their only duty is to eat, grow, and to produce milk and eggs or be converted into meat. We discuss this method and some of its consequences in Box 2.2.

By these various feeding methods, we obtain meat and other products from cows, sheep, goats, pigs, chickens,

(a)

(b)

Figure 2.11 Examples of animal raising: (a) sheep grazing on hillside on the North Island of New Zealand; (b) goats nibbling bushes in Red Sea plains of western Saudi Arabia.

and a variety of other animals that many of you would probably prefer not to consider as a food source (Table 2.6). Pigs can be used only for meat, and chickens for meat and eggs. Cows, sheep, and goats, however, produce meat, milk, cheese, hides, wool, and hair. Variations in the types of animal products used in different areas are partly determined by the agricultural conditions of the area but also, to a great degree, by ethnic, national, and regional variations in people's preferences. For example:

• Moslems and Jews do not eat pork, and countries in which they dominate do not raise pigs. By contrast, the Chinese raise nearly half of all the pigs in the world.

- Americans have always liked beef and never been fond of lamb and mutton, and the percentage of sheep in the United States is very small despite broad areas of the western part of the country that are favorable to sheep raising. Australia and New Zealand, on the other hand, raise a large proportion of the world's sheep, largely for wool (they can't possibly eat that much lamb).
- South Americans also prefer beef. Only in the Andes, where cows cannot survive, is the goat population of much significance.
- Goats are the dominant domestic animal in most of Africa and parts of Asia. They can survive in semiarid to arid regions, and goats can eat virtually any food that relatively poor populations have left over.
- Chickens are another animal that is easy to feed, partly from grain that people cannot eat, and they do not require much space. Chickens are popular in most of the world (there are twice as many chickens as people).

Table 2.6 shows worldwide production of major meat products plus selected information for various areas and countries. We investigate the role of both animals and fish in human nutrition at the end of Section 2.4, after we discuss fish production.

Policy questions

If industrial countries raised cattle, pigs, and chickens in pastures and barnyards instead of in pens, where animals are fed on prepared food, we might have to pay more money for meat that is slightly tougher (more chewy). Are you prepared to do this?

FURTHER READING: Lappe (1982); Rifkin (1992).

2.4 FISHING

Little Tommy Tittlemouse
Lives in a little house.
He catches fishes
In other people's ditches.
— Nursery rhyme (slightly modified)

Virtually all early human settlements were along rivers, lakes, or the seashore, and people obtained some of their food from the water. Although some societies eat almost anything that comes out of the sea (sea urchins, octopus, etc.), the bulk of food from the oceans has always con-

sisted of fish. Also, because the quantity of food taken from the oceans is so much greater than the quantity taken from freshwater, we concentrate on marine fish for this section. First we describe the requirements for growth of healthy fish populations and then the status of the fishing industry.

The nurture of fish

On land, we can see the operation of a food chain. Plants grow by photosynthesis, herbivores eat plants, carnivores eat herbivores, and bacteria and fungi cause dead animals and plants to decay. Although it is less visible, the same food chain operates in oceans and lakes.

The base of the oceanic food chain is single-celled plants. They are part of a group of organisms, known as "microplankton," which float near the ocean surface and drift with the currents ("plankton" means drifting rather than swimming). Because algae need light for photosynthesis, and light can penetrate only a few hundred feet even into very clear water, the plankton must live near the water surface. The two main forms are diatoms, which are surrounded by small shells made of silica (SiO_2), and dinoflagellates, with an organic cell wall. Both varieties have diameters in the range of 100 microns (a few hundredths of an inch). A third variety, more important in the centers of large ocean basins, is coccolithophores, which have a diameter of a few tens of microns and are surrounded by a mosaic of tiny calcite disks. In addition to light, water, and CO_2, algae need the same nutrients that are required by all other organisms, mostly phosphorus and nitrogen. Because these nutrients are largely supplied by freshwater discharged from land by rivers, the algae are more concentrated in the regions of oceans rather than in the centers.

Just above the base of the oceanic food chain are animals that do not photosynthesize, do not move of their own volition, and live by ingesting other organic matter. The simplest of these animals are single-celled planktonic herbivores, including SiO_2-covered radiolarians; dinoflagellates and a few other plankton species alternate between photosynthesis and ingestion. This category also includes sedentary organisms such as clams and corals, which obtain food by filtering it out of the water or similar processes. Most of the food used by animals at this level does not consist of whole single-celled organisms but of particles and dissolved material left from the death and decay of the algae and other plankton. Animal plankton also include large numbers of multicellular organisms. Most of them are arthropods, a diverse phy-

BOX 2.2 MAD COWS AND CAGED HENS

Most of the meat, eggs, milk, or other animal products consumed in industrial countries come from animals raised in cages, small pens, or in very crowded barns. These conditions leave little, virtually no, room for movement. Chickens sit in cages and produce eggs. Cows stand in stalls and produce milk. Pigs grow bacon and ham as they rub together in barns where their waste products are periodically hosed out into holding ponds around the barns. Food is delivered to the cages and pens, the eggs and milk are removed, and when the animals are no longer useful for eggs or milk, or are sufficiently fattened to be converted into meat, they are removed and killed.

These conditions have spawned various individuals and groups that want to change the practice. One of the most effective is Freedom Food, an organization sponsored by the British Royal Society for the Prevention of Cruelty to Animals (RSPCA). Freedom Food has promoted, with limited success, the idea that people should buy meat and other animal products only from organizations that allow the animals freedom of movement while they are alive — that is, a normal life-style. Beef and pork should be bought only from stores that certify that the cows have wandered through pastures and the pigs have rooted through the mud of an outdoor sty. Eggs should be purchased only from places that certify that chickens have been allowed to roam through chicken yards.

Although most people agree that animals should not be confined in tiny spaces during their entire lives, the problem with promoting the idea of freedom is that pens and cages are an efficient form of industrial agriculture. Eggs, milk, and meat obtained from them is generally less expensive than from wider-roaming animals. The low cost to consumers of this industrial agriculture has thus far prevented much change in the way animals are treated in industrial countries, but as with so many other apparent costs discussed in this book (see particularly Chapter 8), we may discover that the real costs are far higher than we think they are. The problem is exemplified by a disease referred to as bovine spongiform encephalopathy (BSE), which literally means holes in the brains of cows and is commonly called "mad cow disease."

This type of brain disease, which is invariably fatal, has been known for many years in a variety of mammals. It is particularly common in sheep, where it is referred to as scrapie. It is caused, in some poorly understood way, when misshapen proteins, called prions, replace or interfere with normal brain proteins and cause the brain to rot away into a "spongy" condition.

The disease can apparently be inherited, and some investigators have proposed that carnivores that eat sheep with scrapie may acquire the disease, but generally there is no evidence that an encephalopathy in one variety of animal can be passed on to another variety. That is, scrapie, BSE, and related diseases have been regarded as a nuisance but not serious. That evaluation may change soon.

The new problem is that a debilitating illness, called Creutzfeldt-Jakob disease (CJD), which normally affects only one person per million worldwide, has recently appeared at slightly higher frequency in Britain. CJD causes death with the same symptoms as BSE, and medical experts have begun to wonder whether cows with BSE might be able to transmit it to people who eat beef. That transmission has not been proved, but the possibility has sent animal experts to search for the reason that BSE seems to be more prevalent in British cows than it used to be. That is where we arrive at the problem of high-efficiency agriculture and animals raised in pens.

Cows foraging in open fields live on grass, or hay harvested from grass, but cows in pens are commonly fed a mixture of hay and food that makes them grow faster. In many countries, this mixture now contains a small amount of meal made from parts of dead sheep (in Britain this meal contains 4% to 5% sheep parts). That is, these cows are no longer purely herbivores but now consume animal products. It seems possible that cows that eat parts of sheep that had scrapie would develop the same brain disease that the sheep had. Then it is possible that people who eat the beef from cows killed before the disease was recognized could develop CJD. No one yet knows whether this chain of events is correct or whether CJD has any relationship to BSE, but the mere chance caused the European Union to ban the sale of British beef and set Britain to the task of culling (killing and disposing) approximately one-third of its cattle herds.

The total cost to British agriculture is estimated to be at least 15 billion pounds ($20–35 billion [U.S.]). This cost may turn out to be far higher than the extra cost of buying beef only from cattle raised according to the precepts of Freedom Food. If so, then industrial agriculture will have cost more in the long run than traditional cattle raising. We examine similar problems with regard to other agricultural issues elsewhere in this chapter and also in Section 7.5.

FURTHER READING: *Economist*, March 30 – April 5, 1996, pp. 25–7.

Table 2.6 *Selected data on meat production*

	Cattle (fresh meat)		Mutton	Goat	Pork	Chicken
	Produced	Imported (+) and exported (−)				
World	5,500		690	250	6,900	3,800
United States	1,200	+33	17		850	3,500
United Kingdom	140	−5	40		110	360
Netherlands	60	+25	2		190	320
Ukraine	140		3		100	290
Japan	65	+56			26	380
Israel	9	+5				30
Egypt	4	+5	6	4		85
Tanzania	20		1	30	1	32
India			18	50	40	27
Vietnam	9				80	130
Peru	11	+3	2		10	62
Guatemala	6	−1.4	0.3		2	15

Notes: All values are in 10^4 tons.
Sources: United Nations Food and Agriculture Organization, *Production* Yearbook (1992), plus additional information from United Nations Food and Agriculture Organization, *World Animal Review* (1993). Calculations by the present writers.

lum that includes shrimp, copepods, and several other groups. Animal plankton also includes the larval stages of fish.

The third, highest, level of the oceanic food chain consists of animals that actively seek food as they move by their own volition (they are "nektonic" rather than planktonic). Most of them are fish, but the category includes mammals (e.g., whales) and invertebrates (e.g., squid). A few of these animals, such as the anchovy, are herbivores, but most are either carnivores that eat other fish or will eat anything (omnivores). The smallest fish, such as sardines, tend to live on organic particles, but larger fish eat smaller fish and so on until the top of the food chain is reached by such fish as tuna. Because of the extraordinary abundance of plant and animal plankton, fish account for far less than 1% of the total biological material (biomass) in the ocean.

Using this information about nutrition, we can discuss the way in which fish reproduce and grow. Fish reproduce (spawn) when the female emits a large stream of eggs, some tens to hundreds of thousands, and the male fertilizes them with a cloud of sperm. Some fish spawn in the open ocean, some in estuaries or coastal wetlands, and some (e.g., salmon) swim up rivers to spawn in the waters in which they were hatched; this remarkable "homing" sense has never been satisfactorily explained. Most spawning and fertilization occurs in the water, with fertilized eggs settling to the bottom, but wide variations occur between species. Female salmon, for example, use their tails to dig shallow holes in stream bottoms, deposit their eggs in the holes, and then cover the holes with sediment after the male fertilizes the eggs.

Fish require some time after hatching before they are fertile adults and can spawn. This time of infertility before fish reach adulthood averages about two years but is highly variable from species to species. Some fish spawn only once, but others spawn repeatedly during their lifetimes, which range from a few years to more than 100 years. In order to preserve spawning stocks, many governmental and other organizations try to prevent catching fish that are too small to be reproductive adults. Anglers may be required to throw fish that are below some minimum standard for the species back into the water.

After spawning, a fertilized egg develops into a larva, which survives for only a short time, perhaps a few days, with energy provided by an attached yolk. At the end of the larval stage, a young fish "hatches" and must find

BOX 2.3 THE COD IN THE FOG _____

The black Labrador padded across the sodden ground, avoiding the deeper water, and weaved through the rain and mist to home, the cottage of a fishing family. Ah, this must be Newfoundland, a place that the residents sometimes refer to as the land of "Cod, fog, bog, and dog." By the late 1990s, the fog, bog, and dog were still there, but the cod was mostly gone.

The fishing riches of Newfoundland, and much of the northeast coast of North America, were based on "ground fish," which live either on and in the bottom sediments or swim in near-bottom waters. These waters are particularly rich on shallow shelves such as the Grand Banks of Newfoundland (Fig. 2.12), where nutrients are supplied by upwelling of deeper Atlantic Ocean water (see text). The bottom-dwelling fish are commonly caught with lines and small drags; the main variety is halibut and the closely related sole and turbot. Slightly above the bot-

tom are cod, at one time widely distributed through the North Atlantic and the main resource for fisherman of the Atlantic provinces of Canada.

The cod were originally caught in small boats using poles and lines but the development of modern techniques made "harvesting" possible. All fishermen had to buy larger and better-equipped boats. The improvement in technology increased the catch, for a few years, but by the 1970s the fishing had been so effective that the spawning population of cod had been reduced below the critical level for replenishment. At that time, the total catch began to decline and continued to do so throughout the 1980s. By 1992, in an effort to allow the cod to recover to sustainable breeding numbers, the Canadian government prohibited cod fishing in the Canadian Exclusive Economic Zone of the Grand Banks. The immediate consequence was to put

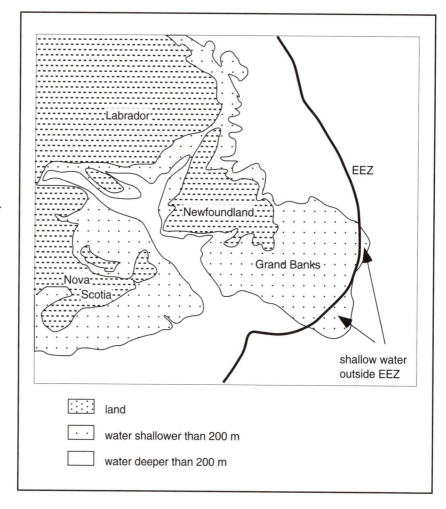

Figure 2.12 Disputed fishing areas on the Grand Banks of Newfoundland. Shallow sea floor around land is delineated from deeper water by the 200-m contour (the deep channel between Labrador and Nova Scotia extends into the St. Lawrence River). The shaded area shows where the shallow Grand Banks extend into the Atlantic beyond the 200-mile Exclusive Economic Zone (EEZ), leading to disputes over jurisdiction in attempts to control fishing. The diagram is generalized from United Nations Food and Agriculture Organization, World Review of Highly Migratory Species (1994).

land

water shallower than 200 m

water deeper than 200 m

much of the population of Newfoundland out of work, and although the Canadian government provided unemployment support and paid the costs of retraining for other jobs, many were still out of work 5 years later.

By the middle 1990s, several varieties of fish in addition to cod were depleted along the whole east coast of the United States and Canada. Federal and local governments in both countries placed additional restrictions on the amount of fish that could be caught. So many fishing boats were for sale that many owners could not recover their original purchase price. Some, but not all, fishermen placed the blame on the governments either for not taking some action earlier or for preventing them from doing enough fishing to make a living. Other criticisms were directed at industrial pollution in coastal areas, tourism, and sport fishing. Right now, there is more blame than fish.

In addition to unemployment, Newfoundland also has a purely geological-geographic problem — the Canadian EEZ includes more than 90% of the Grand Banks, but not all (Fig. 2.12). The part outside the EEZ is theoretically free to the fishing ships of all countries, but a North Atlantic Fishing Organization (NAFO) organized in 1976 has attempted to coordinate fishing in order to prevent depletion of stocks. The NAFO regulations are commonly little more than suggestions, and the outer zones of the Grand Banks continued to be depleted.

Fishing outside of the Canadian EEZ almost caused a miniwar in 1995. Ships chartered in Spain began fishing off Grand Banks, and Canadian naval vessels intercepted them. Because this was outside of the EEZ, Spain claimed that Canada was guilty of piracy. When the Canadians examined the fishing gear on the Spanish boats, they made a singular observation — the mesh size on the fishing nets. In order to spare small fish that had not yet spawned, the standard size of mesh on all cod boats should be set at 18 inches. These large openings are adequate for normal fishing, but by the 1990s they did not provide a sufficient catch as the number of fish decreased. Therefore, some of the Spanish boats were using nets with mesh sizes of 4 inches, which virtually insured depletion of the spawning stock. Because this incident occurred in international water, nothing was officially illegal, war was not declared, and the Spanish simply went home.

We hope that sensible regulations will be maintained and the cod will return to their place in the fog.

FURTHER READING: *Economist*, March 19, 1994, pp. 21–4.

---

food quickly or it will die. Larvae and preadult fish follow a variety of paths depending on the species. Some live wholly in the oceans, but many live in estuaries, bays, and coastal marshes where nutrients are abundant and competition for resources from older fish is not as fierce. The larval and young stages of many animals lower on the food chain (e.g., shrimp) also depend on marshes during their early growth. The fact that a healthy fish population depends so much on coastal wetlands is a primary reason that many local and federal governments do not allow wetlands to be drained for agriculture or commercial use and prohibit sealing coastal marshes off from the sea by building walls around them for marinas or other shore-front property.

Most fish live around the margins of oceans. One reason is this high dependence of many species on rivers and coastal wetlands for their early development. A second major reason is the abundance of nutrients in shallow marine waters. Although some of these nutrients are supplied by coastal rivers, the abundance is primarily caused by "upwelling" of deeper marine water onto shallow continental shelves. Water at depths of several hundred feet in the oceans is particularly rich in nutrients, as most nutrients are produced by decay of plants and animals that live in near-surface water. The particles and dissolved organic materials that result from this decay settle downward, enriching the deeper water. It takes ocean circulation to force this nutrient-rich water onto shallow banks along some continental margins and thus return the nutrients to a depth where they can be consumed. One area of particular wealth is the Grand Banks of Newfoundland, which we discuss in more detail in Box 2.3.

The fishing industry

Commercial fishing occurs both in the oceans and in inland waterways (lakes and streams). Because only 16% of the catch is from freshwater, we concentrate on the oceans for this discussion.

Ocean fish occupy an area that is physically a global commons. It is physically a commons because the only barriers to movement of the fish are oceanographic ones, primarily water temperature and salinity, and not political boundaries. These barriers cause most fish to have a very local distribution, and even those that are highly migratory are restricted to broad temperature zones (cod in cold water, tuna in warm to hot water, etc.). Many fish migrate seasonally as water temperature changes. Salinity shows very little variation from place to place in the open oceans but major changes in near-shore areas. Bays and estuaries range from the 35 parts-per-thousand (ppt) salinity of the open ocean (3.5% salt content) to the near-zero salinity of fresh water; in many places, the variation is tidally controlled.

For most of human history, the oceans were also a commons in the sense of their use for fishing. In theory, anyone who could put a boat into the water could fish anywhere. Conflicts, of course, have occurred throughout the ages. They range from fishermen standing on the decks of their boats yelling and shaking their fists at each other to, hopefully only in more primitive times, attempts to sink or capture a boat and kill or enslave the crew. More serious conflicts have involved different countries, such as disputes between Mexico and the United States, in the North Atlantic between the European Union and Iceland, and in the Pacific between Peru and both of its neighbors. Sometimes, the navies of different countries have been drawn in.

At present, the oceans produce about 85 million tons of fish and other marine products each year and are a source of protein for many of the world's people (Table 2.7). The major categories of marine catch are fish ("finfish"), mollusks, and crustaceans. Small fish such as herring, anchovies, sardines, and pilchard constitute one-fourth of the weight of all finfish. Cod (and the related pollock) and the familiar tuna, sold largely in cans, are other major categories. Mollusks include clams and oysters, but most of the catch is squid (a small percentage of the squid is a variety known as the "jumbo flying squid," a name to excite the imagination). Crustaceans include crabs and lobsters, but the principal tonnage of the catch is shrimp.

Despite this high oceanic productivity, the idea that something was wrong with the fishing commons began to develop in the 1970s and 1980s. More and more fishermen reported trouble making a living, and the stocks of certain major food fish began to decline. For example, Atlantic cod, sometimes called "pigs of the sea," yielded several million tons during peak production in the North Atlantic in the late 1960s, but by the early 1990s the catch had declined to little more than 1 million tons, and some countries had closed their waters to cod fishing (Box 2.3). Cod was one of the standard fish for fast-food restaurants and "fish-and-chips" stores and contributed the major share of the export earnings of Iceland.

Table 2.7 Selected data on world fishing for 1992

Location (%)	
Marine	84
Finfish	80
Mollusks (mostly squid)	10
Crustaceans (mostly shrimp)	6
Fresh water	16
Oceans of catch (%)	
Pacific	60
Atlantic	30
Indian	10
Major marine fish (in million tons)	
Herring, sardine, anchovy, pilchard	20
Mullet, jacks	10
Cod, pollock	10
Bass	7
Tuna	5
Salmon, trout	2
Flounder, sole	1

Notes: Production in kcal per person per year is kcal available in grain produced (at 100 kcal per ounce) divided by population and does not represent total nourishment available to the people of the countries.
Sources: United Nations Food and Agriculture Organization, *Fisheries Statistics: Catches and Landings* Yearbook (1992), with calculations by the present writers.

The development of high technology has been one, if not the principal, contributor to the problems of the fishing industry. Major seafaring countries are able to operate around the world with large "mother ships" (fish storehouses and processing factories) and a fleet of smaller ships to set nets and do the actual catching. Schools of fish are located by helicopters and spotter planes, and sonar (underwater listening) equipment can identify not only the location but, in some cases, the species of fish. Perhaps the most efficient devices, in terms of tons of fish caught, are large drift (gill) nets that hang from floats and extend over miles of ocean (see the quotation at the start of this section). The nets catch everything that comes by, in contrast to the catches of individual species made by older boats that simply kept 20 or more lines in the water at any one time.

The technology joined with growth of the world population and the need for more food resources to deliver one more blow to the fishing industry – growth of the fishing fleets. At present, the world's oceans are fished by more than 5,000 ships that weigh more than 500 tons and an uncounted number of smaller ones, and some 2 million people try to make their living from the sea. During the 1980s, the number of fishing vessels increased twice as fast as the increase in catch.

The problem with world fishing was not immediately apparent because the total fish catch, calculated as tons of fish, increased until the late 1980s. This increase resulted from an increased use of gill nets and other fishing methods that catch a lot of fish that previously were not caught by hooked lines. Although much of this catch is "trash fish" (varieties that would not normally be used for human consumption but which can be processed and incorporated into the mix of fish sold for general consumption), stocks of many valuable species have been drawn down by fishing at a rate greater than replacement by spawning. In effect, this overfishing is an expenditure of fish capital in the same way as overtimbering (excess logging) is an expenditure of forest capital (Section 8.3). A prime example is the increase in the catch of tuna and related species from about 2 million tons in 1970 to 5 million tons in 1990. This increase, however, leaves 10 of the 14 major varieties of tuna at risk because the spawning population has been reduced below effective size. Further reduction of the spawning population was caused by reduction in the mesh size of the gill nets, which reduces the average size of fish caught. For some species, the meshes have been reduced below the size of adult fish, resulting in catching of fish before they have a chance to spawn and replenish the stock.

Clearly, the unregulated use of the oceans as a commons is not working very well. Just as an individual farmer benefits for a short time by putting more cows on a common pasture, even if the pasture ultimately degrades, so an individual fisherman or fishing corporation benefits by extracting more fish from the sea even if the ultimate result is destruction of the stock. When this fact became apparent to local and national governments, several solutions were attempted.

The first, naturally, was hurling acrimonious insults at everybody in sight on the theory that the problem was someone else's fault. In 1976, however, many governments came together on a treaty that gave a country jurisdiction over an area extending 200 miles out from its coastline at any place. This "Exclusive Economic Zone" (EEZ) was reserved to the country for fishing and any other commercial use, such as drilling for oil and gas (Section 5.3). Borders between countries were extended directly outward into the ocean for the purpose of establishing the zones of national control. Along most coasts, the EEZ extended over much of the continental shelf, thus given a country jurisdiction over its own shallow and nutrient-rich water.

The EEZ allows countries to regulate fishing within their water, and some countries delegate further responsibility to states and provinces. Types of regulations include setting quotas, commonly measured as tons of fish that can be landed at a country's ports, and setting allowable seasons in which fishing can take place. For example, for many years Alaska has strictly regulated fishing for salmon and certain other fish by establishing periods of time in which fishing could be done. A fishing season of several months for some varieties of salmon has now been shortened to one day, and the "season" for roe herring has, in some years, lasted only 40 minutes. Establishment of the EEZ provided some control but not very much. Because many of the major commercial fish (tuna, cod, herring, etc.) migrate across EEZ borders, regulation within one zone is seldom sufficient. Furthermore, the ocean is very difficult to police, and fish caught illegally off the waters of one country could be transported to ports elsewhere and sold there.

An additional problem with regulation of fishing is the effects of onshore activities beyond the control of fishing authorities. Industries along coasts commonly discharge pollutants into coastal waters. Large farms and ranches add fertilizer and animal residues to streams. These residues are rich in nitrogen, phosphorus, and potassium and cause excessive growth of plants in coastal wetlands (see the discussion of eutrophication in Section 7.5). This deprives larvae and young fish of the opportunity to develop into adults that can survive in the open ocean (see earlier discussion). Construction of dams along rivers reduces the ability of salmon and other fish to make their way to spawning grounds. For example, by the late 1980s, salmon in the northeastern Pacific that used to spawn in the upper reaches of the Columbia River and its tributaries had been reduced from millions a year to a few hundred thousand.

Animals and fish in human nutrition

All people have some eating habits in common, and in Table 2.8 we show data for selected countries that exemplify these habits. Almost all of us obtain most of our energy from grain, with fruits and vegetables providing vitamins and other necessary nutrients but very little energy. For most people, the only other significant sources of energy are meat, vegetable oils, and various sweeteners (mostly sugar). Other foods tend to be important only in local areas, such as a high use of cassava tubers in some tropical countries. The basic similarities in nutrition, however, mask important variations among the world's people.

The principal differences shown in Table 2.8 are between the diets of people in rich industrial nations, including western Europe and the United States, and those

Table 2.8 *Sources of nutrition energy for selected countries*

	Kilocalories per person provided by food types									
	Total	Grain	Tubers	Pulses	Veg. oil	Sugar	Meat	Milk	Alcohol	Veg/an
United States	3,600	680	95	30	530	570	660	360	190	2
United Kingdom	3,200	640	200	24	300	480	490	320	180	2
Netherlands	3,300	620	170	25	360	390	360	380	180	3
Ukraine										
Japan	2,800	1,200	80	21	280	250	140	100	150	4
Israel	3,000	1,000	64	38	480	370	240	250	36	4
Egypt	3,300	2,000	49	33	340	370	82	61	0	10
Tanzania	2,200	1,000	500	130	92	<20	49	36	93	20
India	2,100	1,300	40	130	140	210	<20	100	<10	20
Vietnam										
Peru	2,200	1,000	190	57	120	340	100	100	75	7
Guatemala	2,300	1,400	13	110	130	370	41	83	<10	10

Notes: Kcal provided by various food types add to less than the total kcal for each country because numerous foods that provide small amounts of energy are omitted from the table. Veg. oil is vegetable oil. Veg/an is the amount of energy provided by vegetable products divided by the amount of energy provided by animal products.
Sources: Data are generalized from United Nations Food and Agriculture Organization (1990), *Food Balance Sheets.* Data for Ukraine and Vietnam were not available in 1990.

of rural people in Africa and southern Asia. We call particular attention to the total daily calories available per person – more than 3,000 in the United States and United Kingdom and barely 2,000 in Tanzania and India. In all poor countries, urban dwellers are, on average, richer and have better diets than the rural people who actually produce the food. In addition to the differences in available calories, the proportions of different foods also vary between rich and poor countries, and we discuss this variability in the remainder of this section.

Everyone needs at least a small amount of animal product (milk, if not meat) as a source of vitamin B_{12} (Section 2.1). People in rich countries obtain a high proportion (about one-third) of their energy from meat, compared with only about 10% in poor countries. The types of animal products used are different – beef, partly fed with excess grain, in rich countries and goats or, to a lesser extent, sheep raised in areas of poor pasture in poor countries. Rich societies also convert a high proportion of their milk to cheese. Cheese is made from milk "curds" (cottage cheese is simply the curds), and the more watery milk "whey" is used for other purposes or, in some cases, just discarded. Poor societies commonly just drink the whole milk from whatever animals are available because they do not have the technology for cheese making and cannot afford to lose any of its nutritive value by discard-

ing the whey (we note that when Miss Muffet sat on her tuffet, she ate both her curds and whey and, thus, behaved in an environmentally ethical fashion).

Fish are nutritionally very useful. The fats in fish tend to be unsaturated (oils) in contrast to the saturated fatty acids of meat, and fish contain a concentrated mix of proteins and vitamins. Most societies, however, use fish only as a minor supplement to normal diets, and consequently fish are not listed in Table 2.8. One exception is the Maldive Islanders in the western Indian Ocean. The Maldive population of 225,000 is scattered over a chain of 1,200 islands with a total land area of only 115 square miles. With little space for corn and none for animals, the islanders harvest most of their food from the sea. Some seafaring countries, such as Iceland and Norway, also choose to eat a lot of fish.

Vegetable oils (unsaturated fats) constitute a significant part of diets in almost all parts of the world. Paradoxically, as people in poor countries want more meat in their diets, people in rich countries are becoming more concerned about the health effects of the saturated fats in meat, milk, butter, and cheese. Thus, industrial countries are progressively replacing butter with partially hydrogenated vegetable (mostly corn) oil to make margarine, and lesser-developed countries are devoting more of their grain production to animal feed (Table 2.3).

Table 2.8 shows a major difference in consumption of sugar between rich and poor countries. Americans and Europeans use about 50 pounds of sugar a year, roughly 10% of their energy intake. Sugar is used not only as a recognizable sweetener in such items as baked goods and soft drinks but as an additive in many foods where we seldom notice it (bread and almost all canned goods). Table sugar, for tea and coffee, and the sugar in soft drinks is invariably sucrose, but industrial societies produce corn syrup (fructose; see Section 2.1) for additives to other food. Poor countries depend almost exclusively on the sucrose from sugarcane, where it is available at all.

Policy questions

A commons is difficult to regulate. In the words of Garrett Hardin (1968), it requires "mutual coercion mutually agreed upon." Thus far, neither the mutual nor the coercion have been effectively agreed upon by the world community. What policies would you suggest to prevent overfishing, either by citizens of your own country or those of other countries?

FURTHER READING: Cushing (1995).

2.5 MONEY AND AGRICULTURE

We are burning the country down. – *Geologist in Moro Goro, Tanzania, as he watched farmers set fire to the stubble from last year's corn fields*

The nutritional, environmental, and financial aspects of food production are intertwined in a maze of relationships. We arbitrarily divide them into environmental and financial categories in this discussion, but the two aspects are really inseparable, particularly if we ever figure out a way to calculate the true cost of environmental degradation (Chapter 9).

Environmental issues

All agriculture modifies the land, particularly the 10% used for crops. The earth's surface, including its soil cover, developed over millions of years as a result of natural processes (Fig. 2.8). Planting of nonindigenous crops, or introduction of nonindigenous animals, must affect the environment (Box 2.4). This does not mean that we must suspend modern agriculture – the human race could not survive if we did, and the environment is not always adversely affected. It does mean that we must understand the changes and limit their deleterious effects as much as possible.

Soil erosion results both from animal and crop production. The world's surface, from lush grasslands to semiarid brush country, is naturally adjusted to grazing and browsing by small numbers of animals. When more animals are introduced, either to feed a local population or to produce meat for export, the vegetation cover is stripped more rapidly than it can grow. The problem is particularly acute with goats and sheep, which strip grass down to its roots. Cows are unable to nibble grass as closely as goats and sheep and are a serious problem only in semiarid areas, such as the western United States. The virtual collapse of the southern Sahara region (Sahel) in the 1960s and 1970s resulted from a 6-year drought imposed on an area where people had greatly expanded their goat and cattle population. In consequence, the Sahara desert essentially advanced southward, compressing desperate inhabitants into lands already occupied and causing widespread starvation.

Replacement of natural vegetation by agricultural crops probably causes more soil erosion than overgrazing. Grass prevents erosion by binding the soil. Wheat, corn, and most other food plants, however, have stalks that are rooted in the soil but leave large areas of bare dirt in between (Box 2.5). Rapid erosion of this dirt is most intense in comparatively dry areas, where the soil is continually dry. Highly variable estimates suggest that replacement of grassland with cropland in the American Midwest, its "breadbasket," have caused nearly one-third of the topsoil to be stripped by erosion in the past 100 years. The most dramatic example of the consequences of modern agriculture was the American "dustbowl" of the 1930s, when drought permitted wind to create dust storms and move dunes of dust and grit across previously fertile fields.

Unfortunately, the world's grassland is currently expanding in the wrong places. Instead of grassland replacing easily eroded cropland, new grasslands are developed largely by cutting down tropical rainforests. Rainforests are generally not logged for their wood, as most of their trees are not as useful as colder-weather pines and hardwoods for the construction industry (Section 8.3). Some deforesting is done in order to clear land for grain crops (mostly corn), but tropical forest soils are so poor that these crops are seldom commercially viable for more than a few years, after which virtually all soil nutrients have been used. Consequently, much of the rainforest destruction is done simply to allow development of grasslands

BOX 2.4 GOLDEN APPLE SNAILS _____

The rice paddies of Vietnam still occasionally yield land mines, deadly relics of the war in the 1960s and early 1970s. They also produce enough rice to feed the Vietnamese and make the country one of the world's largest exporters (see text and Table 2.5). In the 1990s, the rice paddies also became a home and food source for golden apple snails.

A golden apple snail is, as the name implies, about the size of an apple, has a golden yellow color, and needs a lot of food. For those who like snails (escargot), the meat is regarded as a delicacy and snail farming was begun to cater to their tastes. The cultivation of a large amount of snail meat requires little effort. The snails grow to sexual maturity in 2 months, have normal lifespan of 3 years, and can produce 15,000 offspring. Under controlled conditions, they can live in densities of dozens per square yard, and thus, require little space. The snails can eat a variety of foods, but they particularly like rice seedlings. They graze along the floors of rice paddies, breathing through a snail equivalent of a snorkel, and a hungry snail can devour more than 1 square yard of rice seedlings per day. Because they eat the seedlings, rather than the rice itself, the snails can destroy the paddies.

The golden apple snail is native to the Amazon region of South America. It was first removed from there as a business venture by a combination of public and private interests and brought to the Philippines in 1982. Originally they were kept in special tanks, but some of the snails and their offspring escaped to uncontrolled rice paddies. They spread rapidly – that's not hard to do with a potential of 15,000 offspring each. By 1989, they had infested 15% of the Philippine rice land and were causing widespread devastation.

Despite these known problems in the Philippines, a combination of foreign and domestic business interests introduced golden apple snails to Vietnam in the early 1990s. Again some escaped, and by 1995, the snails had destroyed about 100 of the 25,000 square miles that Vietnam devotes to rice growing. That is not a high percentage, but the snails are flourishing in every section of the country, and their depradations can only increase.

Fighting the golden snail is difficult. Special snail poisons (molluscides) are dangerous, particularly to rice farmers who do not have shoes. Furthermore, the poisons are sufficiently expensive that the average peasant farmer cannot afford them. The snails can be "hand picked" out of the paddies, but the labor is intense and also dangerous because broken snail shells can cut bare feet and allow infections to enter.

If the snails are potentially so dangerous, why were they brought to the Philippines and Vietnam? The answer is simple. These countries have cheap land and cheap labor that keep the production costs of exportable goods low. Thus, by raising snails in the Philippines and Vietnam and selling them abroad, some people – a very small number – make a lot of money.

The golden apple snail is only one example of the ecological disaster than can be caused by introducing species to a foreign habitat, or removing them from their native habitat. We mention a few others:

- Wolves and mountain lions have been deliberately exterminated in many parts of the United States in order to protect humans and livestock. Their absence caused such an expansion of the deer population that it now has to be controlled by hunting.
- Australian farmers brought rabbits to the country from Europe. Because the rabbits bred rapidly and had no natural enemies, they destroyed much of the forage both for native wildlife and for imported cattle.
- Extermination of pigs, goats, dogs, and some plants have been carried out in the Galapagos and Hawaiian Islands after their deliberate introduction by European settlers. Because they had no natural enemies to control them, they had bred to such large populations that they were destroying large parts of the natural vegetation and the animals and birds that depended on it for food.
- Highway engineers in the southeastern United States imported a Japanese vine, kudzu, to help stabilize slopes laid bare during road construction. Similarly, tamarisk trees were imported to the southwestern United States to stabilize irrigation canals. Both the kudzu and the tamarisk are now taking over from native vegetation.

We need to keep relearning an important lesson. Natural ecosystems are adjusted to the animals and vegetation that live in them. Predators and prey reach an equilibrium that is occasionally disturbed for short periods of time but usually reestablishes itself naturally. If we disturb that equilibrium by introducing an animal or plant that has no natural control, or by removing a predator, the population of a species explodes and the new equilibrium may exclude some native plants and animals. We discuss this issue in more detail in Section 8.3.

FURTHER READING: Anderson (1993); P. Shenon, *New York Times,* May 17, 1995, pp. A9, A4.

suitable for cattle. Because cattle raising usually benefits only a small number of large landowners and has little impact on the food supply of the local people, we will return to this issue in our discussion of money and agriculture.

Soil erosion can be reduced by proper methods of planting and harvesting. The most extensive erosion is caused by deep furrowing of the soil by tractor-pulled plows. The furrowing opens up the soil for seeding and also reworks the stubble (e.g., wheat stalks) into the soil, thus preserving plant nutrients. The most modern sowing methods drop seed on the ground with virtually no plowing. Furrowing with animal-drawn plows is somewhat less destructive than deep mechanical plowing, but the least disruptive is the primitive method of poking a hole in the ground with a stick and dropping seed corn into it. Unfortunately, poking holes does not produce a large crop (Section 2.2), and farmers who practice this method commonly clear the stubble from their fields by burning it off. During burning, solid nutrients such as phosphorus, potassium, and calcium remain in the ash and are returned to the soil, but the volatile nitrogen is lost, thus reducing the growth of future crops.

Environmental problems arise not only from what is removed from agricultural land by erosion but also from what is added. As we discussed in Section 2.2, both industrial societies and areas of lesser-developed countries taking advantage of the "green revolution" depend to a very high degree on fertilizers, herbicides, and pesticides. We explore the environmental problems caused by these chemicals more thoroughly in Sections 7.3 and 7.5 and can only point out two of the most serious ones here. One is that excess fertilizer, particularly nitrogen, runs into streams and lakes and causes the same degradation that we mentioned for coastal marshes in Section 2.4. The second is that pesticides are generally very stable and contaminate both ground and surface water supplies for many years after they are applied.

Money

The world produces plenty of food for everybody (see the quotation at the start of this chapter). Grain production alone yields approximately 4,000 kcal per day for each person (Section 2.2). Other plants, meat and milk from animals fed on grasslands, and fish all add to the supply of available energy and of necessary nutrients such as proteins and vitamins. Despite this largesse, an estimated one-fourth (more than 1 billion) of the world's people are chronically undernourished. Why? The answer is right before your eyes – if you glance at our section title.

We showed in Section 2.2 that industrial societies could expend about $50 of gasoline to produce 2 to 3 tons of wheat and, thereby, free most of the labor force to do something besides grow grain. The use of energy for food does not stop there. Industrial countries distribute much of their produce in cans or as frozen packages. The canning process uses energy, mostly from fossil fuel (Chapter 4). By the time one pound of corn has been placed in a can, about 6,000 kcal of energy have been expended, mostly in making the steel for the can, but the corn yields only 1,500 kcal of energy when eaten. Also, exceptional amounts of energy are used in transportation in the meat industry. Cows fattened on grain must be brought to feed lots, and/or the grain must be brought to the cows. Once the meat is produced, it cannot be transported or kept in stores without refrigeration, which uses considerable energy. If that energy is available, however, meat is easier to transport than grain because the fat in meat yields about twice as much food energy per pound as grain (Section 2.2), and thus we have to transport less of it.

Clearly, the more energy a nation has, the more advantage it has in the competitive arena of world trade. This advantage is offset somewhat by the fact that people in rich countries earn more than people in poor countries. Just as it is cheaper for rich countries to import goods manufactured in low-wage countries than to manufacture the goods at home, so it would be cheaper for industrial countries to import much of their food. In order to avoid this situation, rich countries have developed elaborate strategies to protect and encourage their agricultural systems. They include subsidies, quotas, and tariffs.

Subsidies work in a variety of ways. In the United States, for example, federal (public) money has been used to build dams and other parts of water supply systems in the arid and semiarid West. These systems supply irrigation water that most agriculture in these dry areas depends on (nationwide, one-third of agricultural crops are grown on irrigated land). Farmers buy water from the federal systems at a rate calculated with great controversy, to reimburse the government for the cost of the dam. The cost does not, however, pay the interest on the money that the government borrowed to build the dams. Thus, the public pays the interest with the intention that the low-cost water will keep the price of agricultural products low. Similar subsidies are used when the federal government rents public land in the West for cattle grazing. The cost of grazing on private land in the western United States is about $10 per acre, but the government charges only about $5.

BOX 2.5 EARLY AGRICULTURE AROUND A MEXICAN LAKE

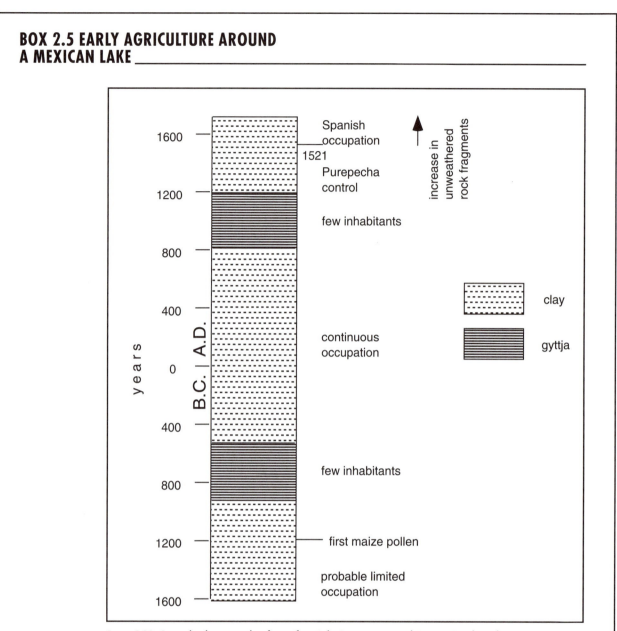

Figure 2.13 Generalized stratigraphy of cores from Lake Patzcuaro, central Mexico. (Depths in the cores are arranged to conform to a linear time scale.) The record of human occupation on the right shows deposition of clays eroded when the drainage basin was used for agriculture and of organic-rich ooze (gyttja) when land was not exposed by farming practices. Deposition rates did not increase after Spanish occupation in 1521, which brought the use of plows and horses, but an increase in the amount of unweathered rock fragments shows increased depth of erosion.

People who believe in rustic simplicity commonly look to native peoples "uncorrupted" by modern civilization as examples of living in harmony with nature. Studies of sediment from a small Mexican lake suggest that these beliefs should be modified.

Lake Patzcuaro lies in a totally enclosed basin with a drainage area of about 500 square miles at an altitude of 6,000 feet in the mountains just west of Mexico City. It receives 40 inches of rainfall annually, and evaporation keeps water depths at an average of 15 feet. Before any human habitation in the area, the lake accumulated the typical lake sediments (officially known as "gyttja") composed largely of organic material with minor component of clays. In the past 4,000 years, accumulation of this sediment has been periodically interrupted by deposition of clays eroded from the surrounding drainage basin. The ages of sediments have been measured by carbon 14 (^{14}C) methods in samples retrieved from cores several feet deep into the lake bottom.

The sequence of sediments, ages, and summarized information about human habitation is shown in Figure 2.13. The earliest evidence of human activity is pollen from maize (corn), apparently planted about 1500 B.C. Prior to this time, the lake sediments are wholly organic, but the advent of agricultural activity in the drainage basin caused erosion of clays into the lake. Erosion has continued to the present except for interruptions from about 900 to 500 B.C. and A.D. 800 to 1200, when deposition of organic sediment correlates roughly with a reduction of human activity in the drainage basin.

Major occupation of the lake area apparently began about A.D. 1100 to 1200, when the area was first occupied by the little-known Purepecha (or Tarascan) Empire. Occupation lasted until the arrival of the Spanish, and there was a population of several tens of thousands at about A.D. 1500. During Purepecha occupation, sedimentation rates of clays increased, and the lake began to contain a high population of diatoms, single-celled plants that live in the surface layers of lakes (and in the oceans: see Section 2.4). The increased erosion apparently was caused by deforestation, partly by clearing land for agriculture, partly to produce wood for buildings, but largely because the Purepechas liked large ceremonial bonfires. The diatoms flourished because of increased nutrients in the lake caused by human effluents, and we discuss it in Section 7.5.

The Spanish conquered central Mexico in 1521. The result-ing interchange between the New and Old Worlds profoundly modified both. In Mexico, the Spanish found corn and shipped it back to Europe, from which it was spread by other colonial adventurers over much of the world. In exchange, the Spanish brought the horse and the plow (also smallpox, which nearly destroyed the indigenous population). After 1521, a reduced population continued to farm in the Patzcuaro basin, using horse-drawn plows. Before the introduction of plows, most of the clays entering the lake had the composition of thoroughly weathered material. After plowing started, the composition of the clays became more characteristic of unweathered material, indicating that the plows bit more deeply into the soil than earlier hand-planting methods. Plowing, however, did not increase sedimentation rates. Degradation of the area was already well advanced by the time plows arrived.

We tell the story of Lake Patzcuaro, not because it is an important part of the history of world agriculture, but because it refutes the idea that the environmental degradation caused by modern agriculture is more severe than that caused by ancient farming methods. In many ways, industrial agriculture is less harmful than ancient techniques. The only reason that we seem to have more problems now is that there are more of us, and we use more of the land to raise crops and animals. For that reason, we must use it better, but we cannot revert to preindustrial methods of growing food.

FURTHER READING: O'Hara et al. (1993); O'Hara, Street-Perrott, and Burt (1994).

Another use of subsidies in the United States has been to support the prices of agricultural commodities. Because the United States produces much more grain and other crops than it needs, it can export the surplus. World prices, however, vary widely depending on world production, and sometimes the prices are considerably below the cost of production in the high-income U.S. society. Consequently, until the program was abolished in 1995 the U.S. government established a "target price" for some crops – for example, $4 per bushel for wheat through much of the 1990s. When the world price dropped below $4, the government used federal money to pay the producer the difference between the target price and the price that the producer can get for the wheat. The formulas for calculating target prices, the amount of

a commodity that the government will pay for, and which commodities were covered by the price-support program generated continual political furor. The policy also generated international tension by making American products more internationally competitive and undercutting sales by such high-efficiency grain producers as Australia and New Zealand. U.S. crop support varied from year to year, but through most of the 1980s and early 1990s the government spent in the range of $10 billion to $15 billion for this program.

Quotas and tariffs have similar effects but operate somewhat differently. Some countries simply do not allow the importation of certain foods. Japan, for example, protects its rice growers by not permitting rice imports, although it does import large amounts of other grains.

The European Common Market allows citrus imports but places quotas on the amounts. Some quotas are on very small commodities; for example the United States prohibits import of avocados from Mexico. Tariffs discourage imports by placing a government "tax" on the commodities. Thus, a tariff places no limit on the amount of import, but it can raise the price to such a high level that consumers will not pay it.

Although industrial countries discourage the importation of some crops, they need other crops that are produced abroad. Because industrial countries are in temperate, relatively cool climates, they cannot produce such "necessities" of modern diets as coffee, tea, pineapples, and bananas. These, and other, crops must be imported from tropical, generally poor, countries. In lesser-developed countries, they are regarded as "cash crops," sold abroad for foreign capital.

At first glance, cash crops seem to be an idea from which everyone profits. People in rich countries can have a cup of coffee with their banana split, and people in poor countries obtain money for the industrial development that will make them rich. For two reasons, however, the benefits to the lesser-developed countries are not always apparent. One is unequal distribution of profits. Most of the land used for export crops is held by a small proportion of the population, thus contributing to the inequality of income distribution. The second, and perhaps more serious, problem with the import of food from lesser-developed countries is allocation of land for various agricultural uses. The choice is between cash crops, which benefit a few owners, and subsistence crops, which provide food for all. We illustrated this issue in Box 2.1 and here will only mention one statistic. In the mid 1990s, some 25% to 50% of the arable land in Central and South America was allocated to production of cash crops that were exported primarily to western Europe and the United States while many local people were inadequately fed.

Numerous other problems arise with land use in lesser-developed countries, and we will mention only two – cattle ranching and monoculture. Although most people in lesser-developed countries cannot afford much meat, large areas are commonly set aside for grazing. The meat is generally raised and processed locally for export, but some countries have permitted land to be rented to foreign organizations for grazing. In some arrangements, the land is sufficiently cheap that U.S. organizations can profitably fly cows to such places as Central America and Haiti. After grazing for a while, the cows are then flown back to the United States for grain feeding before processing. The cattle ranches employ some local people, but most of the profits go to a few landowners, and the loss of land for growing subsistence crops contributes to the scarcity of food for many people who live in the area.

The concept of monoculture signifies raising of a single crop. Some areas, such as the rice-growing countries of southeastern Asia, have little option. Their climate and land do not permit major production of other crops. These monocultures are dangerous, because a disease or pest that affects that single crop may wipe out a nation's livelihood. A prime example is the Irish potato blight of 1852, and we discussed recent examples of the danger to rice production in the Philippines and Vietnam in Box 2.4.

The most pernicious monocultures were those imposed by the Soviet Union on some of its smaller "republics." For example, the Turkmenistan region of central Asia was designated as a cotton-growing area. The announced intention of the Soviet government was to let the Turkmenis grow cotton for distribution to the rest of the country and to have the rest of the country supply food to the Turkmenis. The arrangement worked with customary Soviet efficiency – that is, not at all. By 1993, Turkmenistan was an independent country, stuck with a cotton production of ~1500 million tons, for which there was little market, and a grain production of only 600 million tons. This grain could supply fewer than 1,000 kcal per day to the Turkmen population of 4 million, and depletion of agricultural resources for cotton prevented adequate animal raising or other agricultural production. In the middle 1990s, more than 5% of Turkmeni children died of malnutrition in their first year of life. We can only reemphasize the need for balance in food production.

The future

We finish by asking what the future holds for human nutrition. For the foreseeable future, diets in industrial countries should remain more than adequate. The agricultural productivity of industrial countries, and the wealth that enables them to import whatever they do not raise, will probably not run into insurmountable environmental or economic problems during the lifetimes of most of the people reading this book. Conversely, increasing populations in the nonindustrial world, coupled with already severe environmental degradation, may not be able to sustain their already low average level of nourishment.

The major difficulty in seeing into the future is whether, or how long, the present contrast between industrial (rich) and lesser-developed (poor) countries will be maintained. There are dozens of possibilites, and we present two. One scenario is that poor countries become poorer, driving down the price of agricultural commodities because their people can no longer buy even the food produced locally. In an effort to keep prices up, rich nations then buy food and send it to the poor nations. Part of the apparent largesse would also be an effort to keep people from poor countries from trying legally or illegally to immigrate. A second possibility is that the agriculture of poor countries collapses at the same time as they become richer through industrialization. This will drive up the worldwide prices of agricultural commodities and might even affect the diets of people in the industrial world.

So what is the conclusion? Although we can be confident that some aspects of the world's food situation will change, we don't know what, when, or by how much. Only time will tell.

Policy questions

This section raises a welter of questions, and we suggest pondering the following three:

- Should nations subsidize food production in order to make their food more competitive on the world markets?
- Should treaties promoting world free trade apply to food as well as to manufactured goods and raw materials?
- Should industrial countries import food from lesser-developed countries if those countries are reducing subsistence agriculture for their own citizens in order to produce food for export?

FURTHER READING: Chadwick and Marsh (1993); Rabbinge (1993); Sheldon and Abbott (1996).

PROBLEMS

1 A nutritious food used by backpackers – sometimes called trail mix or "gorp" – consists of chocolate candy, peanuts, and dried fruit. Use Table 2.1 to show how much easier it is to carry trail mix than sandwiches. (Note – about one-half of the weight of bread is grain and the other half water.)

2 A farmer in a lesser-developed country can sell grain to a marketing company for $200 per ton. If the company also sells fertilizer, and it is likely that $300 of fertilizer per acre will increase grain yields from 1.2 to 2.8 tons per acre, should the farmer buy and use the fertilizer? If the farmer must buy the fertilizer 6 months before the grain is sold and must borrow money at 20% annual interest in order to do so, should the farmer borrow the money?

3 The recommended daily allowance (RDA) for vitamin C is 60 mg (0.002 ounces). Propose some combination of foods that will provide the RDA. Contents of vitamin C in various foods is: 1 ounce of potatoes (with skins) – 5 mg; 1 ounce of tomatoes – 8 mg; 1 ounce of peaches – 1 mg; 1 ounce of oranges – 8 mg.

4 Use data in Sections 2.2 and 2.3 to estimate how much water is used so that you can have a one-half-pound (250-gram) steak.

5 A country in which the diet for the average person provides only 1,800 kcal per day also contains enough wealthy people who eat most of the beef, so that the population consumes an average 150 kcal per day of grain-fed beef. Use data in this chapter to calculate the average kilocalories that would be available per person if the resources allocated to beef production were used for production of grain for human consumption.

6 A farmer is trying to grow corn (maize) in tropical soil that has very few nutrients. If edible kernels of corn are approximately 20% of the weight of the typical corn plant, how much phosphate fertilizer must the farmer add in order to produce 1 ton of corn meal? How would this value be affected if the farmer plows the stalks and leaves from the previous year's harvest back into the soil? Use Table 2.2 to answer this question.

7 The table below shows population density (in people per square mile), percentage of arable land, and fertility rate for three countries (A, B, C). Which country do you think has the best economy?

Country	Population density	% Arable	Fertility rate
A	120	50	2.5
B	900	20	1.8
C	30	5	5.5

REFERENCES

Anderson, B. (1993). The Philippine snail disaster. *Ecologist* 23(2):70–2.

Bowes, A. de P. (1980). *Bowes' and Church's Food Values of Portions Commonly Used*. New York: Lippincott.

Chadwick, D. J., and Marsh, J., eds. (1993). *Crop Protection and Sustainable Agriculture*. Chichester: John Wiley and Sons.

Chrispeels, M. J., and Sadava, D. E. (1994). *Plants, Genes, and Agriculture*. Boston: Jones and Bartlett.

Cushing, D. H. (1995). *Population, Production, and Regulation in the Sea: A Fisheries Perspective*. Cambridge: Cambridge University Press.

Farzin, Y. H. (1991). Food aid: Positive or negative effects in Somalia? *Journal of Developing Areas* 25:261–82.

Gershoff, S. M. (1996). *The Tufts University Guide To Total Nutrition*. New York: HarperCollins.

Hardin, G. (1968). The tragedy of the commons. *Science* 162:1243–48.

Lappe, F. M. (1982). *Diet for a Small Planet* New York: Ballantine.

Mather, R. (1995). *A Garden of Unearthly Delights*. New York: Dutton–Penguin Books.

O'Hara, S. L. (1994). Historical evidence of fluctuations in the level of Lake Patzcuaro, Michoacan, Mexico over the last 600 years. *Geographical Journal* 159(1):51–62.

O'Hara, S. L., Street-Perrott, F. A., and Burt, A. T. (1993). Accelerated soil erosion around a Mexican highland lake caused by pre-Hispanic agriculture. *Nature* 362:48–51.

Pimentel, D., and Hall, C. W., eds. (1984). *Food and Energy Resources*. Orlando, FL: Academic Press.

Rabbinge, R. (1993). The ecological background of food production. In D. J. Chadwick and J. Marsh, eds., *Crop Protection and Sustainable Agriculture*. pp. 2–29. Chichester: John Wiley and Sons.

Rifkin, J. (1992). *Beyond Beef: The Rise and Fall of the Cattle Culture*. New York: Dutton.

Samatar, I. A. (1994). *The Somali Challenge: From Catastrophe to Renewal?* Boulder, Colo.: L. Rienner.

Sheldon, I. M., and Abbott, P. C., eds. (1996). *Industrial Organization and Trade in the Food Industries*. Boulder, Colo.: Westview Press.

Touissaint-Samat, M. (1992). *A History of Food*. Translated by A. Bell. Oxford: Blackwell Reference.

United Nations Food and Agriculture Organization (1992). *Fisheries Statistics: Catches and Landings*. New York.

United Nations Food and Agriculture Organization (1990). *Food Balance Sheets*. New York.

United Nations Food and Agriculture Organization (annual). *Forest Products Statistics*. New York.

United Nations Food and Agriculture Organization (annual). *Production* Yearbook. New York.

United Nations Food and Agriculture Organization (annual). *Trade* Yearbook. New York.

United Nations Food and Agriculture Organization (frequent but not annual). *World Animal Review*. New York.

United Nations Food and Agriculture Organization (1994). World Review of Highly Migratory Species and Straddling Stocks. *Fisheries Technical Paper 337*. New York.

Wild, A. (1993). *Soils and the Environment: An Introduction*. Cambridge: Cambridge University Press.

World Resources Institute (1992, 1994). *World Resources*. Oxford: Oxford University

3 | NATURAL HAZARDS

3.0 INTRODUCTION

Civilization exists by geological consent – subject to change at any moment. – Will Durant

On May 18, 1981, a lively group assembled in a tavern in Silver Lake, Washington, to feast on venison and honor the memory of a homespun hero named Harry Truman, who had died a year earlier. It was in the spring of 1980 that the nightly news shows had reveled in the story of this unlikely character. Truman ran a fishing lodge on nearby Spirit Lake, nestled in the shadow of Mt. St. Helens, a dormant volcano in the Cascade Range of the northwestern United States (Fig. 3.1a). That spring, the mountain had exhibited signs of imminent eruption – small earthquakes rumbled across the slopes and plumes of steam rose off the mountain's flanks. Analysis of instrumental data suggested rising lava beneath the mountain. Since Mt. St. Helens was known to be an active volcano and historic eruptions at Mt. Lassen in California provided an example of the devastation to be expected, the United States Geological Survey (USGS) had warned local residents to evacuate. Truman refused to leave, to the great pleasure of the media. He made great copy and gained his hero status by saying "They'll never get me off this mountain.... That's my mountain.... Beautiful, isn't she.... Won't hurt old Truman."

Mt. St. Helens had been Truman's home for 54 years. Home meant safety and security. At home we know what to expect. Home conjures up images of constancy and beneficence. But Truman's home and Truman himself, defiant to the end, were instantaneously buried beneath tens of meters of 600°C ash and mud on May 18, 1980, when Mt. St. Helens blew more than 1 cubic kilometer of rock and ejecta from its southeast flank. Folk hero or not, in spite of "knowing that mountain better than anyone," Truman never had a chance against the geological forces building in the earth's crust for millions of years. Truman has, however, attained a form of immortality. Some geologists refer to the "Truman syndrome" as an unwillingness to face the reality of catastrophic geological events in one's own backyard, leading people to rebuild destroyed structures rather than to move to safer areas.

The earth is the home of the human race as literally as any house or apartment is your home, as his fishing camp was Harry Truman's. This terrestrial home of *Homo sapiens* is deceptively safe and secure. We inhabit a mostly constant realm blanketed by an atmospheric canopy with just enough oxygen to sustain life, enough CO_2 to keep us warm, enough ozone to reduce ultraviolet radiation, and enough total air to burn up incoming astronomical debris that would otherwise bombard the earth's surface with supersonic projectiles. The earth also has an oceanic thermostat to moderate temperature, and it is just the right distance from a stellar energy source to allow water in the form of ice, rain, and clouds to collect on the surface. If it were closer to the sun, water would boil off, and if it were farther away, water would freeze. That is, the earth's location and composition are *just right* to permit photosynthesis and the myriad of reactions and processes that sustain both plant and animal life as we know them (Section 2.1).

We feel secure as a consequence of this perfection, but we must recognize that our planet is truly habitable only because of change. Some of these changes, essential to

(a)

(b)

Figure 3.1 (a) Eruption of Mt. St. Helens, May 18, 1980. Courtesy United States Geological Survey. (b) Eruption of Mt. Pinatubo, Philippines, June 12, 1991. Courtesy R. P. Hablitt, United States Geological Survey.

human survival, are slow and imperceptible; some are fast and terrifying. Among the slow events is the uplift of the earth's crust to form mountains and plateaus. Newly exposed rocks weather to provide essential nutrients to soils and to release other chemicals to rivers and streams to buffer the composition of the atmosphere and oceans. On a more rapid scale, rain and snowmelt water run downhill from these highlands, eroding the mountains and carrying sediments to river floodplains and deltas. Even faster events like storms, tides, or simply the passing of night and day support a complex web of shifting energy and resources essential to life on earth. Thus, we face a quandary – human survival on a scale of centuries or millennia depends on change, even unpredictable change, but the economic, cultural, and social structures of modern society are designed for constancy and stasis.

This changing earth demands thoughtful consideration of the impact of geological processes on human life and well-being. This theme occurs throughout the book; for example, in Section 8.2 we will discuss the effect on the earth's atmosphere of an eruption of Mt. Pinatubo (Fig. 3.1b), a volcano much larger than Mt. St. Helens. We investigate geological change in this chapter, paying particular attention to those events that are rapid and catastrophic, which pose "hazards" to our safety and happiness. We start with a discussion of earth's engines of change (3.1), proceed to a discussion of the concept of risk (3.2), then describe individual hazards (3.3 to 3.8), and finally discuss the ways in which society reacts to hazards and disasters (3.9, 3.10).

3.1 EARTH'S ENGINES OF CHANGE AND THEIR CONSEQUENCES

James Hutton thought that the earth was very old, with "no vestige of a beginning, no prospect of an end." – T. H. Van Andel, New Views on an Old Planet, p. 27

The earth is powered by two heat engines, one external and one internal. The more familiar source is the external heat of the sun. Each day part of the sun's radiative heat is retained within the atmosphere (Section 8.2). Because low latitudes receive more heat than high latitudes, average atmospheric temperatures decrease from the equator toward the poles. This differential heating, plus the rotation of the earth, causes air masses to move around and creates our weather and climates. We discuss this issue more completely in Chapter 8, but here we describe those movements of air, such as hurricanes, that

are rapid enough and cause sufficient coastal flooding to be potentially catastrophic (Section 3.7).

Part of the sun's heat is used to evaporate water, principally from the oceans, thus putting water vapor into the atmosphere (Section 4.1). When the vapor is extracted as rain, it causes erosion and promotes rock weathering (Section 2.2). It also fills rivers and groundwater basins (Section 4.1), and most of the rainwater makes its way back to the sea within a few years of falling on the ground. Erosion by water (and wind) sculpts the landscape, causing some areas to be prone to landsliding and most of the earth to be subject to slow downhill movements (Section 3.8). Fluctuation in the amount of water delivered to river systems may cause floods, which we discuss in Section 3.6.

The earth's internal heat causes earthquake and volcanic hazards. This heat results both from the accumulation of the earth and the radioactive elements that it acquired as it formed. The earth grew from a dispersed set of particles and gas that constituted the primitive solar system. Approximately 4.5 billion years ago this solar system "cloud" condensed into discrete bodies – the sun and planets. The reason for the initial condensation is hotly disputed, but once small nuclei were initially formed, they must have acted as "gravity wells" that attracted material still circulating in the cloud. Most of the particles and gas plunged into the sun, which now contains 99.9% of all of the mass of the solar system. Condensation of material closest to the sun produced the "inner" planets (Mercury, Venus, Earth, and Mars) in a zone that was hot enough to allow the light gases, hydrogen and helium, to escape back into space. Consequently these planets are rocky (the technical term is "terrestrial") and consist mostly of silicate rocks plus a core of iron–nickel (Fe–Ni) that gives them average densities in the range of 4.2 to 5.5 g/cm^3. The outer planets (Jupiter, Saturn, Uranus, and Neptune) were cold enough to retain their light gases and now have densities that range from 0.67 to 1.65 g/cm^3; Saturn, the lightest, would float on water (although the water would freeze).

As the protoearth attracted particles into its gravity well, both its size and temperature increased. The heating resulted from the impact of particles with the accreting earth in the same way that the side of a house heats up when paint is removed by sandblasting. Technically, a particle at some distance from the earth has a "potential" energy that is converted into "kinetic" energy of motion as it begins to approach the earth (or as a grain of sand is blasted against a house). When impact occurs and the particle stops moving, this kinetic energy cannot disap-

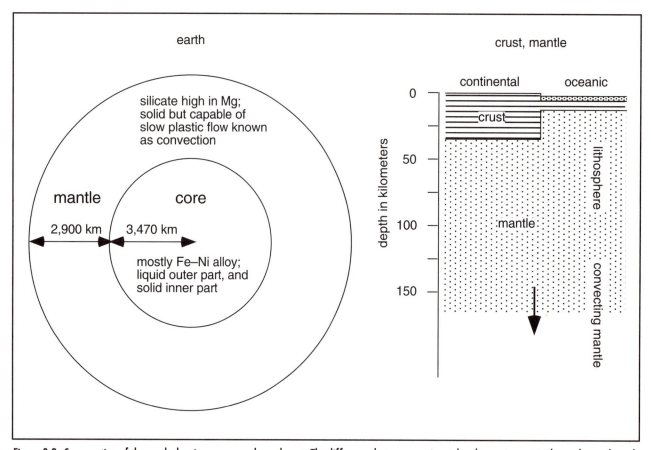

Figure 3.2 Cross section of the earth showing core, mantle, and crust. The difference between continental and oceanic crust is shown, but no boundary is indicated between the rigid lithosphere and the convecting mantle because the depth is so variable. See discussion in text.

pear. This is the so-called Law of Conservation of Energy – if energy disappears in one place or form, it must appear in some other place or form. In the earth (or the wall of the house), the kinetic energy is converted into heat, and the earth heats up. This "heat of accretion" has been augmented by two other heat sources. A relatively small one was the early separation of the earth's high-density core by the sinking of molten iron and nickel toward the earth's center (Fig. 3.2). This resulted in a further decrease of potential energy and release of heat. The more important additional heat source is the continuing decay of radioactive elements, principally uranium, thorium, and potassium (see Section 5.4 for a discussion of radioactivity).

The earth forms an effective blanket for its internally derived heat, permitting only slow escape into space. For this reason, the interior of the earth remains hot, and temperatures increase downward from the surface to the center. The increase is sufficient to make the earth's mantle (Fig. 3.2) thermally unstable. Hot mantle rises from the deep interior and cools below a rigid outer skin

of mantle and crust (known as "lithosphere"), which has a thickness in the range of 100 to 200 miles. This cold mantle then descends back into the interior, and the rising and descending areas combine to form a "convection cell." This convection stirs the mantle in the same fashion that a pot of soup is stirred by a rolling boil as the bottom is heated and the top cools.

Mantle convection cells create stresses in the near-surface, rigid lithosphere of the earth, causing it to deform. The zones of deformation are generally narrow and serve as the borders between broad oceanic and continental areas of continued stability (Fig. 3.3). The stable areas are referred to as "plates," and the concept of stable plates moving around the earth and deforming on their margins is called "plate tectonics." It is a dominant principle of modern geology, but we only summarize it here.

Stable plates can be bounded by three types of margins.

- In some areas the lithosphere is separated (rifted) by rising mantle plumes. Although rifts occur within

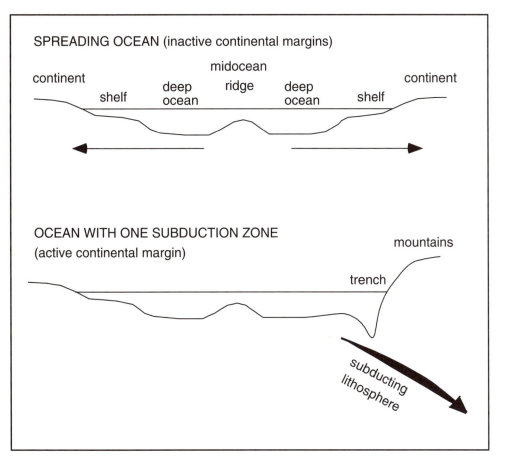

SPREADING OCEAN (inactive continental margins)

continent   shelf   deep ocean   midocean ridge   deep ocean   shelf   continent

OCEAN WITH ONE SUBDUCTION ZONE
(active continental margin)

mountains

trench

subducting
lithosphere

Figure 3.3 Contrast between inactive continental margins around spreading oceans and active margins where oceanic lithosphere is being subducted under the continental margin.

continents, such as in East Africa, most of this separation occurs along "mid-ocean ridges" (Fig. 3.3), where new oceanic lithosphere is formed and the ocean basin grows larger as the lithosphere spreads away from the ridge. For example, the Atlantic Ocean has been entirely created in the past 200 million years by initiation and spreading of the mid-Atlantic ridge.

• Because the earth cannot create new lithosphere in some areas without destroying it elsewhere, in some areas the lithosphere is pulled down ("subducted") beneath other lithosphere. Some of the subduction occurs beneath other oceanic lithosphere, forming "island arcs" within the oceans. Most subduction, however, consists of oceanic lithosphere descending beneath the margins of continental lithosphere, forming mountain ranges. Subduction of oceanic lithosphere reduces the size of ocean basins, and because subduction has occurred on most of the margins of the Pacific Ocean for the past 200 million years, the Pacific Ocean is now much smaller than it was 200 million years ago.

• In some areas, two pieces of the lithosphere are simply shoved past each other, forming faults commonly referred to as "transforms." The best-known example is the San Andreas fault of northern Mexico and California (see discussion of San Francisco earthquake in Section 3.3).

The concept of plate tectonics also permits us to distinguish two types of continental margins, which are important in our discussion of coastal processes in Section 3.7. "Active" margins are plate boundaries, where subduction causes mountain building. "Passive" continental margins are not plate margins; they occur where oceans are growing and pushing continents on either side apart. Almost all of the Pacific Ocean is surrounded by active margins, and almost all of the Atlantic by passive margins.

Mantle convection is the fundamental cause of earthquakes and volcanoes. Earthquakes result from the fact that none of the various types of plate deformation occur smoothly. The lithosphere is strong enough to resist fracturing, and this resistance causes strain to build up in the lithosphere. Thus, when fracturing does occur, the energy is released rapidly and the earth shakes. Most, but not all, earthquakes occur along plate margins, a topic explored further in Section 3.3.

As with earthquakes, most volcanism occurs along plate margins. Volcanoes are the most obvious surface expression of the release of the earth's internal heat. Rising hot mantle either melts itself as it nears the surface or, more commonly, distributes sufficient heat to cause melting of some part of the adjacent lithosphere. Also, descending mantle releases water and other volatile materials that promote melting of overlying lithosphere. Some volcanoes, however, develop above narrow (~100-mile diameter) "plumes" of hot mantle that appear to rise independently of the major convection cells. Hawaii and Yellowstone are prime examples of the surface expression of mantle plumes, which cause the volcanism of Hawaii and the geysers of Yellowstone. Volcanic eruptions not only create local disasters, as Mr. Truman discovered (briefly); they may also affect the earth's climate for periods of several years. We discuss the local effects of their eruption in Section 3.4 and their climatic effect in Section 8.3.

The consequences of change and the recognition of hazards

The earth's internal and external heat sources create a dynamic, rather than a static, earth. Generally, people don't like change but can adapt to it. The need for a people to relocate provides one example of adaptability. Figure 3.4 shows how a changing shoreline over the course of 2,500 years forced the abandonment of the ancient city of Pella and relocation to the present site of Thessaloniki in Greece. Despite this ability to adapt, change makes us nervous, and many rapid earth and atmospheric processes cause changes that are risky – hurricanes, floods, earthquakes, and so on. We do, however, adjust to slower changes and, in fact, benefit from them. For example, we build cities on riverbanks to use the flowing water for transportation, municipal water supplies, industrial purposes, and waste removal. Slow downhill movement of soil supplies rivers with sediment, and if that sediment were not present in rivers, they would erode their banks downstream. The movement of sand along coastlines creates beaches for our enjoyment. The debris from volcanoes develops into rich soils for agriculture.

Our task is not to prevent change, which is usually impossible, or to deny the accompanying risks. What we must do is accept and understand the changes and the dangers that they may pose and then prepare for them.

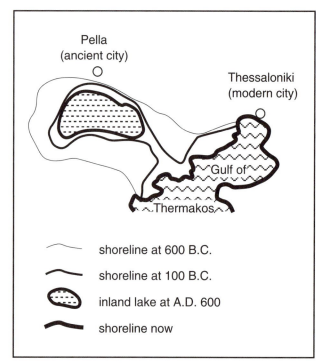

Figure 3.4 Changes in shoreline in northern Gulf of Thermakos, Greece, from 600 B.C. to the present. Two rivers gradually deposited deltas that closed the mouth of the bay, forming a small lake that ultimately was completely filled by small rivers running into it. The closure caused ancient coastal cities to be abandoned. Modified from Eumorphopolous (1963) and Kraft, Aschenbrenner, and Rapp (1977).

For example, when a river floods above its banks, sometimes buildings are destroyed or damaged and lives may be lost (Section 3.6). Were we wise and careful (preferably both) we would understand the possibility of flooding and prepare for it by passing zoning ordinances and/or requiring protective structures to lessen flood damage. Preparations for flooding and all other natural hazards, however, involve political and economic issues that are highly controversial, and we discuss them further throughout this chapter.

Any infrequent natural change that is fast enough to threaten life and limb is referred to as a "natural hazard." Most are episodic, and some are difficult to predict. Examples of such hazards include floods, landslides, earthquakes, volcanic eruptions, tsunamis (often erroneously called "tidal waves"), and meteorite impacts. (We will ignore impact of large meteorites – something that occurs at intervals of thousands to hundreds of thousands of years is too infrequent to worry about, and if the meteorite is really big, there is nothing we can do about it; but see Box 3.1). Also, we defer to Chapter 8 our discussions

BOX 3.1 THE TUNGUSKA EVENT

Geologists like the word "event." It refers to something that actually happened – perhaps a volcanic eruption 2 billion years ago or a fault movement 500 million years ago. We use the term, however, when we are not really sure that a volcanic eruption did occur or a fault did move at the times specified. Within the framework of this definition, what happened on the brightly sunlit morning of June 30, 1908, over the Tunguska River of central Siberia was assuredly an event.

An explosion at an estimated height of 3 km sent an incandescent column of "something" up to a height of 20 km. The explosion was seen at a distance of 500 km, heard more than 1,200 km away, and felt at a distance of 70 km. The sky over much of Eurasia was light for several nights after the Tunguska explosion – the phenomenon is referred to as "light nights." Before the explosion, the "object" appeared to have a gaseous pale-blue aura and left a trail of ionized air along its path as it moved toward the northeast.

Because of the remoteness of the Tunguska area and the Communist revolution in Russia, on-the-ground exploration of the explosion area was not undertaken until about 20 years after the event. The investigation discovered an area of approximately 50 km diameter in which most of the trees had been blown down. They had fallen in a pattern radiating from the center, where the blast occurred, with progressively less destruction outward. Search for other information, such as an impact crater or fragments of a meteorite, produced nothing.

Putting all of the data together led to the conclusion that the "event" was caused by a midair explosion that totally disintegrated the object, whatever it was. Estimates of the energy released have been made possible by use of data from aboveground nuclear tests, in which the energy release was known and the effects measured. On this basis, the Tunguska explosion released the same amount of energy that would have been released by the explosion of about 12 megatons (12×10^6 tons) of TNT – that is, it was a 12 megaton explosion. By contrast, the atomic bomb that destroyed Hiroshima released 20×10^3 tons of energy, roughly 750 times less.

Speculation about the cause of the explosion has been wide ranging (perhaps we should say "far out"). The ionized trail, the incandescent aura, and the height of the explosion clearly indicated the destruction of something that had entered the earth's atmosphere. But what? Theories have included miniature black holes and contra-terrene matter, in which atomic nuclei have negative charges and the place of electrons is taken by positively charged particles (positrons). We particularly enjoy the suggestion that it was an alien spaceship despite the fact that no little green bodies were found among the blown-down tree stumps.

More realistically, proposals center around comets or meteorites. A comet, which consists largely of ice and has a correspondingly low density, could account for the lack of impact crater, and the light nights could be explained by light reflected from ice in the comet's tail before the paths of the earth and the cometary remains completely diverged. Whether a comet could have contained enough energy for the explosion is problematic. An iron (metallic) meteorite could not have been responsible because such meteorites are too strong to break up before they impact the ground. A stony (silicate) meteorite, however, might be weak enough to disintegrate, and the explosion of a stony meteorite with a diameter of about 30 m could cause sufficient energy release.

The Tunguska explosion caused little loss in lives or property, partly because it wasn't enormous and it was in an area where virtually nobody lived. A leading explanation for the end of the Mesozoic era (the age of dinosaurs) 65 million years ago also involves meteorites. That proposed impact, on the coast of what is now the Yucatan peninsula of Mexico, may have so rearranged the earth's surface environment that it caused 80% of living species to become extinct. The estimated size of that meteorite is 20 km, some 600 times the size of the Tunguska meteorite. Fortunately, the really big ones are very rare.

FURTHER READING: Ben-Menahem (1975); Chyba, Thomas, and Zahnle (1993).

of comparatively slow changes in climate and other aspects of the global ecosystem.

How big a problem are natural hazards? The U.S. Geological Survey (USGS) estimates that every year the world endures 100,000 thunderstorms, 10,000 floods, hundreds of landslides and tornadoes, 100 damaging earthquakes, and scores each of hurricanes, volcanic eruptions, and tsunamis. Unevaluated or ignored, each threatens our well-being in different geological and geographical settings. Should you worry? Reading on in this chapter, you would be anything but human if you did not keep asking yourself, "Does this affect me?" You may or may not experience a feeling of relief if you realize that you live in a place where earthquakes or volcanoes are unlikely to effect you or yours. But is that really so? Economically, we *all* pay some price. The same USGS study attaches an annual economic cost of $150 billion per year worldwide from these natural hazards. That is $3 billion dollars a week; a few dollars shy of $300,000 a minute. Worldwide, 150,000 souls perish each year from such causes. It is hard to say that natural hazards are someone else's problem.

Now we turn to risk.

FURTHER READING: Bridges (1990) for a discussion of the landforms of the earth's surface; Rogers (1993); Van Andel (1994).

3.2 THINKING ABOUT RISK

The almost universal feeling of fear that a great earthquake can cause . . . does not arise from the potential damage that can be done. . . . But rather, it is more concerned with the unknown qualities of an earthquake. – R. Iacopi, Earthquake Country, p. 34

Before we look at natural hazards one by one, let us think of how people live with danger. We call exposure to danger or accident "risk." Risk is inherent in life. Plague, pestilence, drought, flood, earthquake, and fire have been our lot since the first hominids roamed East Africa a few million years ago. Stories of such events are deeply rooted in every people's legends and myths. The Bible's Book of Genesis and the Assyrian Gilgamesh legend both tell of a terrible flood that destroyed all of civilization (see Box 3.7). The Book of Hopi describes years of drought that threatened the survival of Oraibi, their ancestral home in northern Arizona. The unwritten history of the Australian Bushmen describes a Dream Time that existed before some uncertain disaster led to their diminished condition that existed when the first British arrived in the 1700s (see Section 8.3).

The anthropologist Sir David Attenborough once characterized people as the "risk-taking animal." He had in mind the human predilection for such activities as running the bulls in Pamplona or bungie jumping, but we accept a risk if we drive a car, walk across a busy street, or use tobacco. In all of these cases, there is a conscious or semiconscious acceptance that life is risky. Table 3.1 shows the risk inherent in a number of everyday activities, but we must distinguish between "individual" and "societal" risk before we proceed further.

Table 3.1 *U.S. annual death rate (in no. of deaths per 100 000)*

Cause of death	Death rate
All causes	879.3
Major diseases	652.8
Motor vehicle accidents	16.3
AIDS	14.9
Suicide	12.1
Homicide	9.9
Falls	5.2
Poisonings	2.5
Drownings	1.9
Fires	1.6
Airline accidents	0.07
Floods	0.06
Lightning	0.05
Landslides	0.02
Earthquakes	0.01
Tsunami	0.001
Volcanoes	0.0004

Notes: Data for illness and death not caused by natural hazards are for 1994. Airline accidents are worldwide average. Lightning, floods, earthquakes, tsunami, volcanoes, and landslides are average annual rates for the late 1900s.
Sources: United States (1994), *Statistical Abstract of the United States;* Smith (1996).

Individual risk is risk that individuals voluntarily accept and acknowledge. When we get a driver's license, we consciously accept the added risk to our physical well-being (especially if we refuse to wear a seat belt while talking on our car phone at 70 mph). Watching television in our living room is surely safer than skate-boarding, but on the other hand, a sedentary life-style and the high-fat snack foods we munch while sitting on the couch significantly raise our susceptibility to heart disease and cancer (Section 2.1). Insurance companies evaluate these risks in order to set premiums for their policies, and public officials attempt to minimize them in order to be reelected, but these are subjects outside the topic of this book.

Societal risk is an involuntary, often unacknowledged, risk that is implicit to simply being alive. It is akin to being struck by lightning or contracting a communicable disease in a public place. It seems frightening and indiscriminate because it affects people who did not consciously place themselves in harm's way. This kind of risk is memorialized in the classic cartoon cliché of a safe falling from a building on a blissful passerby. Many types of societal risk, such as earthquakes and floods, are natural hazards. Exposure to these hazards is generally not voluntary. It is inherent to place rather than to personal choice, and most people have lived with, and been unaware of, the risk all of their lives.

People have different psychological reactions to individual and societal risk. Generally, individual, known risk is less frightening than an equivalent, unpredictable, societal risk. That is, an event that is highly probable but of low societal impact is viewed as less serious than an event with low probability but large social consequences. Consider fatal individual automobile accidents and homicides, which are high-probability risks in many countries but with low consequence to the society as a whole. An example of a low-probability risk with high societal consequence is the bombing of the United States Federal Building in Oklahoma City in May 1995. It killed fewer than 200 people, but 6 months later it still was front-page news. In the same 6 months, 25,000 Americans died in traffic accidents and 15,000 by fatal gunshots. Why the difference in societal attention and concern? Undoubtedly it was due to the dramatic impact, randomness, unpredictability, and concentration of human consequences in a single location of such an event.

Lack of familiarity makes some risks particularly terrifying. For example, people are more frightened of snakes than of cumulus clouds, although annual lightning deaths exceed the total of deaths from snake and spider bites, wild animal attacks, and alien abductions combined. Because geologic hazards are unfamiliar to many people, and also difficult or impossible to predict, a geologic disaster such as a flood or earthquake attracts the same type of attention as an urban terrorist attack. Before we describe various geological hazards, we must consider the relationship between them and societal risks.

Natural hazards as a societal risk

Many aspects of societal risk are exacerbated by the human tendency to live in large and complex groups. A nomadic hunter is exposed to minimal danger from earthquakes, volcanic ash falls, or floods by keeping a wary eye on the surroundings – something necessary for survival anyway. Even if this nomad falls prey to such a hazard, the familial or tribal group placed at risk is small in numbers and limited in possessions. As our ancestors became settled farmers, however, the first permanent villages and cities were located beside rivers, on islands, or along coastlines because city dwellers needed ready and dependable access to water for irrigation, drinking, cook-ing, commerce, and transportation. Then, as our ancestors built larger and more substantial structures, such as granaries, dwellings, and temples, they increased the amount of damage that could be caused by storms, earthquakes, or floods. In a sense, as people became more settled and successful urban engineers, changing the earth to meet their own needs, they increased their societal risk. It soon became difficult, if not economically impossible, to relocate whole civilizations that had made bad decisions about where to settle. Prehistory and written history are full of evidence and legends of cities destroyed by geological events, of states weakened by volcanic eruptions, floods, and earthquakes, and of dire warnings ignored to the detriment of many innocent people (example in Box 3.2).

We do not imply that preindustrial cultures were not aware of natural risk or that they did not adapt to the dangers inherent in their environment. On the floodplain of the Nile Delta, Egyptian farmers built only the most temporary shelters on the riverbanks and then annually abandoned them when warned by their priest-astronomers of the impending spring floods. Thus, they avoided devastation and irreparable loss as the rising waters washed away their frail structures along with the wastes of the previous year and replaced them with a fresh load of fertile silt and mud from the upper Nile. More recently, in southern coastal North America in the late 18th and early 19th centuries, those able to leave the malarial, miasmic swamps of Georgia and South Carolina headed for the "up-country" each summer to avoid the fever. In both of these cases, society adapted to inherent risks in the environment, and the cultures thrived because the hazard was mitigated by an awareness, either conscious or not, that living carries benefits and risks and that to realize the former, one must accommodate the latter.

At the heart of the issue of natural risk are the concepts of "cyclicity" and "predictability." In our previous examples – seasonal escape from disease-bearing insects and hurricanes or abandoning a river floodplain at the annual high flow – the adaptation succeeds because the event is *cyclic* and *predictable*. Accommodation to societal risk is easiest when the life-threatening event occurs like clockwork or is preceded by a warning sign like the changing of the seasons or the rising of Sirius at dawn. Most natural disasters are *infrequent* and *unpredictable* or, using our new terminology, they involve low-probability, high-consequence societal risk. The causes are often hidden from the senses of the nonscientist. With little or no warning, the beneficent earth shakes or huge waves crash

BOX 3.2 HOW TO LOSE A CONTINENT _____

Visitors to Thera (Santorini) depart from their boats into a small port that clings to a narrow ledge on the precipitous side of the island. From the port, they can climb, or be carried by mules, upward past nearly 1,000 feet of volcanic ash and other debris to the top of a curved island about 10 miles long and 3 miles wide. At the top they look west over the Aegean Sea and see small islands that appear to form a ring approximately 15 miles across with the curved Thera plus one island in the middle of the ring. Had they been there 4,000 years ago, they would have been on the side of a landmass with a diameter of at least 20 miles that housed a thriving part of the early civilization of the Aegean region. This larger land, which earlier civilizations might have regarded as a continent, has now been mapped by geological and underwater geophysical methods that clearly show it as a former volcanic edifice (technically, a caldera).

The edifice blew up in 1628 B.C. (the dating is now very accurate). Remnants of the former civilization are preserved on Thera as a city destroyed by violent shaking and fire. At the time of their destruction, the inhabitants of Thera had been part of a broad "Minoan" civilization spread throughout the Aegean and surrounding mainland areas. The principal Minoan development was on the large island of Crete, with its capital at Knossos. The Minoan civilization disappeared about 1500 B.C. (this dating is vigorously disputed), to be replaced by people whose roots were in mainland areas surrounding the Aegean. Many of the Minoan cities were destroyed by shaking and fire, perhaps all at the same time (although this synchroneity is also not clear).

The date of 1628 B.C. is important in parts of the world remote from the Aegean. For example:

- The bristlecone pine is a stunted tree that lives its long life at high elevations in the arid southwestern United States. Bristlecone pines grow, as do most trees, by adding a growth ring each year, and tree-ring scholars (dendrochronologists) have used a large number of living and dead trees to establish a very good sequence of growth rings extending back several thousand years. Starting in 1628 B.C. and for several years afterward, the bristlecone pines produced only very thin rings because of frost damage. Similar stunting has been found in preserved oak wood from Ireland.

- Volcanic ash with ages consistent with an eruption at 1628 B.C. has been found through Turkey, parts of the Middle East south of Turkey, and in the Black Sea.

- Ice obtained from cores into the Greenland ice cap has a small amount of sulfuric acid in layers accumulated about 1628 B.C. (or perhaps slightly older).

- A stele at Thebes, Egypt, has been translated to record that at about 1600 B.C. (data do not permit better dating) "the gods expressed their discontent . . . the sky come with a tempest . . . it caused darkness in the western region . . . each house . . . was floating in water . . . with no one able to light the torch." In other words, high winds, colossal rain, and darkness.

- Chinese annals dated at 1618 B.C. (but they may be older) record "yellow fog, a dim sun, then three suns, frost in July, famine, and the withering of all five cereals." This event marks the beginning of a new (Shang) dynasty in China.

Putting together all of the geological and archeological evidence suggests that the eruption of Thera was the largest in recorded history. By comparison with modern (smaller) volcanic eruptions, volcanologists estimate that about 30 to 40 km³ of Thera was blown into the atmosphere in less than one day. At peak discharge, the column of ash reached more than 30 km into the sky. The distribution of ash into Asia Minor also suggests that the wind was blowing from the west during the eruption.

Did the eruption of Thera destroy the mythical continent of Atlantis? We do not know — there is too much myth surrounding Atlantis to be sure of anything. Did the eruption destroy the Minoan civilization? Possibly, although more accurate dating needs to be done. Did the eruption cause an overturn of Chinese dynasties? That is a bit farfetched, but also possible. But it's fun to speculate, and to conduct further investigations.

FURTHER READING: Hardy et al. (1989); all quotations are from papers in volume 3.

ashore, and destruction is rapid and overwhelming. Then, almost perversely, the earth returns to normal. Thus, while they occur rarely, geologic disasters are dramatic in their impact, often affecting many people in a restricted geographical area who understandably thought themselves at little risk before the event. The event has a fatalistic quality.

Although we have come to understand natural hazards scientifically, because of their infrequency and unpredictability, it is difficult to change behaviors to reduce risk from a hazard that few have ever witnessed. How can people be expected to modify their habits and behavior to avoid building on sites flooded before they were born or along coastlines where powerful cyclonic storms may occur on average only once every 40 years, or when geological events occur only every 100 or even 1,000 years? Too often, our scientific knowledge has postdated patterns of settlement. Until the San Francisco Earthquake of 1906 destroyed the city, few recognized that there was a high probability of seismic risk in the San Francisco Bay area. By 1906, patterns of settlement, economic development, and transportation in northern California precluded rebuilding this major commercial center anywhere else. Furthermore, it was several years before the earthquake risk was understood. In that time, civic leaders and promoters had rebuilt San Francisco on the ruins of 1906 (Box 3.3) and the metropolitan area was once more (and still is) at risk of catastrophic damage (Section 3.3)

Throughout this chapter we emphasize that converting scientific knowledge and engineering design to best practices is not a simple task. A geologist's understanding of the cyclicity and power of a natural event includes uncertainties as to timing, frequency, and severity. These uncertainties may be expressed as statistical probabilities of risk, but public policy makers often find it difficult to translate probabilities into practice. In the absence of direct experience, politicians often find the arguments and enthusiasms of developers, promoters, and boosters more compelling than those of statistics-touting scientists. We ask you to bear these problems in mind as we discuss some of the more destructive sources of risk to health and well-being from natural processes. For each source, we describe the process, the sources of risk, and the current means of evaluating and managing that risk from a scientific-engineering standpoint. Finally, we address the always contentious issue of risk assessment and management in economic and policy terms.

Policy questions

What efforts, if any, do you make to reduce risk?

FURTHER READING: Coch (1995).

BOX 3.3 PLANNING (?) IN THE AFTERMATH OF THE 1906 SAN FRANCISCO EARTHQUAKE

That the human spirit is resilient needs little proof. That we learn from our errors is a bit harder to prove.

San Francisco is earthquake country. Well before the turn-of-the-century, earthquakes had already caused notable damage in settled areas, in particular in 1865 and 1868. The former quake gained some national notoriety through an engraving in the *Police Gazette,* a scandal sheet of its day, showing a well-known senator fleeing a fancy San Francisco bordello clad only in his long underwear. Robert Louis Stevenson visited San Francisco in the 1800s and wrote: "Earthquakes are not only common, they are sometimes threatening in their violence; the fear of them grows yearly on the resident." So no one should have been terribly surprised by the events of April 18, 1906, which destroyed 28,000 buildings and rendered more than a quarter of a million Californians homeless. The earthquake was only the strongest (M = 8.3) of more than 450 seismic events noted by the Smithsonian Institution between 1850 and 1906 in the Bay area. What did not collapse in the quake was incinerated in the fires that burned unchecked for several days.

But it is the aftermath that interests us. What did the survivors learn that could help them rebuild a safer city? In general, knowledgeable observers were impressed by the fact that many of the modern buildings seemed to have survived. George Simpson, the architect who had built the San Francisco Chronicle Building, observed that "all the steel frame buildings in San Francisco withstood the shocks and the only damage done to them outside of fire was the falling out of part of the walls." The City Engineer Thomas P. Woodward noted that areas of the city built over old swamps had settled as much as 4 feet. He also noted that the Mission District, built on fill over two old sluggish creek beds and bay marshes, suffered massive destruction.

Would the residents relocate to Oakland across the Bay where damage was far less or ban construction in areas of unstable fill? Would they learn their lesson and require steel-frame structures? After you write down your answer, read the following from the *San Francisco Earthquake Horror,* a book that appears to have been slapped together for a news-hungry public in 1906:

BOX 3.3 PLANNING (?) IN THE AFTERMATH OF THE 1906 SAN FRANCISCO EARTHQUAKE (continued) _____

The spirit of San Francisco would brook no successful rivalry [from Oakland] and its leading men were united in a determination to rebuild a city beautiful on the ashen site. . . . Smoke was still rising from the debris of one building while the owner was planning the erection of another and still better one. . . .

It should be borne in mind that San Francisco was not destroyed by the earthquake. While old buildings in that part of the city which stood on "made" ground . . . it is true suffered from the shock, it was fire that wrought the great devastation. (chap. 20: The New San Francisco).

San Francisco will be as surely rebuilt as the sun rises in heaven. No earthquake upheaval can shake the determined will of the unconquerable American to recover from disaster. . . . We know not when another shock may come or whether it will come again at all. No matter. The city shall rise again. (introduction by the Rt. Rev. Samuel Fallows)

We may have stumbled on the writings of the first, certainly not the last, public relations hacks in California's history.

This boosterism and unbounded energy quickly took precedence over planning and careful engineering. By coincidence, a year earlier, the noted city planner and architect Daniel Burnham had provided a comprehensive plan for San Francisco that involved broad boulevards, what we would now call greenways,

and vast city parks. It was to be the blueprint that would redefine the city in time for a world's fair to commemorate the opening of the Panama Canal. Ironically, at the time of the earthquake, a bill was pending in Congress to authorize this fair. Instead, Congress passed the first measure in U.S. history to provide federal funds for disaster relief. The Burnham Plan could have been used in combination with the insights of engineers and architects like Woodward and Simpson to avoid another disaster, but the whole idea was ancient history by June 1906. Having no interest in being bound by plans and government restrictions, the "unconquerable" San Franciscans simply rebuilt with what they had wherever they stood.

Undaunted, city leaders, only a year later, began an ambitious project to build a seawall and fill in 184 acres of marshy tidal flat known as the Marina, just east of the Golden Gate. Much of the fill was, in fact, debris from the "ashes" of old San Francisco. In 1911, this became the site of the 1915 Panama-Pacific Exposition, of which the Palace of Fine Arts still survives. In 1989, the Marina District and the Mission District, which had been so badly battered in 1906, were the most seriously damaged residential areas in San Francisco during the Loma Prieta earthquake. As if to mock us all, sand boils in the Marina District, small eruptions of subsurface, liquified sand that often occur on unstable sediments during an earthquake, included pieces of charred wood and tarpaper from the buried debris of 1906.

FURTHER READING: Linthicum (approximately 1906); Baldwin and Sitar (1991).

3.3 EARTHQUAKES

When a French rescue team pulled a man from the rubble [of the earthquake at Yerevan, Armenia, in 1989] after four days, he raised his hands and, thinking that World War III had started, said "I surrender." – B. A. Bolt, Earthquakes, p. 85

Large earthquakes and similar catastrophes have always been gruesomely fascinating. As early as the 13th century B.C., an earthquake that damaged the great Assyrian capital of Ninevah was described in correspondence preserved in clay tablets from excavations. The destruction of Sodom and Gomorrah has been puzzled over by scholars and mystics for centuries and is probably the oral folk tale of a prehistoric earthquake. In a prescientific age, the shaking earth was not understood and was easy to interpret as a warning from the gods or an augury of impending doom. In our modern age, we understand earthquakes better, can locate them with great accuracy, and can estimate the forces released. To a humbling degree, however, we still cannot predict major seismic events, we are only marginally more able to prevent death and destruction from them, and we are wholly helpless to prevent them. The only advantage we have is an understanding of why certain parts of the earth are likely sites of an earthquake – at least we know where the seismic enemy is. In this section, we briefly describe how earthquakes occur. We focus on the risk factors associated with earthquakes and indicate regions at greatest risk. We use historical examples and note the best practices for preventing wholesale destruction in earthquake-prone areas.

Causes and kinds of earthquakes

An earthquake is the ground response to energy released along active faults and transmitted through the earth as seismic waves. The energy ultimately is traced to forces in the earth's interior, including the movement of crustal plates, the rising of magmas, the subsidence of the earth's crust under a load of newly deposited sediments, or the thickening of the crust during mountain building (Section 3.1). Energy generated by these forces is stored in rocks as "elastic strain" (Fig. 3.5). When the rocks can no longer deform without breaking, we say that they have reached their "elastic limit," and they fracture (Fig. 3.6). An analog would be stretching an elastic band. Energy is stored in the band until it reaches a limit of strength, at which time it breaks and the energy is transmitted to your finger tips – ouch! The energy released by faulting passes through and around the solid earth as a complex swarm of waves that may cause the ground to rise and fall with a motion similar to that of a water wave or to fracture and collapse. In either case, human structures sitting on the ground are subjected to motions that most were not designed to withstand.

The amount of energy released by earthquakes is measured on the Richter magnitude scale. The determination of Richter magnitude requires two measurements.

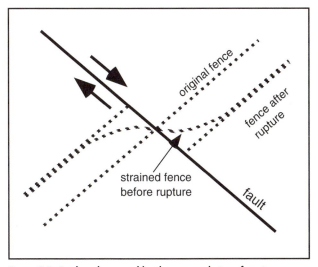

Figure 3.5 Earthquake caused by slow accumulation of strain energy around a fault and sudden release when the fault slips.

The first is made by seismographs, which determine ground motion by the simple process of remaining stationary while the ground moves under them, and the relative displacement of the earth beneath the seismograph increases as the arriving seismic energy increases. The second measurement is distance of the seismograph from the earthquake. Because ground motion decreases with increasing distance, the energy of the quake is determined as a function of both ground motion and distance.

The Richter magnitude varies, theoretically, between zero and infinity. One unit increase in magnitude is a 32-fold increase in seismic energy release, and at any given distance from an earthquake, a difference of one unit in magnitude accounts for a 10-fold difference in ground displacement. The highest Richter magnitude ever measured is 9.2 for the Prince William Sound earthquake of March 28, 1964 (also called the Good Friday quake in the Anchorage area). Any earthquake with a magnitude greater than 8.0 is a monster. Most destructive quakes have magnitudes 7.0 and greater, but very serious earthquakes include one of 6.5 magnitude in Los Angeles, California, in January 1994 and another of 6.9 in Kobe, Japan, in 1995 (both of these quakes are discussed further). At the other end of the Richter scale are tens of thousands of tiny earthquakes every year so small that they pass unnoticed by all but a coterie of ever vigilant seismologists. Table 3.2 shows the average number of earthquakes of different magnitudes worldwide each year since 1900. On average, 19 earthquakes with magnitude ≥7.0 occur each year, but the number has varied between 6 and 41. Table 3.2 also demonstrates that as magnitude decreases, the average annual frequency increases, and a graph of frequency versus magnitude (Fig. 3.7) shows a typical exponential curve (see explanation in Fig. 1.3).

The simple correlation of earthquake locations to active faults, their tendency to occur repeatedly in nearly the same place, plus some information on their magnitudes, provide the single most important predictive tool in seismology. With nearly 100 years of seismic instrumentation behind us and lots of studies to tell us which faults are active, we can recognize general areas of seismic risk (Figs. 3.8, 3.9).

Destruction associated with earthquakes

The toll of death and destruction from earthquakes reads like the aftermath of some of history's most brutal conflicts. The 1923 Tokyo-Yokohama earthquake, 8.3 on the Richter scale, killed 143,000 people, mostly in the

Figure 3.6 Effects of earthquake in Kobe, Japan (more description in text): (a) area destroyed by fire after the earthquake; (b) collapsed residence; (c) landslip caused by earthquake. Courtesy T. L. Holser, United States Geological Survey, and the Geological Society of America.

Table 3.2 *Frequency of earthquakes since 1900*

Description	Magnitude	Average annual frequency
Great	>8.0	1
Major	7.0–7.9	18
Strong	6.0–6.9	120
Moderate	5.0–5.9	800
Light	4.0–4.9	6,200 (est.)
Minor	3.0–3.9	49,000 (est.)

Sources: Data from United States Geological Survey.

great fire that swept Tokyo in the aftermath. In 1976, an 8.0-magnitude earthquake in Tangshen, China, a city less than 100 miles southeast of Beijing, killed 255,000 according to the official China News Agency, although Western estimates of the death toll are as high as 650,000. The 1985 Michoacan earthquake killed as many as 30,000 in Mexico City. A swarm of earthquakes in 1990 in western Iran killed 40,000 to 50,000 people, many in seismic-induced landslides. The 1989 Loma Prieta earthquake, magnitude 7.1, killed only 62 people but caused an estimated $10 billion of damage in the San Francisco Bay area. The 6.7-magnitude Northridge earthquake of 1994 killed 59 people, injured 1,500, damaged 12,500 structures, and caused an estimated $30 bil-

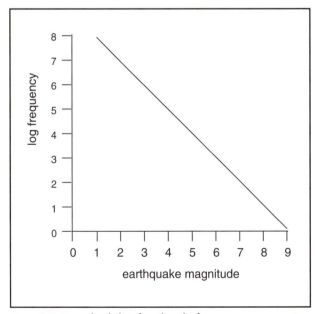

Figure 3.7 Generalized plot of earthquake frequency versus magnitude of quake. The line follows the equation log N = 9 − M, where N is number of earthquakes at each magnitude and M is magnitude.

lion of damage in the northern Los Angeles basin. Portions of eleven major highways linking downtown Los Angeles with surrounding areas were closed.

The two California earthquakes, serious as they were, are dwarfed in human impact by the Hanshin-Awaji (Kobe), Japan, earthquake of January 17, 1995, which serves as a graphic lesson of what a relatively small (magnitude, 6.9) earthquake can do to a modern urban area (Fig. 3.6). Kobe is a modern city in southwestern Honshu, about 50 km north of a major transform (strike-slip) fault between two plates in the western Pacific (Section 3.1). Numerous small faults are active in the Kobe area, and the actual rupture occurred along a near vertical strike-slip fault at 5:46 A.M. local time. Where exposed at the surface, maximum horizontal displacements were measured at 5 feet. Although many of Kobe's larger buildings as well as its transportation system and service infrastructure were built according to the best engineering standards of the times, 5,378 people were killed and 33,189 were injured, almost all in collapsed houses and apartment buildings. In addition, 152,000 buildings were severely damaged or destroyed, including nearly 7,000 in a series of fires started by leaking gas mains and electrical short circuits. Ultimately, the equivalent of 70 city blocks was consumed by fire. Every supporting column of the elevated Hanshan Expressway and all major rail lines and most port facilities were destroyed, essentially isolating Kobe from the outside world for days.

In addition to destruction of buildings in Kobe, many utility services were severely interrupted. While electricity was shut off for 1 million customers, most regained service within a week, but restoration of other essential services took longer. Gas, which most Kobe residents depend on for heat and cooking, was lost to 857,400 individual customers, and only a third had service restored within 1 month. It took 11 days to repair one-half of the water leaks for the 650,000 customers who lost service, and 20% of the leaks were still unrepaired a month later. Sewage treatment facilities were so crippled by the earthquake that Kobe was forced to pump chlorinated, but otherwise raw, sewage into Osaka Bay. Total property damage is estimated at a staggering $100 billion. The psychological damage continued for much longer. Ten months later, public health officials reported an epidemic of depression-induced suicides among those who lost their homes, their possessions, loved ones, and the relative stability of their jobs and neighborhoods.

How does all this mayhem and destruction happen? Contrary to Hollywood movie images, little of the destruction associated with earthquakes is attributed to cracks opening in the earth to swallow people, beasts,

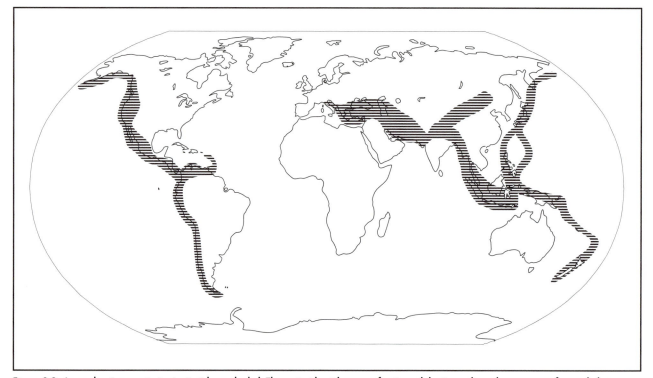

Figure 3.8 Areas showing greatest seismic risk are shaded. They coincide with zones of active subduction or lateral movement of crustal plates.

and property alike. Earthquake destruction is far more devastating, and we describe six processes that are responsible for most of the devastation.

SHAKING AND COLLAPSE OF STRUCTURES   Most structures are built to withstand static, vertical forces. The reason is that, with the exception of an occasional windstorm, the forces on a house, office building, or highway overpass are generated by gradual accumulation of the weight of the building and its contents. When an earthquake occurs, instantaneous forces as great as several times the force of gravity (1 G to 2 G, to use the terminology familiar to pilots and astronauts) are applied in all directions. Unless specifically designed to absorb such forces, human structures are weakened and collapse.

GROUND DISPLACEMENT, LIQUEFACTION, AND SUBSIDENCE   Even if a structure meets codes designed to prevent damage from shaking, there is little hope of survival if the foundation beneath the building gives way during an earthquake. The process by which poorly consolidated muds and other fine-grained sediments become fluid during shaking is called "liquefaction," and it affects many solid-seeming, compact muds or silts. Solid bedrock is the most stable foundation, and buildings on it have a good chance of riding out all but the most severe earthquakes. Where the underlying soils or sediments are

weak and poorly consolidated, however, the story is different. The risk factor from ground displacement is often exacerbated in urban areas where land is at a premium, and many cities have expanded into wetlands and shallow coastal regions by using artificial fill to increase the land area (Box 3.3). For example, Mexico City is built on old lake beds that were deposited atop basaltic lava flows. During the 1985 Michoacan earthquake, few structures in Mexico City collapsed where the underlying foundation was hard, dense lava, but where buildings sat on poorly consolidated, water-saturated muds, the shaking caused chaotic shifts and movements of the muds beneath the buildings with predictable loss of life and property.

FIRE   Historically, fire in the aftermath of earthquakes, has been the greatest danger in urban areas. Prior to modern electrical service, most city dwellers used wood- or coal-burning stoves for heat and cooking and open flames or lanterns for light, all of which were often toppled by the shaking during earthquakes. Even today, the combination of electrical short-circuits caused by destruction of service poles and transformers and the presence of broken gas mains can produce enormous risk from fire. In Kobe, about 10% of the fatalities were fire-related, about two-thirds apparently caused by leaking

Figure 3.9 Approximately 50 feet of surface displacement caused by vertical movement accumulated over several thousand years along fault on east side of Sierra Nevadas, California. The last movement caused a few inches of displacement in 1980 and resulted in an earthquake. Courtesy Allen Glazner.

gas or electrical problems. The problem of fire is exacerbated by broken water mains, loss of water pressure, and the inability of fire companies to negotiate the rubble-strewn streets of an earthquake-damaged city.

DISEASE AND DISRUPTION OF VITAL SERVICES   As with fire, the collapse of the social infrastructure – municipal water supplies, sewage treatment facilities, burial of the dead, isolation of outlying areas from food and medical care – contributes to a general decline of social services. Cholera and other epidemics are common in more remote areas of the less-developed world in the aftermath of earthquake destruction. In 1993, an earthquake centered on Khillari, 300 miles southeast of Bombay, killed perhaps as many as 22,000. In the aftermath, shortages of water and proper sanitation resulted in epidemics of gastroenteritis and malaria, although the far worse spread of cholera and diphtheria was prevented by the rapid response of public health officials. In the last decade of the 20th century, such preventable sources of postearthquake misery should be avoidable by a combination of emergency preparedness and swift governmental and international action. Too often, however, matters of national pride and political concerns make such relief efforts less responsive and efficient than one might wish. Though the incident was resolved, the international response to a 7.7-magnitude earthquake in Iran in June 1990 was hampered by internal political feuds between the more moderate government and the opposition leaders, who wanted no help from Western nations.

LANDSLIDES   As we discuss in Section 3.8, unstable masses of earth and rock can be set in motion by the shaking associated with earthquakes. This process, not unlike that which undermines foundations in areas of unstable fill or sediments, can result in catastrophic destruction in mountainous areas.

TSUNAMIS   Some severe (magnitude ≥ 7) submarine earthquakes produce sea waves of enormous magnitude. They can have devastating effects along coastlines, and we discuss them in more detail in Section 3.5.

Risk analysis and prevention of earthquake destruction

As if the natural risk from an earthquake were not enough, the hazard is severely confounded by habitual patterns of living. Earthquakes that would be of only modest impact in a modern, industrialized nation are devastating in areas where, out of economic necessity, many people live in unmortared stone structures or poorly constructed buildings that could not be better "designed" to collapse and do the maximum damage to inhabitants. For example, in 1988 a swarm of midsized earthquakes killed an estimated 25,000 in Armenia, almost all injuries and fatalities the result of the collapse of prefabricated, urban high-rises built to inadequate Soviet standards. The problem exists in industrial countries, also. Nearly 7% of Americans live in preconstructed (once called "mobile") homes, housing mostly families with few resources at the lower end of the economic scale. Many of these homes sit precariously on poorly constructed piers or jacks and readily fall off when shaken by ground motion.

Some of the risks can be reduced or virtually eliminated if we are willing to spend enough money. For example, the Old City Hall in Kobe, an 8-story building built in the 1960s, pancaked such that the 7th and 8th floors literally collapsed onto the 5th, obliterating the intervening floor. Immediately next door, the 16-story New City Hall, built according to updated codes in the 1980s, was unharmed and was reopened as soon as utilities could be provided. This demonstrates that, technically, we can build earthquake-resistant structures. The problems are that they are expensive and that the expense of making old structures earthquake-proof dwarfs the added cost of applying modern engineering codes to new structures in earthquake-prone areas. The response of most people is to "learn to live with the risk" and trust in good luck, insurance, and disaster relief to lessen the impact.

As long as we must live with the risk, what can we do to minimize it? First, we need to know where the danger is and, therefore, who is in danger. It is easy to say that San Francisco or Tokyo is an area of seismic risk since each has experienced major earthquakes this century, but how many would guess that Charleston, South Carolina, or the Netherlands is an area of historical seismicity? Charleston experienced an earthquake in 1886 that is now estimated to have had a magnitude of 7.5. Between 60 and 100 people died, damage was estimated at $5 million in 1886 dollars, and ground motion was felt as far away as Havana, Cuba, and Boston, Massachusetts. On April 13, 1990, an earthquake of magnitude 5.5 occurred in southeastern Netherlands, just 50 miles east of Antwerp. Although a small earthquake by Pacific basin standards, in this heavily industrialized and densely populated portion of western Europe, one person died (of heart failure) and about 50 were injured, primarily by cuts and bruises from falling masonry and broken glass. Damage is estimated at $100 million, mostly due to failure of masonry chimneys and walls, damage to vibration-

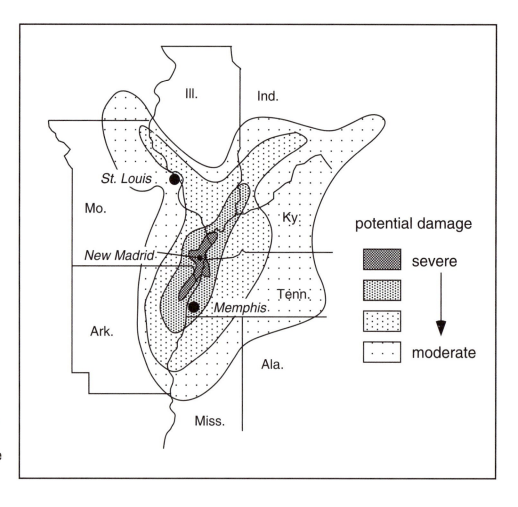

Figure 3.10 Map of seismic risk estimated by the U.S. Federal Emergency Management Agency for the area around the New Madrid seismic zone.

sensitive industrial equipment, and shutdowns of utilities and industrial plants with subsequent production losses due to extensive postearthquake inspections and repairs.

In an effort to help public officials prepare for earthquakes, seismologists have developed maps that indicate which areas of the world are at seismic risk. Such maps are based on mapping active faults, studies of geologic features that allow dating of earthquake-produced scarps, landslides, offsets, and fluidization features, and (very important) the historical record of seismic activity. We showed a generalized map of worldwide seismic risk in Figure 3.8 and now present Figure 3.10 as an example of a more detailed attempt to define areas that have a high probability of a property-damaging earthquake in the immediate future. Maps of this type are useful as a planning device, for developing building codes, and for emergency-management officials.

Although we can develop maps, such as Figure 3.10, that locate major areas of seismic risk, can we predict earthquakes? Attempts to forecast a location, magnitude, and general timing of a seismic event have been based on a wide variety of observations that have variously included statistical patterns of minor foreshocks, changes in local ground elevation, velocities of seismic waves through rocks assumed to be near the limit of stress that they can maintain (their "elastic limit"), drops in the level ("hydrostatic head") of groundwater (Section 4.4), increases in radon gas being emitted from the earth, and the decline in frequency of minor earthquakes along strands of active faults. Some people have even used the unusual behaviors of animals (are the pigs more nervous than usual?). In specific instances, however, all attempts either failed to predict an event or predicted one that never occurred. In understandable desperation, some charlatans and others eager for attention have even ventured into the quagmire of psychic prediction and scientific-seeming, but entirely unproven, theories for earthquakes like the astronomical alignment of planets and the moon. None of these predictors hold up to analysis, and the truth is that anyone who lives in "earthquake country," wherever it may be, must be prepared for the inevitable worst case. It will come without warning.

Policy questions

Regardless of where you live, have you or your government spent enough time and money to investigate the earthquake hazard and to prepare for the possibility of an earthquake? (We ask the same question for Sections 3.3 to 3.8.)

FURTHER READING: Eiby (1980); Bolt (1993).

3.4 VOLCANIC ERUPTIONS

Better put jam in your pocket, general. We're toast. — Airman at Clark Air Force Base before the eruption of Mt. Pinatubo

Remember Harry Truman? It is time we got back to volcanoes and the havoc they wreak. In this section, we briefly discuss how volcanoes erupt, describe the kinds of materials that spew from a volcano, and specifically address those volcanic processes that are hazardous to humans. Finally, we look at what regions are at risk and what can be done to minimize the destruction associated with volcanic processes?

The nature of volcanic eruptions

A volcanic eruption, like an earthquake, ultimately results from the earth's internal heat engine that we discussed in Section 3.1. This heat causes local melting or partial melting of rock below the surface to form a liquid called "magma." The magma, being liquid, is less dense than the surrounding solid rocks and rises because of its buoyancy. Thus it pushes its way upward, often actually lifting the surface, and moves along fractures produced by the upward buoyant forces and other tectonic activity.

Once the magma body comes sufficiently close to the earth's surface, a new mechanism takes over to drive the upward motion and, ultimately, to cause the volcanic eruption. As magmas cool, they crystallize minerals, and their dissolved gases, predominantly water and carbon dioxide (CO_2) are concentrated into the remaining liquid. Then, as surrounding rock pressure is lowered when the magma approaches the surface, these gases expand and come out of solution. This process is similar to opening a partially frozen carbonated soda from your freezer. The soda was under pressure before it froze because the manufacturer had pressurized it with CO_2, and freezing forced the gas into solution in the remaining liquid under even higher pressure. This pressure may exceed the strength of the container, resulting in its bursting with predictable consequences to the interior of your freezer. Similarly, when pressure is released from a rising magma, the same process occurs except that the material that splatters over the countryside consists of molten rock, solid fragments, and gases at temperatures approaching 800° to 1,000°C.

Materials ejected from a volcano include lava flows, tephra, and gases (Figs. 3.11, 3.12).

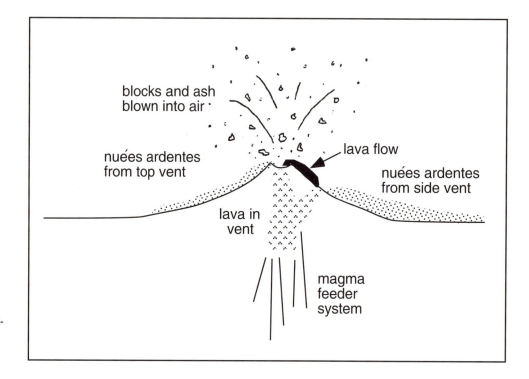

Figure 3.11 Types of materials erupted from volcanoes. See discussion in text.

- A *lava flow* is molten rock – that is, it is magma on the surface of the earth. Some lavas, generally basalts (with low SiO_2 content), are quite fluid and capable of traveling at velocities of 10 to 30 mph for distances as great as 30 to 90 miles from the vent. Other flows, generally with high SiO_2, are as viscous and sticky as taffy.
- *Tephra* are fragments of molten lava and solid rocks blown out of the volcano. They form when gas-rich and viscous magmas come near the surface and explode through the volcano top or nearby fractures, where they sometimes excavate a new crater. Large fragments, called bombs or blocks, generally land near the vent. Fine-grained ash can form clouds that rise as high above a volcano as 10 to 12 miles into the stratosphere. The larger ash particles may rain down like snow on areas downwind, while the finest and lightest ash is often blown thousands of miles, sometimes circling the earth for several years after the eruption. Instead of being blown through the air and cooling somewhat before they settle to the ground, some tephra form a dense mixture of hot ash and steam that rushes down the flank of the volcano as a gas-emitting fiery flow at speeds up to 150 mph; they are referred to as "nuées ardentes," or glowing clouds. In between slow-moving lava flows and racing nuées ardentes are a host of mud and debris flows and debris

avalanches that result from the mixing of lava and ash with lake and stream sediments, glacial debris, and snow and ice on the flanks of alpine volcanoes.
- *Gases* emitted by volcanoes are usually water in the form of steam. The white plume that you often see in photos of erupting volcanoes is just such a cloud of steam. When photos show gray or black plumes above a volcano, they indicate significant quantities of tephra in the rising plume. Many volcanoes, in addition, also emit large volumes of other gases, including CO_2, carbon monoxide (CO), hydrogen sulfide (H_2S), sulfur dioxide (SO_2), and hydrochloric acid (HCl). In addition to being greenhouse gases (see Chapter 8), these gases are, if not poisonous, at least capable of asphyxiating any oxygen-breathing animal.

The result of all of this activity is a volcano, a composite structure built of lava flows, tephra falls, and ash flows. Typically, volcanic structures dominated by lava flows are broad, shieldlike mountains, whereas those characterized by tephra falls are the steep, crater-capped volcanoes that most of us think of when we imagine a volcano (Fig. 3.1). Steam and dry gases may be emitted from vents and fumaroles all along the volcano surface even when the volcano is no longer actively producing lava or tephra. The proportions, temperatures, composition, and nature of lava, tephra, and gas vary from volcano to volcano, from eruptive cycle to eruptive cycle, and even from time to time within an eruptive cycle. While there are patterns to eruptions in terms of periodicity and severity, every volcano can surprise even the most wary and savvy observer.

Volcanic hazards

Flowing lava is generally a minor hazard of active volcanism because the flows move relatively slowly and tend to follow well-defined valleys and stream courses on the flanks of the mountain. Both tephra explosions and nuées ardentes, however, can be very dangerous. Vast columns of tephra that rise as high as 10 miles into the atmosphere in less than 30 minutes consist of rock and mineral fragments, glass shards less than 0.1 inch in size, and noxious gases, all of which quickly spread downwind. The immediate hazard is to commercial jets, whose engines ingest huge amounts of air. On December 15, 1989, a Boeing 747 with 231 passengers bound from Amsterdam to Anchorage flew into a dust cloud produced by an eruption of Redoubt Volcano, near Anchorage and 150 miles upwind of the plane. The pilot tried to climb out of the cloud, but the ash ingested by the engines melted at the high temperatures inside the jet engines to form a glassy coating on the vanes of the turbines. All four engines stalled and the plane literally glided for 8 minutes until, only 2,000 feet above the ground, the crew was able to restart the engines and land.

In addition to dramatic airborne encounters, the downwind dispersal and deposit of ash can be likened to a heavy snowfall of glass fragments. Where thickest, roofs collapse and electrical wires are broken. Ingestion of the fine ash destroys cooling systems of automobiles, diesel engines, and other machinery; sewage treatment plants are clogged and become inoperable; damage to crops and livestock is considerable; pumps are rendered inoperable; windshields are sandblasted by the impact of the tiny grains of glass. The May 18, 1980, eruption of Mt. St. Helens in Washington deposited 1.1 km^3 of windblown ash in layers up to an inch thick over a 30- to 40-mile-wide swath as far east as Missoula, Montana, more than 500 miles away. The nuisance impact of moving this ash from highways and other manmade structures can be estimated by the fact that 300 miles away, the ash amounted to 8 tons per acre. This slightly more than 1-km^3 ash cloud did $2.7 billion in damage in the state of Washington alone. Now consider that this is 1% of the volume of the tephra produced by the eruption some 7,000 years ago at Crater Lake, and 0.1% of the size of the largest his-

torical eruption, Tambora, Indonesia, in 1812, which produced an estimated 100 to 300 km^3 of tephra and put so much ash into the atmosphere that it modified the earth's climate for several years (see Box 8.2).

The most dangerous volcanic eruptions are those that produce nuées ardentes. The weight of the hot ash descending the steep flanks of a volcano provides the driving force, while a diabolical combination of continuous degassing of the ash, air trapped beneath the ash, and gases from the burning of combustible materials and boiling surface water provide a buoyant "air cushion" for the flow. The flows move downslope as ground-hugging, glowing, 300°to 700°C avalanches of steam, ash, and rock fragments at velocities up to 150 mph, and their momentum carries them over barriers in their path as high as several hundreds of meters and for distances as great as 200 km from the vents. They can burn every combustible object in their path and bury hundreds, even thousands, of square miles in tens to hundreds of feet of scalding ash. In Box 3.4 we describe the most famous nuée ardente, which buried the city of St. Pierre, Martinique, on May 8, 1902.

Finally, an insidious and little recognized hazard, at least until recent years, is the venting of toxic and asphyxiating gases that often precedes eruptions. Most of the gas released by eruptions and cooling magmas is simply steam that quickly dissipates into the atmosphere with little harmful effect. However, as we have noted, volcanic gases also contain varying amounts of CO_2 and carbon monoxide (CO), sulfur and chlorine compounds, and ammonia. Near volcanic vents (typically within a few kilometers) these gases are harmful to living organisms and can damage property. Painful acid burns and respiratory difficulties are not uncommon, and the acids in volcanic gases often etch and corrode metals and other materials.

In a few cases, volcanic gases are more deadly yet. In 1986, in a rural district in northwest Cameroon, 1,700 people and more than 3,000 cattle were found dead as if by some silent, fast-acting disease. There was no evidence of violence or of sickness. All lived near a 200-m-deep crater lake, Lake Nyos. There were no signs of explosion or fire; and the few survivors reported only a strange bubbling noise from the lake that lasted less than a minute and then a slightly rotten egg smell and a feeling of inhaling damp, dense air before passing out. For two days following the event, no insects, birds, or small mammals were seen, but plants were unaffected. Later studies showed that CO_2 gas, slowly being released by cooling magmas at depth, collected in the deep cold lake

(a)

(b)

Figure 3.12 Volcanic phenomena: (a) crater of Kilauea, Hawaii; the volcano emits very fluid basaltic lava; (b) basalt flows along the Columbia River, Washington; these flows are part of a volcanic mass more than 1 km thick that covers an area with a diameter of more than 300 km; (c) Crater Lake, Oregon, a largely andesitic volcano showing recent construction of a small volcano inside the 10-km-wide crater; (d) Mt. Vesuvius, Italy, which erupts periodically but is best known for the explosive emission of ash that buried the town of Pompeii in A.D. 79; (e) debris (some blocks as large as small houses) blown more than 10 km from the volcano on the site now occupied by Lake Taupo, New Zealand; the eruption that destroyed the volcano in approximately A.D. 200 is the largest eruption known to have occurred in human history, and Roman chronicles record darkening of the skies by some of the erupted ash.

(c)

(d)

(e)

waters, only to be spontaneously released by some still unknown disturbance. In all likelihood, as the cold CO_2-saturated water rose and depressurized, it released even more CO_2, which is denser than air and hugs the ground as a cloud that suffocates all animal life.

Risk and prediction

With all of this potential carnage from volcanoes, can we do anything to minimize the risk from them? Some risks are obvious. A mountain spewing molten rock and 100-mph ash flows, tossing automobile-sized rocks into the air, and raining incandescent ash on the countryside needs no government-required labels saying "WARN-ING: Volcanoes can be hazardous to your health." The

trick is to know when a volcano is about to erupt and where the resulting flows and falls are apt to be a danger to life and property and how to avoid them. Some volcanoes are active, meaning they are either erupting now or are known to have erupted in the past 10,000 years. Some are unquestionably extinct, meaning that they are no longer linked to a deep crustal source of molten rock, a magma chamber. One problem is that many volcanoes are neither active nor extinct, but "dormant." They are in-between eruptive cycles, appearing deceptively benign to the nonspecialist. Mt. St. Helens was such a volcano before its 1980 eruptive cycle began. It had last erupted in 1857, before any geologist had ever visited it.

On a worldwide basis, more than 1,300 volcanoes have been active in historic times, and about 50 erupt

BOX 3.4 AN ERUPTION OF MT. PELÉE

Martinique is a picture-postcard, tropical island in the Windward Islands of the eastern Caribbean — it has been part of France since the early 17th century. The island lies above the subduction zone formed where the Atlantic plate descends beneath the Caribbean plate, thus developing the Antillean island arc system. The subduction has caused volcanism, and the erupted rocks have provided shallow-water platforms for deposition of coral reef and other limestones, thus forming a complexly alternating sequence of volcanic and carbonate rocks. Martinique contains six principal volcanic centers with ages of activity ranging from approximately 50 million years ago to the present. The particular eruption that interests us here occurred early in this century.

The currently active volcano on Martinique is Mt. Pelée, which probably began to erupt about 2 million years ago, although a well-dated history of eruption can only be extended back to about 20,000 years ago. Massive eruption of fragmental material, including nuées ardentes (see text), has occurred episodically for at least the past 8,000 years, and one deposit less than 1,000 years old covers layers containing artifacts of early Indian inhabitants (Carib Indians). Within historic times, large steam explosions occurred in 1792 and 1851, and then activity began again on April 24, 1902. For about the next week, ash-laden water vapor rose from the cone and began to cover the island with relatively cool ash. Then on the night of May 4–5, the residents heard loud detonations and on May 6 saw a glow above the volcano as the volume of ash began to increase. On May 7, M. Fouché, eager to keep the citizens from fleeing town and missing the upcoming municipal elections, issued a proclamation that said: "In accordance with the Governor, whose devotion is ever in command of circumstances, we believe ourselves able to assure you that . . . we have no immediate danger to fear."

Despite this proclamation, the population of St. Pierre was assaulted by heavy rain that mixed with the new ash to create widespread mudflows on the night of May 7–8. But throughout this period, Mt. Pelee was only getting ready for the enormous nuées ardentes that blew from it at 8:02 A.M. on the morning of May 8 and buried 30,000 people in the city of St. Pierre. Eyewitnesses on surrounding hills described:

A column of fire, which I estimated to be at least thirteen hundred feet in height, descended upon the town. It engulfed the statue of Christ and the cemetery. Then with a great roaring, it encircled the mulatto quarter, leaping over the Pont Basin, moving across L'Centre, where only a short time before I had conversed with the American Consul and his lady. (Father Alte Roche)

[I] beheld the black vapor leap from the side of the mountain. Looking down on it as it rolled on to St. Pierre, it seemed to me as if all Martinique were sliding into the sea. A great tongue of fire seemed to detach itself from the vapor to lick up all the water in the Roxelane River. . . . Only the towers of the Cathedral of Saint Pierre remained untouched, and they only for a brief moment, for the fiery mass enveloped them too, as it spread itself over all of St. Pierre. The mass was being constantly refueled by a huge stream of fire pouring out of the side of the crater to ravage an already devastated town. The cane fields were on fire, as were the plantations around the town. There must be so many victims, hundreds, possibly thousands, and from here there was nothing to be done. (Cure Mary)

We close by pointing out that the eruption of Mt. Pelée brought some measure of recognition to at least two people. The French geologist A. Lacroix (who actually was already famous) used the descriptions of the eruption to introduce the term "nuée ardente" for the type of eruption that destroyed St. Pierre. Also, with an ironical twist, the only survivor in the city was Auguste Ciparon, who was incarcerated in a below-ground dungeon (and who was not already famous). He lived to tour the United States with P. T. Barnum's Circus as the "Survivor of St. Pierre." We do not know whether M. Fouché would have regarded him as a "local boy who made good."

FURTHER READING: Thomas and Witts (1969), from whom all quotations were taken.

each year. This volcanism occurs primarily on plate margins, with a few areas of activity above plumes (e.g., Hawaii) or in unexplained locations (Section 3.1). Consequently, two-thirds of all active volcanoes are in the circum-Pacific "ring-of-fire," the subduction zones around the closing Pacific Ocean. Some 350 to 400 million people, about 10% of the earth's population, live on, or dangerously near, active volcanoes. Because many of these areas are fertile coastal regions, it is unlikely that they can be abandoned, and many circum-Pacific countries have no choice but to adjust to this inherent risk. Among the 1,300 active volcanoes, volcanologists have identified a subset of about 80 high-risk ones that are: (1) known to be active or to have eruptive cycles that lead us to believe they may soon become active and (2) located near populated areas. These high-risk volcanoes are found in about fifteen countries: Italy, Japan, Indonesia, the Philippines, the United States, Mexico, Colombia, New Zealand, Papua New Guinea, and assorted small nations of Central America and the Lesser Antilles. The list is almost certainly incomplete. For example, one of the first lists of high-risk volcanoes, made in 1984, failed to include Nevado del Ruíz in Colombia, which erupted a year later and killed between 22,000 and 27,000 people, the worst volcanic disaster since the 1902 eruption of Mont Pelée on Martinique (Box 3.4).

Fortunately, unlike earthquakes, most volcanic eruptions give warning. The rising magma in the earth's mid- and upper crust typically announces itself by swarms of small earthquakes, by a bulging up of the earth's surface of tens of centimeters to a meter or more per kilometer, and by increases or changes in the temperatures of hot springs and fumaroles. Furthermore, some volcanoes are almost as predictable as the weather (for whatever that is worth). Kilauea volcano on Hawaii, for example, has erupted more than 45 times in the past 200 years, including a nearly constant eruptive cycle that began in 1983. The basaltic eruptions from this crater are relatively passive and nonexplosive. Since 1948, geologists at the Hawaii Volcano Observatory atop Kilauea have been able to monitor its activity constantly and know, perhaps as well as is humanly possible, Kilauea's personality, whims, and behavior. Even so, when Kilauea decides to erupt, all that one can do is warn those in the path of lava flows and watch.

Most volcanoes are not as predictable as Kilauea, and in order to minimize risk, we need to do careful monitoring of high-risk volcanoes combined with detailed studies to allow prediction of the paths of lava and ash flows. All of this requires money and expertise, but the cost is large only if we ignore the potential loss of life and prop-

erty if monitoring is not pursued. Even so, the political will and organization must exist to prevent disasters by careful planning and effective warning and evacuation procedures. Consider, for example, the Colombian volcano Nevado del Ruíz (mentioned earlier). In November 1985, eruption of this volcano generated mud flows that buried the town of Armero, killing 23,000. Armero was built on mudflow debris from the same volcano that were deposited when it erupted in 1595 and 1845. Six weeks before the fatal eruption of 1985, as Nevado del Ruíz awoke from a 140-year "sleep," Colombian geologists notified authorities of the risk to Armero, but the lack of resources to install monitoring devices on the mountain made useful predictions of the time and location of the pending eruption impossible.

Major loss of life and property has been avoided in other instances. In early April 1991, villagers in Patal Pinto on the flanks of Mt. Pinatubo, on Luzon in the Philippines, saw small steamy eruptions on the upper slopes of the mountain and smelled rotten-egg odors (Fig. 3.1b). These reports led to a 10-week multinational monitoring of Mt. Pinatubo by many geologists from the Philippine Institute of Volcanology and Seismology and the U.S. Geological Survey. Over the next few months, these studies allowed the geologists to issue a succession of more and more precise alerts that allowed villages within 10 km of the peak to be evacuated in late April and May. The major U.S. air base at Clark Field was evacuated in mid-June, and Manila International Airport was closed (see the quotation at the start of this section). All of this monitoring and warning limited deaths to 350 despite the fact that the eruption was one of the largest of the 20th century and the volcano had a city of 300,000 people (Angeles) on its flanks.

Can we make predictions about volcanic eruptions where there is no active volcano? Consider the release of CO_2 from Lake Nyos that we described earlier. Well, in 1990, a U.S. Forest Service ranger reported being nearly asphyxiated when he entered a small snow-covered cabin near Mammoth Mountain in eastern California (Fig. 3.13). In this same area, foresters had begun to notice large areas of dying forest. Mammoth Mountain is a large volcano on the western margin of the Long Valley caldera, and a number of shallow magma chambers are known to lie beneath the area around Mammoth Mountain. Although the caldera formed 600,000 years ago, Mammoth Mountain first erupted 200,000 years ago, with activity continuing until about 50,000 years ago. So-called phreatic eruptions, steam and gas emissions that carve out large craters on the flanks of volcanoes, oc-

Figure 3.13 Destruction of trees by carbon dioxide emitted near Mammoth Lake, California. Courtesy Allen Glazner.

curred in the area as recently as 500 years ago, and eruptions of small domes have occurred in the last 250 to 500 years. Subsequent studies in 1994 of the tree kills ruled out insects or pest infestations, but soil analyses showed that the concentration of CO_2 in the soils in the areas of the die-offs was 20 to 90 times greater that is normal for healthy forests. Forest biologists think that the high CO_2 inhibits normal root function in these mature conifer forests and, since the oldest trees being killed are more than 250 years old, that this CO_2 emission is a recent and infrequent phenomenon.

Some geologists who study the volcanic history of Long Valley have noted that the onset of tree kills coincided with seismic and other evidence of the shallow emplacement of magma in the area. Will these dikes be feeders to new eruptions such as those that occurred some 250 to 500 years ago? Is the CO_2 emission signaling an onset of phreatic eruptions? We do not know but should mention that 30,000 visitors a day come to the Mammoth Mountain ski resorts during the height of the season. The U.S. Forest Service has closed a campground several kilometers south of the mountain in an area of tree kills (Fig. 3.13). The CO_2 content of air samples in tents and rest rooms (>1% by volume) exceed national health standards. Underground utility vaults in the same region have had CO_2 contents as high as 89%. Now, if you think Harry Truman was unusual, consider these quotes from the Associated Press (September 17, 1995) about Mammoth Mountain:

It's just too beautiful here to worry about the volcano. – Danielle Lane, resident

The volcano is part of the attraction. – Alexandra Campbell, a recent visitor

It's not a requirement or a feeling on our part that it [the volcano] should affect land use planning. – Mammoth Lakes City Manager Glenn Thompson [echoes of Mayor Fouché]

We leave the rest to the reader.

Policy questions

Regardless of where you live, have you or your government spent enough time and money to investigate the volcanic hazard and to prepare for the possibility of an eruption? (We ask the same question for Sections 3.3 to 3.8.)

FURTHER READING: Francis (1993).

3.5 TSUNAMIS

Oeiras Bay [Lisbon, Portugal] all but emptied itself, exposing rocks never before seen by man to a distance of more than a mile out. . . . few comprehended its meaning. They watched,

fascinated. Only minutes passed before the wave was first seen [description of the tsunami that completed the carnage started by an earthquake and caused the deaths of approximately 60,000 people in Lisbon on November 1, 1755].
– D. Myles, The Great Waves, p. 124

Popular folklore holds that ships in the open ocean can be overturned by "tidal waves." The problem is that people are not really referring to tidal waves, and what they are referring to is not large enough to do damage in the open ocean. Tides are driven by the moon and are predictable and cyclic. In a few places in the world and at some seasons of the year, they are larger than others; but they are never noticed in the open ocean, and they do not reach amplitudes that can overturn ships. What people usually mean by a tidal wave is a "tsunami," which is a giant, rogue wave that, in the open ocean, has amplitudes of only a meter or less, distances between wave crests of tens to hundreds of kilometers, and speeds of hundreds of miles per hour, enabling a tsunami to cross the Pacific Ocean basin from Chile to Japan in 20 to 24 hours. On reaching shallow coastal waters, these waves slow to 30 to 40 mph, and they can attain heights (amplitudes) of 300 feet (100 m) and run up on shallow coastal plains and bays for distances of a mile or more inland.

The mechanism that produces a tsunami is simple enough. A sudden change in the volume of the sea floor – a landslide, a volcanic eruption, an earthquake-related shift – produces an impulsive force that drives a huge wave just as kicking your feet will cause the water in your bathtub to oscillate back and forth. Most tsunamis are caused by earthquakes, but we are never sure which ones will produce a tsunami. Although there have been more than 2,000 earthquakes with magnitudes greater than 7 this century, there have been only a few hundred tsunamis, the vast majority of them in the Pacific Basin. Despite the enormous number of earthquakes in the Pacific area, Japanese geologists estimate that only 106 major tsunamis struck the Japanese coast between 416 and 1978, a period of more than 1,500 years. These studies also show that the coast of Honshu about 200 miles north-northeast of Tokyo, known as the Sanriku coast, is the most tsunami-prone area in the world. Conservative estimates are that large tsunamis occur there with an average frequency of once every 40 years.

Tsunamis are fickle not only in their production but in their behavior. Bays and headlands can focus or diffuse the energy, depending on the direction of approach of the wave, and the topography of both the offshore and nearby coast influences the amount of run-up onto the land. The number of aftershocks and the actual frequency of the seismic surface waves at the earthquake epicenter, as well as the magnitude of the seismic event, influence the size and number of tsunamis. On rare occasions there is only one wave crest; typically, there are successions of several, even a dozen, waves separated by minutes or hours with no predictable relation between order in the succession and height. Some tsunamis arrive so gently that the sea level slowly rises to inundate coastal areas passively and then just as quietly subsides. Others arrive with huge crashing wave crests the height of a 20-story building and run across the land faster than a person can run.

The destruction caused by tsunamis can be enormous. The Sanriku coast, mentioned already, has seen approximately 150,000 deaths from tsunamis since records have been kept. The largest tsunami was in A.D. 869, and another one on June 15, 1896, killed 27,122 people and destroyed thousands of homes. A set of tsunamis associated with as many as 50 earthquakes in Chile from May 21–22, 1960, some with magnitudes as strong as 7.8 to 8.2, killed between 1,000 and 1,500 in Chile, 61 in Hawaii, and 80 to 90 in Japan, caused $350 million in damage to coastal communities and port facilities in Japan, and also minor damage to the ports of Los Angeles and San Diego.

Preventing damage from tsunamis is like preventing damage from any coastal flooding. It relies first and foremost on sound coastal management, zoning, and construction practices. Literally any Pacific coast area within 50 feet of sea level and 1 mile from the coast is at risk of a tsunami; although it has been more than 30 years since the last major tsunami in North America caused the destruction of Crescent City, California, after the Prince William Sound earthquake (Section 3.3). It is also possible to save lives and some destruction of mobile property by issuing tsunami warnings. This is done in the Pacific Ocean basin through oceanographic observatories in Alaska, Hawaii, and Japan. They monitor large earthquakes, and when they detect an epicentral location beneath the Pacific Ocean, they query a series of stations around the basin for evidence of large coastal waves. If a tsunami were generated, its velocity and trajectory can be predicted and appropriate warnings issued.

Policy questions

Regardless of where you live, have you or your government spent enough time and money to investigate the tsunami hazard and to prepare for the possibility of a tsunami? (We ask the same question for Sections 3.3 to 3.8.)

FURTHER READING: Myles (1985); Dudley (1988).

3.6 RIVER FLOODS

In general, flood control in the United States should place much greater emphasis on restriction of development of floodplains, flood-proofing of individual sites and local areas, an insurance program in which premiums are proportional to risk. — L. B. Leopold, A View of the River, p. 121

The kitchen sink floods when we add more water than can run down the drain. For exactly the same reason, a stream or river floods whenever the volume of water exceeds the channel's capacity to carry water away. The "surplus" stream water rises, overflowing the stream's bank, and inundates the surrounding countryside. This section describes floods and the features of rivers related to them.

Floods cause variable amounts of damage. At one end of the spectrum are damp basements and waterlogged automobiles of those who built or parked too close to a flooding stream. At the other extreme are undermined building foundations and roads, entire towns destroyed, transportation and communication disrupted on a regional scale, wholesale crop and livestock destruction, loss of human life, and disease and famine in the flood's aftermath. The 1993 floods in the upper Mississippi River basin in central North America cost an estimated $15 billion dollars in lost and damaged crops and flooded urban areas. Floods on a minor tributary of the Danube in north-central Romania in 1970 destroyed 225 towns and villages, and in 1887, floods on the Yellow River (Huang Ho) in Henan Province, China, are reported to have killed 900,000 people.

All of this flood damage results from the close relationship between people and rivers. Roughly one-third of the world's population obtains its food from crops grown on river floodplains, which means that hundreds of millions of farmers and hundreds of thousands of farm communities are located in vulnerable areas. Many cities are wholly or mostly on floodplains. For example, the famous seaport of New Orleans, Louisiana, is entirely on the floodplain of the lower Mississippi River; the entire city of Cairo, Egypt, is on the floodplain of the Nile; large parts of several cities along the Rhine River are flooded during particular high-water stages; and nearly 90% of Phoenix, Arizona, the desert capital of the arid southwestern United States, is on the floodplain of the Gila River.

The rivers that cause this flooding commonly flow in "meander loops" through valleys that are broader than the swath occupied by the present river. The cross sec-

tion in Figure 3.14 shows a river valley deepened by erosion, with two former river levels preserved as ledges ("terraces") along the valley walls. Because of this progressive deepening of the valleys, the higher the terrace the older it is, a fact that archaeologists use to establish the history of people who have inhabited a river valley for a long period of time. A steep bank occurs on the outside of the meander loop, where erosion is occurring, and a more gradual bank on the inside, where the river is depositing sand and gravel in relatively quiet water. In its normal position, the river is slightly below the level of its floodplain, which is named for the fact that it is covered during flooding. Floodwaters commonly deposit sand and silt along the sides of the river channel, forming natural levees that rise slightly above the level of the floodplain. Riverbank residents often build up and strengthen these levees to provide protection from floods. The floodplain commonly rises to higher elevations away from the river.

Normal and flood stages of a river are shown in the lower part of the cross section of Figure 3.14 and further illustrated in Figure 3.15. Flooding causes a river both to rise, which we see, and to deepen its channel by erosion, which we do not see. As water level rises, the water covers progressively more of the floodplain and, in large floods, may rise over the banks of the normal valley and flood higher terraces and areas outside the river's valley. Floods can be categorized into three types:

- Seasonal or cyclic floods come in response to climatically induced wet and dry seasons or to spring thaws which melt the snow pack in the stream's headwaters. These floods are relatively predictable.
- Infrequent and unpredictable (or "random") floods occur when rapid runoff is caused by a singular event such as unusually heavy rainfall or an abnormally persistent storm pattern. Although difficult to predict far in advance, floods of this type often give enough warning to allow measures to be taken to save lives and some property.
- Finally, some floods are caused or exacerbated by human activity.

We now treat each of these categories.

Seasonal floods

People who live near rivers are easily lulled into thinking that the natural state of the river is one in which the waters flow within the banks. This pastoral view of a river is true about 98% of the time, but every stream floods periodically unless prevented from doing so by human inter-

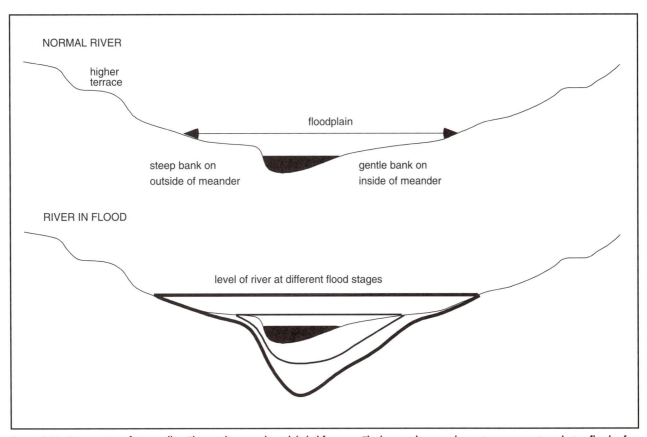

NORMAL RIVER

higher
terrace

floodplain

steep bank on
outside of meander

gentle bank on
inside of meander

RIVER IN FLOOD

level of river at different flood stages

Figure 3.14 Cross section of river valley. The top diagram shows labeled features. The bottom diagram shows river cross sections during floods of different magnitudes.

vention (see the sample hydrograph in Figure 3.16). Seasonal flooding is so common and so predictable that geologists can estimate the frequency, height above the river bottom, and area likely to be inundated by periodic and seasonal floods. For example, we cannot say for sure that the Rhône River will flood in 1999, but we are fairly certain that it will rise above its bank-full stage several times in the period from 1999 to 2009. In the eastern United States, streams overflow their banks on average about once every 2 years, usually only flooding the adjacent areas to a depth of about 1.2 times the vertical distance from the stream bottom to the rim of the bank; thus, a 5-foot-deep stream would normally rise about 1 foot above its banks every second year.

In areas of cyclical or predictable flooding of this kind, planners can use recent history to make robust predictions of the risk to any landowner. Height above bank-full level and the timing of flood crests are easily predicted because they are keyed to observations such as snowpack in the mountains, whose runoff feeds the streams, or long-range weather patterns in a watershed. The regularity of monsoon rains in the Indian subconti-

nent or of the spring floods on the Nile makes construction and management of flood control structures to regulate annual floods a matter only of the availability of capital, engineering skill, and political will (none of which are commonly in great supply).

Randomly occurring floods

Floods that occur with random frequency are far more difficult to manage than seasonal ones. Many catastrophic flood events are the result of unusual sets of conditions, such as a massive cyclonic storm that blows inland over a watershed on average only once a century. In Box 3.5 we illustrate the effects of such a storm in Khartoum, Sudan (where the whole concept of rain is virtually unheard of). Another example is the so-called Great Flood of 1993 in the Mississippi River basin of North America, which produced the highest river discharges for the months of August and September in 63 years. These floods were caused by no single storm but by a midsummer weather pattern unlike any seen before. Day after day, moist air from the Gulf of Mexico was drawn over

Figure 3.15 Potential flood areas of a river valley: flood-plains on both sides of Grey River, New Zealand.

the Midwest by a fixed upper-atmosphere jet stream and poured rain during the driest time of the year.

To predict infrequent and rare floods, hydrologists resort to statistical analysis. Many streams and rivers throughout the world have instruments known as "stream gauges" along the length of their channels. Typically maintained and operated by public agencies, these stations provide long-term data on stream discharge and water levels. In North America, western Europe, and other regions with major rivers, stream discharge records go back as far as 75 to 150 years. Using these data, we select the highest discharge event for each year. Then we place

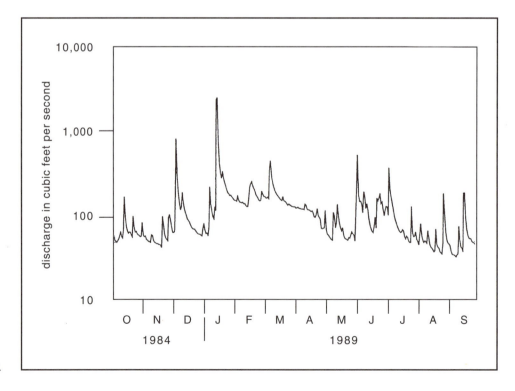

Figure 3.16 Example of hydrograph. Courtesy Charles Daniel, U.S. Geological Survey.

BOX 3.5 FLOODS AT KHARTOUM, SUDAN _____

"One cannot but praise the genius of the man who founded the city and foresaw the importance of its position. On the other hand, there is a drawback, for when the river rises, the town is flooded. This is because Khartoum is below the level of the Blue Nile." This statement, which we requote from an article by Walsh, Davies, and Musa (1987), was written in 1858 by Sr. Dal Bosco, an Italian traveler in the Sahara. It is accurate but minimizes the flood danger.

Khartoum and adjoining towns sit at the junction of the Blue and White Nile Rivers (Fig. 3.17). The Blue Nile rises in the highlands of Ethiopia and runs rapidly westward to join the White Nile, which wends its way sluggishly through broad swamps from its many branches in central Africa. Rainfall in Ethiopia causes the annual summer flooding of the entire Nile River to its outlet on the Egyptian coast, an event now largely controlled by the Aswan Dam, which we discuss in Box 4.4. In some years, the floodwaters of the Blue Nile have been so strong that they backed up the White Nile upstream, causing overbank flooding in and south of Khartoum. Overall climate change over north-central Africa (Section 8.2), however, has reduced the amount of water entering the Blue Nile and thus reduced flood levels. Floods in Khartoum in 1988, however, were among the worst on record. Here is why.

Khartoum has gone through various cycles of population growth and reduction since its founding in the early 1800s, but obviously it is much larger now than it has ever been. From a town of a few thousand when visited by Sr. Dal Bosco, Khartoum reached a population of 500,000 at the time of Sudanese independence and now contains about 4 million people. Older parts of the city were situated on relatively high parts of the flood-plain away from the river banks (see Fig. 3.15 for an explanation of floodplain elevations). An almost 10-fold increase in population in the past 40 years, however, has been accomplished largely by construction of homes and other buildings closer to the river, and hence lower. These areas had formerly been restricted to agricultural use, which could be resumed after the short-lived floods had passed.

A further difference between the older, and wealthier, parts of Khartoum and the newer areas is the roads and drainage systems. Old parts of the city have paved roads and adequate drainage conduits that remove water rapidly and prevent damage from flooding. New parts of the city, however, have neither paving nor drainage systems. The new parts of Khartoum were constructed on river sediments (alluvium) that have been com-

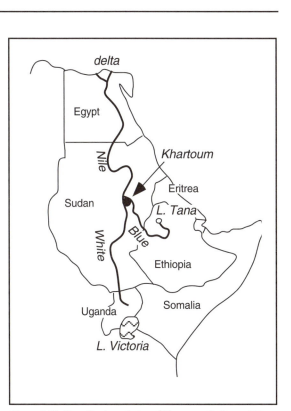

Figure 3.17 Map showing relation of Khartoum, Sudan, to Nile River.

pacted to a hard surface by the constant passage of people and animals. This unvegetated, packed surface is virtually impermeable in much the same way that paved parking lots around shopping malls in rich countries are impermeable.

The concentration of flimsy structures on impermeable ground in low-lying areas led to the tragedies of the 1988 floods. They were not caused by the Blue Nile, which received very little rain that year, but by an unusual 8-inch downpour over the Khartoum area itself. This much water with no place to go ponded in low areas, ran along normally dry watercourses (wadis), and finally flowed over into the Nile River and was carried downstream. In the course of the flood, approximately one third of the homes sustained some damage, much of it severe, and 2.5 million inhabitants were affected either in their homes or their places of work. The flood was simply another example of having too many people in the wrong place.

FURTHER READING: Toniolo and Hill (1974); Walsh et al. (1994).

these events in a rank order from highest to lowest and calculate T_r, the recurrence interval (average time between floods), using the formula $T_r = (N + 1)/m$, where N is the number of years of record, and m is the rank order of each discharge event. With T_r calculated for each discharge, we plot T_r versus discharge to obtain the "flood frequency curve" in Figure 3.18. This curve tells us the recurrence interval of each discharge event, including floods. For example, for the Euphrates River at Hit, a small town on the river almost due west of Baghdad, Iraq, the curve shows an average 50-year interval between measured discharges of 250,000 cubic feet. In probabilistic terms, the curve shows a 1 in 50 (2%) chance of such an event occurring in any given year, commonly referred to as the "fifty-year-flood." As we will see, this is a potentially misleading phrase.

Flood frequency diagrams are widely used by people interested in floodplain management. They might include real-estate developers, land-use planners, the local chamber of commerce, or disaster relief officials. With their varying interests and motivations, these and other interested parties often use the same data to come to different conclusions. For example, some people think that

a "50-year flood" will occur only once every 50 years. That is, if a discharge event of 250,0000 cubic feet per second occurred last year on the Euphrates River at Hit, such a flood will not occur again for 49 years, or the probability of such an event occurring this year is far less than 1 in 50. This line of analysis would seem to justify the construction of a facility with a predetermined 45-year practical life-span on the river floodplain. Or would it?

We have entered the slippery realm of statistics here and, to help navigate this terrain, we must consider some of the pitfalls in the construction of the flood frequency curves. We ranked the peak annual discharges and made our calculation without considering the *actual* intervals between the highest peak discharges. In other words, this analysis does not, nor does it claim to, take into account the *pattern* of discharge events. We do not worry whether the highest and second highest discharges occurred in sequential years or not. The method gives the same recurrence interval if the highest and second highest discharges occurred in two subsequent years or 60 years apart. For climatic reasons, however, years with high peak discharges may cluster together – a long wet spell. Conversely, years with low peak discharges may also clus-

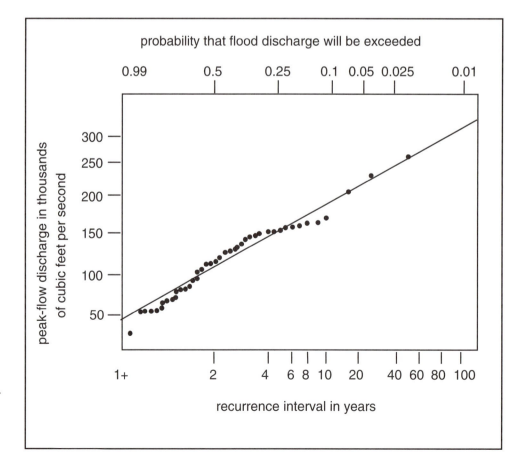

Figure 3.18 Plot of flood frequencies for the Euphrates River. Modified from Kolars and Mitchell (1991).

ter – a multiyear drought. If clustering does occur, then the fact that last year we had a 50-year flood might actually mean that this year or next we are at a significantly greater than a 1-in-50 chance of having a 50- or even a 100-year flood.

Two other difficulties arise with the use of flood frequency curves. One is that a brave (and not very careful) person might project them to determine the discharge of 100- or 500-year events, long beyond the sampling interval. Such a projection encounters three problems. First, the record is derived from a relatively short sampling period, and it is obviously problematic to predict the discharge of a 100-year flood based on a 75-year record. Another problem in basing predictions on data that are 75 to 100 years old is that flood-management projects, land development, and urbanization have changed the patterns of stream flow and runoff in most industrial nations.

This change may be a gradual or an incremental one, but in either case it means that data on annual discharge that predate the human interference with stream flow are no longer applicable. This restriction shortens the useful sampling interval and further reduces the statistical utility of long-term flood recurrence predictions of the type we are discussing.

With the caveats noted here, recurrence intervals *are* useful for planners and engineers who need to know what conditions and eventualities to design and plan for. Using flood recurrence calculations and topographic maps, a hydrologist can convert discharge in cubic feet per second to feet of river height above the channel bottom. This allows us to transfer historic discharge data to maps that show average probabilities of flood risk on a river floodplain (Fig. 3.19). These curves and maps are not intended to predict next year's flood risk based on last

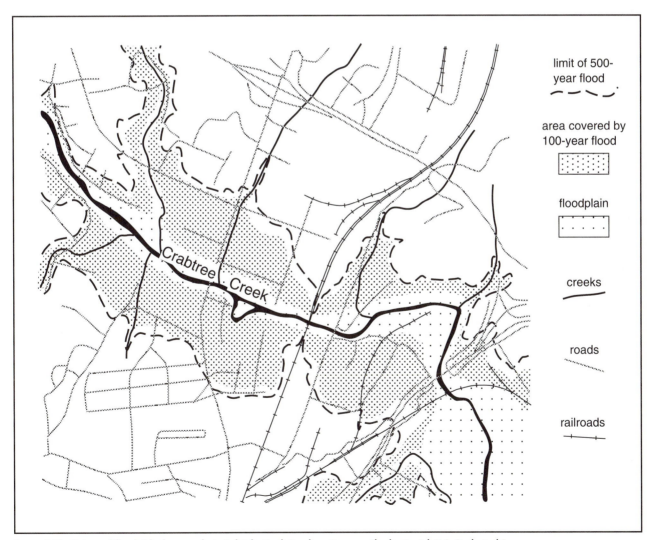

Figure 3.19 Estimated flood risk along creeks in Raleigh, North Carolina. Courtesy Charles Daniel, U.S. Geological Survey.

year's. They simply show those areas that theoretically have a 100% probability of being inundated during the chosen period of time, if average conditions prevail.

Floods induced by human activities

Some engineering structures along riverbanks are important and effective. Evidence from the previously mentioned 1993 floods in the North American Midwest indicates that those areas that suffered the most crop and structural damage were either unprotected or imperfectly protected from flooding. Many levees that gave way were known well before 1993 to be weak or too low to be effective. Areas such as St. Louis, where modern, well-engineered, hardened levees and channels existed, suffered little or no damage.

Nonetheless, human interference in natural river basins often causes unintended results. As we discuss in Section 4.3, the percentage of any rainstorm that flows overland with the possibility of becoming part of a flood varies with many factors, including the nature of the soil, the duration and intensity of the rainstorm, vegetation patterns, the time of year, and the level of the groundwater table. The percentage is also greatly affected by urbanization. Typically as much as 50% of any rainstorm soaks into the ground in undeveloped forested areas. Following urbanization, rain that falls on buildings is channeled by gutters and downspouts into storm drains, and water on the streets and parking lots runs directly into drainage ditches or storm sewers. Furthermore, the water arrives in the stream channel much faster through concrete gutters and pipes than it would if it ran across the irregular ground surface and periodically ponded in depressions. Thus, urbanization increases peak discharge two to four times for any given rainfall, reduces the time between peak rainfall and peak discharge by half, and causes flooding from relatively minor storms that, prior to development, caused no flooding at all. Seasonal or storm-induced precipitation events that would have caused flooding before urbanization become far more severe and damaging. To prevent these larger floods from occurring, developers in many urban areas may be required to build catchment basins that retain and hold water long enough to ensure that the eventual runoff of surface water will equal the rate and amount under natural conditions.

Urbanization is not the only human activity that can cause floods. Straightening river channels and building high concrete levees on either side can protect adjacent areas from floods; but downstream of the straightened channel the more rapid throughput during periods of high discharge and the prevention of even minor flooding onto the floodplain upstream can result in far more severe flood events. Destruction of natural wetlands (swamps) by farmers or developers in upstream areas can eliminate natural catchment basins that act as a brake on river discharge during periods of high flow (also see discussion of wetlands in Section 2.4). A common consequence of wetland destruction in uplands is the need to build flood-control dams to catch, hold, and allow gradual release of seasonal high discharge that was once naturally controlled by flooding of swampy areas. Even the routine destruction of beaver dams can result in local flooding where none occurred before.

A final example of human-produced floods is a terrifying one – dam failure (Box 3.6). Among the most catastrophic floods in the industrial era are those caused by the collapse of poorly engineered or inadequately maintained dams. Is this a serious problem? No one really knows. Certainly some dams in relatively remote areas might fail with little or no loss of human life or property, and we do know that new dams are safer than old ones and concrete dams have a better safety record than earthfill dams. We also know that the United States, for example, has about 20,000 dams with water depths of 20 feet or more that would cause significant loss of life or property if they failed. Furthermore, some dams hold more than water, such as by-product solutions and sludge from industrial manufacturing, tailings from mining activities, or agricultural wastes from large swine or cattle feedlots. These dam failures have serious short- and perhaps long-term environmental impacts beyond those caused by floodwaters alone.

Policy questions

Regardless of where you live, have you or your government spent enough time and money to investigate the flood hazard and to prepare for the possibility of flooding? (We ask the same question for Sections 3.3 to 3.8.)

FURTHER READING: Beven and Carling (1989).

3.7 COASTAL FLOODING AND EROSION

The seawall worked well until it failed. – *Orrin Pilkey, quoting the U.S. Army Corps of Engineers*

An estimated 260 million people (4.6% of the world's population) live within 2 m (7 feet) of mean sea level.

BOX 3.6 THE JOHNSTOWN FLOOD _____

Late 19th-century America was awash in unfettered growth and greed. The Gilded Age was a time of industrial expansion, accumulation of vast wealth, and an attitude toward the common good immortalized in the words of William Vanderbilt, "The public be damned." It was the perfect era for a human-created catastrophe that should never have happened.

The river town of Johnstown lies in the once-sylvan valley of the Little Conemaugh River, about 60 miles east of Pittsburgh in the rugged Allegheny Mountains of western Pennsylvania. This remote part of Pennsylvania would have remained little known to the world had not the Western Division of the Main Line of Pennsylvania's barge canals followed this river valley. In the 1830s and 1840s, the major, east-to-west commercial lifeline of the growing United States was this canal system, which allowed durable goods to criss-cross the mountain barrier between the Ohio River, the largest eastern tributary of the Mississippi, and the east coast. To compensate for insufficient water during dry summers, the canal-builders constructed a 72-foot-high earthen dam, said to be the largest in the world, on a tributary of the Little Conemaugh River. More than 250 feet thick at its base, South Fork Dam was built to the best design standards of the day. It created a 425-acre lake, the Western Reservoir.

Dams are not inherently dangerous, and even 150 years ago, engineers knew how to build safe and durable structures. Among the common practices used then (and now) were the installation of discharge pipes in the base of the dam to allow water to be released from the lake in a controlled manner during periods of high water. In addition, the core of the dam was made of carefully tamped and packed clay to prevent water seepage that might erode an underground channel and undermine the dam's interior. Huge blocks of stone, known in the business as riprap and so large as to require three teams of horses to move, were piled on the clay core as ballast. Finally, the lowest point on the dam's crest was a spillway cut into solid rock in the valley wall. At a height of 9 feet below the top of the dam, the spillway was a safety valve such that overflow was channeled from the reservoir without eroding the earth-fill dam. Like cars, however, dams have to be maintained. Unfortunately, the South Fork Dam was completed at just about the time that the Pennsylvania Railroad built its main tracks along the Little Conemaugh, forever ending the need for the canal and reservoir. In 1857, the railroad acquired the dam along with the canal right-of-way, and over a period of years drained the lake and allowed some 200 feet of the dam to wash away. None of this caused much concern; no one lived in the immediate area.

Thirty years later, the railroads and the booming iron and steel industry had turned Johnstown, 14 miles down the narrow and twisting valley of the Little Conemaugh from the dam, into an industrial center of nearly 20,000 residents with several large steel mills sprawled along the floodplain of the river. Railroads, iron, and steel had also made Andrew Carnegie, Henry Frick, and Andrew Mellon, among other residents of Pittsburgh, very wealthy. In 1879, the old South Fork Reservoir was acquired by the South Fork Fishing and Hunting Club, with Carnegie, Frick, and Mellon among its members. With no engineering expertise and total disregard for standard practice or safety, the club proceeded to plug the old discharge pipes at the base of the dam, to fill in the washed out portions of the dam with loose and uncompacted fill, and to lower the top of dam so that a carriage road could be built to provide access to a 47-room clubhouse and the 17-room "cottages" for members. As the lake refilled, they also obstructed the spillway so that the 1,000 imported black bass could not escape into the river below. The newly christened Lake Conemaugh covered 500 acres and was about 60 feet deep at the dam. The club, with Gilded Age arrogance, ignored warnings that their new dam was a threat to those downstream, warnings that included a detailed critique of the reconstruction by John Fulton, a geologist and engineer employed by the Cambria Iron Works in Johnstown.

On May 30–31, 1889, a massive storm dropped perhaps as much as 8 inches of rain on southwestern Pennsylvania. The rivers flooded and Lake Conemaugh rose to the top of the dam. Efforts by workers at the club to dig diversion ditches and clear the spillway failed to lower the lake. By 11:00 A.M., May 31, water was spilling over the dam's center, where there was no spillway. For the next 4 hours, the water cut a channel and carried fill off the dam face until the dam was so thinned and weakened that it literally burst. A 420-foot-wide section of the dam gave way (an eyewitness said, "The dam didn't burst. It moved aside") and in 45 minutes, more than 20 million tons of water drained into the narrow river valley, gaining velocity as it rushed 14 miles and dropped 400 feet in elevation toward Johnstown. It took only 10 minutes for the first wall of water to hit Johnstown. The town was literally torn to shreds, and more than 3,000 people died in the flood and the fires and disruption that followed.

The Johnstown Flood caught the attention of the American people like few other disasters before or since. It was a time of aggressive and free-wheeling journalism, and reporters and photographers descended like a plague. It had its villains: Andrew Carnegie idly fishing and playing whist while placing inno-

---

BOX 3.6 THE JOHNSTOWN FLOOD *(continued)* _____

cent lives at risk. It had heroines: Hettie Ogle, a telephone switchboard operator, who died at her post trying to warn the town. It had miraculous escapes: Michael Roneson, swept away by the flood, who held fast to a lightning rod on top of a building for 2 hours until the flood subsided. It was Victorian melodrama at its best: with widows and orphans, and the ruthless violation of home and hearth by evil incarnate.

No legal responsibility was ever established. Editorials, coro-

ners' inquests, and a blue ribbon report by the American Society of Civil Engineers cited the negligence of the South Fork Fishing and Hunting Club, but juries in Pittsburgh (!) ruled that the flood was "a visitation of Providence." Andrew Carnegie, however, did give $5,000 to the relief efforts.

FURTHER READING: O'Conner (1957); Degen and Degen (1984).

Some of them have little choice in where they live, but others are there because they want to be. We discuss these groups briefly before proceeding to coastal geology and engineering.

Many people live near sea level because the lower reaches of major river systems, including the deltas, provide the richest soils and most abundant freshwater supplies for subsistence farming. The annual floods that come with seasonal change provide new topsoil and remove the agricultural and animal, including human, wastes. The earliest developments of major civilizations were along the lower reaches of the Nile River of Egypt and the Tigris and Euphrates Rivers, now in Iraq (Box 3.7). Bangladesh is the eighth most populous nation in the world, and almost all of its 125 million citizens live on the floodplains and deltas of the Ganges and Brahmaputra Rivers, much of which is within 10 feet of sea level. The deltas are frequently swept by monsoon storms that blow northward from the Indian Ocean, bringing flooding, high winds, and damaging waves, both along the coast and to areas up to 20 feet above sea level. In a great monsoon of April 1991, 131,000 people were killed on the Bangladesh coastal plain and on outlying islands that were totally covered as the wind-driven ocean swept completely over them. This single storm caused $2.7 billion damage to property and agriculture in one of the poorest nations of the world, a country whose annual gross domestic product is only $23.7 billion. This amount would be the economic equivalent of a $600 billion disaster in the United States.

Some people who live near sea level because they have no other choice are in minimal danger because they not only understand their situation but can partly control it. For obvious reasons, the major centers of the fishing industry and commercial ports are located in low-lying coastal regions. Rich countries provide protection

for these areas. For example, parts of the city of New Orleans are below sea level but are guarded from flooding by the Mississippi River and the Gulf of Mexico by manmade levees. A second major example is the Netherlands, a rich nation with a long history of industrialized agricultural practices (Chapter 2), where 60% of the population live below sea level. With 38% of the country below sea level, the Dutch have no options in where they live, but they protect themselves by a sophisticated system of more than 1,500 miles of dikes and sea walls (some as high as 16 to 20 m above mean sea level) and pumping stations.

At the opposite end of the spectrum from the peasant farmers of the Ganges delta and the industrialized farms and factories of the Netherlands are people who choose to use the coast for recreational purposes. In industrial countries, most coastal regions within a few hours drive of metropolitan areas are highly developed by those seeking fun on the beaches and dunes. Perhaps, at best, these people are unaware of the dangers of coastal areas or, at worst, are victims of self-delusion. They may have never lived along the coast, may fall under the persuasive influence of a smooth-tongued developer, or may have that unfailing human ability to assume that bad things always happen to someone else.

Coastlines at risk

The risk of flooding along coastlines is highly variable. Areas not at significant risk include continental margins undergoing active mountain building ("active" margins along subduction zones; Section 3.1) and young volcanic islands (Fig. 3.21). In these areas, the active uplift of the coast or the constant addition of hard, resistant rock by volcanic eruption keeps all but the narrowest beach strand well above the level of coastal flooding.

BOX 3.7 THE GREAT FLOOD

O man of Shuruppak, son of Ubartutu:
Tear down the house and build a boat!
Abandon wealth and seek living things!
Spurn possessions and keep alive living things!
Make all living beings go up into the boat.
— Tablet XI, *The Epic of Gilgamesh* (trans. Kovacs)

Nearly 150 years ago, British archaeologists found pieces of 12 clay tablets in the ruins of the library of Ninevah, the pride of King Ashurbanipal in the 7th century B.C. Not until about 1900 were these cuneiform writings translated and found to tell the rich story of *The Epic of Gilgamesh*. The Ninevah version, it turns out, is only one of the youngest of perhaps a dozen, the oldest of which was written in Babylon about 1800 B.C. from legends dating back to the earliest Sumerian peoples of about 5000 B.C. Just as all English-speaking peoples know their Shakespeare, apparently for more than 4,000 years all Mesopotamians knew their Gilgamesh.

The story of Gilgamesh, a Mesopotamian hero who lived about 2700 to 2500 B.C. along the fertile banks of the Euphrates River, is the age-old search for immortality. To find it, Gilgamesh (perhaps a king or priest) sets out to find the one man known to be immortal, Utanapishtim. Utanapishtim won his victory over death by leading a virtuous life and having heeded the word of his god, Ea, when Ea, in anger, had sent a flood to destroy mankind. The chronicle says:

Six days and seven nights
came the wind and the flood, the storm flattening the land.
When the seventh day arrived, the storm was pounding,
the flood was a war [then]
The sea calmed, fell still, the whirlwind flood stopped up.
I looked around all day long — quiet had set in
and all the human beings had turned to clay!
The terrain was flat as a roof . . .

On Mt. Nimush the boat lodged firm,
Mt. Nimush held the boat, allowing no sway.

Sound familiar? Change the names and numbers and we have Noah's flood as told in Genesis (7:11–12):

. . . were all the fountains of the great deep opened . . .
and the windows of heaven were opened
and the rain was upon the earth forty days and forty nights.

Fifteen cubits upward did the waters prevail . . .
upon the earth a hundred and fifty days.

(A cubit is the distance from the elbow to the tip of the finger, about 18 inches for average people.)

The legend of the "great flood" forms the basis for the more than 500 such tales in more than 750 different cultures. After resonating among the river peoples of the Tigris and Euphrates, the story apparently was carried to the Nile valley, where cuneiform tablets have been found with the marks of hieroglyphic translators highlighting various passages. But was there a great flood? Why did this story seem so plausible? And what about all the sightings of the wreckage of the Ark on Mt. Ararat?

The Tigris and Euphrates Rivers constitute the floodplain that is Mesopotamia, now mostly Iraq (Fig. 3.20). The earliest cities on the floodplain were built of available clay with elaborate canals, dikes, and irrigation systems to control the waters. They were also built closer to the sea than Baghdad and other major cities of modern Iraq, partly because the rivers had not extended their deltas as far south as they have in the past few thousand years. These cities generally were no more than 25 feet above sea level and were terribly vulnerable to flood. Could there have been a time when a great flood suddenly occurred? We propose two possibilities — coastal floods and river floods. A coastal flood is suggested by the swiftness of onset and the tales of local rains lasting for weeks. Current monsoons commonly bring rain that lasts for a month or more to the northern edge of the Indian Ocean, and typhoons sweep onto low-lying coasts for many miles (the effects in Pakistan and Bangladesh are described in the text). Typhoon-induced flooding could cover much of the lower, settled, reaches of the Tigris and Euphrates floodplain, which 300 miles from the sea has an elevation less than 100 feet.

Conversely, a river flood is suggested by the history of climate in the Middle East. The Tigris and Euphrates Rivers are only 700 miles long and drain the relatively arid highlands of the Taurus Range of southern Turkey. Thus, the annual flow is irregular and unpredictable — they flood in some years and bring drought the next. We can investigate the possibility of a sudden flood or floods by looking at Lake Van in the Turkish highlands near the headwaters of both rivers. During the summer, streams flowing into the lake deposit sand and silt in a thin layer over the lake floor. The finest silt and clay remain in suspension, however, and settle to the lake floor only in winter, when the streams freeze and the lake water is quiet. This alternation produces annual layers of fine and coarse sediments, called

BOX 3.7 THE GREAT FLOOD *(continued)* _____

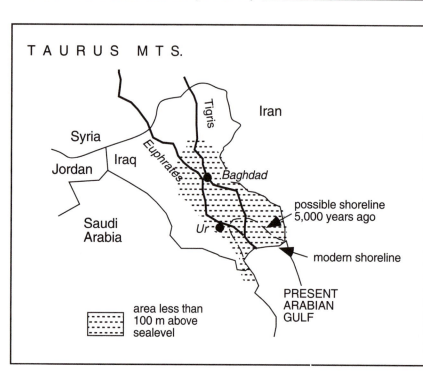

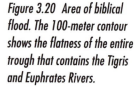

Figure 3.20 Area of biblical flood. The 100-meter contour shows the flatness of the entire trough that contains the Tigris and Euphrates Rivers.

"varves," which record the passing of the years just as tree rings record annual cycles of growth and dormancy. When geologists core the sediments in the bottom of Lake Van, they find almost 14,000 varves that record the variation in temperature and precipitation in the headwaters of Tigris and Euphrates since the end of the last glaciation (14,000 years ago; see Section 8.1) and the entry of the first settled peoples onto the Mesopotamian plain. The record is clear; 5000 B.C. saw a sudden transition from dry to wet, from low to high rainfall. If this transition was ushered in by a season of extraordinary rainfall, the Mesopotamians must have seen their clay cities and all they knew swept away in a climatically induced flood or series of floods of proportions unprecedented in their cultural memory.

Regardless of the source of the Great Flood, what about the wreckage of the Ark on Mt. Ararat? Mt. Ararat is a 16,000-foot-high, snow-capped peak near the border of Turkey and Iran.

During the last glacial maximum, it was covered with alpine glaciers that picked up soil and loose rock as they swept down from the peak and, at the bottom (toe) of the glacier, plowed the debris into a mass of unconsolidated earth and rock known as a moraine. In this moraine, sticking out like the masts of an ancient vessel, are great pine logs plucked from the forested flanks of the mountain. To the curious eyes of prescientific wanderers, this debris looked like a flood deposit in which the remnants of an old sailing vessel are preserved. Unfortunately for the legend, however, flood waters 15 cubits (approximately 25 feet) high could not cover Mt. Ararat. The legends must have had a small local hill in mind, but this discrepancy does not weaken the compelling evidence for a great flood in the early history of Mesopotamia.

FURTHER READING: Keller (1956); Kovacs (1985).

Coastlines that are most at risk are along continental margins where mountain building is not occurring ("inactive" margins; Section 3.1) and where human habitation is densest. They include most of the margins of the Atlantic Ocean in North and South America and Eu-

rope, virtually all of the African coast, and much of south and east Asia. The margins of the Arctic Ocean are also inactive, but they have never been developed as beach resorts. Features along these coastlines include (Figs. 3.22 and 3.23):

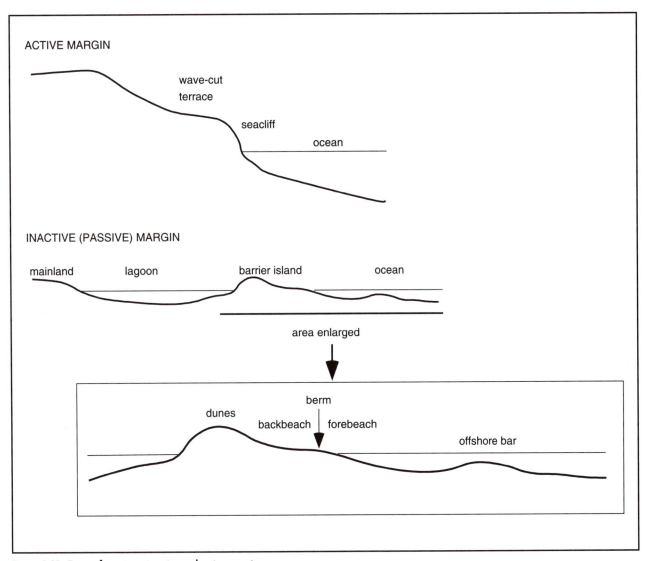

Figure 3.21 Types of coasts on inactive and active margins.

- barrier (offshore) islands, commonly with sandy beaches facing the ocean and swampy areas on the landward side;
- lagoons with areas of shallow water and swamp between the barrier islands and mainland;
- flat mainland areas that rise very slowly away from the shore and may have elevations of only a few feet above mean sea level inland as far distant as several miles;
- shallow bays and complex estuaries extending far inland.

Even in many of these areas, the risk is limited to small coastal embayments and the narrow strand of actual beach. Where rocky headlands produced by erosionally resistant rocks form the coastline, few areas are at risk.

The reason that risk is high along inactive margins is that these coastlines are adjusted to the supply of sediment and to its removal by normal processes of erosion. All of these areas are dynamic systems that depend on rivers to provide a steady supply of sand and silt to deltas and steady longshore drift to supply sand to bars and barrier islands. Beaches are built and destroyed, islands migrate, channels fill, and marshes flood and silt in with the predictable short-term changes in tides and weather and longer-term changes in sediment supply and land subsidence.

Coastal storms accelerate the pace and amplify the impact of changes on coastlines. With storms come surges of fast-breaking waves that wash over old beaches and sand bars, often cutting new channels and silting in old ones. The magnified erosive power of

Figure 3.22 Features of shorelines: (a) notch cut by wave erosion on coast of Kenya; (b) healthy beach on west coast of Florida; the photograph is taken from the top of the dune ridge and shows unharmed dunes, beach, and several offshore bars (light color in water); (c) damaged coast at Swakopmund, Namibia; erosion resulted from destruction of dunes during construction of the beach houses.

(a)

(b)

Figure 3.23 (a) Consequence of constructing breakwaters along a beach in Southern California; the longshore current coming toward the viewer has been interrupted, preventing sand deposition and allowing erosion of the shoreline. (b) Destruction of building and removal of beach along North Carolina coast caused by hurricane in 1996; courtesy C. Brush.

3-to 20-foot waves breaking along and overrunning a beach strand that normally has 2- to 5-foot waves causes major destruction. The storm surge scours the beach, cuts new channels though sandy bars and islands, overwashes muddy coastal wetlands, and may carry away most man-made structures in its path. While people who live smugly inland may scoff, we should note that, on average, 65% of the insured prop-

erty losses in the United States are caused by hurricanes, much of it related to coastal flooding and erosion (Section 3.9).

As an example of such a storm, consider Andrew, a category 4 (winds between 130 and 155 mph) tropical storm. On August 26, 1992, Andrew swept across the delta of the Mississippi River in southern Louisiana after causing billions of dollars in wind and water destruction in south Florida. The on-rushing wind- and pressure-driven surge of water effectively raised water level during the 2-day duration of the storm as high as 20 feet above mean sea level. The erosive power of the high waves combined with the storm surge removed 120 feet of shoreline on some of the hardest hit beaches. In the low-lying coastal marshes and shallow bays behind the beaches, the receding storm surge left a 1.5-foot-thick layer of mud over tens of thousands of acres of land – about 1.5 billion cubic feet of mud was deposited by one storm in 2 days! The most striking changes were on Isles Dernieres and Timbalier Island, a short distance into the Gulf from the mainland. They were completely inundated by the storm surge, and, when the storm passed and the islands reemerged, 70% of the beach sand had been removed and 15 new channels had been cut in the islands, in effect creating 16 small islands out of 2 large ones. Geologists who had studied the Louisiana coast had predicted that these islands would, in the natural course of events, disappear by the year 2020. Hurricane Andrew probably shortened their life by 20 years.

Human activity and the coastline

The lesson of coastal storms is that living in a dynamic geological environment is a gamble. Beaches erode; islands grow, are inundated, and migrate; wetlands flood; estuaries, channels, and bays become silted in and no longer navigable by deep-draft ships. Those who deliberately live along coastlines at risk have two choices: (1) adjust to the dynamics of active coastlines by building structures only in areas at less risk – for example, behind protective vegetated dunes or on relatively high parcels of land at the widest part of an island; or (2) attempt to use engineered structures like seawalls and breakwaters to prevent erosion. The first option is available only if land development has not preceded understanding or if there are, in fact, options. For many subsistence farmers whose livelihood is dependent on fertile coastal lands, it is not possible to move. For many residents along developed portions of coastlines in the United States and western Europe, it is too late to move.

The temptation to protect a coastline against flooding and erosion thus becomes great. In fact, building protective structures (Fig. 3.24) is a public political-economic decision that requires both an engineer's expertise and a geologist's understanding of earth processes. As geologists, we emphasize the following generalizations that bear on the decision to engineer a coastline to prevent flooding or erosion:

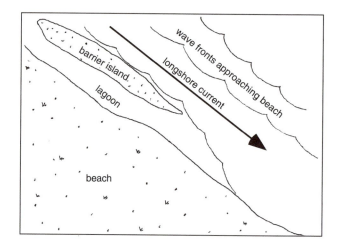

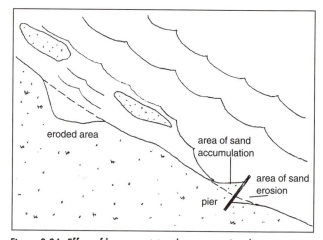

Figure 3.24 Effect of human activity along coasts. See discussion in text.

- Eventually the ocean will prevail. This maxim might seem unduly fatalistic, but it simply means that no single structure, no single coastal modification, will last forever. If we build a sea wall to protect, for example, the foundations beneath a bridge, then we must expect to have to reconstruct and redesign that sea wall on a regular basis.
- Engineered structures often lead to new problems, sometimes for the landowner, sometimes for someone else.

We conclude with two examples of the effects of efforts to modify or protect beaches.

First, before a detailed understanding of beachfront processes was common knowledge, it was a normal practice to create or extend sandy beaches by pumping offshore sand onto them (Fig. 3.24). The immediate result is a broad sandy beach to replace one that had been eroded or a new beach where none had existed before. Because the cost of beach replenishment projects is highly dependent on how far we must pump or transport the sand, the temptation is simply to pump from offshore bars seaward of the beach being replenished. This process has several unintended results. The waves you see breaking offshore of a sandy beach are breaking on offshore bars; without these bars, the new beach will be exposed to the full energy and erosive force of incoming waves. Furthermore, the removal of the bar and adjacent sandy banks typically steepens the shoreline profile, with predictable slumping of the oversteepened beach front. And, to add insult to injury, beach replenishment is expensive as well as temporary. The minimum cost of replenishment is $1 million per mile and some projects cost ten times that much. While some have been highly successful – for example, the 16-km-long Miami Beach replenishment, which is still intact some 15 years later – others are not. In 1982, a $5-million replenishment project at Ocean Beach, Maryland, was washed away within 2 months, and Wrightsville Beach, North Carolina, has been replenished at least 10 times.

A second common example of an engineered structure resulting in more, rather than less, erosion is the building of barriers perpendicular to the shoreline along the coast or on lakes. Longshore currents move sand along the shore (Fig. 3.24). As sand is eroded from up-current beaches, it is transferred to down-current beaches. We can sympathize with the owner of up-current beachfront property who watches sand wash away to replenish the beach of a neighbor. We can understand the desire to build a breakwater along the extension of the property line to trap that drifting sand. The unintended result, however, is that sand is washed away from the beach of the down-current neighbor because the breakwater did nothing to stop the incoming waves. Consequently, in the absence of a continual sand supply to stabilize the beach, the waves simply erode the shoreline.

Because of the inevitability of shoreline movement, we should recognize that the problem is not the control of shorelines but control of people's use of them. If people, generally affluent ones, did not flock to coasts and build homes and businesses on them, there would be no shoreline problem. Coastal movement does not affect people who only visit the shore during the day or camp there in temporary shelters. As a society, we must decide not only what type of use we will permit along shorelines but also whether we should provide public support of permanent

structures through government-supported engineering projects and property insurance not obtainable privately.

Policy questions

Regardless of where you live, have you or your government spent enough time and money to investigate the problem of coastal erosion and to prepare for any difficulties that it will cause? (We ask the same question for Sections 3.3 to 3.8.)

FURTHER READING: Kaufman and Pilkey (1983); Bird (1985).

3.8 LANDSLIDES AND RELATED PROCESSES

The landslide event came as a surprise to the community.
– F. B. Leighton, commenting on the Bluebird Canyon landslide in Laguna Beach, California, where 23 residences built on top of an earlier landslide became uninhabitable when a new slide occurred, in Coates and Vitek, Thresholds in Geomorphology

Gravity is like death and taxes – it is always there. It acts on rocks and soils as it does on apples and hang-gliders, pulling all earth materials toward the earth's center. If mud, gravel, or boulders happen to be on a sloping surface or, even better, on the edge of a cliff, they may move downslope, creating a hazard to any person or structure when the land gives way. Some downslope movements are slow, some fast. We recognize three categories based largely on speed of movement (see the illustrations in Fig. 3.25).

- At the slow end of the spectrum are all kinds of gradual migrations of soil and unconsolidated materials downhill. They are known collectively as "creep," and rates of movement are commonly a few inches a year. Anyone who has ever lived near an oversteepened bank, perhaps near a roadcut or a construction project cut into a hill, has seen the effects of creep. Landscapers and engineers often build retaining walls and various kinds of buttresses to hold off this steady, slow encroachment of slumping earth. We will not discuss creep further because, although much effort is expended on controlling creep, it is a geological nuisance and not a hazard to life and limb.

- At the opposite end of the velocity spectrum are "debris avalanches" and "rock falls" that fall off of a mountain slope and then race at velocities upwards of 200 mph along the ground surface (see Box 3.8).

(a)

(b)

Figure 3.25 Types of earth movement: (a) creep in Antigua, Caribbean. Downslope creep has caused the palm trees to rotate toward the left, but their tendency to grow straight upward has caused them to grow in a curve. Similar curvature in pine and hardwood trees provided ready-made prows and ribs for sailing ships in seafaring nations; (b) rock pile (talus) along mountain front in Alberta, Canada; (c) rotated slump block in northwestern Argentina; (d) landslide at Lake Tahoe, California. The debris on the lower part of the slope came mostly from the scar on the upper part of the slope. The highway is being constructed across the debris.

(c)

(d)

• Between creep and debris avalanches are a host of mass movements that are generally referred to as "landslides" and "slumps." A landslide moves suddenly downslope and creates a jumbled mass of debris at the foot of the "scar" on the cliff from which the debris came. A slump is a block of soil and/or rock that moves coherently (without breaking up) along a slip surface that is commonly exposed above the slump block.

Landslides and other downslope movements occur frequently. In mountainous regions or along eroding river bluffs and coastlines, slides are an important part of the natural erosional cycle. Evidence of large prehistoric landslides may consist of landslide scars along ridges and steep slopes, and in the valleys below we may still find the landslide debris in the form of hummocky deposits of poorly sorted, chaotic debris with angular cobbles and

BOX 3.8 THE LUCKY SHEEP OF HUASCARAN

A shepherd (name unknown to us) was startled to look up and see a sheet of boulders, mud, and snow pass above him and the sheep that he was tending near the Andean town of Yungay. With neither the time nor the inclination to make measurements, the shepherd would not have known that the airborne mass of debris was a detached part of a rapidly moving landslide triggered by an earthquake 85 miles offshore from the Peruvian coast. The earthquake, which directly killed 80,000 people, caused part of 22,000-foot Mt. Huascaran to break off and fall into the valley below. This falling rock picked up snow, ice, and glacial debris and amalgamated them into a coherent mass that raced down the valley of the Rio Santa (Santa River). In the valley:

- The slide mass contained enough air trapped in the snow and loose mud that its density was low enough to glide over the ground.
- As the slide moved faster, it trapped air under the front edge and "hydroplaned" down the valley at speeds of about 250

mph, losing 14,000 feet of altitude and covering 9 miles in 3 minutes. In physics terminology, the slide had encountered almost no friction.

- At a sharp curve in the valley, one branch of the slide detached itself, leaped over a ridge (and the previously mentioned sheep), landed nearly intact on the other side, and then plowed into the village of Yungay. Yungay was covered by a pile of debris 5 to 100 feet thick, almost completely destroying the village and all of its inhabitants.
- The main branch of the slide fanned out over the town of Rangrahirca (most of which it obliterated), made a right-angle turn along the Rio Santa, and proceeded for another 30 miles along the valley at a speed of "only" 15 mph.

In the 3 minutes that elapsed between the initial rockfall from Mount Huascaran to the destruction of Yungay and Rangrahirca, an estimated 20,000 people died.

FURTHER READING: Browning (1973).

---

boulders enclosed in finer sediments. A landslide origin of the debris is confirmed if the boulders are similar to rocks in place upslope on the landslide scar, and, if the slide is recent enough, we may even find evidence for the passage of the slide in devastated forests and scoured soils.

The risk from landslides is obvious, for a few seconds, to anyone in the path of one, and it is useful to know how to evaluate whether an area is at risk of future landslides. The most logical precaution is to avoid building in areas of prehistoric or historic landslides, using the time-tested logic that what has happened in the past can happen in the future. In areas where ancient slides are not easily recognized, we must learn to predict the danger by understanding the basic causes of sliding and then see how poor land-use practices and bad engineering decisions exacerbate the risk.

Causes of rapid downslope movement of earth materials

Whether material will move downhill by any of the mechanisms described here depends on whether it is on a

stable or unstable slope. Slope stability is determined primarily by the type of rock under the slope and the angle of the slope. Consolidated, hard rock resists downslope movement better than unconsolidated sand or mud (Fig. 3.26). Also, different materials are stable at different slope angles. Unconsolidated materials typically lie at a characteristic "angle of repose," which is the angle that a pile of a particular size of sediment makes with a horizontal surface (Fig. 3.27). The angle of repose depends on grain size (coarser material makes a steeper pile), angularity (angular grains make a steeper slope than rounded ones), and presence or absence of moisture (wet sand, for example, does not hold as steep a slope as dry sand).

Downslope movement is promoted by any natural process that increases the slope angle, for example, undercutting a bank by stream erosion or adding new sediment to the slope by deposition of windblown sand. A slope that is normally stable may fail catastrophically when conditions that affect the strength or cohesiveness of the material change or a sudden force acts to disturb the slope. In regions without much rain, for example, a sudden rainstorm can initiate slides by infiltration of groundwater beneath steep cliffs of unconsolidated mate-

(a)

(b)

Figure 3.26 Photos showing (a) low slope angle on loose sediment in Death Valley, California, and (b) vertical cliff in resistant rock at Half Dome, Yosemite, California. Courtesy Catherine Flack.

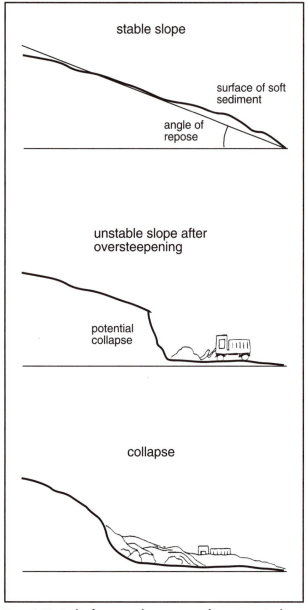

Figure 3.27 Angle of repose and consequence of oversteepening by construction activity. See discussion in text.

turbed areas along coastal bluffs show no evidence of past downslope movement, they are underlain by a sequence of unconsolidated glacial deposits in which an impermeable, slippery-when-wet clay (locally called the "blue clay") lies beneath coarse glacial outwash sands and gravels (Fig. 3.28). Consequently, the attractive waterfront bluffs provide a recipe for slope failure. In heavy rains (which are the norm for western Washington), the rainwater infiltrates the permeable sands and gravels, ponding above the blue clay bed. The blue clay, once wetted, becomes a layer of grease – with predictable and unhappy results along unsupported cliffs and sea bluffs.

Human influences on downslope movement

By now, it should come as no surprise that human intervention has a way of increasing the potential for landslides. Consider again Fig. 3.27, which shows a stable slope in a natural state of equilibrium. It will fail only if more sediment is added to the surface or the lower part is eroded, steepening the slope, or if excessive water infiltrates the slope, lessening the sediment's cohesive-

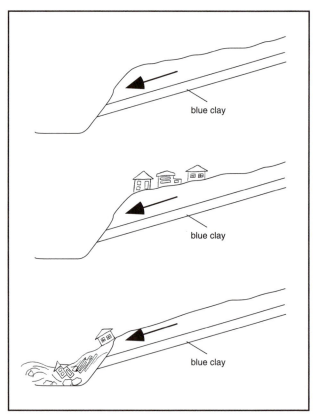

Figure 3.28 Diagram showing slippage on blue clay near Seattle, Washington. See discussion in text.

rial. Sometimes sudden ground movement, typically associated with earthquakes, can set off a slide just as we might knock over a house of cards by tapping on a table.

Inhomogeneities in the rock mass beneath a slope may also play a role in downslope movement. Planar weaknesses in seemingly hard, consolidated rock that parallel a slope allow an otherwise competent rock mass to fail along the planar surface. An example of planar weakness is along the coast of much of Puget Sound in the northwestern United States. Although many seemingly undis-

ness. Human activities can increase the risk of slope failure by playing into these natural means of increasing the potential for slope failure. Examples include:

- oversteepening the slope by undercutting for roads, grading to make level building sites, or regrading, perhaps, to produce additional building lots;
- increasing water infiltration either through the use of sanitary septic systems or by altering the natural runoff and infiltration patterns with landscaping or drainage modifications (this moisture not only lessens the cohesiveness and strength of unconsolidated materials but may cause some clays and silts to behave like liquids);
- overloading an otherwise stable slope with dense fill as a foundation material for roads or structures (this would be like placing a loaded drink cooler on top of a sand castle at the beach).

Many of these human-induced problems can be, and are, avoided by careful zoning restrictions to prevent building in hazardous areas or by strict building codes that legislate maximum slope angles or thicknesses of fill on unstable or unsupported slopes. Expensive engineering modifications after building, such as retaining walls and soil anchors, may be used to counterbalance the weight of the unsupported slope materials or to provide a buttress against the downhill forces. Elaborate and costly slope drainage systems may be required to prevent infiltration or to remove groundwater. Even more elaborate chemical, electrochemical, and thermal treatment of slide-sensitive clays is possible, if we have enough money.

Policy questions

Regardless of where you live, have you or your government spent enough time and money to investigate the earth-movement hazard and to prepare for any difficulties that it will cause? (We ask the same question for Sections 3.3 to 3.8.)

FURTHER READING: Coch (1995).

3.9 WHO PAYS FOR NATURAL DISASTERS?

The introduction in 1971 of insurance cover against landslip and subsidence, on domestic policies in the UK, has put additional pressure on the legal aspects. This reflects the desire to

Table 3.3 *Costs of some recent natural catastrophes (in billions of U.S. dollars, 1992 value)*

	Value	Year
Kobe earthquake, Japan	140	1995
Hurricane Andrew, United States	30	1992
Northridge earthquake, United States	20	1994
midwestern floods, United States	15	1993
Loma Prieta earthquake, United States	10	1989
Hurricane Hugo, United States	6	1989

Note: Estimates for the same event by different people and organizations are highly variable. The values shown here are approximately the median from a variety of sources.

recover losses from others involved in the property, whether as builder, developer, designer or professional advisor.
– *J. C. Doornkamp, in McGregor and Thompson,* Geomorphology

Someone sitting back in a cozy flat in London or an apartment in Des Moines might be thinking, "Hey, tough luck all you beachcombers and mountain dwellers. I'm doing fine here. It's not my problem." Or is it? Human compassion aside, who really pays for all of this damage and how much is the bill? The numbers are hard to come up with because averages are so deceiving. Natural disasters of all kinds probably cost about $50 billion per year worldwide. Perhaps about 40% of this total, or $20 billion, is caused by the geological agents we have been discussing – the rest would be mostly weather-related or the result of biological catastrophes like insect infestations or algal blooms. In the United States during an average year, the toll from geological hazards is about $5 billion in destruction to property and perhaps 1,000 deaths, but it would not take too many Kobe-style earthquakes to shift these averages significantly.

The average figures cited above mask an important aspect of the cost of geologic catastrophes – geological events are independent of arbitrary time divisions like months and years. Thus, there are good years and bad years, even good centuries and bad centuries. We illustrate the costs of individual disasters with just a few recent examples in Table 3.3, which clearly shows that the past few years have been bad years (possibly we are in a bad century). More likely, the worldwide cost of disasters is increasing as the population grows and societies become more complex. Perhaps, we have been lucky. Insurance experts estimate that a Northridge-style earthquake in central Los Angeles would cost $50 billion; in central Tokyo, a staggering $800 billion.

And, speaking of insurance, how *do* we pay for all this carnage? There are really only three options:

1 Individuals pay.
2 Insurance companies pay.
3 Governments pay.

In most of the lesser-developed world, option 1 is the only one available. In the past, say prior to 1945, most people in industrial countries also took option 1 and went on with life after disaster struck. Today, our behavior is more in the realm of what insurers call "risk-averse." This means we would rather accept a high probability of paying a little (insurance premiums) than a low probability of paying a lot (losing everything in an earthquake or coastal storm). The insurance industry used to like this very human trait of risk-aversion because they could easily convince us to part with small amounts of money for premiums in return for, say, sleeping well at night.

In recent years, the insurance industry has been re-thinking its approach. In constant 1992 dollars, in the decade of the 1960s, insurance companies in the United States paid out an annual average of $1 billion for natural disasters. In the 4-year period from 1990 to 1993, they paid out an average of $12 billion. This period includes the worst year in disaster insurance history, 1992, when they paid out $20 billion. The question has been asked, "Are there more disasters or are people insuring themselves more?" The answer is probably yes. Some actuaries who study the risk of natural disaster point to global warming, drought in the Sahel, and other weather-related causes. Others simply point out that for a long time, insurers were simply lucky. All agree that more people are buying insurance. Along the hurricane-sensitive east coastal regions of the United States, for example, insurance companies insure twice the property value in constant dollars as they did a decade ago. How do they do this? In theory, insurance companies operate by pooling risk, which is done by having a large enough portfolio of risk spread over a large enough population that their probability estimates will be sufficiently accurate, and the likelihood is that most of their clients' risk will be independent of each other. Thus, if an insurer has a loss due to floods in Illinois, the premiums will keep rolling in from policyholders in Florida and California.

Aside from just bad luck, insurance companies have two nightmares, moral hazard and adverse selection.

Moral hazard means that once a policyholder has insurance, he or she acts carelessly or behaves badly in some other manner that increases the insurer's risk. Insurance companies handle this one by inspecting properties, setting standards and codes, and generally working to lessen the risk exposure of its policyholders. This is good for the insurer and good for the policyholder, unless the company decides that your house is an unacceptable risk – so old, say, that it will fall down in any earthquake or storm. Adverse selection means that only people who are really vulnerable to risk decide to buy insurance. In the realm of geological hazard, this aversion would mean that only people who live on barrier islands or river floodplains would buy flood insurance, while the only people buying earthquake insurance live in seismically active areas. The end result for the insurer is either: (1) not to write insurance for that hazard, which is bad for policyholders, (2) to raise premiums so high that few can afford them, which is bad for policyholders; or (3) seek government help, which may be bad for taxpayers.

Recently in the United States, government help is the approach taken most frequently by the insurance industry. The government, either by insuring the insurers against catastrophic risk with public funds, known as reinsurance, or by legally setting a limit on net liability, steps in to hold premiums artificially low and to allow high-risk policies to be written. The unintended result of government programs of this kind is that many of the built-in negative incentives that would lead individuals to avoid risk – that is, to be risk-averse – are removed. Where one might think long and hard about building that dream house on a sandy spit projecting into the Atlantic Ocean for fear of losing everything in the next storm, a federally reinsured flood policy with artificially low premiums might just be the incentive to go ahead. The result is that the owner of the house shares the risk of losing it with a host of unwitting taxpayers, which removes any economic disincentive that might keep a person with choices from accepting a risk.

Governments commonly attempt to reduce deliberately risky behavior when they step in with reinsurance guarantees or with government-run insurance of one kind or another. They often insist upon changes in zoning, building codes, and other regulations in a commendable effort to lessen risk and improve the survivability of structures. This process is an efficient means of bringing government expertise in risk analysis and best-engineering practice to bear in rural or out-of-the-way communities. Thus, for example, the U.S. National Flood Insurance Program, run by the Federal Insurance Administration and the Federal Emergency Management Agency (how's that for a bureaucratic potpourri), makes flood insurance available to more than 2.7 million homes and businesses who would otherwise, due to risk, not be able to buy flood insurance.

The National Flood Insurance Program requires that eligible communities develop appropriate floodplain management practices and building codes so that, in the aftermath of a flood disaster, reconstruction follows appropriate guidelines to lessen damage and destruction from future floods. In some cases, this includes *not* providing future insurance to properties that are not in compliance with recommended land-use practices. Critics, however, have pointed to one problem with the requirements for obtaining flood insurance – construction and management codes are set by local communities rather than by the federal government. Thus, if a local community allows inadequate construction in risky areas, the federal government is required to provide insurance anyway – up to $350,000 for premiums of a few hundred dollars per year. Efforts by the Flood Insurance Program to deny insurance have been overturned in local courts, thus placing federal money at risk without federal oversight.

Why does the U.S. government provide inexpensive insurance for houses built next to houses that have been repeatedly damaged by storms and floods and, in some places, on ground where the previous house has been washed away? Simply because the insurance saves government money (also known as taxpayer dollars). The reason is that many people living, or planning to build, in those areas would live there anyway without insurance. Then, when their houses were washed away, the U.S. government would provide relief funds to compensate them for their loss. Consequently, by collecting at least some insurance premiums, the government has a reserve against the costs of disaster and need not take the full cost out of general tax receipts. Some people view this policy of reimbursement as an act of compassion and others as a policy of government welfare. What do you think?

Federal generosity is further exemplified by the payment of nearly $10 billion to the Los Angeles area to compensate for destruction after the Northridge earthquake of 1993. In answer to a question concerning whether the state of California should raise taxes to pay for the damage, a state assemblywoman for Los Angeles stated, "We are taxed to death already. . . . This is a federal problem." Enthusiasm of people in other parts of the United States to pay was not recorded.

Policy questions

Who do you think should pay for disasters? Here are two possibilities:

- Federal or local governments pay for everything out of tax money.
- Individuals and businesses pay for everything, either out of their own funds or from private insurance policies. (In this case, should insurance companies be required to write policies for risks that they do not want to cover?)

FURTHER READING: Griggs and Gilchrist (1983); Alexander (1993); McGregor and Thompson (1995), Palm (1990).

3.10 CONCLUSION

We can learn four lessons from our study of geological hazards and the issue of who pays for disasters? The first lesson, and one that is almost trite, is in the quotation at the start of Section 3.0 – "Civilization exists by geological consent." This shorthand statement nicely summarizes the realization that the forces of nature are immense and often seem capricious. As with any risk in life, we can and should make adjustments to geological hazards. This is best accomplished when we understand the hazards and the forces that produce them, when we insist on vigilance where vigilance is called for, and when we make careful plans to deal with natural disasters and their aftermath.

A second lesson, one for planners and politicians, is that the Truman syndrome, the denial of risk often in the face of undeniable evidence, is a fact of life – it may even be a survival strategy. Educating a populace that does not want to hear bad news is difficult enough. Motivating the Harry Trumans of the world to change the way they live or to invest scarce resources in preventive measures is often a political impossibility. The truth is that there is often but a brief "window of opportunity" for changing behaviors and practices. That window is the aftermath of a geologically induced disaster.

We might call the final two lessons the Ozymandias Effect and the Law of Unintended Results. The Ozymandias Effect alludes to a poem by Shelley in which he describes encountering, half buried in the sands of Egypt, a broken and askew pedestal on which, at some time, a statue must have stood. Engraved on the pedestal are the words:

My name is Ozymandias, king of kings!
Look on my works, ye Mighty, and despair!

Thus it is with all our engineered efforts to hold off the forces of nature. It is inherent to any project whose intent is to lessen or modify the impact of geological processes on human structures that they will always require maintenance, repair, and likely redesign as we learn more about the processes we are encountering. Indeed, some geological forces in some places may well be unopposable.

The Law of Unintended Results is simply the inevitability that as one modifies any natural process, those very modifications will change the process, perhaps in ways that were not foreseen. Confining a river to a straight, concrete channel will change the way the river behaves downstream of the channel. A coastal breakwater to trap sand will deprive other beaches of sand needed to replenish and maintain themselves. In the context of geological hazards, unintended results are only avoided by a full scientific understanding of the geological process being controlled.

That is the reason we have written this chapter.

PROBLEMS

1 How much more powerful (in terms of energy release) is an earthquake of magnitude 5.9 than one of magnitude 4.4?

2 A magma chamber beneath a volcano contains 10 km³ of magma before eruption. The top half of the chamber contains so much water that the eruption produces airfall ash. The bottom half of the chamber contains just enough water to produce glowing clouds (nuées ardentes). If the typical ignimbrite deposited from a nuée ardente is 10 m (approximately 30 feet) thick, how much ground will be covered by ignimbrites?

3 The world insurance industry estimates that about $50 billion will be paid annually in claims for damage from natural disasters in the 1990s. Virtually all of this money will be paid in OECD countries (Section 1.4). If insurance premiums to cover this damage were spread equally among all residents of these countries, how much would each person (including children) pay each year? (See population information in Chapter 1.)

4 A coastal plain rises from the shoreline at a slope of 0.05%. If a hurricane causes a high tide of 15 feet above normal, how far might the water penetrate inland?

5 A house builder wants to place a house on the property shown in the accompanying illustration, where the ground slopes at an angle of 10°. If the angle of repose is 20°, how much level surface can the builder develop by bulldozing into the slope? (Note – cotangent 10 = 5.67; cotangent 20 = 2.75.)

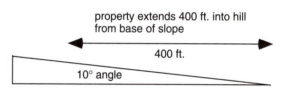

6 Assume that an earthquake of magnitude 7 does 32 times as much damage as one of magnitude 6 (because it releases 30 times as much energy) and that earthquakes of different magnitude occur with the frequencies shown in Table 3.2. If an insurance company decides that insurance for quakes of magnitude 6 will require annual premiums of $2,000 for a $100,000 house, what should the premiums be if the insurance is to cover damage from quakes of magnitude 7?

7 Flow records are available for the past 87 years for a small stream in the steep-sided valley shown in the accompanying illustration. On seven occasions the discharge has reached 74 cubic feet per second, causing flooding up to 4 feet above the normal river level. Use the equation on p. 102 to calculate the recurrence interval for floods of 74 cubic feet per second.

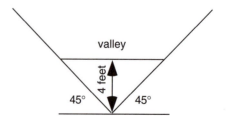

Calculate the height that would be reached by floods with discharges of 148 cubic feet per second.

8 Assume that a hurricane cut a large channel into the beach at the point marked A in the accompanying figure. If the longshore current continued to operate after the hurricane, roughly sketch what you think would be the configuration of the beach a few years later.

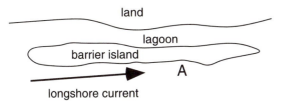

REFERENCES

Alexander, D. (1993). *Natural Disasters*. New York: Chapman and Hall.

Baldwin, J. E., II, and Sitar, N., eds. (1991). *The Loma Prieta Earthquake: Engineering Geologic Perspectives*. Sudbury, Mass.: Association of Engineering Geologists Special Paper 1.

Ben-Menahem, A. (1975). Source parameters of the Siberian explosion of June 30, 1908, from analysis and synthesis of seismic signals at four stations. *Physics of the Earth and Planetary Interiors* 11: 1–35.

Beven, K., and Carling, P., eds. (1989). *Floods – Hydrological, Sedimentological and Geomorphological Implications*. Chichester: John Wiley and Sons.

Bird, E. C. F. (1985). *Coastline Changes – A Global Review*. Chichester: John Wiley and Sons.

Bolt, B. A. (1993). *Earthquakes*. New York: W. H. Freeman.

Bridges, E. M. (1990). *World Geomorphology*. Cambridge: Cambridge University Press.

Browning, J. M. (1973). Catastrophic rock slide, Mount Huascaran, north-central Peru, May 31, 1970. *American Association Petroleum Geologists Bulletin* 57: 1335–41.

Chyba, C. F., Thomas, P. J., and Zahnle, K. J. (1993). The 1908 Tunguska explosion: Atmospheric disruption of a stony asteroid. *Nature* 361: 40–4.

Coates, D. R., and Vitek, J. D., eds. (1978). *Thresholds in Geomorphology*. London: George Allen and Unwin.

Coch, N. K. (1995). *Geohazards*. Englewood Cliffs, NJ: Prentice-Hall.

Degen, P., and Degen, C. (1984). *The Johnstown Flood of 1889*. Philadelphia: Eastern Acorn Press.

Doornkamp, J. C. (1995). Legislation, policy and insurance aspects of landslip and "subsidence" in Great Britain. In D. F. M. McGregor and D. A. Thompson, eds., *Geomorphology and Land Management in a Changing Environment*, pp. 35–49. Chichester: John Wiley and Sons.

Dudley, W. C. (1988). *Tsunami!* Honolulu: University of Hawaii Press.

Eiby, G. A. (1980). *Earthquakes*. New York: Van Nostrand Reinhold.

Eumorphopolous, L. (1963). *Geographica Helvetica* 18: 269–75.

Francis, P. (1993). *Volcanoes – A Planetary Perspective*. Oxford: Oxford University Press.

Griggs, G. B., and Gilchrist, J. A. (1983). *Geological Hazards, Resources, and Environmental Planning*. Belmont, Calif.: Wadsworth.

Hardy, D. A., Keller, J., Galanopoulos, V. P., Flemming, N. C., and Druitt, T. H., eds. (1989). *Thera and the Aegean World*. 3 vols. London: Thera Foundation.

Iacopi, R. (1971). *Earthquake Country – How, Why and Where Earthquakes Strike in California*. Menlo Park, Calif.: Lane Books.

Kaufman, W., and Pilkey, O. H. (1983). *The Beaches Are Moving: The Drowning of America's Shorelines*. Durham, N.C.: Duke University Press.

Keller, W. (1956). *The Bible As History*. New York: William Morrow.

Kolars, J. E., and Mitchell, W. A. (1991). *The Euphrates River and the Southeast Anatolia Project*. Carbondale, Ill.: Southern Illinois University Press.

Kovacs, M. G., trans. (1985). *The Epic of Gilgamesh*. Stanford, Calif.: Stanford University Press.

Kraft, J. C., Aschenbrenner, S. E., and Rapp, G., Jr. (1977). Paleogeographic reconstruction of coastal Aegean Archaeological sites. *Science* 195: 941–7.

Leighton, F. B. (1978). Bluebird Canyon landslide, Laguna Beach – A geomorphic threshold event. In D. R. Coates and J. D. Vitek, eds., *Thresholds in Geomorphology*, pp. 387–400. London: George Allen and Unwin.

Leopold, L. B. (1994). *A View of the River*. Cambridge, Mass.: Harvard University Press.

Linthicum, R. (ca. 1906). *The San Francisco Earthquake Horror*. Chicago: Hubert Russell.

McGregor, D. F. M., and Thompson, D. A., eds. (1995). *Geomorphology and Land Management in a Changing Environment*. Chichester: John Wiley and Sons.

Myles, D. (1985). *The Great Waves*. New York: McGraw-Hill.

O'Conner, R. (1957). *Johnstown: The Day the Dam Broke*. New York: J. B. Lippincott.

Palm, R. (1990). *Natural Hazards: An Integrative Framework for Research and Planning*. Baltimore: Johns Hopkins University Press.

Rogers, J. J. W. (1993). *A History of the Earth*. Cambridge: Cambridge University Press.

Smith, K. (1996). *Environmental Hazards*. 2nd ed. London: Routledge.

Thomas, G., and Witts, M. M. (1969). *The Day the World Ended*. New York: Stein and Day.

Toniolo, E., and Hill, R. (1974). *The Opening of the Nile Basin*. London: Hurst.

Van Andel, T. H. (1994). *New Views on an Old Planet – A History of Global Change*. Cambridge: Cambridge University Press.

Walsh, R. P. D., Davies, H. R. J., and Musa, S. B. (1994). Flood frequency and impacts at Khartoum since the early nineteenth century (Sudanese capital). *Geographical Journal* 160(3): 266–79.

4 WATER

4.0 INTRODUCTION

We must play the hand nature dealt us – we must learn to use wisely the water we have, and not think we can go to some other country or some other region and bring in new supplies when we already waste enormous amounts of water.
– Stewart Udall, as quoted by J. Waterbury, Hydropolitics of the Nile Valley, *p. 116*

There is nothing like a drink of ice water to satisfy your thirst. But here are a few questions to ponder next time you have a drink. Where does that water come from? We don't mean from the faucet. Where is it before it enters the pipes of your local or municipal water system? If it originated as rain (and at some point it undoubtedly did), when and where did it fall? Are you sure that this water is safe to drink? Is there enough of it left wherever it came from for you to count on a drink tomorrow, next year, in fifty years? Are you the first person to drink this water? And, while we're asking impertinent questions, where does the water go that runs down the drain?

People living in the 20th century find these questions far more difficult to answer than our ancestors would have. Our great-grandparents likely drew their water from a well that their parents dug behind the house, but now, as we discussed in Sections 1.6 and 2.3, industrialization and urbanization have made us lose immediate contact with the land. In the United States, for example, fewer than 2% of Americans now live on farms compared to 20% only 50 years ago, despite more than a doubling of total agricultural productivity during that time. We have become street-smart and urbane but unaware of the natural world, and issues like the sources of the water that we depend on are assumed and mysterious. And be-

cause we are out of touch with nature, we place many new, conflicting, even excessive demands on water resources. As this chapter describes, we are quite capable of overusing and degrading the water we need.

This is the first of three chapters considering earth resources – Chapter 5 will focus on energy and Chapter 6 on mineral resources. A natural resource is any material found in nature that is held in reserve for future use. This is a nice way to think of resources since it suggests that the totals of such resources are finite and that we are only the current husbanders of resources, which we hold in reserve for future generations. This chapter is about a single resource, water, the one resource truly essential for survival. Why essential? You can go weeks without food, a lifetime without titanium or natural gas, but less than a few days (depending on the temperature and your activity level) without water (Section 2.1). The average adult must have several liters (quarts) of potable water every day just to maintain the essential chemical balances in tissue and blood, to control body temperature, and to allow elimination of toxic wastes from metabolism. We can make a list of all the things people do with water: drink it; cook with it; wash with it and then use it to carry away our wastes; irrigate crops with it and water livestock; generate power and do work with it; transport heat and cool the human environment; travel on it and use it to transport goods, especially heavy items like crude oil and stone; play in it and on it; gain pleasure and solace in its beauty and grace.

Even when water is inappropriate for many of the uses listed here, usually because it is too salty or is solid (ice), water still makes life possible. It moderates the climate, sculpts the landscape into pleasing forms, acts as the universal solvent in converting rocks into soils and trans-porting vital nutrients to plants, is balm to the soul and inspiration to our creative spirit, and provides homes to fish, fowl, and myriads of invertebrates and plants upon whom we are utterly dependent for survival on earth. Al-though, as we saw in Chapter 3, people have a long, close, and sometimes fatal relationship with bodies of water, we are, to put it simply, at the mercy of an abun-dant and available supply of water.

This chapter is organized in six sections. First we do a bit of bean-counting to see where water is (4.1) and how we use it (4.2). Then, from a geological perspective, we discuss where water is found and how it behaves, looking in detail at the two major freshwater repositories – the surface water in the network of interconnected lakes, streams, and rivers (4.3) and the groundwater in rocks and soil beneath us (4.4). We also discuss the quality of water that we need for various uses (4.5), and, finally, we look at how society structures laws and institutions to ap-portion this critical resource (4.6). As you might imag-ine, anything this important, this critical to survival, car-ries around an enormous amount of economic, social, and political baggage. Throughout, we try to emphasize that people have it in their power to use water wisely or carelessly.

4.1 ABUNDANCE OF WATER

Water, water everywhere, nor any drop to drink. – *Samuel Taylor Coleridge,* **Rime of the Ancient Mariner**

How much water is there on the earth? In order to an-swer that question we first examine the global availabil-ity of all water, then explain the concept of a water "bud-get," and finally discuss regional distribution of fresh water.

Global considerations

In a world of limited resources you may be surprised that, at the global level, there is no absolute shortage of water on the earth's surface. About 71% of the earth's surface is covered by seawater, another 3% by ice, and enough wa-ter to cover the entire earth to a depth of 1 inch is held in an atmosphere that contains only 0.001% of the earth's water (Section 8.1). Our concerns about water as a resource are obviously not based on a global insuffi-ciency, but unless we want to desalinate salt water (which is being done in some areas), or melt ice (by people who live in the Arctic and a few Antarctic explorers), or con-

Table 4.1 *Water on the earth's surface*

Reservoir	Volume (in 1,000 cu km)	% of total
Atmosphere	13	0.001
Oceans	1,320,000	97.54
Glaciers and ice caps	29,200	
Salt lakes	104	0.007
Freshwater lakes[a]	125	0.009
Rivers	1	0.000002
Soil moisture	40	0.001
Groundwater[b]	8,350	0.63

[a]Includes reservoirs.
[b]Shallower than 4 km.
Sources: van der Leeden, Troise, and Todd (1990).

dense dew from the atmosphere (also done by a few hardy souls), we must draw our water supplies from fresh water on land.

Looking somewhat more closely at fresh water, we show the location of water on the earth's surface by vol-ume in Table 4.1. It shows that only 0.644% of the earth's upper crustal and surface water is in the form of liquid, fresh water, and 99% of that is in the ground. That is still a lot of water, a total of 2.25×10^{18} gallons (8.5 million km[3)], about 400 million gallons for each of us. Just the amount in rivers and freshwater lakes alone is 6 million gallons for every adult and child on earth. It's enough water to submerge the United States beneath a little more than 40 feet (12 m) of water and all of Europe under about 16 feet. Knowing all of this, however, is like knowing how much gold is in Fort Knox. It's reassuring to know it's there, but it doesn't help pay the rent unless we can take it out and use it.

Taking the fresh water out poses a problem. It occurs either as standing water (wetlands and swamps, lakes and ponds), as flowing water (rivers and streams), or in the soil and upper crust as groundwater, but neither the water nor the people who need it are uniformly nor rationally distributed on the 29% of the earth's surface that is dry land. Too many people live in arid regions such as south-ern California or the Middle East. One million people live in the greater Phoenix, Arizona, metropolitan area, which gets less rainfall per year (7.66 inches) than Dam-ascus, Tripoli, or Tehran. At the other end of the spec-trum, people avoid places with lots of rainfall, such as Waialeale, Kaua'i, or the Isle of Skye. The former gets a whopping 444 inches of rain a year, and the latter has so much water vapor in the air that it is covered by clouds most of the time. We need to examine this freshwater resource in greater detail to understand where it is and how it moves from place to place.

The (mostly) freshwater budget

Tallying water by volume seems to suggest that water is fixed in its various reservoirs, such as the ocean or atmosphere. We know this isn't true simply because it rains on some days and not on others. Also water moves. It flows from underground springs, sinks into the soil, flows from streams to lakes to rivers to the ocean, evaporates from pavement surfaces after a hard rain. Lawyers, as we will see, call such a resource "fugitive" because it is capable of moving across property lines and jurisdictional borders.

The total volume of this fugitive resource, 2,250 quadrillion gallons (Table 4.1), is relatively constant, but the amount of water that moves through the system on an annual basis, the "annual water flux," is very large. For example, about 106,250 cu km of water falls on the land

surface each year, but Table 4.1 shows that only a little more than 125,000 cu km of water resides in freshwater lakes and rivers at any time. This means that the equivalent of 85% of the fresh water contained in catchments on land falls onto the land in any given year, and unless some water moves off of the land, in 2 years new catchments would have to store an additional 90,000 cu km of water (2 × 106,250 −125,000). Of course this is not necessary, because some of the water that falls on land evaporates back into the air, some runs off into the oceans, and some soaks into the ground. Because the volume of fresh water is constant as it moves from one part of the earth to another, the process is cyclical, and we describe it as a hydrologic cycle, or freshwater budget (to clarify this concept and the later concept of residence time we use the analogy of household budgets in Box 4.1).

We start our description of the hydrologic cycle (Fig. 4.1) with the only renewable source of fresh water on the earth, the water vapor put into the atmosphere by oceanic evaporation. Evaporation is a distillation process

---

BOX 4.1 WATER BUDGETS AND PERSONAL BUDGETS

We spend a lot of time in this chapter discussing water reservoirs (lakes, the ocean, etc.) and rates of movement between them. In order to clarify these concepts, we offer an analogy based on a subject that you are probably very familiar with — your bank account.

Think of your bank account as a reservoir, or "catchment" in the terminology of hydrology, that you put money into and take money out of. If you are financially stable, you receive income from a variety of sources (wages, scholarships, parental subsidy, interest earned) and this equals your expenses (tuition, food, room, clothes, car payments, savings, interest paid, this book). From year to year, your net worth (all your cash and other assets) may remain unchanged and disappointingly small, but the total resource (dollars) that moves through your checking account can be surprisingly large. If, for example, the average daily balance in your checking account was $800 last year but your income was, perhaps, $12,000, on some days, you might have had $2,500 in your account; on others, $50. Water is moving through various catchments on the earth's surface just as money is moving through your checking account.

Because the total amount of fresh water on the earth is virtually constant through time, its rates of movement and "residence times" in reservoirs can be likened to transfers among your total assets. Suppose your net worth, everything you own — cash, checking accounts, savings certificates, your car — is $10,000. As a student, you are lucky to be holding your own, so last year you also had a net worth of $10,000 as well. During the year, you spent all your income, $12,000. Obviously, money is moving through this system, but if you look at individual dollars, you will find they move through different parts of your budget system at different rates. For example, a dollar invested in a savings certificate or in your car hasn't "moved." We would say that it has a residence time in your savings or automobile equity of greater than 1 year. On the other hand, if you get a weekly wage of $150 on Friday and have it all spent by next Friday, then we would say that a dollar in cash has a residence time in your pocket of less than a week.

We hope these analogs are useful as you read our description of hydrologic processes.

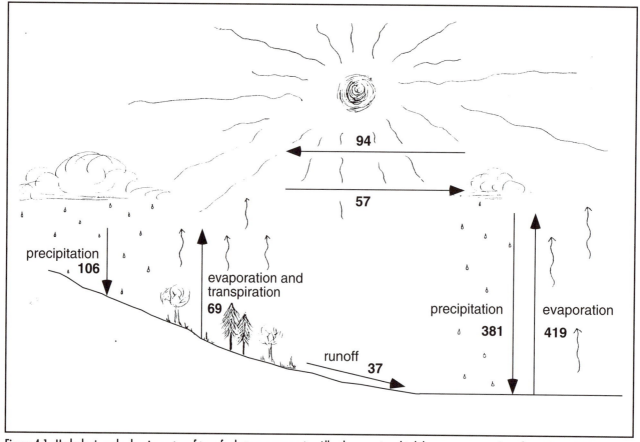

Figure 4.1 Hydrologic cycle showing rates of transfer between reservoirs. All values are in cubic kilometer per year. Data from Moore (1989).

that leaves the ocean's salt behind and puts only freshwater vapor, which we call humidity, into the atmosphere. Humidity is reported either as absolute humidity, the total amount of water in a mass of air (in units like grams of water per cubic meter of air), or relative humidity, expressed as a percentage of the total amount of water that an air mass could contain if saturated (that is, if it could hold no more water). Air that feels dry is undersaturated for its temperature, perhaps with a relative humidity of 30%, and heavy, damp air may have a relative humidity of 90% or more and be close to saturation. The amount of water that air can hold at saturation, the absolute humidity at 100% relative humidity, is dependent on temperature (Figure 4.2). A cubic meter of air at 35°C (let's say on a very hot, steamy summer day) can contain almost 40 g of water (that is pretty close to three tablespoons in a large suitcase of air), while the same cubic meter of air at 10°C (a crisp autumn afternoon) can hold less than 10 g of water.

As air cools, it approaches its saturation value, and when it is saturated its contained moisture condenses into droplets of liquid or crystals of ice depending on temperature. If you had poured a glass of ice water when you began this chapter, it would, by now, have left a ring of condensed moisture on your desk top as it cooled the air around it to 0°C at the glass–air interface. Take the glass back to the kitchen, and by the time you finish this chapter the water droplets on your desk will have evaporated back into the warm and water-undersaturated air that slowly moves across your desk top.

The processes of evaporation and condensation on the earth are analogous to what happened to your cold glass. As we will see in detail in Section 8.1, sunlight penetrates the atmosphere, warming the earth's surface and causing the air to be warmest just at, or immediately above, the ground or water surface. As the air warms up, it becomes progressively more water-undersaturated and extracts water from the surface in the same way that the draft in your room evaporates the ring of water from your desk top. The diurnal (daily) change in temperature causes changes in humidity driven by evaporation during the day and condensation during the night (Fig. 4.3). By itself the diurnal cycle of evaporation and condensation leads to no net removal of water from a body of surface water, and transfer of the moisture requires a second process. That process results from the fact that warm air, even laden with moisture, is less dense than cool air and consequently rises from the ocean or a lake as a plume of warm air while cold, dense, and water-undersaturated air descends to the surface. Depending on the time of day, this descending, cold dry air may warm, expand, rise, and thereby transfer evaporative water to upper levels of the atmosphere.

The moisture-laden air undergoes two changes as it rises in the atmosphere. One is an expansion resulting from the upward decrease in atmospheric pressure, which is simply the weight of the overlying column of air (for reference, pressure at 18,000 feet is one-half that at sea level), and we commonly call this expanded air "thinner." The second change is cooling caused by the expansion (see the discussion of air conditioners in Section 8.4). The expanding and cooling air eventually becomes water-saturated, and its moisture condenses to form clouds or precipitation. More than 75% (326,000 km³) of the almost 420,000 km³ of water that evaporates from the oceans each year simply falls back into them as precipitation, forming a closed loop of evaporation and precipitation that is critical to climate and the thermal balance of the earth but adds nothing to the fresh water on land. The 94,000 km³ of the water evaporating from the world's oceans that is carried aloft and blown by strato-

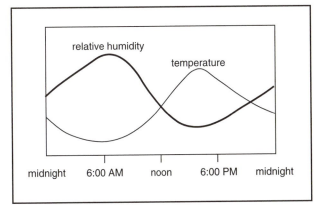

Figure 4.3 Typical diurnal (daily) change in relative humidity and temperature.

spheric winds over the continents falls to the surface where we can use it. A similar evaporative process above lakes, rivers, and wet soils, plus emission from the leaves of plants (transpiration) puts about 69,000 km³ of water into the air each year from the land surface, with about 12% of that returning to the land as precipitation and the rest falling into the oceans. These transfers and the fresh water that flows in rivers to the sea close the loop and account for all of the water in the freshwater budget.

Using numbers in the preceding paragraph plus Table 4.1 and Figure 4.1, we can now determine both the total amount of water available for human consumption and the rate at which it becomes available. The total evaporation is 488,000 km³ (419,000 + 69,000), with precipitation on land of approximately 106,000 km³. About one-third of the precipitation (7.6% of the total evaporation) flows into rivers and lakes and is directly available for consumption, but the rest can be used only if we trap it before it soaks into the ground, evaporates, is taken up by plants, or runs off the surface to the sea. The water that infiltrates the ground is referred to as groundwater, and, although it is not shown in Figure 4.2, on a global basis it appears that as much fresh water enters the groundwater system by percolation of surface water as leaves the groundwater system through springs and unseen feeders to lakes, rivers, and the oceans. In other words, this is a vast resource in a state of dynamic equilibrium, and we discuss it in Section 4.4.

We can further clarify the rates of transfer between reservoirs by calculating "residence times," which are the periods of time that a volume of water spends in each catchment (Table 4.2; for further definition, see Section 8.1). Atmospheric water cycles every 9 days, which means that on average water that evaporates from the

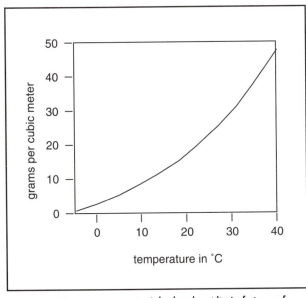

Figure 4.2 Water-vapor capacity (absolute humidity) of air as a function of temperature.

Table 4.2 *Global residence times of water*

Reservoir	Residence time
Atmosphere	9 days
Oceans	2,500 years
Glaciers and ice caps	9,700 years
Freshwater lakes	17 years
Rivers	16 days
Soil moisture	1 year
Groundwater	1,400 years

Sources: McDonald and Kay (1988, table 2.4).

earth's surface returns as precipitation in 9 days. This means that if we rely on a steady source of atmospheric water, we can expect rapid replenishment of the resource since water spends such a short time in the atmosphere. On the other hand, the average residence time for groundwater is more than 1,000 years, and replenishment may require 100 years or more. Water held in glaciers is also an important resource, but the residence time of water in polar ice is on the order of 10,000 years, and the only other time glacial ice enters the discussion of freshwater resources is when some visionary decides that we should tow icebergs from high latitudes to some water-needy municipality like San Diego or Kuwait City to provide abundant, cheap fresh water. The economics of this never seem to work (although one study in 1972 suggested that, at that time, the cost of harvesting icebergs and towing them to Australia or Chile was about 1% of the cost of a comparable desalination plant).

Regional considerations of freshwater availability

Thus far we have discussed only global averages, which are good for understanding processes but carry little consolation to individuals wrestling with specifics. Fresh water is not evenly distributed about the globe, and there are people in West Australia or Namibia to whom discussions of average annual precipitation give little solace. Many of these inequalities in water availability have local geological and climatic explanations, but a few are global in the sense that they derive from worldwide climate patterns that fundamentally control and limit the water availability. We discuss these patterns first and refer to some local issues later in this section.

The most important control of water availability on the time scale of human requirements is the relative humidity of the air at the surface (Fig. 4.4). A plot of humidity versus latitude shows a wavelike shape, with humidities high at the equator, where the earth is warmest, lowest at 30° north and south of the equator, and then high at 60° north and south. This curve results from the general decrease in air temperature away from the equator and the lower density of hot air than cold air, which combine to create three major convective movements, named Hadley cells for their discoverer, the 18th-century meteorologist George Hadley (Fig. 4.5).

One Hadley cell is caused by cold polar air moving under warm air away from the poles. In the Northern Hemisphere, the southward-moving air causes the development of "polar fronts," which break out southward as blasts of Arctic air during the winter. A second convection cell develops when hot air rising at the equator cools, loses its moisture as rain condenses from the cooler air, and moves toward higher latitudes. The now dry and cooler air moves north of the equator in the Northern Hemisphere and south of the equator in the Southern. It eventually becomes dense enough by cooling to descend at about 30° to form what meteorologists call the subtropical convergence. It is called a convergence because the circumference of the earth is much smaller at 30° than at the equator, and consequently the air moving to higher latitudes is pushed into a smaller area. (An analogy would be to place your fingers on adjacent meridians at the equator of a globe, slide them north keeping each finger on its meridian, and watch them converge, even bunch up, as they move north.) As the air converges at 30° north and south, it produces a thick wedge of cold and dry air, a high-pressure zone where the air must descend. As the air descends it heats up by compression,

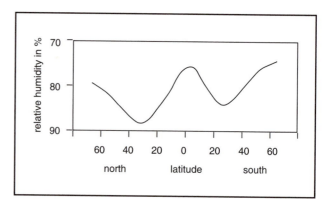

Figure 4.4 Latitudinal variations in relative humidity.

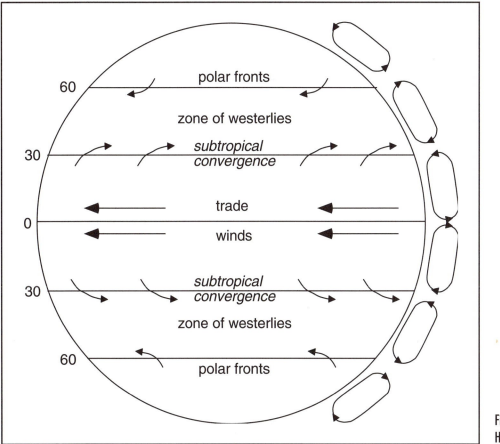

Figure 4.5 Distribution of Hadley cells on the earth.

and if the convergence is over the land, this hot dry air has an effect like blasting a potted plant day and night with a hair dryer. Between the subtropical convergence and the polar front lies a third convective cell dominated by prevailing westerly winds (blowing eastward).

Figure 4.6 demonstrates the global impact of the alternating Hadley cells on evaporation and precipitation by latitude. A surplus of precipitation over evaporation occurs from the equator to 20° north and south – the tropics – and a small surplus north and south of 40°. Conversely, from 20° to about 40° north and south, the subtropical convergences, evaporation exceeds precipitation and creates deserts (Fig. 4.7). The Sahara, Arabian, Gobi, Sonoran deserts in the Northern Hemisphere and the Atacama, Namibia–Kalahari, and western Australia deserts in the Southern faithfully line up along the subtropical convergences. The people who live there can do nothing about it, for the deserts beneath the subtropical convergences are built into the earth system.

FURTHER READING: Mather (1984).

4.2 HUMAN USES OF WATER

There are plenty of people . . . who approach land, water, grass, timber, mineral resources, and scenery as grave robbers might approach the temple of a pharaoh. – Wallace Stegner

We now know where fresh water is found and how it moves, and before we go on to details of the surface freshwater system we have to consider where people, in particular large groups of people, get the water they need and what they actually do with it. In this section we confine our discussion to "offstream" use, which relies on the removal of water from its source and subsequent transport to its place of use by bucket, ditch, pipe, canal, tanker, or cupped hands. "Instream" use includes hydroelectric power generation, which we discuss in Section 5.1. Trans-

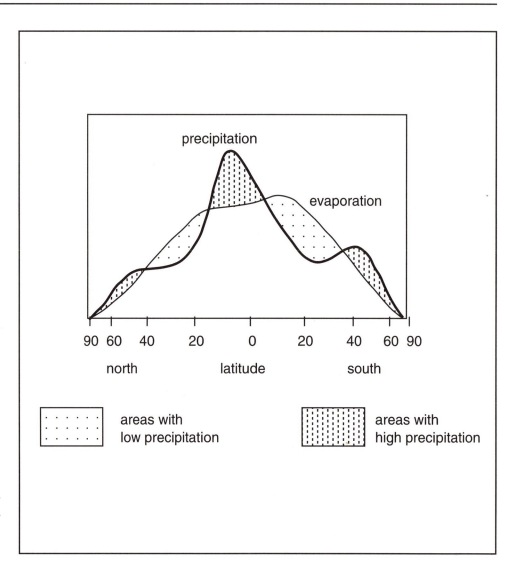

Figure 4.6 Latitudinal variation in evaporation and precipitation, showing areas of precipitation deficit and precipitation surplus.

portation and recreation are beyond the scope of this book, but we discuss one transportation issue in Box 4.2.

Where do we get water?

There are really only a very limited number of ways to get significant amounts of offstream fresh water. They include:

- direct capture of precipitation;
- desalination of saline water, most often seawater;
- recycling (reclamation) of used water (this category includes water that is put back into the water system but does not include treated waste water that is put back into streams and the ground);
- diversion of water in streams, rivers, and lakes, called offstream surface water;

- extraction of groundwater, commonly by the drilling of wells and pumping, but also by diversion of groundwater-fed springs.

We discuss the first three methods first.

Desalination and direct capture of precipitation are insignificant for somewhat different reasons. Desalination is so expensive that it is used only where there are no other options, and although capturing precipitation is cheap, it is only done on a small scale and rarely provides a sufficiently steady supply of water that people can depend on it as their sole source. In many rural communities and isolated farms, particularly in the 1800s, rain barrels and cisterns were used primarily to capture water for washing purposes. This was desirable because in many parts of the world groundwater is "hard," meaning that it contains low levels of dissolved salts, typically calcium

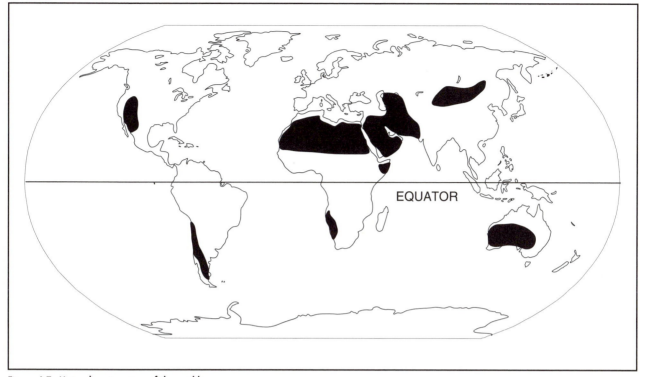

Figure 4.7 Major desert regions of the world.

carbonate (CaCO$_3$) or caclium sulfate (CaSO$_4$), which interfere with the cleaning properties of natural soaps (further discussion in Section 4.5).

Reclaimed water is the single fastest growing source of water in most developed nations, but it is still a minuscule fraction of water consumed. In the United States, for example, offstream and groundwater use declined between 1980 to 1990 while reclaimed water increased by 50% (Table 4.3). In spite of that increase, reclaimed water still only accounts for 0.02% of water use and, at its current growth rate, would not account for even 1% of total U.S. water consumption for almost 100 years.

So, where do people get their water? It is either from groundwater, most often wells though occasionally from springs, or from diverted surface runoff in lakes and rivers. Table 4.3 shows that in 1990 an average of three-quarters of all offstream fresh water in the United States came from surface waters, but the range of variation is from a high of 98.2% surface water in the upper Colorado River basin (southwestern Wyoming and the Four Corners region of western Colorado and New Mexico and eastern Utah and Arizona) to a low of 51.2% in the Arkansas–Red River basin (southeastern Colorado, Oklahoma, the Texas Panhandle, southern Kansas, and western Arkansas). We will discuss these two areas, the

high surface and groundwater consumers in the United States, in greater detail. Worldwide, the importance of surface water is even more pronounced than it is in the United States, and few nonindustrial nations use groundwater at the level of the United States. Large-scale extraction of groundwater generally uses more energy, money, and technology than people have (Box 4.3), so most people rely on surface water even though it is a very imperfectly distributed resource.

No matter where people live, water-rich Canada or water-poor Australia, the places where most people live are cities, and all cities import water. A simple calculation will demonstrate this. The London metropolitan area is home to about 9 million people. Data on water consumption in the industrialized world indicate that the average person uses about 10 gallons of water per day for domestic uses. This means that the London metropolitan area would require 90 million gallons of water daily just to meet the everyday needs of each Londoner. Typically, the total public water supply need for industrial, utilities, and commercial uses is at least 10 times the domestic need. Let's say, therefore, that London needs a minimum of 900 million gallons of water every day. If we consider the size of metropolitan London and capture all the rain that fell over London each year, we would still be short

BOX 4.2 THE HIDROVIA PROJECT

The waterways of the Hidrovia Project will strike like an arrow for 2,000 miles into the heart of South America (Fig. 4.8). Starting from the La Plata estuary, which passes by Montevideo and Buenos Aires, it will follow the Parana River inland to its junction with the Paraguay River and on nearly to the town of Caceres, Brazil. The purpose is to create an inland waterway capable of passage by large oceangoing vessels and, in the hopes of its proponents, bring economic boon to the economically distressed interior of South America. If it becomes fully operational, the river system will affect the lives of 200 million people who live in an area of approximately 12 million sq km. The project, which will not start until 1997 at the earliest, is supported both politically and financially by the governments of all five countries that border the river and the La Plata estuary — Argentina, Bolivia, Brazil, Paraguay, and Uruguay. It is also partly supported by European companies, which expect to recover their investment by charging tolls to ships that use the waterway.

In order to convert the river system so that ships can use it, the river channels must be greatly modified. Both rivers now move sluggishly across very flat land, passing through numerous swamps and meandering past reed-covered banks even where the land is not swampy. Consequently, modification to a channel at least 150 feet wide and 150 feet deep will require a lot of dredging, diversion of the rivers into different channels in several places, and construction of some dams that will contain locks large enough to handle the large ships. At the head of the system, the channel dredging will pass the Pantanal, one of the world's largest wetlands, and plans are being considered to continue dredging canals that will connect across the Amazon River and link the Hidrovia Project with the drainage of the Orinoco River.

The issue raised by the Hidrovia Project is, as usual, considerably more complex than the traditional conflict between development and environmental preservation. First, we can agree that development is good, but whose development? The interior of South America will supply iron ore, construction materials, and lumber that can be transported down the waterway, but who will profit from their exploitation? Will the development bring a higher standard of living to the poor people who already live in the area, or will they work at low wages so that profits can be made by people already wealthy, including investors in foreign companies underwriting part of the cost? And how much will modern industry affect the lives of people who subsist on

Figure 4.8 Location of the Hidrovia Project.

fishing and modest agriculture along the banks of the present winding streams?

Second, we all agree that environmental protection is good, but whose environment? Should the wetlands that provide a home to animals and plants not found anywhere else on earth be preserved while people who live around them need jobs? How important is it to preserve them? Can a bargain be arranged in which the wetlands and way of life for some local people are preserved at the same time as the channel is cut to provide better incomes for people who need them? Possibly some arrangements can be made, and possibly not. Because this is an issue that dominates so much of worldwide discussion, we discuss it more completely in Section 8.3.

FURTHER READING: Heath (1995).

Table 4.3 *Major sources of offstream freshwater in the United States*

Source of water	% change 1980 to 1990	1990 % of total water withdrawn
Surface water	−10.7	76.4
Groundwater	−4.3	23.4
Reclaimed water	50.0	0.02

Sources: Selley et al. (1993, table 31).

about 300 million gallons per day. In other words, London must import much of its water, and it does so from throughout the Thames basin and from other nearby river basins, including the Trent, Severn, and Ouse.

Even in the best of times, surface water is a quirky resource. In a 1986 study, the World Resources Institute estimated that of the total 41,000 cu km of surface runoff from continents each year – this might be called the global runoff – only about 9,000 cu km (or 22%) is sufficiently stable in the sense of remaining constant through annual or longer variations in precipitation to justify the financial investment in large-scale water diversions. And this 9,000 cu km is so inequitably distributed that water-rich nations may literally have thousands of times as much water per capita as water-poor nations (Table 4.4). In general, the Western Hemisphere is awash in water, while Africa and southern Asia are impoverished and parched.

We now have a sense of how much water there is and how it is distributed. Before considering what geology can teach us about surface and groundwater – our two largest freshwater resources – we need some idea of the nature of the worldwide water demand.

What do we do with water?

How much water do we need? That depends on who we are. As noted, at the fundamental level each of us needs one-half gallon (2 liters) of water every day just to stay alive. If we were active in a hot, dry climate, the daily requirement is as great as 5 gallons (20 liters). Often, how much water a person uses depends on how easy it is to get. A hiker who has to carry water (at 8 pounds to the gallon) probably drinks less than an equally hot, but certainly less active, rider on a train who can simply walk to the water cooler at any time. A study of water consumption in the Oyo state of Nigeria in 1980 found that fami-

lies who had to walk several kilometers a day to obtain water used 5 to 6 gallons a day for all domestic purposes, while those living in nearby, small villages with piped-in water consumed an average of 20 to 25 gallons per day.

Water is not used only for drinking. If it were, most of us would use only about 1 or 2 gallons per day, but, as we discussed, the average person worldwide uses 10 gallons per day (almost 40 liters) for domestic purposes alone, and in a modern urban area such as London the total need for municipal water is typically as much as 10 times the domestic need, or 100 gallons per day. In 1990 in the United States, the daily per capita offstream freshwater use was about 1,360 gallons per day, more than 600 times the amount needed to sustain life. We had better take a careful look at who is using all of this water.

Table 4.5 shows the principal offstream users of water in the United States in 1990. The worldwide figure of 100 gallons a day for municipal use is less than U.S. municipal needs by about 70%, and most of the major water users are not municipal. They are principally industrial and agricultural consumers and the power industry, which uses almost as much water as farmers, but the figures in Table 4.5 hide important differences in total consumption. Less than 3% (14 gallons per day) of the water withdrawn by power generators is actually consumed by evaporation in the cooling process, and the remainder is returned unaltered, though usually warmer, to the surface water system for other uses. By contrast, irrigation returns only a fraction of the water withdrawn, and that fraction may be contaminated with inorganic salts and chemicals like pesticides, herbicides, and fertilizers (Sections 7.3 and 7.5). Almost all other water uses are also consumptive in the sense that water is either evaporated, transpired, incorporated into products, or so modified (polluted) by its use that it cannot be returned to the immediate water supply without treatment that may range from simple filtering to elaborate purification (Section 4.5).

As you might now expect, since irrigation is by far the major consumptive use, per capita water consumption varies geographically depending on the economic base of a region. Worldwide, about 75% of daily water consumption is for agriculture, predominantly irrigation, but the range is enormous. In India, a populous nation that is nearly self-sufficient in food, about 96% of water is consumed in agriculture, 3% in industry, and only 1% for domestic uses. Arid nations, like Saudi Arabia and Mexico, also direct 96% and 95% of their water consumption to agriculture, respectively, in order to produce some fraction of their domestic food needs. Chile, a nation with a relatively small population and little agriculture, on the

BOX 4.3 THE LIBYAN ARAB PEOPLE'S JAMAHIRIYA AND THE "GREAT MAN-MADE RIVER"

In early 1996, over a desert construction camp hundreds of kilometers south of the Libyan capital of Tripoli, hung a sign proclaiming "278 days to Tripoli." This sign was not informing travelers of how long it takes to get to Tripoli. Rather, it was announcing the estimated time of arrival of a 900-km long, 4-m-diameter water pipeline that Libyan leader, Colonel Muammar Qaddafi, hopes will save his nation's economy and his political future. Not a man noted for rhetorical restraint, Colonel Qaddafi called the Great Man-Made River "the world's largest civil engineering project" and the "Eighth Wonder of the World." Why is he spending perhaps as much as $43 billion and 15% of his nation's oil revenues on this immense venture — especially when nearly $20 billion in contracts will be let to the Dong Ah Construction group of South Korea, a close ally of Libya's archenemy, the United States? In fact, much of the pipeline for the current phase of the project is manufactured under license from an American company. Are these the actions of a desperate man?

Perhaps. Along the coast of North Africa, a narrow band of rain-fed highlands is the only significant arable land between the coast and the southern margin of the Sahara more than a thousand kilometers to the south. Traditionally, food for the sparse population of this strip was supplied by small farms that relied wholly on rainfall, narrow intermittent wadis, and wells fed by local aquifers. These aquifers are recharged by the limited, seasonal precipitation from the Mediterranean, supply only a minimal amount of water, and in many places have been contaminated by saltwater incursion. This problem of water supply became more serious as Libya's economy grew in the last half of the 20th century, largely through profits generated by the sale of petroleum produced in the Sahara, and Libya began to experience a 25-year population doubling rate.

Thus, even though about 80% of Libya's water resources are used for agriculture, Libya has been forced to seek food elsewhere, mostly by trading oil for food in international markets. This would not be bad were it not for Colonel Qaddafi's periodic brushes with state-sponsored terrorism, which have led to boycotts and embargoes on goods in and out of Libya by the United States, many western European nations, and the United Nations. Other sources of food also have begun to dwindle. The collapse of the Soviet Union cut off one traditional market. The People's Republic of China has its own internal economic difficulties and needs to maintain close trading ties with the West. And Colonel Qaddafi's periodic disagreements with his Arab neighbors also limit his options.

Seeing himself isolated from international food suppliers and wishing to gain agricultural independence, Colonel Qaddafi engaged an American engineering company to design a water project to bring deep groundwater to the coast. It proposed a series of well fields to extract water discovered during oil exploration above several large basins in the Libyan Sahara. The water, locked in aquifers above the oil reservoirs, could feed a 3,000-km network of pipelines leading to the coast and ultimately connecting the water with areas along the entire Mediterranean seaboard of Libya. In addition to the pipelines, the project will build reservoirs covered to prevent evaporation in the desert heat; hundreds of wells, pumping stations, roads, railroads, irrigation systems, even municipal water systems will be built. The project is to be completed in segments between 1991 and 2007 and will ultimately deliver millions of cubic meters of water per year, irrigate hundreds of thousands of acres of land, and provide a stable, potable water source for both Libya's urban population and farms for the first time in decades.

Will it be worth it? That is not an easy question to answer. The project has consumed much of Libya's expendable oil revenue surpluses, even forcing Libya to seek loans from the African Development Bank in 1992. Experience in other arid nations suggests that the cost of providing water will exceed the market value of any crops grown, thus committing Libya to subsidizing agriculture indefinitely. Saharan groundwater is being mined. Geological studies indicate that most of the water is 15,000 years old, having flowed into the aquifers at the glacial maximum when the North African climate was temperate and wet. Estimates for the longevity of supply at projected extraction rates vary from 50 to 100 years. Finally, no one is quite certain what the effect of this massive water withdrawal will be on groundwater in neighboring states like Egypt, the Sudan, Chad, and Tunisia. On the other hand, Colonel Qaddafi will no doubt achieve his short-term goal of greater, if not total, independence from world food markets.

It remains uncertain what the long-term economic impact will be. Some insight is suggested by the fact that the Dong Ah Construction Group has insisted that payments for much of the project be made in Japanese, German, and U.S. currency.

FURTHER READING: Gardner (1995).

Table 4.4 *Some water-rich and water-poor nations*

Nation	Rank	Annual per capita surface water supply (in cu m)
Water-rich		
Canada	1	121,930
Panama	2	66,060
Nicaragua	3	53,480
Brazil	4	38,280
Ecuador	5	33,480
Sweden	7	22,110
Former Soviet Union	10	16,930
Austria	12	12,002
United States	13	10,430
Water-poor		
Malta	1	70
Libya	2	190
Barbados	3	200
Oman	4	540
Kenya	5	720
Egypt	6	1,200
Belgium	7	1,270
South Africa	8	1,540
Poland	9	1,570
India	12	2,430
China	13	3,530

Sources: McDonald and Kay (1988, table 4.1).

Table 4.5 *Annual per capita water consumption in the United States*

Use	Per capita consumption (in gallons/day)
Public and private domestic and commercial water	176
Irrigation	543
Livestock	18
Industry	76
Mining	13
Power station cooling water used	519
Power station cooling water consumed	14
Total	1,359

Sources: Selley et al. (1993).

other hand, uses 76% of its water for industrial purposes, while the more populous United Kingdom, at another extreme, uses 66% of its water for domestic purposes and only 3% for agriculture. The United States uses 57% of its water for agriculture and has a range in daily per capita water consumption from a high of about 19,000 gallons per day in Idaho to 180 in Rhode Island. This 100-fold difference does not mean that Idahoans are thirstier or that Rhode Islanders drink more beer. It means that people in Idaho use enormous amounts of water to irrigate fields of potatoes and sugar beets, while in Rhode Island, a tiny populous state with a much smaller agricultural base, almost all water consumption is for municipal uses.

Choices versus necessities in water use

It is clear that industrialized nations use more water than less-developed ones, rich people use more than poor, and those who live where water is abundant are more profligate than those who routinely experience water shortages. This is human nature, but there are instances where options exist and people have made choices for one reason or another that can have profound, often unintended, consequences in the allocation of water. We give two examples.

The first example is based on the fact that Americans love beef, and the average American eats about 100 pounds each year. While the 1990 U.S. water use data (Table 4.5) suggest that only a paltry 18 gallons per day is used for watering livestock, in fact about 45% of all irrigated lands in the United States are used to raise forage crops for cattle (Section 2.2). As much as 25,000 gallons of water are needed to produce a pound of beef in the American Southwest as contrasted with 250 gallons to make a pound of bread. This choice of beef protein over vegetable protein obviously has profound implications for our allocation of a scarce resource. We will return to this in Section 4.3.

The second example involves a purely economic decision, rather than a dietary preference based on history and culture, that had entirely unexpected results on water availability. In the semiarid regions of West Africa – Niger, the Sudan, Mali, Burkina Faso, and Chad – soils are poor. Subsistence farmers have long used a time-honored and successful strategy for soil conservation that allows them to survive in a harsh and difficult land. These farmers allow their croplands to lie fallow for several years between plantings of grains and other crops that have a high demand on soil nutrients. By tradition, nomadic pastoralists use these fallow areas for seasonal grazing of goats, sheep, or camels and, thereby, refertilize the soils through the natural application of animal wastes (Section 2.2).

This agricultural strategy in West Africa worked for thousands of years, but in the 1960s, in a seemingly unrelated series of events, agribusiness interests in the United States began to capture the European market for vegetable oils by expanding the production and marketing of oils from soy beans. American farmers, needing a market for crop surpluses, took markets away from French economic interests which, for over a hundred years, had imported oils from the West African countries where subsistence farmers raised groundnuts (peanuts) by traditional means. Not wanting to lose their market share to American producers, the French decided to provide a subsidy by artificially guaranteeing high prices to African groundnut farmers. In the late 1960s French interests bought high-priced oils at guaranteed prices, sold them at competitive prices in the European market, and absorbed the loss to preserve market share. Of course, the African farmers knew an opportunity when they saw one. Groundnut production doubled, tripled, and even quadrupled in some areas of the Sudan and Niger. Unfortunately, this was accomplished by increasing the area under cultivation and decreasing the amount of land left fallow. The decline in fallow acreage drove the nomadic herders northward to marginal grazing lands along the southern margin of the Sahara Desert. In the early 1970s, drought hit these areas, and the water demand for the increased lands under groundnut cultivation combined with overgrazing by displaced herds resulted in an ecological crisis that some feel caused irreparable damage to marginal croplands in this region of central Africa.

These examples show that we make choices on how to use water and, in an increasingly crowded and thirsty world, these choices will become more difficult. Americans, who place enormous demands on water supplies to raise fodder for beef may either have to raise less beef or allocate water away from other uses (Section 2.3). International agribusiness interests that inadvertently cause expansion of the Sahara Desert may either have to relinquish land back to farmers or take water away from other people. The answers to these questions are easier to determine than who answers them. In the United States and other industrial countries, choices can be made by free-market pricing or by local and federal governments. In lesser-developed countries, the choices will not only be made locally but will be affected by powerful international organizations and governments. We pursue these issues through the rest of this chapter.

Policy questions

This section raises two sets of policy questions:

- Do you or your community use more water than you should? If so, how might you or the community reduce this use?
- If water resources are scarce, should they be allocated by the free market, with people and businesses who have the most money getting the water that they want, or by government allocation (rationing)?

FURTHER READING: Mather (1984); McDonald and Kay (1988); Speidel (1988).

4.3 THE SURFACE WATER RESOURCE

We are the Arabs of water. – *David Cliche, environment minister for the Province of Quebec, commenting on floods that had heavily damaged northeastern Quebec in July 1996*

When rain falls on the land, it either soaks into the soil (infiltrates), evaporates, or runs off the land surface as overland flow. Overland flow is the principal contributor to the surface water reservoir, and because surface water is by far the most important source for water (Section 4.2), we are interested in what controls how much water flows off the land. We divide this section into four parts: the amount of surface runoff; a discussion of drainage basins and watersheds; a brief description of how we extract water from the surface, including the construction of dams; and a case study of the Colorado River of the western United States.

Amount of surface runoff

The amount of water that a rainstorm contributes to surface runoff is the result of numerous factors, and we discuss four.

One factor is the duration and intensity of rainfall. Slow, steady rains allow more infiltration into the ground than short, torrential storms. Thus, slow rains result in less runoff, but intense storms often cause flash floods when excessive rain falls too fast for the land to absorb water or for stream channels to accommodate the abundant runoff.

A second factor is the frequency and patterns of recent precipitation. Abundant rainfall in advance of a storm (antecedent precipitation) may so saturate the soils and raise the level of groundwater that little or no additional infiltration can occur. Thus, more runoff will occur during the latter stages of a storm or at the end of the rainy season than earlier. Also a warm front that produces rain after a heavy snowfall will result in rapid melting and far greater runoff than a gradual melting of snowpack as the spring thaw arrives.

Another factor that determines the amount of surface runoff is patterns of vegetation. Trees are the best vegetation to slow runoff. Meadows and grassy fields are better than bare ground, but no ecosystem helps slow and control runoff like a mature deciduous forest. Thick canopies of leafy trees intercept rainwater and allow it to flow slowly along leaf stems, branches, and tree trunks to the ground, where it infiltrates into the soil by slow percolation through the leaf litter beneath the tree. In such thick canopies of deciduous trees, a large volume of rain water also wets the leaves and branches so that more evaporation occurs when the storm ceases. The cumulative effect of stemflow and surface wetting is why standing under a tree in a rainstorm (not recommended if lightning is active) can keep you dry. A graphic example of the impact of vegetation patterns on runoff comes from a study of clearcutting of forests in the highlands of Kenya. Where all the trees are harvested, in contrast to selective cutting that removes only one-third of the trees, there is a fourfold increase in the magnitude of annual surface runoff.

Human activity is a fourth factor. Buildings with gutters feeding into storm sewers, asphalt parking lots, forests replaced with meadows (including golf courses) or lawns all increase the runoff from any precipitation event. Experiments show that surface runoff can increase five to eight times as an area changes from forest to pasture to a developed and urbanized site. If you live in a city, look about you the next time you go outside in the rain. How much of that rain is going to soak into the soil; how much will evaporate? All of the rest will be diverted into storm sewer systems consisting of interconnected pipes, tunnels, and ditches that carry storm water from hardened surfaces directly to the nearest stream, river, or catchment basin. These storm sewers should not be confused with sanitary sewer systems, which carry waterborn wastes to sewage treatment plants. While these systems are independent of one another, it is common for high runoff events to overwhelm the capacity of the storm sewer system and flood the sanitary sewer system.

This usually results in either too much input to the sewage treatment plant, forcing the release of untreated or partially treated sewage into streams and rivers, or a backup of sanitary sewers and release of raw sewage in populated areas (further discussion in Section 7.5).

As a result of these variations, the amount of runoff to the surface water system from any storm varies. As little as 15% of the rainfall in a slow, steady rain onto a heavily forested plot of land runs off as contrasted with almost 100% in urban areas. Ultimately, runoff ends up in channels and catchments of one kind or another, and, since surface water is our most important water resource, we need some insight into this network of interconnected catchments that ultimately leads to the ocean (further discussion of rivers in Section 3.6).

Drainage basins and watersheds

One of the great fascinations, one might say distractions, of 19th-century geographers and explorers was to find the headwaters of the great rivers of the world. Desire to spread the faith and colonial interests aside, there seems to have been some deep yearning on the part of these hardy souls to find the source of the highest tributary that fed the Nile, the Congo, the Amazon, or the Colorado. In this search these explorers were defining, step by step, the drainage basin (watershed) of each of these rivers. Like the veins of a leaf that connect the leaf margin with the central stem, each creek feeds a larger stream, each stream a river, and so on and on until the main river eventually empties into the ocean. Along the way there may be marshes, ponds, and lakes, but they are only temporary way stations for water on its way from the headwaters to the ocean. If we map out every single stream that leads into a river system, we have geographically defined the river drainage basin (Figure 4.9a). The boundary between drainage basins is an imaginary line along topographic ridges and saddles called a drainage divide, and every drop of runoff within a drainage basin must, unless it is diverted by people, find its way into the main channel of the river. Thus, watersheds are the fundamental unit of surface runoff systems, and to varying degrees, every person living within a river drainage basin is affected by, or can affect, the water supply of other occupants of that basin. This effect occurs regardless of political boundaries, which, all too often, ignore drainage basins and divides.

We discussed in Chapter 3 some of the reasons why rivers are important in human history. River valleys and interconnected lakes were the cradles of most great civi-

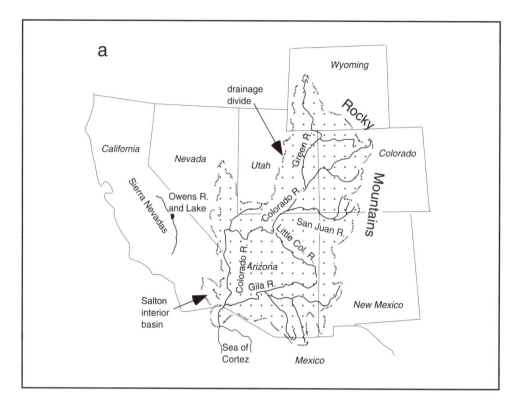

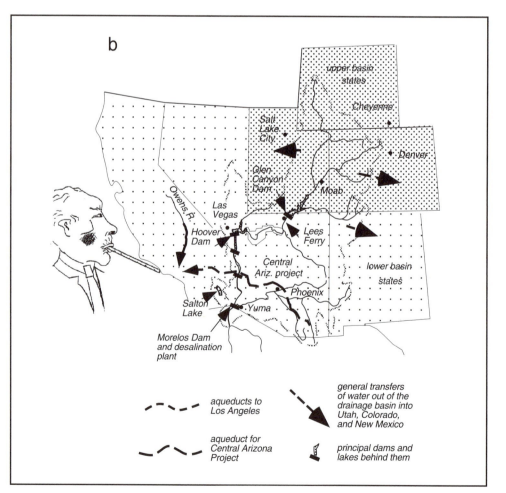

Figure 4.9 Colorado River drainage basin: (a) before human modification, showing drainage divide and principal rivers; (b) after human modification. This figure shows major dams and lakes behind them; transfers of water out of the basin; pipeline from the Colorado River into southern Arizona; pipeline to a very thirsty Los Angeles; and the basin filled accidentally to create the Salton Sea. The pipeline from Owens Valley to Los Angeles (Box 4.5) uses water outside the Colorado River drainage basin.

lizations, the routes of access and trade to the interior of continents (Box 4.2), and the providers of our most essential resource, water. To understand, protect, and wisely allocate this resource, we must recognize that the fundamental unit of the surface water system is this stream or river drainage basin. This may seem trivial to you. It is like saying that to understand the blood in your index finger you must understand all the interconnected vessels and arteries, veins, and other organs that effect the transfer of blood within your body. A physician would not remain in practice for long who decided to treat a blood disease by only applying topical medicines to your index finger. Yet, surprisingly, river systems are often treated in just such a reductionist manner. Local political subdivisions or economic interests may act as if the surface water at their disposal is somehow disconnected from everyone up- or downstream. This misfit between the natural boundaries of river watersheds and human jurisdictions causes many problems that endanger the quality and quantity of the surface water resource. After a brief discussion of the ways in which we extract water from rivers, we elucidate this problem of misfit by describing a major river system where history, tempered now and again by a little greed, has resulted in a nightmare of surface water allocation, diversion, contamination, and misappropriation – the Colorado River in the western United States.

Extraction of water for human use and the construction of dams

In order to use water, we must get it and, for many purposes, purify it. We discuss extraction of groundwater in Section 4.4, water quality in Section 4.5, and here simply discuss the extraction of water from surface reservoirs.

The simplest way to extract water from a river is to put a pipe into it or cut a canal from it. We generally use pipes for municipal or industrial purposes and canals for irrigation. Canals from rivers commonly branch into an intricate series of irrigation ditches that ultimately lead small rivulets of water along individual rows of plants. If we can put the pipe or canal into the river upstream from where we want to use the water, then the water will flow downhill and come out where we want it (further discussion of water flow occurs in Section 4.4). In most situations, however, we must remove the water from a river or lake near the place where we use it. Then, because we live above the level of the river (which keeps us dry), the water must be pumped up to an elevation at which it can be used.

Piping or pumping water directly from rivers and lakes works well if the river flows continually and the lake is continually recharged. Flow and recharge rates are generally adequate in areas of high precipitation, such as Europe, eastern North America, and most of the tropics (see the discussion of rainfall distribution in Section 2.2). In arid regions, however, water must be stored during periods of high river flow so that it can be distributed when the flow is low, or perhaps even drops to zero in some desert areas. Lesser-developed countries and individual ranchers in the dry parts of some industrial countries commonly fill natural depressions or constructed holding ponds during wet seasons and hope that the water will last through the dry season. A more sophisticated storage method, and the one that accounts for the majority of held water, is the construction of dams. Because of their convenience, they are used in areas with high rainfall almost as much as in dry areas.

Dams are constructed for a variety of purposes in addition to storing water. One is that placing a dam across a river upstream from the city or other area that will use the water keeps the water at a high enough elevation that it will flow downhill through pipes connecting it to a water-supply system. Two other functions of dams are to control floods (Section 3.6) and to generate electrical power by letting water flow out through turbines (Section 5.2). Finally, construction of a dam for any, or several, of the other purposes also creates a recreation area for people who like to boat, fish, or simply walk along the shore of a lake.

All construction of dams begins with diversion of the river out of its normal channel around the place where the dam will be. This requires cutting a channel through the side of a valley and building a gate between the river and the channel that can be opened to let the river flow into the channel and then closed to return the river to its normal course after the dam has been constructed. For swiftly flowing rivers in deep valleys, the diversion channel may be a large part of the cost of the dam.

After the river has been diverted, the dam can be built in either of two ways. Most dams are made by filling the river channel with some mixture of rock and dirt quarried elsewhere and hauled to the river; these dams are referred to as "rockfill" and "earthfill" respectively. Some dams are made by pouring concrete either directly into the dry riverbed or, more commonly, on top of a base of rock and earth fill. Dams that are to be used for power – and all large ones are – have turbines placed near the base. All properly constructed dams must have overflow channels (essentially safety valves) so that water can be

BOX 4.4 THE ASWAN HIGH DAM OF EGYPT

The Nile River begins in two main branches in central and eastern Africa, and after they join at Khartoum, Sudan (Box 3.5), the Nile makes its way through Sudan and Egypt to the Mediterranean. Because most of its journey is in the Sahara Desert, it is possible to stand with one foot in the fertile soil of the Nile valley and the other foot on a steep bank of inhospitable sand and rock. Thus, people who live in Egypt have been almost totally restricted to the Nile valley since they first began to occupy it well before 5000 B.C. The population of Egypt is now (in the mid 1990s) approaching 70 million, the great majority of whom live between the steep cliffsides of a valley ranging from 3 to 5 miles wide and 600 miles long.

The early settlers of the Nile valley learned that each year, just after the peak of summer heat, the level of the Nile would rise and, in many years, it would overflow its banks. They also learned that when the Nile receded they could plant crops that would grow from the water left by the flood, and soon they also learned that they could haul water out of the Nile in baskets and put it into irrigation systems that increased their crop yields. Modern agricultural experts now also know that the floods of the Nile delivered not only water but also silt and clay laden with nutrients that the crops needed for their growth. For thousands of years this annual delivery of water and nutrient allowed Egyptians to plant one crop per year and live on the harvest until the next flood. Naturally, there was famine in years when rain in central Africa was inadequate, and displacement of people and destruction of fields when very high rainfall pushed the Nile far over its normal flood level.

One crop per year and alternation of famine and flood was acceptable when the population was small. With the population in the tens of millions and growing rapidly after World War II, however, this subsistence agriculture was no longer adequate. Something had to be done. Egypt proposed to build a dam at Aswan, one of the narrow parts of the river, and create a lake that would store water in bountiful years until it could be used in dry years, would permit nearly year-round removal of water so that more than one crop could be planted each year, and would generate all of the electricity that Egypt could use, thus freeing the country from the necessity of importing fuel for its power system.

This proposal led to a bizarre sequence of international negotiations. Egypt first approached Western countries and organizations, who rejected the idea for a variety of reasons, and finally the Soviet Union agreed to advance the money to build the dam, provided, naturally, that it would get its money back and that Egypt would support Soviet interests in the area. The Egyptian leader throughout much of the argumentation was Gamal Abdel Nasser, and he accepted the arrangements. Mr. Nasser had died by the time the dam was completed in 1971, but the Egyptians named the lake behind it Lake Nasser. The Aswan High Dam now has a height of 230 feet and floods a lake 300 miles long, extending back into northern Sudan. The complexities of financing, building, and using it are described in lucid detail in J. Waterbury's book *The Hydropolitics of the Nile Valley*.

The argument is whether this dam is good for Egypt. On the positive side are its provision of year-around water, leading to a rapid increase in production of food and also of cotton for export, its control of floods, and its generation of many times more energy than Egypt needs. There are three negative effects, however. One is that the dam traps the nutrient-rich silt that the floods used to bring to Egypt's fields, thus requiring heavy use of fertilizer and turning Egyptian agriculture away from family holdings to larger industrial operations (see Sections 2.2 and 2.5). A second negative effect is that the low flow of the Nile permits seawater from the Mediterranean to infiltrate into the groundwater of the Nile Delta, which is one of the most important agricultural areas of the entire Nile River system. The third effect is that the valley no longer dries out between floods, and the continual presence of pools allows spread of parasites that used to die in the dry season, thus increasing health problems for many Egyptians.

So, is the Aswan High Dam good for Egypt? On balance, probably yes. The present population could not possibly survive without it, but we should not minimize its drawbacks.

FURTHER READING: Waterbury (1979).

discharged during floods, because water flowing over the top of any type of dam weakens the structure and may lead to complete dam failure (Box 3.6).

Dams provide both benefits and risks. Because of this combination, they are sources of major controversy both during their planning and their operation. All dams create lakes upstream, flooding some people who are accustomed to live along the river banks. All dams block transportation up and down rivers unless they are designed with locks that allow shipping to bypass them. On rivers that are used for spawning by ocean fish such as salmon, dams reduce access to spawning grounds even if they are constructed with "fish runs" (Section 2.4). Dams that are used to control floods place their operators in the position of deciding whether to retain water during a flood, and thus flood out people upstream and risk losing the dam by flow over the top, or to release water and risk flooding out people who live below the dam. We illustrate these problems in Box 4.4 on the Aswan Dam of Egypt.

The Colorado: A parable of a river

The Colorado River in the southwestern United States was the last great river to be explored in North America. Its late discovery, however, has not shielded its valuable water from use, reuse, and abuse. The main branches of the Colorado River rise on the western slopes of the Rocky Mountains in Wyoming and Colorado at elevations in excess of 14,000 feet above sea level (Fig. 4.9). It flows down alpine valleys, across high plateaus, and in deep canyons to the desert basins of the northern Sonoran Desert. Its tributaries flow through portions of seven U.S. states and two Mexican ones before "emptying," some 1,700 miles from its headwaters, into the Sea of Cortez. We put "emptying" in quotation marks because water no longer flows to the mouth of the Colorado River. Where paddlewheel steamers once plied their way up the river almost as far as present-day Las Vegas, Nevada, there is only a dry channel. Since 1960, the once-plentiful water of the Colorado has been so heavily used by farmers, urban dwellers, and industry or extracted for export to neighboring river basins that not a drop of Colorado River water has made its way to the ocean in 35 years.

This is the tamed river that was once called the American Nile since, like the Nile, it flows almost exclusively through arid lands. The average rainfall in most of the watershed is less than 10 inches per year. The Colorado drains 8% of the area of the lower 48 states, an area larger than any nation in western Europe, yet its natural discharge is one-third that of the Rhine, one-half that of the Rhône, and barely equal to that of the Delaware River. About 4.5 million people live in the Colorado River watershed. A whole lot of cattle do too. Many books and monographs have been written on the Colorado River and its water. It has been the subject of novels, movies, lawsuits, international treaties, and more government studies than we care to contemplate. Our purpose here is to use the river as an example of how the quality and integrity of a river watershed can be placed in jeopardy when there are competing and incompatible uses for the water and when political jurisdictions do not match nature's prior topographic arrangements.

The story of this river is complex. Since 1922, the Colorado River drainage basin has been formally, but arbitrarily, divided into two parts (Fig. 4.9b). The Upper Basin lies upstream of Lee's Ferry, Arizona, where the Glen Canyon Dam was completed in 1963. The Upper Basin includes the three main tributaries – the Green, the main Colorado (once called the Grand), and the San Juan Rivers – and includes most of the river interests in Colorado, Wyoming, Utah, and New Mexico. The Lower Colorado River Basin includes the Grand Canyon and most of the water interests in Arizona, Nevada, California, and Mexico.

The explanation for this arbitrary subdivision of one watershed into two basins begins at the turn of the century. It seems that land-hungry real-estate promoters in southern California were eager to lure naive settlers into the arid basins east of San Diego. One such starry-eyed enthusiast christened a dry desert basin on the California–Mexico border with an impressive-sounding name, the Imperial Valley. The Imperial Valley was, in fact, a vast dry interior basin with no throughgoing rivers and wholly insufficient water for farming, grazing, or any other agricultural use (it is not a bad place for geothermal power plants; see Chapter 5). This land promoter and his fellow boosters decided to set up a land development company to sell, at no small profit to themselves, we imagine, farms to unsuspecting immigrants. Read their own honeyed words:

> This vast plain of opulent soil – the mighty delta of a mighty river – is rich in the potentialities of production beyond any land in our country which has ever known the plow. Yet here it has slept for ages, dormant, useless, silent. It has stood barred and padlocked against the approach of mankind, What is the key that will unlock the door to modern en-

terprise and human genius? It is the Rio Colorado. Whoever shall control the right to divert these turbid waters will be the master of this empire. "Without the right and ability to use water, nothing is possible." (William E. Smythe in *Sunset* magazine, ca. 1900, in Fradkin [1996], p. 269)

So what did they do? In 1901, they built a canal to divert a portion of the flow in the Colorado into the Imperial Valley. A land boom was on. Unfortunately, in 1905, spring floods on the Colorado River washed out the headgate (the main diversion channel that controlled the flow of a small fraction of the river into the Imperial Valley) and the entire flow of the Colorado River was diverted into the Imperial Valley. Since the Valley has no outlet, it began to fill like a bathtub without a drain, and by the time the Colorado River was returned to its natural channel by a major engineering effort 2 years later, the Salton Sea had formed (it is still there though shrinking by evaporation).

Disasters like the accidental diversion of the Colorado River to form the Salton Sea convinced many users of Colorado River water that some federal government control and regulation were required. Otherwise, upstream users in one state, say Colorado, could use all the water needed by downstream users in California, or strong political interests in California could, by appropriate legislative and lobbying activity in Washington, limit the amount of water that upstream users could access. Farmers in the Upper Basin were diverting large amounts of water for irrigation and returning these waters to the river so enriched in salts leached from the soils that they were saltier than seawater. (It may surprise some readers that this diversion of a river for private use would be legal – we will discuss the unique aspects of water allocation law in the western United States in Section 4.6.) As a result of growing concern over the fate of Colorado River water, the U.S. secretary of commerce, a mining engineer by the name of Herbert Hoover who is more remembered for subsequent political missteps, brokered a settlement between the four states in the Upper Colorado River Basin and the three in the Lower Colorado River Basin. The agreement was simple. Since stream gauging stations on the river at Lee's Ferry had measured an average annual flow of 16.8 million acre-feet of water between 1896 and 1921, then the upper basin would be allotted 7.5 million acre-feet each year and the lower basin would get an equal amount, with Arizona receiving a bonus of 1 million acre-feet because the Gila River, the only significant tributary to join the Colorado below Lee's Ferry, flows entirely in Arizona.

Now this seems like a rational approach. It was, however, based on some basic assumptions that we might wish to look at. The agreement clearly assumes that the best use of every drop of water in the river (save a paltry 0.8 million acre-feet left to be allocated) is to remove it from the river and somehow employ it for human, livestock, or agricultural purposes. Such an assumption is somewhat at odds with conservationist views or, as a matter of fact, with those of the 15,000 people a year who raft through the Grand Canyon. This allocation formula also leaves little margin for error, allocating 95.3% of the water. The 25-year baseline of 1896–1921 was an unusually wet period, and the primitive stream-gauging equipment of the time was also in error. From 1922 to 1976, however, the average annual flow at Lee's Ferry was only about 13.9 million acre-feet. This means that Mr. Hoover allocated 16 million acre-feet of water from a river that normally produces less than 14. Also, a few interested parties were left out of the 1922 negotiation, such as the government of Mexico and the interests of many Native American tribes who had long-standing water rights based on custom, land deeds, and treaty obligations. These forgotten obligations came back to haunt this agreement over the next 70 years.

In addition to the problems of total supply of water, the 1922 agreement says nothing at all about water quality. By the 1960s, the salinity of the Colorado was 17 times higher at Lee's Ferry than it was upstream of the farms in the Upper Basin. In fact, by the early 1960s, the little water being delivered to Mexico was so saline that it killed plants that were irrigated with it. And finally, as if all these problems were not sufficient to assure a lack of trust and cooperation among water users, the 1922 agreement took no notice of the impact of the Hoover and the Glen Canyon Dams, both of which were in the planning or design stages. When completed, however, two large lakes (Lake Mead behind Hoover Dam and Lake Powell behind Glen Canyon) resulted in huge water losses. Impounding river water in the desert produces spectacular recreational lakes enjoyed by water-skiers and fishermen alike, but the lakes significantly increase the evaporation rate as well as allowing water to leak into aquifers by infiltration through the lake bottom. Lake Powell loses 450,000 acre-feet of water a year to evaporation and Lake Mead, 800,000. Lake Powell also loses 1 million acre-feet a year to groundwater infiltration (dam builders like to call this "bank storage" since this water is still there, though this euphemism dodges the issue that this is water lost to downstream users). It turns out that the water lost from Lake Powell each day is equal to the amount needed by the million or so residents of the Phoenix metropolitan region.

With nine states, two federal governments, various native tribes, a dozen or so major cities, two artificially defined water basins, millions of residents, farmers, and vacationers, and the powerful mining/agricultural/industrial interests squabbling over a river whose vital water is already overallocated, we have a problem. In the Upper Basin, a 1949 agreement gives each state a formulaic proportion of the water allocated, while still obligating the states to deliver 7.5 million acre-feet each year under the terms of the 1922 agreement. Colorado gets about half, Utah about a quarter, and Wyoming and New Mexico split the rest (with a small dividend given to Arizona as compensation for its small contribution to Upper Basin flow). No restrictions are placed on how this water is to be used. Thus, it turns out that while much of this water is used for dry-land farming in the Upper Basin and returned to the river diminished by evaporation and infiltration and increasingly saline, much of it is diverted completely out of the basin to the eastern slope of the Rockies (Denver and vicinity), to the upper Rio Grande basin, and to Salt Lake City. The Lower Basin, after the huge losses to evaporation and infiltration in Lakes Powell and Mead, diverts water back up the Gila River to Phoenix through the Central Arizona Project, and sends more water out of the basin to Los Angeles (see also Box 4.5), the Imperial Valley, and Tijuana. Mexico gets its treaty-based quota, but not a drop makes it to the ocean.

Is this the best use of the river? Does it make sense to drain a river dry to grow forage for cattle? One estimate is that 90% of the water used for irrigation within the Lower Colorado River Basin is used to grow hay for beef cattle. Is it rational to use this water to irrigate arid lands? The salinity of the river is so high from this process that desalination plants were built at a cost of hundreds of millions of dollars at Yuma, Arizona, near the Mexican border so that the water delivered by treaty obligation to Mexico can be used for irrigation and animal consumption. Is it an appropriate use of public funds to build dams in the desert that result in scarce water being lost to evaporation and infiltration. A steadily increasing amount of the power generated from Hoover Dam is used to light up the night sky over Las Vegas's casinos. The answers to these questions may be, "Yes, this is appropriate." The point of this discussion is that because the river runs through independent political jurisdictions of a dizzying variety, such questions were never asked. Upstream consumers and downstream users (and vice versa) who share only residence in the same watershed are the unsuspecting victims of decisions they never participated in. We will consider some of the political and legal remedies to these problems in Section 4.5.

Policy questions

What should people do if they discover that their community, state or province, or country is using more river water than is really available?

FURTHER READING: Stegner (1953); Shaw (1983); Reisner (1986); Manning (1992); Gleick (1993); Fradkin (1996).

4.4 THE GROUNDWATER RESOURCE

The first work to systematize the knowledge of groundwater was that of the Persian naturalist M. Karadi (died 1016). This work was called *The Search for Waters Hidden under the Earth*. . . . Unfortunately, this work became known to Europeans only in the second half of the twentieth century.
— E. V. Pinneker, *General Hydrogeology, p. 6*

The second most important source of water for human needs is groundwater, water that has soaked into the earth and can supply springs or be extracted by digging and drilling wells. We divide this section into three parts: the geology of groundwater, issues dealing with its extraction, and a case study of the High Plains aquifer of the United States. We start the discussion by defining five terms.

- Porosity is the amount of open space in a rock, sediment, or soil.
- Permeability is a measure of the ability of water to flow through the ground. It is related to the size and shape of pores and the degree to which the pores are isolated or connected to each other Permeability is roughly equivalent to hydraulic conductivity, which we will describe in our discussion of Darcy's Law.
- The water table is the upper surface of the part of the ground in which all of the pores are filled (saturated) with water. Above the water table, water that soaks into the ground from rain or a river filters downward through soil and rock containing open spaces (pores) until it reaches the saturated zone.
- An "aquifer" is any rock or sediment beneath the earth's surface that holds and transfers water. The term is generic, but many important aquifers are named, and we will return to one very famous one, the Ogallala aquifer, later in this section.

BOX 4.5 THE OWENS VALLEY WATER WAR

Los Angeles County is home to more than 9 million Californians, and more arrive each day. It is also a desert basin with an average of 23 inches of rain each year, falling in a small number of heavy storms. The adjacent San Fernando Valley is one of the richest agricultural regions in the world. All of these people, all these farms need water — lots of it. This need is not new. For at least 100 years, the Los Angeles basin has needed more water than it could find locally. And, for that 100 years, the lifeline of the city and surrounding areas has been the Los Angeles Aqueduct System, hundreds of kilometers of pipelines, siphons, reservoirs, and canals that bring water from the east side of the massive Sierra Nevada mountains to sunny, southern California (Fig. 4.9b). The construction of that aqueduct pitted Los Angelenos against local farmers in the Owens Valley, a narrow but rich valley in Inyo County nestled between the Sierras on the west and the White-Inyo Mountains on the east. In small towns like Lone Pine and Bishop, California, settlers raised crops with stream water that drained the snow-capped, eastern slopes of the Sierras.

About the turn of the century, as Los Angeles began its growth, public officials responsible for water supply had to look beyond the Los Angeles Basin for water, and they eventually settled on the Owens River as the only practical source. The problem was that they did not own the water, and the farmers and ranchers in the valley were unlikely to give it to them. In fact, the U.S. Reclamation Service, the federal agency responsible for developing federal lands for farming and grazing, had just begun to acquire land and water rights for the Owens Valley Project, a series of dams, reservoirs, and connectors on the Owens River to enhance agricultural development in Inyo County. Knowing that local residents were not going to assent willingly to Los Angeles's taking this water and operating under pressure from droughts in southern California, a number of colorful and marginally scrupulous men — a senator or two, several newspaper magnates, public officials, and even a cabinet secretary under President Theodore Roosevelt — concocted and colluded in a scheme whereby representatives of Los Angeles posed as U.S. Reclamation Service agents and private citizens to buy up huge tracts of land with rights to the water. About 1910, by the time the citizens of the valley realized what had happened, it was too late. Los Angeles controlled the water in the valley.

For the next 20 years, by offering inflated land prices and taking advantage of bank failures, the city acquired enough land and water to justify an aqueduct system more than 500 km long that would drain Owens Lake and send some 600 million m^3 of water to Los Angeles each year. But this was not before the 20-year long, bloodless Owens Valley Water War had ebbed and flowed across the valley. That it was bloodless may be due more to luck than intent. At various times, portions of the aqueduct were destroyed by dynamite, a representative of Los Angeles was once kidnapped from a restaurant and nearly hanged by angry citizens, and a band of local residents even took over portions of the aqueduct and diverted water to the Owens River. On several occasions, trainloads of armed deputies from Los Angeles were faced down by rural sheriffs and their deputies and, on one occasion, roadblocks on all the highways and spotlights trained across the valley gave the Owens Valley the appearance of an open-air penitentiary. At least one of these armed confrontations had a charming denouement. As locals and armed deputies faced off along the aqueduct, Tom Mix, a silent-movie star who was filming one of his many oatburner westerns nearby, sent over his orchestra and fired up the pits for a huge outdoor barbecue. The face-off became a picnic and the evening is supposed to have ended with everyone singing "Onward Christian Soldiers" around a bonfire under the stars.

Ultimately tempers cooled. The political inevitability of Los Angeles's claims, as well as the irresistible temptation by many valley residents to profit from land and water sales, led to the collapse of local opposition. As the years went on, most people turned to the courts for relief, and by the 1970s, a series of state court rulings in favor of Inyo County led to negotiation such that residents of Owens Valley are now more secure, whatever water rights remain to them. In addition, intense national pressure from conservation and wildlife groups further restricted the ability of Los Angeles to extract unlimited amounts of water from the scenic and ecologically unique Mono Lake at the north end of the valley.

Nonetheless, the problem persists after 100 years. The competition for a limited but vital resource between a populous and politically powerful urban water consumer, Los Angeles, and the environmental and economic concerns of a sparsely populated but scenic region, the Owens Valley, present a seemingly insoluble quandary. Matters might have been handled better, but the insufficiency of water supply in southern California is a product of finite resources rather than of bad manners or unscrupulous politicians.

FURTHER READING: Walton (1992).

• An "aquiclude" is the opposite of an aquifer. Its name shows that it "excludes" water and is a rock that prevents the flow of groundwater.

The geology of groundwater

Every water witch or dowser amazes clients because they seem to assume that groundwater must be in an underground channel and, thus, is difficult to find from the surface. These channels exist only in a few areas in the world, however, where near-surface weathering turns limestone into Swiss cheese–like warrens of caves and tunnels, which are sufficiently unusual that many are major tourist attractions (e.g., Carlsbad National Monu-

ment, New Mexico). All other groundwater simply permeates the near-surface rocks, sediments, and soils. In fact, water is everywhere beneath us. What varies is the depth of the water table, the rate at which water flows, and the ease of its extraction. So, what controls groundwater flow?

The first point to remember is that water runs downhill. As geologists, however, we would like to take this obvious fact and make it a little more complicated by discussing potential energy, pressure, and hydraulic head (Fig. 4.10). Potential energy is energy imparted by the earth's gravity that could be converted into kinetic (movement) energy. The equation for potential energy is PE = mgh, where m is mass, g is the force of gravity, and

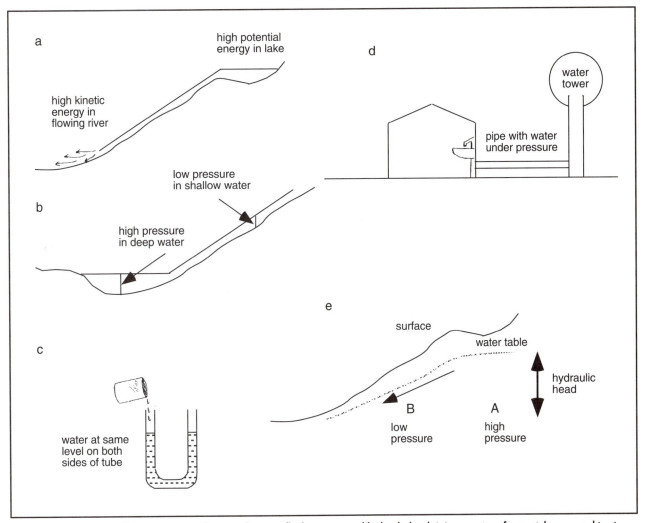

Figure 4.10 Diagrams to illustrate concepts of potential energy, fluid pressure, and hydraulic head: (a) conversion of potential energy to kinetic energy; (b) pressure proportional to depth; (c) U-tube showing same level of water in both sides; (d) water at high elevation in water tower supplying water to home; (e) pressure beneath sloping water table, showing water flowing from high to low pressure in the direction in which the water table slopes downward. See further discussion in text.

h is the height above some reference level, commonly the surface of the earth. Thus, potential energy increases with increasing elevation for objects of equal mass. We can illustrate this definition by dropping two balls of equal mass from two different elevations. Because the one at the higher elevation has higher potential energy when it is dropped, it is moving faster when it hits the ground than the ball dropped from the lower elevation.

By this definition of potential energy, water at the top of the hill in Figure 4.10a has more potential energy than water at lower elevations, and the water in a river converts this potential energy into kinetic energy as it flows downhill. Throughout human history, people have converted this energy into mechanical energy by putting a waterwheel in the river, which then does work for us as the river transfers some of its energy to the wheel. In industrial societies, we are more likely to store the river's energy temporarily behind a dam and then let the water transfer its energy to electricity as it flows out through turbines near the base of the dam (Sections 4.2 and 5.1).

In addition to imparting potential energy, gravity generates pressure in fluids. The pressure at any point is proportional to the height and density of the fluid above it. For example, the one atmosphere of pressure generated by the earth's atmosphere at sea level is equal to the pressure generated by 32 feet of water. Thus a person diving into the ocean to a depth of 32 feet is under two atmospheres of pressure (one at the surface plus one more for the 32 feet of water). We illustrate this concept in Figure 4.10b, where the pressure on the bottom of a river is much less than the pressure at the bottom of a deeper lake.

Water that is free to move will always flow so that the top surface is flat. For example, if we pour water into one side of the bent tube (U-tube) in Figure 4.10c, it will rise up the other side and maintain the same level in both sides. In doing so, the water maintains the same potential energy at its top surface on both sides of the tube, and the water on both sides generates the same pressure on the bottom of the tube. For example, most public water-supply systems deliver water to faucets by pumping water into a tank at an elevation above the level of all of the faucets and then simply letting the water run downhill (Fig. 4.10d).

Water filling connected pores in the ground exerts the same pressure as water in a lake and attempts to flow so that its top surface becomes flat. Consequently, in Figure 4.10e, the pressure at point A is greater than the pressure at point B because A is farther below the surface than B, and the water flows from A to B, that is, from high pres-

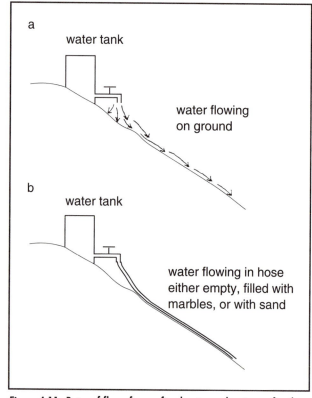

Figure 4.11 Rates of flow of unconfined water and water confined to media of different permeabilities. See discussion in text.

sure to low. This leads to the general rule that water below a water table moves in the direction toward which the water table is sloping downward. The difference in elevation of the water table at points A and B is known as the "hydraulic head," a common term used in describing the driving force for fluid flow.

Now that we have taken the simple fact that water runs downhill and made the subject more complicated, we are ready to conduct a thought experiment with a water tank at the top of a hill (Fig. 4.11). If we open the valve, the water runs down the hill as a stream if a channel is present or as a sheet of water if there is none. If we connect the valve to a hose, the water still runs down the hill, but now it is confined to the hose. If we fill the hose with marbles (don't ask why – we are simply trying to illustrate a principle) and put a screen on the end so the marbles can't come out, the water still runs down hill, but the flow will be more restricted and slower. Finally, pack the hose somehow with fine sand, and water will flow even slower. All of these experiments have been conducted with the same hydraulic head; but simply by changing the medium through which the water is flowing, we have greatly changed the rate.

From this simple experiment to the real world of groundwater is an easy jump. Instead of sand packed in a hose that leads from a water tank, think of the hill itself as a pile of sand or sandstone and of rainwater falling on that hill and soaking into the pores between the grains of sand. The water that soaks into the sandy hillside and reaches the water table develops a hydraulic head and moves downslope. The rate at which it moves in the ground is described by an equation known as Darcy's Law:

$$V = (h_1 - h_2)/d \times K,$$

where V is the velocity of flow of the groundwater; $(h_1 - h_2)$ is the hydraulic head; d is the distance between the two points that determine the head; $(h_1 - h_2)/d$ (vertical distance over horizontal distance) is the hydraulic gradient, which is equal to the slope of the groundwater surface; and K is the hydraulic conductivity.

The hydraulic conductivity, K, is a constant for any rock, sediment, or soil and can be measured in the laboratory. We illustrate it with three different rocks with approximately 20% porosity in Figure 4.12. Case 1 is a rock of equally sized, roughly spherical sand grains that are cemented to each other only where the grains are in contact. This rock has a high permeability because the pores are open and interconnected. Thus, it has a high hydraulic conductivity because the pores or voids are interconnected and water can readily flow around and between grains. In case 2, the total porosity is still 20%, but the holes in the rock are "frozen" gas bubbles in a lava. Since each bubble is independent of each other bubble and the material filling the interstices is a dense volcanic

glass with no porosity, the rock has no permeability, water cannot flow, and the hydraulic conductivity is effectively zero. Case 3 is intermediate – a rock with total pore space of 20% but a wide variety of grain sizes. This means that the pores are smaller, the pathways very tortuous, and as water flows it may even entrain the smallest clay-size particles and clog the tiny connections between pores. Thus, hydraulic conductivity is a physical property that depends on a wide variety of features such as grain size, grain shape, grain-size distribution, degree of isolation of pores or fracture density, and also on dynamic factors like flow rate since faster flowing water could entrain larger particles that might jam the flow pathways.

Several examples of measured hydraulic conductivities of geological materials are shown in Table 4.6. The flow rate of groundwater in sand is 4 times faster than in sandstone, 13 times faster than in limestone, 60 times faster than in a metamorphic schist, and 60,000 times faster than in clay. Regardless of the medium, all groundwater flow is very slow compared with surface water. With a gradient of 0.5, for example, groundwater would require nearly a week to travel the distance of a football field in sand and more than 1,000 to travel that distance in clay. Most groundwater gradients, however, are much lower, and because flow rates are proportional to gradient (Darcy's Law), they are also much lower. Thus, for example, groundwater with a gradient of 0.05 in sand would need 10 weeks to flow the length of a football field.

The depth and shape of the water table and the nature of the ground in which the water occurs are all important in determining the behavior and availability of groundwater. In some areas the ground is relatively homogeneous, and all of the rock in them is essentially an aquifer. By

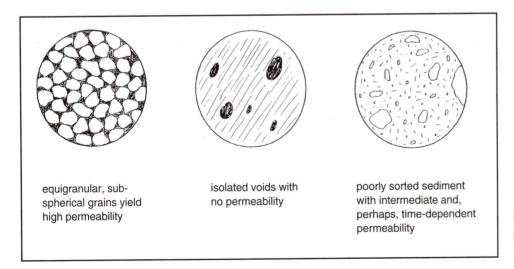

equigranular, sub-spherical grains yield high permeability

isolated voids with no permeability

poorly sorted sediment with intermediate and, perhaps, time-dependent permeability

Figure 4.12 Examples of permeability variation in rocks of 20% porosity.

Table 4.6 *Hydraulic conductivity of some common sediments and rocks (in m/day)*

Type	Hydraulic conductivity
Well-sorted sediments	
Gravel (16 mm in diam.)	270
Sand (0.5 mm in diam.)	12
Clay (0.004 mm in diam)	0.0002
Selected rocks	
Sandstone	3.1
Limestone	0.84
Schist	0.2

Source: Shaw (1983).

contrast, many areas are inhomogeneous, with groundwater flow impeded by a mixture of aquifers and aquicludes.

Homogeneous ground in which all of the rocks are aquifers can occur both in humid and arid environments. In a humid region rain is continually infiltrating the ground, and the water table will generally be shallow, although it will rise and fall as the amount of rain varies. The continual supply of rainwater maintains a slow rate of groundwater flow that keeps the water table above the elevation of the nearby valley and supplies springwater to the creek. Hydrologists refer to such streams as "gaining streams" since they gain water from the underground aquifers, and except in long-continued drought this flow keeps streams running year-round whether it has rained recently or not (Fig. 4.13). Another consequence of this continual flow is that, if rock permeability is homogeneous, the water table will resemble a subdued version of the land surface. By contrast, in an arid region the lack of rain over long periods means that water tables are very low, and some streams are above the water table (Fig. 4.13). They will only run when there is overland flow after a rainstorm, sometimes called flash floods, and when they run they actually lose water through their bed to the underlying sediments and rocks. These are "losing streams," called arroyos in the southwestern United States or wadis in North Africa, and people who live in arid regions may have seen water flow down them and literally disappear along the channel as it soaks into the underlying undersaturated sands and gravels.

Relationships are more difficult to understand in ground that includes both aquifers and aquicludes. Hills are rarely simple piles of sand or mounds of sandstone,

and real accumulations of rocks or sediments are layered into beds that vary in grain size, sorting, thickness, and permeability or porosity. Figure 4.14 shows two of the most important configurations of aquifers from a geological point of view. Case a is a series of flat-lying sedimentary rock layers that have been dissected by a stream valley, with the top of the hill underlain by a sandstone. Because sandstone is usually a good aquifer, rainwater readily soaks into the ground and finds its way to the water table, but an impermeable shale bed acts as an aquiclude that prevents the percolating rainwater from penetrating through to underlying aquifers. The result is an aquifer "perched" above the aquiclude. Recognition of a perched aquifer is very important to any groundwater user on top of the hill since this aquifer has a significantly smaller holding capacity than would an aquifer that makes up the entire hill.

Case b in Figure 4.14 is a little more complicated, and it is very important to millions of people who depend on artesian water for consumption or, more commonly, for irrigation. Here an aquifer is sandwiched between two aquicludes, and one end of this package of sediments has been uplifted along a mountain front. Rainfall and snow meltwater that flow off the mountain can percolate into the upended aquifer just as one might fill a pipe from a mountain lake. The impermeable beds on either side of the up-turned aquifer confine the groundwater as it flows downhill. The aquifer can literally be filled to capacity, and someone seeking water hundreds of kilometers away can tap this underground water by drilling a well through the upper impermeable aquiclude. Furthermore, since the water in that deep aquifer is pressurized by being recharged far away at much higher elevations, it will rise up any pipe inserted into the well and become a flowing or artesian well in the same way that water in the U-tube of Figure 4.10c rises to the same elevation on both sides of the tube.

Extraction of groundwater

When we extract water from an aquifer, we need to take two precautions. One is that it should be pure enough for our purposes (Section 4.5), and the second is that we should extract water at a rate that keeps the well or wells flowing. If we pump too fast or pump too much over a long period of time, the aquifer may stop yielding water and we will have to find another water source. We use two examples to illustrate these problems – drawdown of the water table around pumped wells and saltwater incursion into coastal aquifers.

Because water percolates slowly through aquifers, the water that is pumped from a well comes mainly from a re-

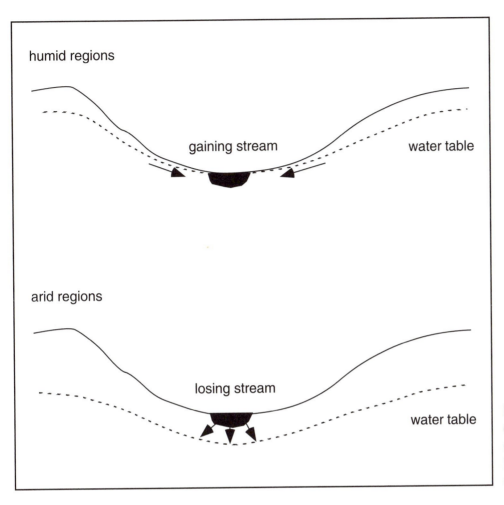

Figure 4.13 Differences in depth to water table in humid and arid regions, showing gaining and losing streams. See discussion in text.

gion immediately around the well. As this water is removed, more flows in to replace it, creating a water table that tilts downward from the level of the water in the ground to the level of water in the well (Fig. 4.15a). Faster pumping causes a steeper tilt on the water table as water flows faster into the region around the well. If the well is pumped too fast, the inability of water to flow fast enough into the region around the well may cause a variety of problems. One is simply that the level of water at the well may be below the depth that the well casing extends into the ground, thus stopping all flow. A second problem is that, even if the well keeps flowing, the water must be pumped from deeper and deeper levels, making the pumping process more expensive. Finally, and possibly most serious, is the possibility that some rocks collapse upon themselves and close their pores when they dry out. If this happens, then the well may have to be completely abandoned because the ground around it is no longer an aquifer.

The problem of saltwater intrusion is illustrated in Figure 4.15b. The boundary between fresh water and salt water is controlled by their relative flow rates. Fresh water flows toward the ocean because of the hydraulic head generated by the water table further inland, and salt water flows inland because it is denser (1.035 g/cm^3) than fresh water (1.0 g/cm^3). Some mixing occurs along the boundary, and most coastal wells encounter fresh water near the ground surface and water with increasing salt content (salinity) further down. If the fresh water is pumped too rapidly, its hydraulic head is reduced, and the salt content of water in the well rises as salt water from the ocean percolates further inland. Many coastal areas with increasing water demands for industrial or tourist use are encountering this problem of increasing salinity.

A case study of groundwater use: The High Plains aquifer

If you have ever flown in an airplane across the United States and sacrificed comfort for curiosity by sitting in the window seat, you may have noticed that large areas of the central United States look as if it had measles (Fig.

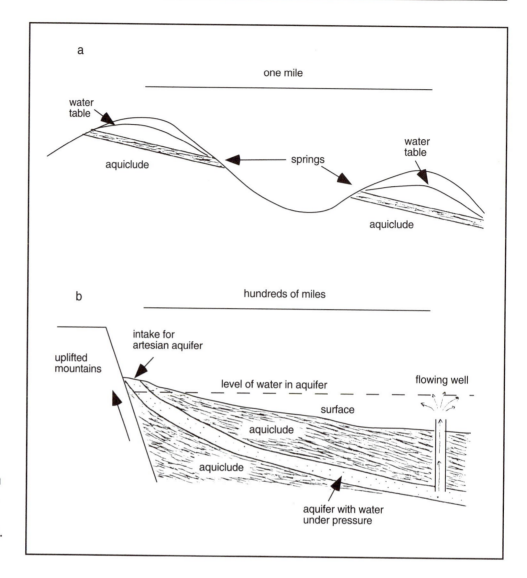

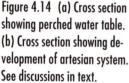

Figure 4.14 (a) Cross section showing perched water table. (b) Cross section showing development of artesian system. See discussions in text.

4.16). Mile after mile over Nebraska, western Kansas, Oklahoma, and the panhandle of Texas, you see large circular patches of green (in summer) centered on a checkerboard pattern of brown. What is going on here?

What is going on is our eternal search for water. This area of the United States is known as the High Plains, a broad, flat, featureless region underlain by thick sequences of sediments deposited by rivers draining the eastern slopes of the Rocky Mountains in Wyoming, Colorado, and New Mexico. The High Plains are to the east of the broad Rocky Mountains, which intercept air flowing eastward from the Pacific Ocean, thus preventing the ocean from moderating the climate and also wringing moisture out of the air as it expands and cools. Thus, the High Plains are uniformly hot in the summer and also very dry. Comparable areas in Europe and Asia are known as the steppes, areas either semiarid or subhumid

depending on your point of view. From the point of view of human habitation, the important fact is that there is insufficient rainfall to raise grains or support large herds of animals, yet some 2 million people have settled here and make their living off agriculture. To do this, they must use groundwater to irrigate crops from what has been called "the land of the underground rain."

Beneath an area of about 175,000 square miles (three-quarters the size of France) lies the High Plains aquifer, the source of life-sustaining water for tens of thousands of farms and hundreds of farming communities (Fig. 4.17). The High Plains aquifer itself is not one porous bed, but many. The most important is the Ogallala Formation, an unconsolidated, poorly sorted mixture of silt, sand, and gravel laid down about 10 million years ago as a thin blanket of fluvial deposits shed eastward from the rising mountain front to the west. Along its western margin, for

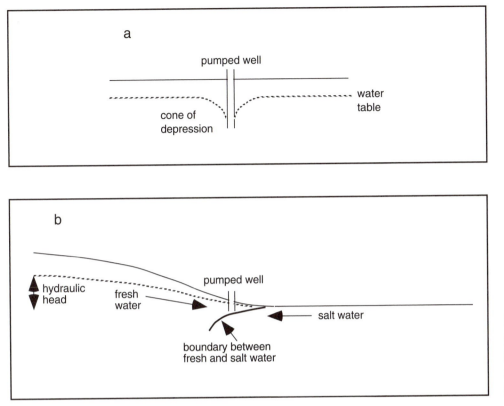

Figure 4.15 (a) Cone of depression around pumped well. (b) Intrusion of salt water into coastal aquifer. See discussions in text.

the past 10 million years surface runoff, rain, and meltwater from the mountains have entered that Ogallala Formation and other beds in the High Plains aquifer and flowed the hundreds of miles eastward to saturate the confined aquifers beneath the High Plains. In some places, wells drilled into these beds flow with artesian water – Artesia, New Mexico, for example; in other places, the water must be pumped, but pumped or flowing, there seems to be an inexhaustible supply of groundwater there for the taking. And take it we do; more than 170,000 irrigation wells pump tens of millions of acre-feet of ground water each year to irrigate millions of acres of rich farmland. In the center of each of the green circles seen so vividly from the air is a well, which, when actively pumping, drives deep groundwater into enormous rotating sprinkler systems (a process called central pivot irrigation, first introduced in the High Plains in the early 1950s) that can send water over a circular field a quarter mile in diameter.

We said that there seems to be an inexhaustible supply of groundwater, but let's look at this in detail. A study by the U.S. Geological Survey estimated that in 1986 the High Plains aquifer contained approximately 3.25 billion acre-feet of water, about the amount of water in Lake Huron. In a typical year some 20 to 25 million acre-feet

of water are extracted, suggesting that it would take more than 100 years to pump the aquifer dry. And anyway, new water must be flowing in to replace the pumped water, so why worry? As usual, statistics can be misleading. Even if there is 3 billion or so acre-feet of water, this is still only about one-half of the original resource available when the first wells were drilled more than 100 years ago. Then there is the issue of recharging of the High Plains aquifer as farmers pump. The average velocity of water through the High Plains aquifer is about 30 cm per day, 6 miles a century, which means that rapid pumping is capable of draining the aquifer. The best way to estimate the rate of removal of groundwater is to measure the decline in the level of the water table within a specific aquifer, the drawdown in the cone of depression (discussed earlier). In many places in the Ogallala aquifer, the saturated thickness, the depth of water in the aquifer, has decreased as much as 50%. Wells in which the saturated thickness was once 200 or 300 feet may now have only a 100 to 150 feet of water and levels are dropping on average across the region at 1 to 2 feet per year.

There are other considerations in the health and future of the High Plains aquifer that make simply dividing the total capacity of the aquifer by the annual depletion deceptive. First, the aquifer is neither of uniform thick-

Figure 4.16 Photo of High Plains region of West Texas. The perfect circles are areas irrigated by rotating sprinklers. The more irregular patches are sinkholes in the limestone terrain.

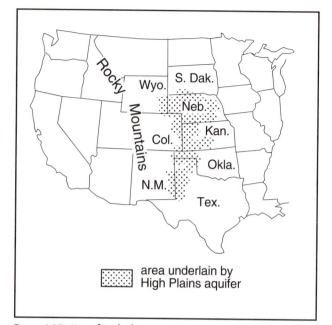

Figure 4.17 Map of High Plains reservoir.

ness nor uniform saturation. In central Nebraska the aquifer's saturated thickness exceeds 1,000 feet, but in west Texas it is less that 100 feet. Economically, it is not feasible to drill very deep wells to an aquifer with a saturated thickness of less than about 35 feet; so many parts of the High Plains aquifer are not likely to be tapped for groundwater in any significant way. In addition, not all of the water in an aquifer can be extracted under the best of circumstances. Best practices are fortunate to yield 15% of the total water in a saturated rock. Add to these limitations the fact that the depth to the top of the water table varies significantly across the High Plains. In some places like central Kansas it is near surface and even feeds gaining streams like the Arkansas River, but in west Texas it is more than 200 feet down. The wells that must be drilled deeper have both higher pumping and a much higher original investment. Consequently, those farmers who had to invest the most to extract groundwater are also the ones least able to cease pumping groundwater since they must amortize their significant capital invest-

ment. This is the problem in west Texas, where the High Plains aquifer is deepest and thinnest and undergoing the most rapid depletion.

The use of the High Plains aquifer also generates significant environmental problems. Lowering of regional water tables by heavy demand for irrigation dries up municipal wells and even changes the local hydrology of river systems. In central Kansas, for example, the regional drop in near-surface water tables has cut off groundwater sources for spring-fed portions of the Arkansas River, with the result that the river often dries up in the summer. Finally, recharge rates from local infiltration increase as water tables drop, causing serious deterioration in the quality of groundwater in areas with significant applications of fertilizers, herbicides, and other agricultural chemicals or where industrial or feed-lot sources of pollution are a problem.

We spent so much time on a local problem of water use because it is a useful example of how a resource that seems limitless can, in actuality, have an ultimate or practical limit. Farmers in the High Plains need groundwater from the High Plains aquifer and extract it at a rate greater than natural processes can replenish it. This may or may not be good, and opinions differ. You can get a sense of the difference of opinion in the language alone. Those who think we are extracting groundwater at too fast a rate talk about "mining" groundwater; those who are less concerned, discuss "overdraft" production of groundwater. In some places like west Texas, this mining means that the Ogallala aquifer will be essentially pumped dry by 2025; in other places like central Nebraska, overdraft production can continue at present rates well into the 22d century.

The problems of the High Plains aquifer are not just local. The High Plains states account for about 20% of the U.S. corn production and 30% of wheat; and the United States is the world's leading exporter of these two grains. Also the High Plains aquifer is but one of too many examples of overzealous use of a seemingly endless, but in fact remarkably fragile resource (two more examples in Boxes 4.6 and 4.7). In most parts of the world, water is being withdrawn from the ground more rapidly than the ground is being recharged. This mining of water has reached its absolute limit by total depletion of the aquifer in only a few places, but in many places the lowered water tables are already causing rivers and lakes to dry up, salt water to encroach into coastal aquifers, and extraction costs of remaining water to increase.

Policy questions

Which of the following policies on groundwater extraction would you prefer?

- Ignore environmental degradation and continue groundwater extraction until there is nothing left to pump.
- Let the price of groundwater rise, thus presumably reducing water use, and use some of the money to restore the environment and to do research on methods of water conservation.
- Require action by local and national governments to restrict groundwater use and, hopefully, begin to replenish the aquifers.

FURTHER READING: Shaw (1983); Mather (1984); Smith (1989); Kromm and White (1992); Price (1996).

4.5 WATER QUALITY

Without a plentiful supply of water the whole structure of our society would collapse. – P. N. Cheremisinoff, Water Management and Supply, p. 1

In industrial countries and some, but not all, parts of the lesser-developed world, the quality of water that we use is closely controlled. Most groundwater is fairly pure and needs only a small amount of treatment, although some is so contaminated with natural salts that societies have

always identified springs that yield potable water (they were called "sweetwater" in the western United States, a name that dots maps to this day). Some groundwater also is locally contaminated by widespread application of toxic chemicals to the land surface, both intentional and unintentional, by routine disposal of noxious substances in deep wells (out of sight, out of mind), by inappropriate disposal of human and animal wastes, and by the slow, steady, and insidious leakage of chemicals from tanks, landfills, waste dumps, junkyards, industrial sites, feed lots, and who knows what. Surface water that is used for drinking must almost always be purified, partly because of natural contamination but largely because almost all surface water is used repeatedly as the wastewater (treated sewage) from one water district is discharged into rivers and then taken by the next water district downstream. It has been estimated that a glass of water drunk in New Orleans, at the mouth of the Mississippi River, has passed through at least six sewage systems en route down river.

The allowable concentrations of impurities in water varies with the different purposes for which the water is used. We start by discussing contaminants in drinking water, which must be the most pure, and then briefly mention the quality of water used for other purposes.

Drinking water

Regulations covering hundreds of possible contaminants in drinking water are published by the national governments of all industrial countries, numerous state, province, and local governments, and such international agencies as the World Health Organization (WHO) and the United Nations Food and Agricultural Organization (FAO). We discuss only a few contaminants of major importance here.

BACTERIA AND PARASITES (MICROBES) Bacteria are single-celled organisms that do not contain a nucleus (they are "prokaryotic") and live by drawing nourishment from other organisms. Sometimes when they attack organisms, they transmit diseases. They generally cannot survive in the presence of oxygen (which is why they enjoy life in your intestines) but some bacteria can survive long enough in surface and groundwater to cause disease when the water is drunk. The principal illnesses caused by waterborne bacteria include typhoid fever, transmitted by *Salmonella* bacteria, and various gastrointestinal problems such as cholera and dysentery.

Parasites are single-celled or multicellular organisms with nuclei (they are "eukaryotic") that live by sucking nourishment from their living hosts. They include amoe-

BOX 4.6 SHRINKING THE ARAL SEA

Until the 1960s, the land-locked Aral Sea of central Asia was a body of water with an area of nearly 70,000 sq km and the home of boats used for fishing and transportation. Its size was maintained because the rate of evaporation from the sea was closely matched by the inflow from two rivers, the Syr Darya in the northeast and the Amu Darya in the south. The evaporation had brought the salinity (salt content) to about 1%, approximately one-third the salinity of seawater but within the range that could be used for agriculture around the edge of the sea.

In the 1960s, life around the Aral Sea began to change rapidly. The Soviet Union decided to expand agricultural production in the area of the sea and in the watersheds upstream. In order to do so, it extracted groundwater at a rate much greater than could be replaced in this land of sparse rainfall, causing rivers to lose water to the ground, and also took so much surface water throughout the river watersheds that the rivers dried up for part of most years. At this point, the Aral Sea began to shrink (Fig. 4.18). Because it no longer received an adequate influx of fresh water, evaporation began to cause a rise in salinity, which reached that of seawater in the 1990s and is projected to become much higher in the future. With the shrinking of the sea came a drying up of the lands around it. These lands and the now-dry sea bottom, lacking any adequate vegetation or soil cover, began to blow away in the fierce winds that howl through central Asia. Plumes of salt and dust rose as high as 4 km and stretched as far as 500 km downwind, and when the wind died down, these particles settled down on the land and began to stifle vegetation there. Even the Soviet government admitted that nearly two-thirds of people in the area of the sea suffered from some illness, and the infant death rates were about 10%, compared with less than 1% in most of the industrial world.

Now the question is what, if anything, can be done about the Aral Sea and its watershed. It is no longer a Soviet problem because they are both wholly within five independent countries created as Russia withdrew within its own borders (we discuss in Box 5.6 a similar Soviet "gift" of the nuclear accident at Chernobyl to two other newly independent countries). The Aral Sea is on the border between Kazakhstan and Uzbekistan. The Syr Darya flows mostly through Kazakhstan but has its headwaters in Kyrgyzstan, and the Amu Darya forms much of the border between Uzbekistan and Turkmenistan and receives some of its

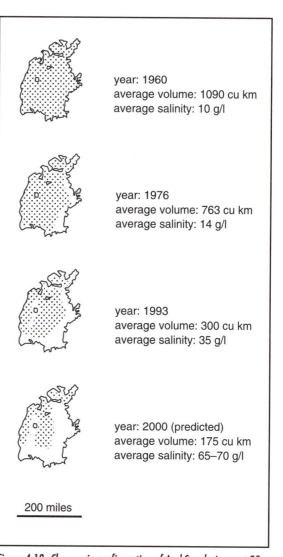

year: 1960
average volume: 1090 cu km
average salinity: 10 g/l

year: 1976
average volume: 763 cu km
average salinity: 14 g/l

year: 1993
average volume: 300 cu km
average salinity: 35 g/l

year: 2000 (predicted)
average volume: 175 cu km
average salinity: 65–70 g/l

200 miles

Figure 4.18 Changes in configuration of Aral Sea during past 25 years. Adapted from Micklin (1993).

water from the mountains of Tajikistan. That means that any restoration of the Aral Sea and the lands around it will require the cooperation of the governments of all five countries.

While awaiting results, we suggest staying upwind of the Aral Sea.

FURTHER READING: Micklin (1993).

bas (single-celled) and tapeworms, and they can remove so much of the infected person's vitality that the person literally wastes away. Principal diseases include amoebic dysentery and bilharzia, a disease prevalent in Africa and the Middle East. Bilharzia is caused by worms that breed in snails, enter surface and groundwater, and then bore through people's skin and infect the urinary system.

The allowable limit for all organisms that can cause bacterial or parasitic diseases is ideally zero. Complete disinfection of drinking water in a water-distribution system is obtained primarily by adding chlorine (chlorination) but also partly by filtering to remove the large parasites. Chlorination is effective because chlorine gas (but not chloride ion from salt) kills all microbial life. Early methods of chlorination used cylinders of gas, but the more common method now is to add some chemical that releases chlorine more slowly, a common one being sodium hypochlorite (NaClO), which is also used as an ordinary laundry bleach. Chlorination and other treatment commonly occurs at the same time as small amounts of fluorine are added to the drinking water in order to harden bones and teeth (see the discussion in Section 2.1).

When water systems temporarily break down after storms or other disasters, and in areas that do not have effective purification systems, avoidance of illness requires individuals to disinfect their own water. The most common method is boiling the water, which kills all parasites and virtually all bacteria, but people living or traveling in areas of chronically bad water carry packages of disinfectant chemicals that release chlorine or, less commonly, bottles of iodine, which is as poisonous as chlorine to microbes.

CHLORINATED HYDROCARBONS   In Section 7.3 we discuss the worldwide proliferation of chlorine-bearing hydrocarbons in the last half of the 20th century (also see Section 2.2). Because of their toxicity to animal and insect pests, they have been used in virtually all countries, but that same toxicity also makes them hazardous to people. Most of these pesticides have now been banned in industrial countries, but because they do not break down easily in soil and water (they are not biodegradable), some of our supplies of surface and groundwater contain residues of pesticides sprayed on fields tens of years ago. The concentrations that render water unfit to drink vary from one compound to another, generally less than 1 part per million (ppm; $10^{-4}\%$). Water with this extent of contamination can be rendered drinkable only by the very expensive process of distillation (boiling the water and condensing the pure steam), and for that reason some water sources simply have to be marked undrinkable.

VOLATILE HYDROCARBONS   Hydrocarbons that are easily vaporized (i.e., they smell) mostly include gasoline and lubricating and heating oil, but a significant source is also the cleaning industry (Section 7.3). As with pesticides, toxicities vary from compound to compound, commonly only a few parts per billion (ppb; $10^{-7}\%$). Soil and water are contaminated with these hydrocarbons by oil spills, leaking gasoline tanks at filling stations, cleaning fluids that are flushed down the drain, and simply by householders throwing away leftover gasoline for lawnmowers, charcoal starters, oil-based paint, and lighter fluids. Because most of these hydrocarbons are so volatile, they can be removed from contaminated water by spraying the water repeatedly in the air, where gradually the most volatile compounds simply evaporate away. The process is extremely expensive, however (see Box 7.3), and consequently water that contains more than the allowable amount of any volatile hydrocarbon must be marked undrinkable in the same way that pesticide-contaminated water is designated. For this reason, many communities now have special collection of toxic chemicals so that households and small business will not simply throw them away.

LEAD   Because it is so dangerous to the nervous system (Section 8.4), the permissible amount of lead in drinking water is very low, with most regulations allowing not more than 0.01 to 0.02 ppb (0.01–0.02 $\times$ $10^{-7}\%$). Because lead compounds are so insoluble, most natural water does not approach those concentrations, but human activity has created four sources.

- Drainage from lead mines and smelters. Lead ore (PbS) can oxidize to soluble lead ions and sulfuric acid, and this source is very dangerous in the vicinity of mines and smelters.
- Landfills that contain discarded lead batteries. Many modern disposal operations either recycle lead from these batteries or make special arrangements to isolate it (or ship it abroad; see Section 7.3).
- Soil and water widely contaminated by exhaust from burning leaded gasoline. This source has now been stopped in most of the world, but lead contamination persists from older use (Section 8.4).
- Lead-soldered pipes. Solder was originally an alloy of lead and various other metals that permitted melting at low temperature. For this reason, many older water pipes have lead solder at their connections, and where this solder begins to etch, the lead can enter drinking water. Partly for this reason, most solder used now contains no lead, and as older pipes are replaced, the problem should disappear.

BOX 4.7 TOO MANY FLORIDIANS AND THE FATE OF THE EVERGLADES

Beginning in the early part of the 20th century, when railroads and real-estate developers began to tout Florida as the "Sunshine State," Florida's population has increased more than 25-fold, making it the nation's fourth most populous state. Every year there are a quarter of a million more Floridians, and armies of tourists who come to see Florida's man-made attractions as well as its few remaining natural regions. They all need water. They are joined in this need by huge agribusiness concerns that raise citrus fruit, vegetables, and sugarcane and which constitute the second largest contributor to Florida's economy after tourism. Florida has become the 10th biggest consumer of water in the United States, and many of the states that use more are sparsely populated and dry places like Idaho and Colorado, which consume huge amounts of water for dry-land irrigation.

The problem in Florida is both too much water and too little water. Florida gets lots of rainfall, more than 50 to 60 inches a year, but that is too much for Florida's topography to handle. The highest point in the state, less than 100 m above sea level, is in the north, and southern Florida is low and flat, with elevations that rarely exceed 20 m above (Miami is at 3 m). South Florida's drainage is dominated by the Kissimmee River, which flows south to Lake Okeechobee (Fig. 4.19). The natural outflow of Lake Okeechobee, combined with the heavy rainfall and the shallow water table, once covered nearly all of interior South Florida to form a vast wetland, dotted with dry isolated hammocks (cypress islands), and sluggish rivers that drained the swamps into Florida Bay. In the wetland, mats of brown-yellow algae floating among the swamp grasses stabilized and held the plant debris, allowing it to settle to the bottom as natural fertilizer, and also served as the bottom link in a complex food chain that included snails, frogs, alligators, and even the endangered Florida panther (Box 8.5; for more on eutrophication, see Section 7.5).

Less than 40% of this great wetland remains, and half is in the Everglades National Park. As Florida's population and economy grew, developers and farmers outside the park, needing more land, cut ditches and deep canals to drain the swamps. These canals lowered the water table, rushed any runoff to the ocean, and produced arable land and desirable sites for homes and communities. Unfortunately, canals dug outside the park also lowered the water table beneath the park. Visitors today see vast dry, grassy prairies that once were beneath perennial sluggish water. On the newly created dry land, houses, condo-

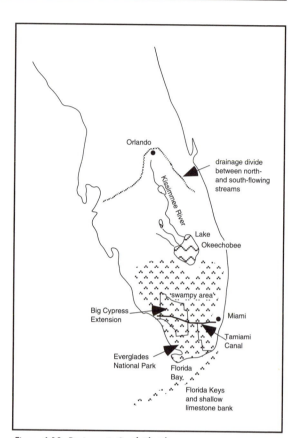

Figure 4.19 Drainage in South Florida.

miniums, and shopping centers are surrounded by lawns and asphalt parking lots. The runoff now acquires nutrients from fertilizers and pollutants from human activity and delivers them to Florida Bay. The decrease in runoff and increase in nutrients has made the bay saltier, causing salt-intolerant sea grasses to die, blooms of nonnutrient algae to occur with recurring severity, and commercial shrimp hatcheries to disappear.

The situation is the response of south Florida to too much water. Despite this amount of water, the lack of large rivers requires that Floridians make do with groundwater for municipal supplies, and in this sense they have too little water. Lowering of the regional water table to create new land and heavy groundwater extraction for municipal purposes resulted in seawater incursion into aquifers. In addition to incursion below the water table, the twice-daily change from low to high tide produces a surge of salt water up the rivers. This saltwater wedge, being denser than the fresh water in the river, travels up stream along the river bottom. This wedge usually doesn't get very far since it has to force its way against the oceanward flowing river water, and if the river is a gaining river, the wedge simply sloshes up and down the river and has minimal effect on the groundwater. But, if the coastal river is a losing stream, if its channel is above the water table, then the wedge will seep into the underlying aquifer from the river bottom. In some parts of south Florida, salty groundwater is creeping inland 100 feet or more each year, and communities have been forced to abandon wells that became salty or install reverse-osmosis plants to remove the salt.

Overall wetland loss has probably now been stabilized by building dikes around portions of the park and pumping water from nearby drainage canals into rivers that run through the park. But to return the park to its more pristine state, to replace the many dry areas with wetlands, will require a more sustained effort. Some have suggested that the state of Florida buy adjacent farmlands, fill the drainage ditches and canals, and breach the dikes surrounding the park so that natural recharge can be reestablished. Farmers clearly oppose this. They suggest a "curtain" wall along the east side of the Park to form a vertical groundwater barrier that would shield the aquifers inside the park from the groundwater withdrawals outside that are needed to preserve the farmlands. Florida is considering creating man-made marshes north of the park to allow natural processes to filter out excessive nutrients that enter the Everglades from runoff in the highly developed agricultural areas around Lake Okeechobee. The U.S. Army Corps of Engineers, has a $372-million project to take a 52-mile, straight stretch of the Kissimmee River and, over a 10- to 15-year period, divert it into a 100-mile meandering channel that will create 26,000 acres of new wetlands and allow natural processes to clean nutrients from the river and recharge aquifers. This project will essentially return the river to its original channel from which they diverted it into its current straight channel in the 1960s.

All of these projects have price tags of hundreds of millions of dollars, and we can easily imagine the reaction of various groups to the question of who should pay. While we may have learned a good deal about wetland ecology in the past 30 years, we haven't learned a whole lot about conserving the public's tax dollars.

FURTHER READING: Cohn (1994); Finkl (1995).

Water that has unacceptable levels of lead can be cleaned up by running the water through water purifiers, which contain various chemicals that absorb ions, but ultimately the lead contaminates the purifiers and they have to be replaced. Replacements are expensive, but the only alternative is distillation, which is even more expensive. Thus, water that is contaminated before it enters pipes must be regarded as unusable.

SALT   Salt (NaCl) is not harmful except at very high concentrations, but any level above about 0.05% makes the water taste bad. This problem generally arises only in the groundwater of coastal areas (Section 4.4) and can usually be resolved by reducing the rate at which the water is pumped or by installing expensive purification systems.

HARDNESS   Hardness is a term that refers primarily to the concentration of dissolved calcium carbonate ($CaCO_3$). Most of this calcium carbonate comes from the solution of limestone that rivers run over or, more commonly, that groundwater filters through. As with salt, calcium carbonate is not harmful when drunk, but it causes problems in two other ways. One is that the calcium carbonate does not evaporate when water is boiled. Instead, it precipitates on the side of the pot or coffee percolator and may ultimately clog openings in spouts. The second problem is that calcium ion reacts with some types of soaps and detergents, making insoluble scums that stick to clothes and dishes and clog drains, although modern varieties of cleaning agents are far less susceptible to this type of reaction. People who live in areas of hard water either have to accept their problems, use water purifiers to absorb the calcium, or buy distilled water.

IRON   Iron is so insoluble that virtually no water contains enough iron to be a health problem. The problem with iron is that its dissolved ions oxidize to iron oxides (commonly Fe_2O_3) with a reddish color. This iron oxide is insoluble and tends to coat sinks and clothes and clog pipes and drains. In areas with iron-rich water, people either buy water purifiers or accept the fact that their sinks will be stained.

Other uses

We can distinguish four major uses for water other than drinking, and each of them has different requirements.

- Water used for cooling or other industrial purposes may be contaminated with microbes or most hazardous chemicals. It cannot, however, contain iron because of the ability of iron oxides to coat pipes and filters.

- Water used in boilers to produce electricity (Section 5.1 and Fig. 5.5) should contain only very low levels of iron, but the contaminant that causes the most problems is calcium carbonate. Precipitates of calcium carbonate, which are referred to as "boiler scale," may be a minor annoyance in a tea kettle, but in a boiler generating electricity for several counties they can be a disaster. Generating plants try not to use very hard water, and those that do must temporarily shut down and clean the scale out of the boilers.

- Irrigation water may contain moderate levels of microbes and even some toxic chemicals. Plants generally do not absorb these contaminants and can be safe to eat or to be used as cattle feed. The most serious contaminant for irrigation water is salt, because most plants have only a limited tolerance for salt, and its buildup in soil can render large areas virtually unusable.

- Water for animals and poultry has to pass many of the same specifications as drinking water. It cannot contain microbes or lead, and the levels of chlorinated and volatile hydrocarbons, although higher than permitted for people, must be as low as possible. The major difference from drinking water is that hard and somewhat salty water is fine for animals, who appear to be less finicky about taste than we do.

Policy questions

Do you know whether the water that you use at home or at work is pure enough for drinking or some other intended purpose? If not, are you going to inquire?

FURTHER READING: van der Leeden, Troise, and Todd (1990); Cheremisinoff (1993).

4.6 WHAT DO SOCIETIES DO TO CONTROL AND APPORTION WATER?

In the international river basins we have studied, . . . outcomes reflect the distribution of power. Cooperation is not achieved unless the dominant power in the basin accepts it, or has been induced to do so by an external power. — *M. R. Lowi, Water and Power, p. 203*

Water, like air, belongs to the commons, to that part of the terrestrial environment which seems that it should belong to everyone (see Sections 2.3 and 2.4). It is a mobile resource that moves over, across, and through private property, and a resource whose misuse or misappropriation can adversely affect others who depend on it for a livelihood or for survival. Therefore, it should not come as a surprise to you that water is, and has been, the subject of duels, litigation, treaty negotiations, and war. After all, if wars can be fought over who won a soccer football game, wars can certainly be fought over rights of access to a vital resource like water.

Some historians and anthropologists are convinced that providing a steady and reliable supply of water was critical to the rise of urban civilization and consider hydraulic engineering to have been as critical as astronomical observations and domestication of animals and grains in our rise from preagricultural nomadism to the complex postindustrial society we are today. In most parts of the world, and certainly in the fertile river valleys of the Middle East where Western civilization evolved, a preindustrial farmer dependent only on rain and what water the family could carry from the nearby river and shallow wells was barely able to raise enough food to sustain a family. What few surplus crops and animals the family produced went to support priests and soldiers, a few tribal leaders and their families, and perhaps an itinerant merchant or two who traded critical materials not locally available. Thus, our simple farmer got salt to preserve meat, spices to cover the rancid flavor of old meat, stone and flint for knives, implements and weapons, and pigments and metals for arts and adornment by exchanging the few surplus agricultural goods.

This simple trading might have been the beginning of settled civilization but still not the takeoff point for urban civilization. However, in areas with soils sufficiently rich and bountiful, like the lower Nile Valley or the fertile crescent of the lower Tigris and Euphrates Rivers, more and more people who did not raise food or livestock became more and more dependent on those who did. By increasing this dependence, they also became dependent on the regular and adequate supply of water, and those who could design catchments to hold water in reserve for the dry months or years and could build irrigation systems to allow farmers to grow more crops on the available land became critical to the growth and development of cities. These unsung heroes were hydraulic engineers. We know that hydraulic engineering started at least 6,000 years ago when water was lifted by early Sumerians using giant waterwheels into canals that watered the fields of ancient Mesopotamia. In fact, it is likely that the first interbasin transfers of water, from the Euphrates to the Tigris basin, occurred in the 4th century B.C.

With the engineers, came the lawyers. The great Babylonian lawgiver, Hammurabi, worried a good deal in his codification of laws about such matters as who was responsible for maintaining the life-supporting canals and floodgates, who should pay for materials, and who should be required to work on these public projects. Irrigation civilizations require laws and central administration, engineers and public works departments, political stability and regulation to survive. This lesson was learned by the Sumerians and Babylonians and it is a lesson we must remember, since most of the world today is dependent on food and fiber that derive from irrigation-based agriculture. It is categorically impossible for our billions to survive in any state even remotely approaching sustainability without massive governmental water projects and international cooperation on uses of water and maintenance of its quality.

Who owns the water?

At first this seems to be a silly question. Water is common property – everybody owns it. But is it quite that simple? Remember when we discussed the Colorado River in Section 4.3 how complex things became because rivers have this bad habit of not confining themselves to one political jurisdiction. Suppose you own a piece of land and a small creek runs across it. Can you build a pond on that creek even if it shuts off the water supply to your downstream neighbor while the pond fills? If you have a septic system that leaks untreated sewage into the creek, can your neighbor sue you even if it leaked into your creek, on your land? Could you use that creek water to irrigate your vegetable garden even if the evaporation and infiltration in your garden decreases the creek flow to such an extent that the fish all die? Whose fish are they? If the edge of your property is bounded by a good-sized river that flows 1,000 miles to the sea, can you prevent ugly barges from floating by your river-front property, can big shipping companies prevent you from building a dock out into the river so you can go fishing, can an upstream user divert water into another drainage basin so that your lovely river view is now a mud flat? And what about the groundwater beneath your land? Is it yours? Can you pump as much as you want even if you end up removing groundwater from beneath your neighbor's land and making his well run dry?

Ownership of water – or, perhaps more correctly, control of water use and misuse – is not as simple as it may

have seemed at first. This book is not long enough for a discussion of this topic, which lawyers call water rights – search of the phrase "water rights" turned up 283 references on our library's computer – and we simply address some of the critical issues and traditions that will allow you to understand many of the problems that underlie the apportionment of both surface and groundwater in a modern populous society. Perhaps the most important lesson to carry away from this discussion is that the difficulty of the issue of who owns water depends largely on: (1) how scarce water is; (2) how many claimants there are for the water; and (3) how powerful the claimants are. In areas of abundant rainfall and clear flowing rivers, water rights are not the most contentious issue in the public eye. Where rain is infrequent, rivers run dry, and groundwater is deep and expensive to pump, who owns the rights to the water is a life-and-death issue. Not surprisingly, since people first settled and wrote their laws and constitutions in places where water abounds, these legal doctrines often do not fit well in arid regions or political jurisdictions where too many people are putting too much demand on a limited resource (Box 4.8).

SURFACE WATER RIGHTS   In the following discussion, we will first focus on U.S. water rights since the two main principles that operate represent the policy extremes in the apportionment of surface water. Other nations have their own traditions, but the two we describe will give you a good flavor of the options that exist in deciding who gets to use surface water and for what purposes. The first general principle with respect to rivers and streams, which applies in most of the nonarid world where private property rights are paramount, is the "riparian doctrine." This concept distinguishes two classes of streams: (1) navigable, meaning that a person in a flat-bottomed boat could reasonably expect to ascend the stream; and (2) nonnavigable. The riparian doctrine assumes that no individual can own a navigable stream, neither the water flowing in it nor the river bed beneath it. For this reason, no one charges tolls on a river (although they may charge tolls to cross it), nor can anyone be prevented from boating along a river even if the land on either side is privately held. Riparian owners who own land adjacent to such a navigable river can not be denied access to the river or its water; but equally, they can do nothing to hinder navigation (like build a huge dock out into the main channel) or withdraw water in excessive amounts. For a navigable river or stream, it is clear that the water and the watercourse are part of the commons.

Issues are trickier on nonnavigable streams. Here, a riparian owner can own the stream bed that crosses the

BOX 4.8 THE JORDAN RIVER

Eighteen million years ago, as the Arabian peninsula continued to move away from Africa and toward Asia it also began to slide directly northward (see Section 5.3). Its northward movement crumpled the mountains along the southern border of Turkey and also left a zone of crustal weakness extending from those mountains all the way south to the Red Sea. The earth's crust has sagged along that zone, and now much of it is below sea level. Because it is so low, most of the water in the area runs toward this zone of weakness, creating a broad watershed that extends through four countries (Fig. 4.20). All water ends up in the Jordan River, which is wholly below sea level and runs down through Lake Tiberias (Sea of Galilee; 700 feet below sea level) to the Dead Sea (1300 feet below sea level).

Prior to World War II, the entire area was controlled by Britain (Jordan and Palestine) and France (Lebanon and Syria), but after the war the European authorities left. At that time, Lebanon, Syria, and Jordan became independent countries, and the British designated the new country of Israel in part of the former area of Palestine. The borders of Israel were drawn along intricate lines that were hoped to cause the least international friction in the future. The eastern border of Israel is along the Jordan River in some areas but west of it in others, leaving a dominantly Palestinian area between the river and Israel that is now known as the West Bank. The drainage divide on the western edge of the Jordan watershed is within this West Bank area, thus officially leaving Israel access to water from the river only upstream along the Israel–Syria border or along a small stretch of border just below the junction of the Jordan and Yarmouk Rivers (Fig. 4.20).

In this book, we cannot hope to discuss the religious and diplomatic issues and resulting conflicts in this part of the world, but we can point out the contribution of water to them. The headwaters of the Jordan River drainage are in Lebanon and Syria. Lake Tiberias is within Israel, but much of its drainage comes from other countries. Water from the Jordan River along the Israel–Syria and Israel–Jordan borders should be shared by mutual agreement. The largest input to the Jordan River is from the Yarmouk River, which is wholly in Jordan except for a small stretch along the Jordan–Syria border. When we add to these complications the fact that the whole area is semiarid and not a drop of water can be wasted, we can see geological roots to a conflict already deeply rooted in cultural and economic differences. Perhaps the necessity of solving the problem of water distribution can contribute to the solution of the other problems.

FURTHER READING: Lowi (1993).

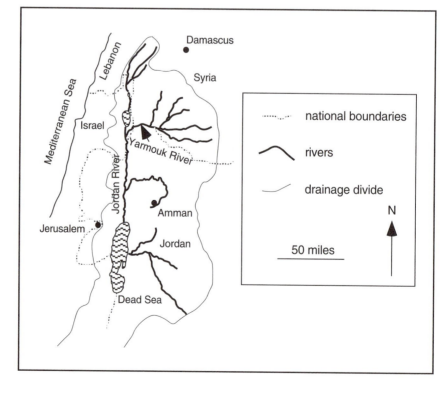

Figure 4.20 Watershed of Jordan River.

owner's property, but the riparian doctrine holds that the landowner across whose land the stream flows, and all riparian landowners whose land is crossed by or abuts the stream, have the right to have the water in that stream remain in a natural state. No one, the owner included, can impair the quantity or quality of that water. The owner may use the water in that stream – build a pond, drive a mill wheel, water a garden, go fishing – but only if that use in no way affects the equal right of every riparian owner, large and small, on that stream or river. In its strictest sense, application of the riparian doctrine would mean that cities, for example, could not dam streams to provide drinking water since that would, in the most literal sense, change the natural state of the water. So courts take a more lenient view of this in modern society and only support riparian complainants who can show that an upstream user has interfered with their "reasonable" use of the water. There is clearly lots of room for litigation in that word "reasonable."

As we noted, the riparian doctrine is a legal principle derived in humid climates. It developed in England, eastern North America, and western Europe in the early 19th century when population densities, with a few notable exceptions, were low and there was a sufficiency of water for all public and private uses. By the 1870s, emigrant settlers from Europe and the east coast of America had begun to penetrate into the semiarid lands west of the one-hundredth meridian in North America. They settled on rivers and creeks and began to farm. They quickly found, as the Sumerians in Mesopotamia and the Native Americans and Mexicans in the American Southwest had before them, that people in semiarid lands must divert surface water to irrigate crops if they are going to survive. So, with little or no thought to riparian rights, they diverted streams and rivers with some abandon. It did not take long, perhaps the first summer in residence, for downstream settlers to realize that upstream farmers and ranchers were depriving them of vital water for their crops and livestock. Angry people and eager lawyers descended upon the territorial and state capitals to seek court injunctions and rulings and to pass laws protecting their new economic interests.

Out of this process, which is best described as making laws up as you go along, the western United States threw out the riparian doctrine and came up with another, rather extreme, idea. It is called the "appropriation" doctrine, and it was a radical departure from our legal traditions. It basically says that whoever gets to water first and uses it in some way has the right to that water. Put simply: "First in use; first in right." The radical part of this

doctrine is not the idea that water in a creek or river has no prior owner; the riparian doctrine says that. What is radical is that the appropriation doctrine holds that this water can be claimed and used regardless of anyone else's rights simply on the basis of who gets there first. The appropriation doctrine also invented something else. It said that water must be appropriated for "beneficial" use (remember, the riparian doctrine talked about "reasonable" use). The courts quickly decided that beneficial uses implied out-of-stream uses, not fishing or rafting, and did not limit the use to riparian lands – that is, lands adjacent to the body of water in question. Thus, the first person to divert a river to irrigate lands, no matter how far away, even if they were in a different drainage basin entirely, would be the person with the best case in court. Since the courts also ruled that to maintain your right, you had to exercise it ("Use it or lose it"), it became customary for those possessing water rights, but lacking the means or interest in using them, to sell or lease the rights. Out of this tangled web of appropriation comes a vast bureaucracy that regulates, and a host of lawyers who litigate, the competing surface water rights of private citizens, cooperatives, corporations, states, the federal government, and Native American tribal groups (to name only the most prominent participants).

In other parts of the world, where market economies are not as highly developed, the extremes of the riparian and appropriation doctrines are usually replaced with governmental ownership of the surface water. Such centrally planned economies rely on government experts and agencies to decide who gets water and who does not, to approve all riparian improvements and projects, and to penalize those who fail to abide by recognized priorities and standards of use. Where the situation gets particularly nasty is where rivers cross international boundaries. To the extent that neighboring nations who share a river watershed get along on other issues, we may come to understand how well they get along on riparian issues. This is not an area where logic and rationality prevail.

There are considered to be four basic types of principles that govern the control of international rivers:

1 The *principle of absolute territorial sovereignty* says a state can do whatever it wishes with the water in any international river regardless of where these waters may subsequently flow. Clearly, this would be favored by upstream states, so it should come as no surprise that this is called the Harmon doctrine after a U.S. attorney general who took this position in negotiations with Mexico in 1895 over utilization of the Rio

Grande River. This position has been taken by others: India over rights to the Indus, Turkey over rights to the Tigris.

2　The *principle of absolute territorial integrity* is the opposite of territorial sovereignty and holds that no state can utilize the waters of any international river in any way that *might* cause detrimental impacts on coriparian states. While not stating it outright, this, in effect, implies that no nation can alter the natural state of any international river since that *might* have a harmful effect downstream. As you might imagine, this is a bit touchy since it represents a state's surrender of its sovereignty over some of its rivers. It could not build a dam or, in theory, even divert waters to cool a power station. This principle, not surprisingly, is favored by downstream states like Bavaria in the past century in a dispute with Austria over the River Inn and the Egyptians with respect to the Nile.

3　The *principle of common jurisdiction* limits a nation's freedom of action by requiring prior consent from coriparians before any use of an international river that might have a detrimental impact. Usually arbitrated by an international body, such consent agreements have been obtained between the United States and Canada over the Columbia River, the Netherlands and Belgium over the Meuse, and Israel and neighboring Arab states over the Jordan (Box 4.8).

4　The *principle of equitable utilization* allows a nation to use an international river's waters as long as its use does not do harm to any coriparian state. This is a more reasonable approach to territorial integrity since it requires a nation to use best-engineering practices and possess a full scientific understanding of the conditions in a watershed before venturing into a project on an international river. This principle has become formalized in 6 chapters and 37 articles of what are often called the Helsinki Rules, which govern most international legal actions in the area of surface water allocation and use today.

If only the pesky water would stay in one place, all these problems would be solved. It would also help if rivers and lakes hadn't been such barriers to transportation and communication since that attribute makes them so often the boundaries between jurisdictions thereby assuring conflict over the rights to the water therein. But water moves and the problems inherent to surface water are only compounded in the case of groundwater by its invisibility and by most people's lack of understanding of the physical laws that govern its behavior.

GROUNDWATER RIGHTS　Who owns a fugitive resource that is hidden from view and often requires significant labor or capital expenditure to obtain? This is the question that underlies groundwater appropriation. It is somewhat like asking who owns the moon? Until the resource becomes accessible and is exploited at a rate that makes others with equal access begin to worry about future availability, such questions of ownership are interesting debating points but hardly as important as day-to-day questions like the weather and who won the football match. But, as we have seen in our discussion of the High Plains aquifer, we can and do exploit groundwater at rates that ultimately jeopardize the resource, and thus issues of ownership and control are just as important in subsurface as in surface water.

As one student of the subject has put it, "For most of history . . . the major problem concerning groundwater has been how to get it out of the ground" (Smith, 1989, p. 6). In effect, as long as animals and wind were the only energy sources for pumping groundwater, no one worried about whose water it was. In the Anglo-Saxon legal tradition (born in a water-abundant land in a time before the science of hydrology was conceived of), use of groundwater was classified as a property right no different from the right to the soil, timber, or minerals on or beneath a parcel of land. Landowners had virtually a free hand to do as they wished with any groundwater they could extract from springs and wells on their land. This principle worked moderately well until inventions like the steam-driven centrifugal pump and improvements in pump design employing cheap, usually government-subsidized, electrical power increased the volume of groundwater extraction in most industrialized nations by 10-, 100-, 1,000-fold in a century. Suddenly, a single-property owner could, by aggressive pumping, begin to reduce the availability of a neighbor's groundwater resource.

The legal recourse in such situations is the court system and, over a period of years, the courts argued that all property owners above an aquifer had coequal rights to a "reasonable" amount of groundwater – that word "reasonable" again is fodder for many a lawsuit. In general, the courts rule that reasonable uses are those essential to the landowner's survival and economic well-being. Unreasonable uses would be those that polluted, wasted, or transferred water off of the property in question. Of course, no one challenges an unreasonable use until that use has some negative impact on others. In some regions of the United States, regulatory limits are placed on reasonable use by regional groundwater agencies in order to guarantee the coequal rights of all landowners above a common aquifer, especially during periods of drought. Thus, for example, the "3-mile, 40%, 25-year rule" in the High Plains of Colorado allows a farmer to drill a well as long as no one within 3 miles is pumping at such a rate as to consume more than 40% of the water during any 25-year period – fairly arbitrary, but a starting point at least. Not surprisingly, the "appropriation" doctrine that is applied to surface water in many water-deficient parts of the United States has also been applied to groundwater. In these states, some central agency or office grants permits for all wells based primarily on who got there first and who continues to use the water for beneficial purposes.

Prior appropriation of groundwater raises many of the same problems as prior appropriation of surface water does. Proof of priority in time and of continuing use can be difficult to obtain. Competing claims are inevitably going to end up in litigation since, in the arid western United States, survival literally may depend on access to groundwater. Definition of what is a beneficial use is a nagging problem. Tradition holds that domestic use, irrigation, stock watering, and manufacturing are beneficial. Is protection of wetlands or scenic areas a beneficial use? What happens when the needs of a large number of citizens run counter to the needs of a few? For example, a large volume of groundwater may be needed to provide water to transport coal as a slurry in a pipeline to feed a power plant for 1 million people in Arizona. Suppose that all the groundwater rights have been appropriated and are used by a few ranchers and farmers to raise cattle. What happens? Who has the highest and best right, if not the earliest? Who is going to win in the state legislature? And, finally, just to make matters worse, we should remember that groundwater commonly feeds surface water streams, rivers, and lakes. This means that in many cases, extraction of groundwater by a property owner could conceivably reduce the amount of water flowing in a stream whose water is owned by someone else.

Policy question

Do you know of any water-allocation laws affecting you, your community, or your country that you think should be changed?

FURTHER READING: McDonald and Kay (1988); Moore (1989); Smith (1989); Lowi (1993); Young, James, and Rodda (1994).

PROBLEMS

1 A river flowing through a flat ranchland has a surface 10 feet below the land surface (a bank 10 feet high). A well drilled on the ranchland intersects the water table at a depth of 20 feet. Is the river gaining water from the ground or losing it to the ground?

2 The accompanying diagram shows a 1-mile stretch of

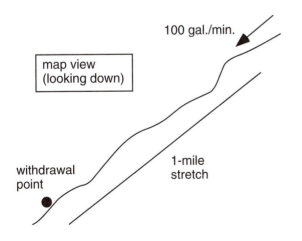

stream in a humid area. The stream above this stretch delivers a discharge of 100 gallons per minute, and groundwater seepage adds 20 gallons per minute along the 1-mile stretch. If you were in charge of a water supply system that withdraws water at the lower end of the stretch, how much water would you withdraw?

3 The mass of ocean water is 1.4×10^{21} kg. The salt (NaCl) content is 3.5%. Rivers deliver 4×10^{14} grams of NaCl to the oceans each year. Based on these data, what is the residence time of NaCl in the oceans? Does your calculation require that oceans maintain the same concentration of NaCl at all times?

4 Groundwater is moving through the area shown in the diagram at a rate of 6 feet per day. If heavy rains cause the water table to rise 4 feet under the hill, how fast will the groundwater flow be?

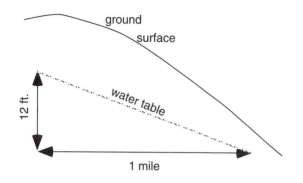

5 The populations (in millions) of the seven states in the Colorado River Compact are: Arizona, 3.67; California, 29.76; Colorado, 3.29; Nevada, 1.20; New Mexico, 1.52; Utah, 1.72; Wyoming, 0.54. If the water in the river were allocated to each state based on its total population, what percentage would go to each state?

6 A power company is considering building dams to produce hydroelectric power on the river shown. In the diagram, as shown, they might build one dam or two. Using the concepts in Section 4.4 and Figure 4.10, decide whether one dam would produce more, less, or the same power as two dams.

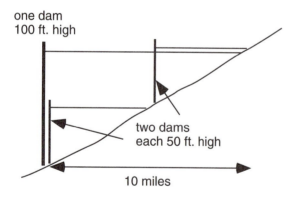

7 Saudi Arabia annually uses about 4 cu km of water but receives only 2 cu km of rain. The remaining water is obtained either by desalination of seawater or by importing. A potential source is icebergs towed from the Antarctic across the Indian Ocean to ports in Saudi Arabia. Antarctic icebergs commonly float about 10 m above sea level. Using the density of ice as 0.9 g/cm^3 and water as 1.0 g/cm^3, and assuming that 75% of the ice volume will melt as it is towed northward, calculate the total surface area of icebergs that would have to be towed to Saudi Arabia to supply its needs.

8 A lake with a surface area of 2 square miles maintains a relatively constant depth of 20 feet. Each year, it discharges about 30 million cubic feet into a river. Calculate the residence time of water in the lake.

9 The average person in the United States uses about 50 gallons of water each day for domestic purposes (bathing, washing clothes and dishes, flushing toilets, watering lawns, etc.; for more information, see Table 4.5). Estimate the percent reduction in total U.S. water use that could be obtained by replacing present toilets, which use about 3 gallons per flush, with modern toilets that use about 1 gallon per flush.

10 A well is drilled in each of the locations shown in the diagram as cross sections. In which well or wells will the water rise highest, and in which will it be the lowest?

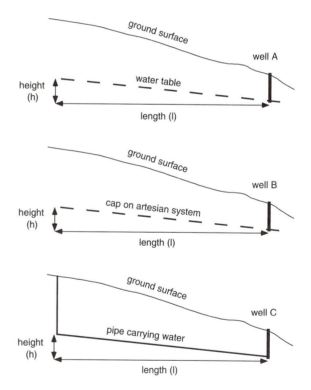

11 The diagram shows a profile of the top (water–air surface) and bottom (water–sediment surface) of a river. In which direction is the river flowing?

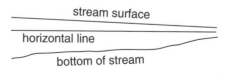

REFERENCES

Cheremisinoff, P. N. (1993). *Water Management and Supply.* Englewood Cliffs, N.J.: PTR Prentice-Hall.

Cohn, J. P. (1994). Returning the Everglades. *Bioscience* 44: 579.

Finkl, C. W. (1995). Water resource management in the Florida everglades: Are "lessons from experience" a prognosis for conservation? *Journal of Soil and Water Conservation* 50: 592–600.

Fradkin, P. L. (1996). *A River No More: The Colorado River and the West.* 2nd ed. Berkeley: University of California Press.

Gardner, G. (1995). From oasis to mirage: The aquifers that won't replenish. *World Watch* 8(3): 30–6.

Gleick, P. H., ed. (1993). *Water in Crisis – A Guide to the World's Fresh Water Resources.* New York: Oxford University Press.

Heath, R. (1995). Hell's highway: Conservationists campaign against Hidrovia Project converting Paraguay–Parana River system into a shipping lane. *New Scientist* 146, no. 1980: 22–5.

Kromm, D. E., and White, S. E., eds. (1992). *Groundwater Exploitation in the High Plains.* Lawrence: University of Kansas Press.

Lowi, M. R. (1993). *Water and Power – The Politics of a Scarce Resource in the Jordan River Basin.* Cambridge: Cambridge University Press.

Manning, J. C. (1992). *Applied Principles of Hydrology.* 2nd ed. New York: Macmillan.

Mather, J. R. (1984). *Water Resources: Distribution, Use, and Management.* New York: John Wiley and Sons.

McDonald, A., and Kay, D. (1988). *Water Resources: Issues and Strategies.* Harlow: Longman Scientific and Technical.

Micklin, P. (1993). The shrinking Aral Sea. *Geotimes* 38(4): 14–18.

Moore, J. M. (1989). *Balancing the Needs of Water Use.* New York: Springer-Verlag.

Pinneker, E. V. (1983). *General Hydrogeology.* Cambridge: Cambridge University Press.

Price, M. (1996). *Introducing Groundwater.* London: Allen and Unwin.

Reisner, M. (1986). *Cadillac Desert: The American West and Its Disappearing Water.* New York: Penguin Books.

Selley, W. B., Pierce, R. R., and Perlman, H. A. (1993). *Estimated Use of Water in the United States in 1990.* Washington, D.C.: U.S. Geological Survey Circular 1081.

Shaw, E. M. (1983). *Hydrology in Practice.* Berkshire: Van Nostrand Reinhold (UK).

Smith, Z. A. (1989). *Groundwater in the West.* San Diego, Calif.: Academic Press.

Speidel, D. H. (1988). *Perspectives on Water: Uses and Abuses.* New York: Oxford University Press.

Stegner, W. (1953). *Beyond the Hundredth Meridian.* Cambridge, Mass.: Riverside Press.

van der Leeden, F., Troise, F. L., and Todd, D. K. (1990). *The Water Encyclopedia.* 2nd ed. Chelsea, Mich.: Lewis Publishers.

Walton, J. (1992). *Western Times and Water Wars: State, Culture, and Rebellion in California.* Berkeley: University of California Press.

Waterbury, J. (1979). *Hydropolitics of the Nile Valley.* Syracuse, N.Y.: Syracuse University Press.

Young, G. J., James, C. I. D., and Rodda, J. C. (1994). *Global Water Resource Issues.* Cambridge: Cambridge University Press.

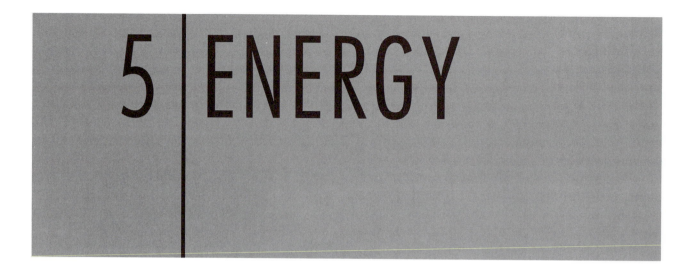

5 ENERGY

5.0 INTRODUCTION

By the early 1980s, KGB reports declared that the cushion of oil profits was all but gone. The abyss awaits us. — D. Remnick, Lenin's Tomb, p. 24, citing a report by the Russian secret service predicting decline and failure of the Soviet Union

At twilight in a rural village in India, a resident walks home along a path through fields of grain, around some fruit trees, and then follows a dusty road past a few cows to a hut of mud-daub walls and thatched roof tenuously supported by a precious wooden beam. At home, the cooking must be started before the daylight completely disappears, and some of the bread, commonly unleavened, was probably baked in small earthen ovens during the day. Other parts of dinner consist of vegetables and grain fried or boiled over fires made with animal dung, scraps of wood painfully collected from dwindling forests, or small kerosene stoves in the more affluent homes.

As night sets in, the village darkens. In most of it, the only light may come from small oil lamps carefully lit before family shrines in alcoves in the walls of the huts. In some villages, the residents make their way to a central area in front of a public building along the main road where a local generator or a wire from the country's power grid supplies electricity. The electricity provides light, perhaps a radio for the villagers to listen to or even dance to, and, increasingly in modern times, even a television set. After a short time, all villagers return home to bed, warming themselves with small blankets on those uncommon occasions when the nights are cool. Because of the lack of light, most people will sleep nine or more hours each night.

Meanwhile in the United States, a resident of suburbia drives home from work, a commute of perhaps 15 to 20 miles. Arriving home exhausted after spending the last half hour dodging maniacs on the freeway, the resident realizes that the evening meal will not be complete without some frozen egg rolls, which have been forgotten in the furor. The resident does not even enter the house but backs the car out of the driveway, drives 3 miles to the local grocery, dodges more maniacs with shopping carts, buys the rolls, and drives another 3 miles home.

At home, the family turns on all of the lights and two television sets (one for junior, who won't watch anything that the rest of the family likes). The egg rolls are popped in the oven (electric), and the rest of the dinner is taken out of the icebox (electric) and cooked on the stove (gas). This evening both dishes and clothes must be washed, requiring use of the dishwasher (electric, with hot water from a gas water heater), clothes washing machine (electric plus gas-heated water), and dryer (gas). Because temperatures may drop to 60°F (about 15°C) overnight, the family turns on the heat (air heated by gas and distributed by an electric fan).

Before going to bed, to sleep an average of 6 to 7 hours, the family leaves all of the lights on and opens the day's mail. It contains three bills. One is from the oil company whose credit cards the family uses to charge purchases of gasoline for their cars. A second bill is from the electric company, and a third is from the company that supplies natural gas to the home. All three bills are high. Intolerably high! Protests must be filed – to the companies, to the mayor, to state agencies, to congress, possibly even to the president. After all, the family uses a minimum amount of energy. The family cars are only driven for necessary trips, lights are sometimes turned off in unoccupied rooms, thermostats are never set above 75°F in the winter or below 65°F in the summer, and sometimes the family uses paper plates and doesn't even have to run the dishwasher. There is absolutely no way that the bills for gasoline, gas, and electricity should be this high!

Well, it's a fact that American suburbanites use a lot of energy. They drive too far, use too many electrical gadgets, and keep a very large house too warm in the winter and too cool in the summer. And this energy used personally is only a small part of the per capita energy that is used by America's industrial society to provide the food and goods necessary to lead a "comfortable" life-style. The per capita energy consumption in the United States is approximately 25 times that in India. This energy does work that residents of India do for themselves (Fig. 5.1).

The rest of this chapter is devoted to energy – where it comes from, how it is transported, and what we use it for. We summarize the basic concept of energy (5.1) and the sources of energy (5.2), and discuss fossil fuels (5.3), nuclear power (5.4), and the present and possible future sources and uses of energy (5.5).

5.1 TEMPERATURE, ENERGY, HEAT, AND POWER

All of our discussion requires a knowledge of the concepts of energy (heat) and temperature and the units that we use to measure both.

The temperature of an object is a measure of the rate of vibration of the atoms and molecules in it – as the vibration rate increases, the temperature rises. One indication of temperature increase is an increase in volume, which permits us to construct ordinary mercury thermometers. As temperature increases, the mercury expands upward along a scale calibrated in degrees, and if the thermometer is at the same temperature as the object being measured (perhaps the area under your tongue), the calibrated scale tells us the temperature of the object (perhaps you). Temperature scales are calibrated arbitrarily. The centigrade scale is used by virtually everyone in the world and is set so that the freezing point of water at sea level is zero (0°C) and the boiling point is 100°C. An older scale, used only in the United States, is the Fahrenheit scale, set with 32°F as the freezing point of fresh water and 212°F as the boiling point. Translation between these scales is provided by the equation °C = 5/9(°F − 32).

Although we commonly refer to temperatures as hot or cold, the term "heat" actually has quite a different meaning. Whereas temperature is a property of an object, heat is a quantity in the same sense as pounds, grams, gallons, or other measures of weight and volume. We can demonstrate the difference with a simple experiment (Fig. 5.2). Place two pots on a stove, one containing 1 quart of water and the other containing 2. Start both pots at the same temperature and bring the water in them to boiling. The 2 quarts of water require longer to boil

Figure 5.1 Consequences of living in an energy-poor country like India include walking instead of driving along roads.

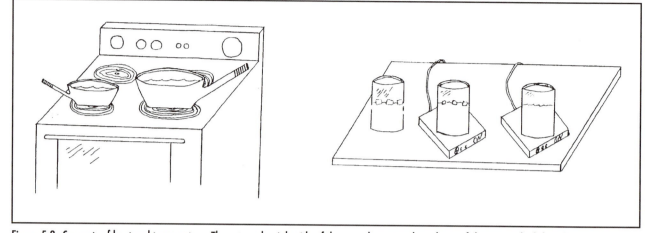

Figure 5.2 Concepts of heat and temperature. The pot on the right side of the stove has twice the volume of the pot on the left and requires twice as much heat for the same temperature increase. On the table, the glass with ice on the hot plate will remain at 0°C until all of the ice melts, but the glass without ice will begin to heat up as soon as the hot plate is turned on.

than the 1 quart, indicating that the stove has had to add more of something – we call it heat – to the larger volume of water than to the smaller one. The temperature change for both pots is the same, however, from a starting room temperature to the boiling point (100°C). In the simple situation of heating or cooling something, the amount of heat that must be added or subtracted is proportional both to the temperature change and the mass of material undergoing the change. More precisely, heat equals mass times temperature change.

The relationships between heat and temperature are slightly more complicated when the object being heated or cooled undergoes a change of condition (change of "phase"). We can illustrate this relationship with another simple experiment (Fig. 5.2). Start with a glass of water containing ice cubes. At "thermal equilibrium" both the water and the ice are at a temperature of 0°C, the freezing point of water. If we put the glass on a hot plate and add heat, the ice will melt, but the temperature will not rise until all of the ice is melted. In this case, the added heat is used to change the solid ice to the liquid water, and we conclude that water contains more heat than ice at the same temperature. After all of the ice has melted, the temperature of the water rises in proportion to the amount of heat that we add until boiling starts. At 100°C, the water undergoes another phase transition to a vapor, and heat must be added to water to produce that vapor even though the temperature remains constant.

The amount of heat needed to raise the temperature of a substance or to cause it to melt or boil can be measured in various units (Table 5.1). The simplest is the calorie, which is the amount of heat needed to raise the temperature of 1 gram of water by 1 degree centigrade. On this scale, the heat needed to melt 1 gram of ice (heat of fusion) is 80 cal/g, and the heat of vaporization is 540 cal/g. Because the calorie is a small unit, we commonly measure heat in kilocalories (kcal), equal to 1,000 cal (in Section 2.1 we note that food calories are actually kilocalories). Equivalent units of heat have been defined in various ways (Table 5.1), but only one need be discussed. The British thermal unit (Btu) is defined as the amount of heat needed to raise the temperature of 1 pound of water by 1 degree Fahrenheit, and 1 Btu is equivalent to 252 cal. A commonly used unit is a quadrillion Btu (10^{15} Btu), which is referred as a "quad."

Energy is essentially equivalent to heat and is measured in the same units. For example, we say that a barrel of oil contains approximately 5.8 million (5.8×10^6) Btu (about 1.5 billion; 1.5×10^9 cal) of energy because burning that barrel yields enough heat to raise the temperature of 1.5 billion g of water by 1 degree centigrade. Similarly, electrical energy can be used to operate a stove, which releases heat in proportion to the amount of electrical energy provided to it. Electrical energy is commonly measured in terms of kilowatt-hours (kWh), which is simply the number of kilowatts of power transmitted times the number of hours (or parts thereof) over which the current is maintained. As shown in Table 5.1, 1 kWh equals 3,413 Btu.

Power is simply the rate at which energy becomes available. For example, an engine that generates 1 horsepower is doing work (supplying energy) at the rate that

Table 5.1 *Units of energy and power and fuel equivalents*

Calorie (cal)	1 calorie of heat will raise the temperature of 1 gram of water 1°C; also known as a gram calorie
Kilocalorie (kcal)	1,000 calories; commonly referred to as a calorie in books and articles on food (Section 2.2); also known as a kilogram calorie
British thermal unit (Btu)	252 calories (gram calories)
Quad	1 quadrillion (10^{15}) Btu
Joule (j)	0.24 cal (gram calorie)
Kilowatt hour	Energy produced by 1 kilowatt of power in 1 hour; 3413 Btus
Watt (w)	Unit of power that delivers 1 joule per second
Kilowatt (kw)	1,000 watts
Horsepower	746 watts; the amount of power delivered by a standard British horse
1 barrel (bbl) of oil	$\sim 5.8 \times 10^6$ Btu
1 cu. ft. of gas	$\sim 1.3 \times 10^3$ Btu
1 lb. of coal	$\sim 1.3 \times 10^4$ Btu
1 g ^{235}U fissioned in a reactor	$\sim 80 \times 10^6$ Btu
1 sq. mi. of solar cells at 10% efficiency	$\sim 10^{13}$ Btu/year

could be provided by one horse. Thus a horse that works for 1 hour generates 1 horsepower-hour of energy, and one that works for 5 hours generates 5 horsepower hours. Similarly, an electrical generator that transmits a signal of 25,000 kW (kilowatts) is generating 25,000 kWh (kilowatt-hours) of energy per hour, equivalent to 85 million Btu.

5.2 TYPES OF ENERGY SOURCES

There is no panacea among these energy options.
– *L. Schipper and S. Meyers, Energy Efficiency and Human Activity, p. 34, commenting on technologies such as nuclear, solar, geothermal, and wind power*

We obtain energy in three basic ways. One is direct heating, which occurs whenever we sit by a fire, turn on a gas-fired heating system, lie in the sun on a cloudless day, or soak in a natural hot spring. A second, less-direct, energy source is electricity. Electricity is secondary in the sense that it must be produced from more direct sources such as boiling water by burning fossil fuel and then transmitted to the place where it is used. The third energy source is even less direct than electricity. It consists of stored energy that can be transported elsewhere and then released. A typical example is the battery in your car. We discuss all of these sources in this section.

Direct energy

Most of the direct energy that people use is produced by burning organic carbon, which is the carbon that constitutes all animals and plants. Organic carbon exists on the earth because of photosynthesis, the process that enables plants to grow. As we discussed in Section 2.0, chlorophylls in plants permit them to take carbon dioxide (CO_2) from the atmosphere and water from the atmosphere and soil and convert them into organic molecules (mostly sugars). The carbon in the organic molecules is in a reduced form, with an effective valence of -4, in contrast to the oxidized carbon in CO_2, which has a valence of $+4$. This reduction requires energy, which is supplied to the chlorophyll by sunlight, and consequently the organic molecules contain more energy than the CO_2 and other components from which they were formed (i.e., photosynthesis is referred to as "endothermic," requiring an input of energy).

The energy stored in organic molecules can be released by reversing the results of photosynthesis by oxidizing the organic carbon back to its $+4$ valence in CO_2, a process described by the following general equation:

$$\text{organic compounds (C,H)} + O_2 \rightarrow CO_2 + H_2O$$

Because the organic molecule contains more energy than the CO_2 plus H_2O, the reaction occurs spontaneously and emits heat (i.e., it is "exothermic"). Most of the oxidation of organic matter on the earth occurs naturally by the decay of dead plants and animals, but we can accelerate the process by deliberately burning them. Nonindustrial societies commonly use recently growing vegetation as their energy source. The best vegetation is trees, which are cut down both for their timber and as a fuel source (Section 8.3). Lacking trees, which many societies now do, anything else that burns can be put into the fire – twigs, leaves, corn stalks, dung from herbivorous animals, whatever is available if you are cold and need to cook a meal.

Only the poorer societies obtain most of their energy by the burning of recently dead vegetation, and industrial societies obtain very little by that method (Section 5.5). For the world's richer people, the principal energy source is fossil fuel, the remains of plants and animals that died and were buried up to several hundred million years ago. This preserved organic carbon now occurs at sparse locations around the earth in the form of oil, natural gas, coal, and other less-used deposits (Section 5.3). We burn these fuels for a variety of purposes. Gas is used primarily for direct heating of buildings or industrial processes. Some coal is used for direct heating, but most is consumed in the production of electricity. Oil is used for direct heating as fuel oil, for petrochemicals, but primarily as gasoline in cars and other vehicles.

When burned as gasoline, the direct energy supplied by oil is used to run internal combustion engines (Fig. 5.3). These engines, which run virtually every automobile in the world, are designed with a series of cylinders that contain individual spark plugs. The cylinders are arranged around a rotating shaft, and when the cylinders fire, they ignite a mixture of gasoline and air that expands explosively and pushes the shaft around its axis. By arranging the cylinders so that they fire in sequence, instead of all at once, the shaft can be rotated continually at speeds that depend on the rate at which the gasoline is fired. The rotating shaft then is connected to the axles that drive the wheels of the car, and, when all is working well, we can head off for another day on the track (i.e., highway).

Passive solar energy refers to a use of the sun's radiation in a way that is, well, passive. Anything exposed to sunlight will absorb some of the radiation and become hotter because of the addition of heat. This heat is usually radiated back out of the object when the sun disappears at night or behind a cloud, but we can design systems that retain as much heat as possible. Most of these systems operate on two basic principles. One is that different materials have different heat capacities, which means that they must absorb different amounts of solar radiation for the same amount of temperature rise. The second major principle is referred to as the "greenhouse effect." As we discuss in Section 8.1, solar radiation arrives at short wavelengths that penetrate glass, but it is radiated back into space largely at longer wavelengths that do not pass back through glass. We could make use of both principles by designing a house in which a large swimming pool is ringed by curtained glass doors on the

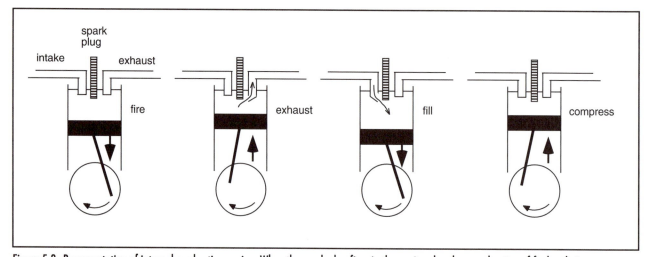

Figure 5.3 Representation of internal combustion engine. When the spark plug fires in the engine chamber, combustion of fuel and air creates power by forcing the piston and drive rod down. The rod turns the drive wheel until the spark plug fires again, causing a sequence of three events. First the rod is pushed back up, forcing exhaust out of the engine chamber; second, the wheel continues to turn and let fuel and air into the firing chamber; and, third, the turning wheel compresses the fuel and air for the next ignition.

Figure 5.4 Geothermal water emitted as steam and geysers at Yellowstone. Courtesy Stuart Rojstaczer.

south side of the house (in the Northern Hemisphere). The water in the pool has a high heat capacity, as do most liquids, and it absorbs heat when the curtains are open during the day. At night, we close the curtains, and the water loses heat sufficiently slowly that it keeps the house warmer than it would have been if we had not allowed the water to absorb energy during the day.

Besides burning and sunlight, the only other direct energy source is hot water from inside the earth. Because temperatures increase downward in the earth at almost all locations, water at some depth must be warm enough to provide heat if brought to the surface. The problem is that, in most places, temperatures of interest to us (perhaps about 50°C) are reached only at depths greater than several kilometers. Because drilling wells to these depths and pumping the water out is much more expensive than other means of heating water, geothermal energy is used

only in a few very favorable locations. The best known is Iceland. It sits directly on the mid-Atlantic spreading ridge, and hot water from geysers and fumaroles generated by the high temperature of rising lava is used by the Icelanders to heat most of their buildings (Fig. 5.4).

Electricity

Two of the most fundamental principles of electromagnetism are that a rotating electrical current generates a magnetic field and, conversely, that a rotating magnet generates an electrical field. We make use of the first principle by wrapping coils of wire around iron bars, and when a current is passed through the wire, it rotates around the bar and converts the bar into a magnet. We use the second principle to generate electrical power by turning magnets inside coils of wire. The only other method of generating electricity is active solar power cells, which have not yet been developed as a major energy source.

Machines designed to rotate magnets in coils of wire are called generators (Fig. 5.5). The magnets are attached to a circular bar that rotates at high speed within as many coils of wire as can be packed into the generator. Because the densest packing is usually achieved with packets of wire that are wound in opposite directions, the rotating magnets generate a current in which the electrical field changes ("cycles") rapidly from one direction to another. These rapid changes produce an alternating current (AC) that is more stable than the direct current (DC) that would be produced by rotating a magnet in one coil of wire. Generators deliver this alternating current with frequencies that vary from country to country (examples are 60 cycles per second in the United States and 50 in Canada). The rate at which electricity is supplied by a generator is commonly measured in kilowatts (Table 5.1).

The preceding discussion shows that the basic problem in the generation of electricity is providing a mechanism to turn the magnets and the energy to run it. Several mechanisms are available, but the one used for about 90% of the world's electricity is boiling water (Fig. 5.5). Steam released as the water boils flows past flanges attached to the bar holding the magnets, causing the bar to turn. As long as the water tank is kept filled with water and energy is supplied to boil the water, electricity can be generated up to the capacity of the system. The ~10% of the world's electricity not generated by boiling water is generated almost entirely by flowing water, referred to as hydropower or hydroelectric systems (Fig. 5.6). These systems operate by letting water at the base of a dam flow

Figure 5.5 Representation of electricity generator. Some type of heat source boils water in the boiler, forcing steam out of the nozzle. The steam uses flanges attached to the rotating shaft of the generator to rotate iron bars wound with wires carrying electric currents that induce magnetism in the bars. The rotating magnets then induce electric currents in coils of wire wound around the outside of the generator. There is net generation of electricity because only a small percent of the amount generated is used to make the iron bars magnetic.

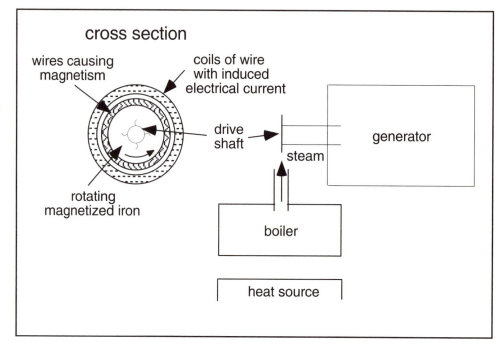

out through turbines, which are generators designed so that the flowing water passes flanges that rotate the bar containing magnets. The only other method for rotating magnets is an internal combustion engine (Fig. 5.3), which is relatively expensive and is used only for very small, perhaps private, generators.

By now, we hope that it has become apparent that electricity is not free. It is available only because some other energy source is used to rotate the bar that contains the magnets. If that energy source stops, the bar stops rotating, and the electricity stops. The costs vary with the type of generating system used.

Systems based on boiling water must build generators and distribution systems and pay for a continual input of energy to make the water boil. In most of the world, this energy is supplied by coal, which is relatively cheap but has large environmental costs that we discuss in Chapters 7 and 8. Oil and gas are more expensive than coal but have fewer environmental costs. Nuclear reactors (Section 5.3) have low fuel costs, but construction is expensive, and environmental hazards are generally regarded as high.

Systems that use hydropower must build dams, install turbines, and maintain an adequate head of water in the dam to turn the turbines (see Section 4.4 for a discussion of hydraulic head). Maintenance of water in the dam commonly requires negotiations with people who want to withdraw the water for agricultural or other use or those who want to retain the water to prevent downstream flooding (one example of the problem is discussed

in Box 5.1). On balance, however, hydropower is cheaper than the more commonly used boiling-water systems, and it is restricted only by the fact that most of the industrial world is already close to the maximum capacity that can be generated by this process.

The costs of construction, purchase of fuel, potential for environmental damage, and negotiation with competing economic interests all invite the development of a cheap and environmentally safe alternative source of electricity. The one with the most promise is the sun, which provides its energy absolutely free. In theory (but see our subsequent discussion), the sun's energy could be used in various ways. One simple way is to construct a set of mirrors that concentrate the sunlight on a steam-operated electrical generator. Mirrors that are arranged in a parabola (Fig. 5.8) that is continually moved so that it points toward the sun will reflect the sunlight to the focus of the parabola and can generate very high temperatures.

Because parabolic mirrors, although simple in concept, are cumbersome to operate on a large scale, the most promising method of using solar energy is by converting sunlight directly into electricity. This process is known as active solar in contrast to the passive solar methods discussed already. It uses banks of photoelectric (photovoltaic) cells of various designs similar to the one shown in Figure 5.9. The cells contain a material, such as amorphous SiO_2, that becomes unstable when struck by light and emits electricity as the crystalline lattice reorganizes itself. The percentage of sunlight that is convert-

(a)

(b)

Figure 5.6 Harnessing water power: (a) Gulfoss waterfall, Iceland; (b) dam across Columbia River, British Columbia, Canada; the picture shows turbines in the face of the dam.

ible by a cell (the efficiency of the cell) has been rising as research is done on solar power, and in the mid 1990s, good solar cells performed at ~10% efficiency. Cells triggered by light have been used for many years for such mundane devices as automatic door openers, which operate by placing a photoelectric cell on one side of the approach to a door and a light beam aimed at the cell on the other side; when someone approaches the door, the

beam is interrupted and the cell triggers the door-opening mechanism.

As door openers and similar devices, photoelectric cells have been convenient and not too expensive. On a broad scale, however, the construction of an electrical power system based on photovoltaics has run into severe cost problems despite the fact that the sunlight is free. The photovoltaic cells are expensive to manufacture un-

BOX 5.1 CHINA'S THREE GORGES DAM _____

It will contain enough crushed rock to rise 610 feet above the valley floor and span 6,864 feet across the valley. It will impound a lake approximately 400 miles long and will contain locks that will raise oceangoing vessels to the level of the lake and let them visit ports as much as 1,500 miles inland from the coast. It will control floods in one of the most heavily flooded valleys in the world. It will generate twice as much power as the Grand Coulee Dam on the Columbia River, the largest dam in the United States. The power will permit rapid industrialization of one of the poorest regions of the country. It may never be built.

The Three Gorges Dam takes its name from three deep gorges of the Yangtze River that will be partly flooded if the dam is built (Fig. 5.7). They are only a few of the gorges formed by the rapidly downcutting river as it journeys from its headwaters in the mountains of northern Tibet to the China Sea near Shanghai. The Yangtze receives so much rain and snowmelt that it floods frequently, erodes incessantly, and carries the eroded sediment out into the fertile fields of eastern China. Many Chinese along the river feel that they have adjusted their lives to these natural processes, and they fear that the dam will cause irreversible damage to them and the places where they live. Their efforts to resist the government decision to build the dam, however, have been thwarted by the repressive Chinese government, in some cases resulting in imprisonment of the dissenters. We present a few of their arguments here.

The lake behind the dam will be 400 miles long, displacing more than 1 million people, who will have to move to higher and less agriculturally productive ground. Because cities and communities along the lake do not have adequate sewage treatment plants, the lake will soon be heavily contaminated because the sewage can no longer wash down the river (see Section 7.5 on eutrophication). Because of the heavy sediment load carried by the Yangtze, the lake will soon be choked with sediment, thus preventing oceangoing ships from going upriver despite the locks at the dam. If the sediment is trapped by the dam, it will not resupply the fields of eastern China (see similar discussion of the Aswan Dam in Box 4.4). The dam's estimated cost of $15–20 billion (U.S.) would divert money from other projects that China needs and might require massive foreign borrowing.

Some of the environmental concerns have affected foreign interest in the project. Financially, this interest arises from the partial Chinese dependence on foreign firms to do some of the engineering work on the dam, and because they are uncertain that the Chinese will ever pay them for the work, several companies have asked their governments to provide them with guarantees of payment if the Chinese default. In the United States, these arrangements are sometimes made by the Export–Import (Exim) Bank, which encourages private companies to work abroad by guaranteeing to pay them if they cannot collect from the government or organization that they worked for. These requests, consequently, place governments in the position of answering several unanswerable questions. Will the Chinese pay, or will foreign governments be stuck with reimbursement of their companies? And even if the Chinese don't pay, is the money well spent in maintaining good relations with the Chinese government? And finally, do environmental concerns of people

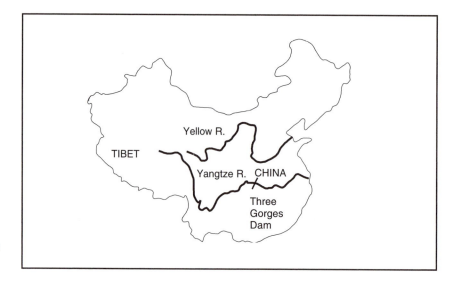

Figure 5.7 Location of Yangtze River and Three Gorges Dam.

living along a river in China outweigh the benefits of profits for companies that would work on the dam? As of the middle of 1995, construction is scheduled to begin, and most governments (including the Exim Bank) have not reached a conclusion.

FURTHER READING: Pearce (1995); Sullivan (1995); Tyler (1996); and P. E. Tyler, *New York Times*, January 15, 1996, pp. A1, A4.

less they can be made in large numbers, but large numbers can be justified only if large numbers of people, perhaps an entire country, decide to convert much of their electrical production from their present system to a solar one. That conversion requires dismantling some of the present power system and redesigning it to use the production of solar grids located in places other than the present power plants. Also, because photovoltaics cells work only in direct sunlight, a solar power system must have either conventional electric systems or electrical storage systems to provide electricity at night and on cloudy days. We discuss all of these issues more specifically when we consider the future of various forms of energy in Section 5.5.

Stored energy

The ability to store energy for future use provides modern people with a freedom that earlier civilizations did not

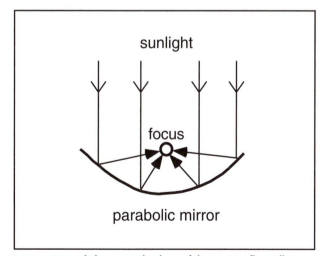

Figure 5.8 Parabolic mirror. The shape of the mirror reflects all sunlight to the focus. For individual use as a solar oven, the focus would contain food to be cooked. For commercial use, electricity could be generated either by placing photovoltaic (PV) cells or a steam boiler at the focus.

possess. People who live in a society that has only fireplaces must burn more wood whenever they want to cook food or warm themselves. With the exception of closing the doors and windows of their houses to retain heat, they have no way of using energy from today's fires to warm themselves or cook food tomorrow. If today's fires produce more heat than is actually needed for today's cooking and warming, the excess is simply lost. In order to avoid this loss, we have made continual efforts to improve our ability to store excess energy. The principal one is lead storage batteries, but other methods such as hydrogen fuel have been proposed, and we discuss them in the rest of this section.

Lead storage batteries operate on the principle of maintaining lead in two different valences, one oxidized and one reduced (Fig. 5.10). A bar of lead (Pb) metal has a valence of 0 (reduced), and the lead in a meshwork of lead dioxide (PbO_2) has a valence of +4 (oxidized). In a solution of sulfuric acid, both of these varieties of Pb are in chemical equilibrium with a small amount of Pb^{+2} ion dissolved in the acid. The Pb and PbO_2 are referred to as electrodes, and if they are kept separate, nothing happens. If the electrodes are connected by a wire, however, electrons will flow from the PbO_2 to the metallic Pb, converting both forms of Pb into dissolved Pb^{+2} and generating an electrical current in the wire. In case this process seems complicated or obscure, we should inform you that it is not. It happens every time that you turn on the ignition switch of your car, thus closing the wire connection and providing the current that starts your car's internal combustion engine.

The energy stored in your automobile battery is placed there during manufacturing and renewed as the car is driven. The manufacturing process starts with natural lead, which has a valence of +2, mostly in the ore mineral galena (PbS; Section 6.2). Extracting the lead from the ore, reducing some of it to metallic Pb, oxidizing some to PbO_2, and constructing the battery around the electrodes requires energy. This energy cannot escape (be discharged) until the wire connection between the elec-

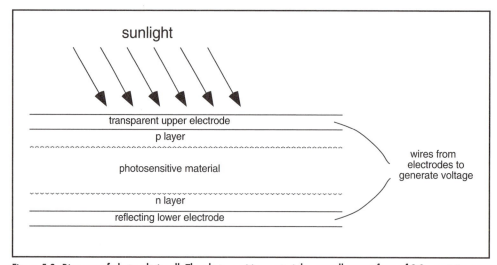

Figure 5.9 Diagram of photovoltaic cell. The photosensitive material, generally some form of SiO$_2$, contains an upper p layer and a lower n layer. The p layer has electron "holes" caused by replacement of some of the Si^{+4} by trivalent ions such as boron, aluminum, and gallium. The n layer has excess electrons caused by replacement of some of the Si by pentavalent ions such as phosphorus, arsenic, and antimony. When activated by light, flow of electrons from the n layer to the p layer creates an electric current because of voltage difference in wires attached to the upper and lower electrodes.

trodes is closed, essentially reversing the manufacturing process and converting the Pb back to its natural state. When the battery is in use, the energy lost by the brief current that flows when you turn the ignition on is replaced as you drive the car. Part of the energy generated by the engine is used to reverse the electron flow in the battery, thus taking Pb^{+2} out of solution and forming

more metallic Pb and PbO$_2$. The battery is thus recharged and stores its new charge until the car has to be started again. Of course, if you leave the lights on when the motor is off, the battery will continue to discharge and will be "dead" when you want to use it again.

Although all batteries operate on the same principle as the lead storage battery in your car, they come in an enormous variety of designs and use an enormous variety of metals and solutions. An ordinary flashlight battery, for example, consists of an outer sheath of zinc (Zn) and an inner core of graphite packed into a paste of manganese dioxide (MnO$_2$), ammonium chloride (NH$_4$Cl), and graphite powder. When the flashlight makes a connection between the graphite core and the Zn sheath, electrons from the dissolving Zn flow through the NH$_4$Cl to reduce the MnO$_2$ and ultimately flow out through the graphite core to complete the circuit. Because a flashlight battery contains a paste rather than a pool of sulfuric acid, it is referred to as a dry cell.

In theory, an adequate system of batteries could give the world a virtually limitless supply of energy. We could, for example, charge batteries with solar power and then run electrical systems from them at night or on dark days. By expanding the electrical systems, we could replace heating systems that use natural gas and oil. We could also reduce gasoline consumption (perhaps to zero) by replacing internal combustion engines in cars with batteries. It sounds utopian, and unfortunately

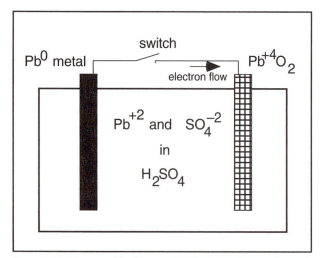

Figure 5.10 Diagram of lead battery. The lead metal has a valence of 0 and emits electrons when it dissolves in the acid, and the lead in the PbO$_2$ has a valence of +4 and gains electrons when it dissolves. Thus the battery stores energy until the switch is closed and the electrons can flow along the wire.

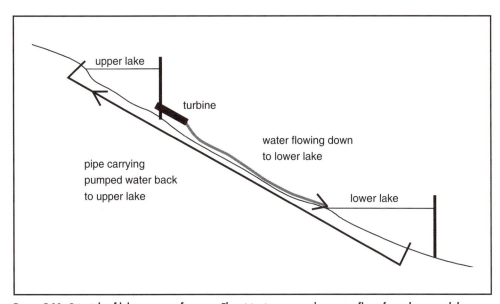

Figure 5.11 Principle of lake storage of energy. Electricity is generated as water flows from the upper lake through the generator to the lower one. Some of this electricity is stored until the generator's power is not needed, probably at night, and is used to pump water back to the upper lake. This system loses energy because it cannot pump as much water back up to the lake as flowed down to generate the electricity used for the pumping. Consequently, lake storage is only useful as part of a broader electrical generation system.

it is. With present technology, batteries do not store enough energy and do not discharge rapidly enough to replace all of these other energy sources. Perhaps in the future, we will develop adequate solar technology, but until then the material discussed in this chapter remains pertinent.

Although batteries are now the only significant method of energy storage, two other methods may become important in the future. The most promising is manufacture of hydrogen gas (H_2). Hydrogen is formed by electrolyzing (passing a strong electrical current through) water, which breaks water down into oxygen and hydrogen gases. The oxygen can be retained as a pure gas but usually is allowed to escape harmlessly into the atmosphere. The separated hydrogen, however, is more commonly stored in pressurized cylinders, and when it is let out of the cylinder's nozzle it can be burned to produce energy in the same way as natural gas and other fossil fuels. Thus the hydrogen stores some of the energy that was used to electrolyze the water. Some observers have suggested replacing gasoline with H_2 in many of our cars, but the technological difficulties of this transfer have not been overcome.

One last method of energy storage should be mentioned, not because it is significant at present but because it exemplifies the important concept of potential energy. In our discussion of hydraulics in Section 4.4, we defined mechanical potential energy as energy that could be converted into kinetic energy of movement. We saw that the potential energy of an object at a high elevation became kinetic energy when a rock fell off of a cliff or water ran downhill. In theory, then, if we pump water uphill and store it, perhaps in an artificial lake, we have created a potential energy source that can be converted into electrical (kinetic) energy if we let the water run downhill through a turbine (Fig. 5.11). Because energy is required to pump the water back uphill, this process cannot create more total energy, but it allows us to use energy for pumping at a time when public energy demand is low, commonly at night, and then release it when demand is high. Unfortunately, this storage system is probably less viable than batteries and H_2, but it illustrates one more possibility for our energy future. At present, that future is still highly dependent on fossil fuels, and we turn to them in the next section.

Policy questions

Have you assessed the amount, type, and source of energy that you use?

FURTHER READING: Howes and Fainberg (1991); Scientific American (1991); Johansson et al. (1993).

5.3 FOSSIL FUELS

The twentieth century rightly deserves the title "the century of oil." – D. Yergin, The Prize, p. 15

As we discussed in Section 5.1, all organic matter ultimately derives from carbon that has been reduced photosynthetically, and it cannot survive unless the dead plants and animals are preserved from natural oxidation. Under anoxic (no oxygen) conditions, organic matter decays in various ways depending on the type of animal or plant that forms the starting material, the temperature and pressure to which that material is exposed, and the length of time during which decay occurs. Most of this section describes the geology and occurrence of various types of fuels, but in order to provide a background for the discussion, we first summarize the chemistry and results of the decay process and the types of combustion that the fuels can undergo.

Products of decay of organic matter

Organic matter accumulates in two principal types of rocks – thick sequences of marine sediments and coastal swamps (the description of geology follows). Animals and plants buried in marine sediments originally consist largely of a mixture of carbohydrates, fatty acids (fats and oils), and minor proteins and other organic compounds (Section 2.1). In addition to carbon and hydrogen, almost all of these compounds contain oxygen, and some contain nitrogen, sulfur, and other elements. One of the principal results of anoxic decay is removal of the oxygen and consequent formation of different varieties of hydrocarbons, compounds that consist almost exclusively of hydrogen and carbon. Hydrocarbons include some with ring structures (benzenes) and some that consist almost entirely of long chains of carbon atoms, each attached to two other carbon atoms and to two hydrogen atoms on the side of the chains (Fig. 5.12). Compounds range in complexity from those of high molecular weight that are virtually solid to the lightest and simplest, known as natural gas (methane, CH_4).

Continued decay in buried sediments causes hydrocarbons to "mature," generally becoming lighter and simpler. Some of this simplification produces methane, but much remains as a mixture of very complex hydrocarbons. This mixture commonly accumulates in organic-rich sediments known as oil shales, which can be mined as a source of oil but more commonly become the source rocks from which oil and methane migrate to reservoirs The oil pumped out of reservoirs is referred to as crude oil; it consists of a mixture of the long-chain hydrocarbons, referred to as paraffin oil, and ring structures containing small amounts of oxygen and referred either as naphthenes or asphaltic oil (Fig. 5.12). This crude oil burns, but it is very sticky and cannot be used for most human activities. Consequently, crude is refined ("cracked") to simpler molecules and then distributed from refineries in a variety of forms suitable for use as heating oil, kerosene for stoves, gasoline for cars, diesel fuel, lubricating oil, and numerous others. The gasoline designed for automobiles consists largely of long-chain hydrocarbons with six or seven carbon atoms (hexane and heptane).

The organic matter that accumulates in the anoxic waters of coastal swamps consists largely of plant remains. Under these conditions, decay caused by burial and time causes loss of oxygen from the dominantly carbohydrate plant material and converts the loose plant debris to an increasingly denser and harder sedimentary rock known as coal. Loss of oxygen and hydrogen by elimination of water and breakdown of carbohydrates results in a rapid increase in the percentage of carbon, and almost all of the energy derived from coal comes from burning carbon. Coals also contain other elements, principally sulfur, left from the plants plus a small amount of clay or other sediment trapped in the initial deposit. The best coals contain the highest carbon and lowest sulfur contents and the smallest amount of clays and other noncombustible material, commonly referred to as ash. (The types of coals are discussed later in the section on geology and distribution of coal.)

Combustion

Combustion is a natural process that consists of a reaction between a chemical substance and oxygen. Because combustion occurs naturally (chemists would say spontaneously), it generates heat (i.e., it is exothermic). Table 5.2 shows the amount of heat generated by burning three organic materials: carbon, which is virtually the only component of coal that burns; methane (CH_4), the principal component of natural gas; and hexane (C_6H_{14}), a major component of gasoline. The table shows the amounts of heat generated by burning 1 gram molecular weight of each material, but we can ask how much of each material needs to be burned to yield the same amount of heat. Thus, to obtain an arbitrary 1,000 kcal of heat we must burn 10.6 mols of pure carbon (coal) because $10.6 \times 94.4 = 1,000$. Similarly 1,000 kcal can be obtained by burning 4.74 mols of CH_4 ($4.74 \times 210.8 = 1,000$) and 1.01 mols of C_6H_{14} ($1.01 \times 989.8 = 1,000$).

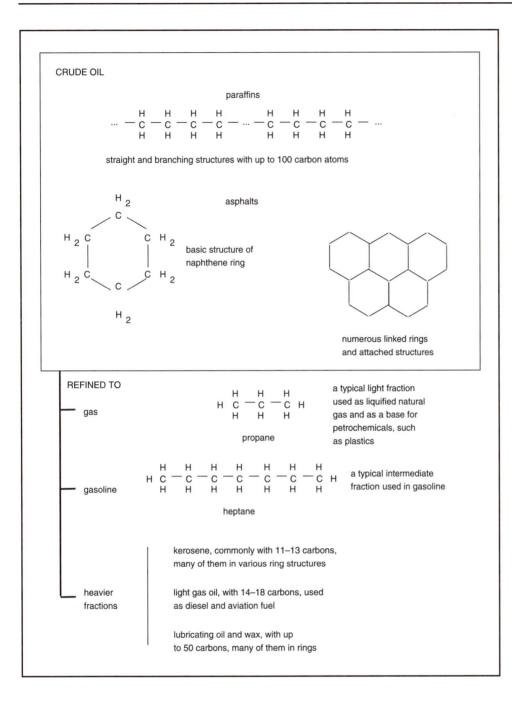

CRUDE OIL

paraffins

straight and branching structures with up to 100 carbon atoms

asphalts

basic structure of
naphthene ring

numerous linked rings
and attached structures

REFINED TO

gas

propane

a typical light fraction
used as liquified natural
gas and as a base for
petrochemicals, such
as plastics

gasoline

heptane

a typical intermediate
fraction used in gasoline

heavier
fractions

kerosene, commonly with 11–13 carbons,
many of them in various ring structures

light gas oil, with 14–18 carbons, used
as diesel and aviation fuel

lubricating oil and wax, with up
to 50 carbons, many of them in rings

Figure 5.12 Typical hydrocarbons in crude oil and in gasoline after refining.

The equations for this burning are:

$$10.6C + 10.6\,O_2 \rightarrow 10.6\,CO_2$$

$$1.01C_6H_{14} + 9.6\,O_2 \rightarrow 6.1CO_2 + 7.1H_2O$$

$$4.74CH_4 + 9.49O_2 \rightarrow 4.74CO_2 + 9.49\,H_2O$$

The important lesson to be gained from these data is that different substances release different amounts of CO_2 for the same amount of heat generated. In the example that we have used, coal releases more than C_6H_{14}, which releases more than CH_4, and generally fuels with lower C–H ratios release more H_2O and less CO_2 for the same amount of heat generated. The significance of this observation will not be clear until we discuss global warming in Section 8.2, and here we simply mention that release of H_2O is not a problem, but many atmospheric scientists feel that continued release of CO_2 is changing the earth's temperature. If that is true, then

Table 5.2 *Heats of combustion (in kcal)*

To burn 1 mol (12 g) of carbon	
$C + O_2 \rightarrow CO_2$	94.4
To burn 1 mol (16 g) of CH_4	
$CH_4 + 2O_2 \rightarrow CO_2 + 2H_2O$	210.8
To burn 1 mol (44 g) of hexane (C_6H_{14})	
$C_6H_{14} + 9.5O_2 \rightarrow 6CO_2 + 7H_2O$	989.8

coal is the most serious environmental hazard and natural gas the least among fossil fuels. (Also, we should point out that nuclear and solar power release no CO_2 at all.)

Now, finally, we are ready to discuss the geology and worldwide distribution of fossil fuels.

Geology of oil and natural gas

The anoxic environment required for the preservation of animal and plant debris (see earlier discussion) in marine sediments requires either rapid burial, deposition in a basin with restricted water circulation, or a combination of conditions (Fig. 5.13). Rapid burial because of high sedimentation rate can take the organic debris below the zone of surficial oxidation that commonly occurs in the upper layers (perhaps only a few centimeters) of sediments at the bottom of oceans with currents that bring oxygen-bearing water to the sediment-covered floor. Even where sedimentation rates are low, basins with restricted circulation may not be able to bring oxygen-bearing surface water to the bottom of the basin, thus preventing oxidation of the newly deposited organic matter. Either of these conditions permits a start of the decay process that leads to development of oil and gas (as

already discussed), and whether actual petroleum reservoirs develop depends on circumstances that we describe in two categories: (1) local geologic processes that bring oil and gas into trapped reservoirs and form oil and gas fields; and (2) regional geologic conditions that lead to the formation of areas within which the local processes can operate.

Oil and gas fields develop where organic debris that has matured to oil and gas can migrate from its original sediments (source rocks) to reservoirs of sufficient size to be commercially useful. Migration is caused by the simple fact that both oil and gas are lighter than water. Because almost all of the outer part of the earth is saturated in water, the oil and gas from source rocks move upward in the same way that oil moves to the top of a bottle of salad dressing when you shake it. If unimpeded, the oil may seep out onto the surface and the gas percolate upward into the atmosphere. In favorable circumstances, however, the oil and gas enter a reservoir rock that has enough porosity to store significant amounts of petroleum and enough permeability for it to move freely (these are the same criteria for water-bearing aquifers discussed in Section 4.4).

A reservoir by itself does not generate a petroleum field, but if the reservoir is below some layer of rock that is impermeable to oil and gas, then they can be trapped. Figure 5.14 shows three common types of traps. The simplest is an anticline, an upward-arching fold of the sedimentary layers that catches the oil and gas as they move up through the water-soaked rocks. A well drilled on the top of the anticline will first recover gas at the top of the reservoir. As the gas is exhausted, oil is forced further upward by water pressure, and the well can recover oil. After the oil is pumped out, water occupies all of the reservoir, and the well shuts down. A well drilled slightly to the side of the anticline may miss the gas pocket and re-

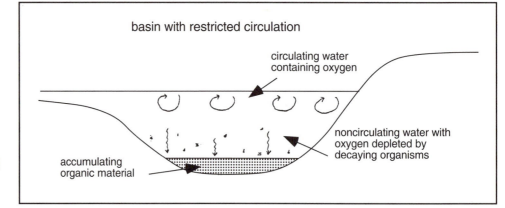

Figure 5.13 Preservation of organic material by rapid burial in sediments below the level of oxidation. See explanation in text.

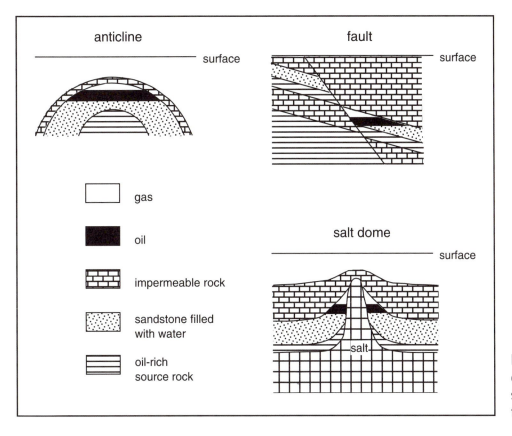

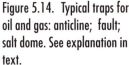

Figure 5.14. Typical traps for oil and gas: anticline; fault; salt dome. See explanation in text.

cover only oil. And a well drilled too far to the side of the anticline will encounter only water.

The two other traps shown in Figure 5.14 are only slightly more complicated than anticlines, and the process of trapping oil and gas at the top of the reservoir is the same as in an anticline. In some places movements along faults seal off the top of a reservoir against impermeable rocks on both sides of the fault. Salt domes originate from two properties possessed by sedimentary salt deposits – salt has a lower density than other sediments (principally sand and clay) and is capable of flowing under the pressure of overlying sediment. Consequently salt tends to rise and may form long "fingers" (domes) that extend upward from a salt horizon and deform the sediments into which they intrude. Where these domes deform a sequence containing both porous horizons and overlying impermeable rocks, they can form reservoirs geometrically very similar to those formed by faults. Salt domes tend to form reservoirs in a ring around the center of the dome, leading to a circular pattern of wells on the surface.

Regions that contain petroleum fields are those in which geologic processes have provided all of the necessary ingredients – source rocks, reservoir rocks, and traps. The fundamental requirement is a large basin containing a thick sequence of marine sediments. In this type of basin, rapid accumulation of sediments preserves hydrocarbons, reservoir and impermeable rocks are interspersed in patterns that can form traps, and the thick pile of sediments is sufficiently "weak" (easily deformed) that anticlines, faults, and other traps are likely to develop.

Basins with thick sediments commonly form in areas of crustal extension, both continental and oceanic, where thinning (stretching) of the crust permits it to subside as sediment is piled on top (Section 3.1). Extensional basins within continents commonly form above a basement cut by one or more small rifts. Some continental basins are wholly on land, but extension near some continental margins may be sufficient to cause the basins to subside far enough that shallow seas can extend over them from the open ocean. The more complex extension that causes splitting of continents leads to the formation of continental margins that border open oceans. As separation begins, the margins begin to subside and accumulate sediments very rapidly. Continental-margin basins are very productive for two reasons. One is that the pile of sediments makes a wedge that is relatively unconfined on its seaward margin, and the sediments slide toward the ocean, forming faults and other structures that can become traps. The second reason is that the first sediments deposited along a continental margin are likely to be salt precipitated from seawater just as it begins to encroach into the newly developing ocean basin, thus forming a horizon that can later produce salt domes.

We clearly cannot discuss even a small part of the world's oil-bearing basins in this book, and we offer brief descriptions of six of them in order to provide a clearer understanding of how oil and gas occur (locations in Fig. 5.15).

• *Arabian (Persian) Gulf.* Starting about 500 million years ago, the present Arabian peninsula became attached to present Africa and formed the northern edge of a large continent (Gondwana) that mostly included Africa and other continents now in the Southern Hemisphere (Fig. 5.16). With only minor interruptions, sedimentation occurred on this margin until the present time, forming a sedimentary pile more than 10 km thick. About 30 million years ago, the Arabian peninsula began to split away from Africa, opening the Red Sea and beginning to collide with the southern margin of Asia. Collision forced the northeastern edge of the Arabian crust down under Asia, permitting ocean waters to form a shallow sea-

way along a narrow belt that originally extended almost to Turkey but has now been partly filled in by deposits from the Tigris and Euphrates Rivers (see Box 3.7). Whether on land or in the Gulf, the entire area has the thick sediments, mild deformation, and underlying salt that combine to give it by far the largest accumulation of oil and gas in the world.

• *Gulf of Mexico.* The Gulf coast of the United States and Mexico is a typical continental margin. Rifting in the Gulf began about 200 million years ago as South America moved away from North America. The Gulf shows no effects of continental collision, but the same thick sediments and underlying salt as the Arabian Gulf combine to make it second only to the Arabian Gulf in both reserves and production.

• *West Siberia.* Continental crust under the West Siberian basin began to thin about 150 million years ago, following the collision that formed the adjoining Urals. This thinning formed a network of small rift valleys that are now overlain by several kilometers of sediment. A great complexity of reservoirs and traps formed in these sediments, and they now contain the world's largest reserves of gas.

• *Williston basin.* The Williston basin of western North America is underlain by one rift, but the entire crust

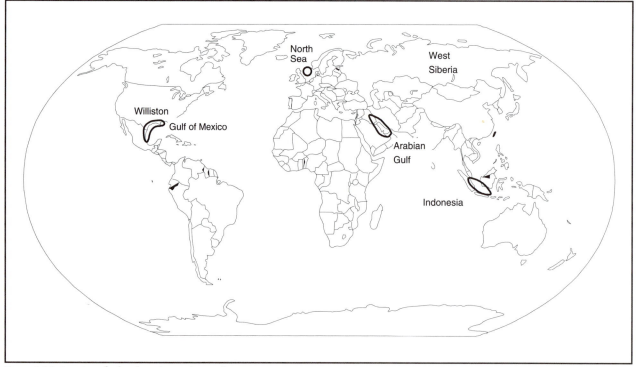

Figure 5.15 Location of oil and gas basins discussed in text.

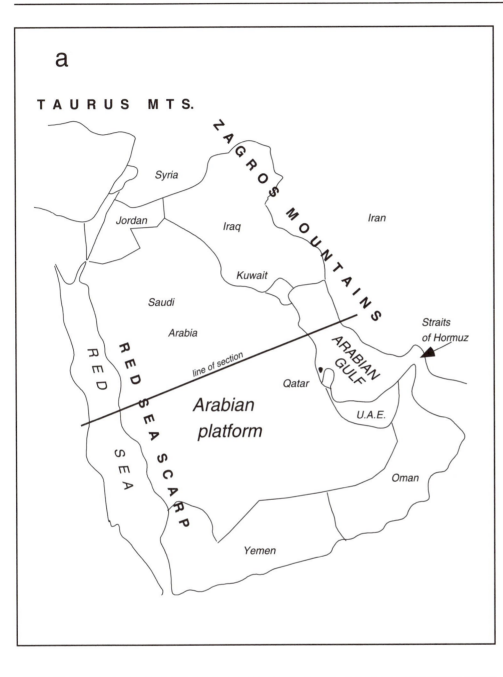

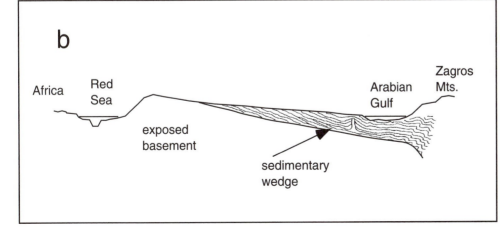

Figure 5.16 Evolution of oil deposits of Arabian Gulf: (a) the location of the Gulf and the line of the cross section of the accompanying diagram; (b) the cross section showing accumulation of thick sediments on the northern edge of the Arabian-African continent from approximately 500 million years ago until the present, followed by compression of these sediments as the Arabian peninsula collided with the Iranian part of the southern edge of Asia. This collision formed oil traps (including salt domes) in the entire Arabian Gulf area. See further discussion in text.

of the area has subsided in an almost circular pattern to accumulate more than 5 km of sediment.

- *North Sea.* The North Sea is underlain by a series of connected rifts similar to those of the West Siberian basin. Thinning in the North Sea, however, has been sufficient that an area that would normally be continent has subsided below sea level despite accumulation of nearly 10 km of sediment. The oil wealth beneath this shallow sea is shared by the bordering countries (mostly the United Kingdom and Norway).

- *Indonesia.* The oil and gas fields of Indonesia appear to be wholly in the ocean, but they are actually on thin continental crust. The Indonesian islands are constructed on continental crust that is barely covered by shallow marine water (Sunda Sea) and was actually dry land during the last glacial period, only a few tens of thousands of years ago. Scattered basins within this area have accumulated more than 5 km of sediment and developed an oil wealth that had a major effect on the conduct of World War II (see the discussion in Box 5.2).

BOX 5.2 OIL FOR GERMANY AND JAPAN IN WORLD WAR II

The German and Japanese war machines needed oil. They both used it for planes and trucks; Germany needed a lot for the tanks that it used in its blitzkrieg attacks and a little for its navy (mostly submarines); and Japan needed oil to fuel the navy that maintained its supply routes through the western Pacific. They got their oil in different ways.

Before the war, Germany had begun a "synfuel" program making oil out of coal by converting the carbon into hydrocarbon. This supply was not adequate for the whole war, however, and Germany found four other sources of supply. A temporary one was capturing supplies left behind by armies retreating in occupied countries, and this oil helped fuel the initial pushes to the east and west. The second, but fairly small, source came in the attack westward, when the Nazis occupied the Netherlands, a country that they had bypassed in World War I. In order to secure longer-term supplies, Germany had obtained a third source through an alliance with Rumania, whose oil fields at Ploesti provided Germany with nearly 60% of its prewar oil needs. Then, in order to provide further security, Germany seized Greece, putting the Ploesti oil fields out of the range of the Allied air force.

The fourth, and largest, potential source of oil for Germany was the rich fields in the Caspian Sea area, centered around Baku (Box 5.4). When Germany invaded Russia, the major push of the Nazi armies was southward, intending to cross the Caucasus to Baku and seize enough oil to supply the German armed forces virtually forever. They never got there. Partly they failed because some of the invasion force was sent toward Moscow, expecting the fall of Moscow to put Russia out of the war and let the Nazis annex whatever they wanted. The army came within 20 miles of Moscow but finally stalled in the face of the winter

and unexpectedly heavy Russian resistance. Another diversion was Stalingrad, a city bearing the name of a man the Nazis hated, so they decided to take it and then mop up the rest of southern Russia later. The defenders of Stalingrad, however, put up a resistance that is now legendary, and the few survivors of the German army finally had to creep away. As a final blow, the attempt to cross the Caucasus also ground to a halt in an area known as Chechnya.

With Russian oil denied and no more supplies of conquered countries to capture, by the end of the war the German war machine was forced back to dependence on the synfuel program and on dwindling production in Rumania. Even with stricter civilian rationing, these resources were inadequate to maintain the heavy use of planes and tanks that Germany had made the backbone of its military strategy. With the exception of one last thrust in the Battle of the Bulge in Luxemburg and Belgium in the winter of 1944–5, the Germans reached the end of the war almost completely without offensive capability and could only sit and resist as long as possible as the Allied armies closed on them from both the west and the east.

The oil situation in Japan was even more precarious than in Germany. With the exception of a very limited synfuels program, the Japanese had to import all of their oil. Before widespread attacks in December 1942, Japan had already obtained a small oil resource by invading China and securing its fields in Manchuria. They were not enough, however, and the Japanese started to assemble a huge stockpile of oil in preparation for what they felt would be a short war leading to complete hegemony over the western Pacific. Much of their oil came from the United States, and when the American government finally decided to stop oil shipments the Japanese knew that they must quickly secure the only major oil fields in the Pacific area, which were in the Dutch East Indies (now the country of Indonesia).

The plan to get the Dutch oil fields was very simple – lightning strikes. By moving the army quickly down through southeast Asia and the navy to Indonesia, the oil would be available to Japan. It would not be available, however, if oil tankers were intercepted by a hostile fleet, so that fleet had to be immobilized. That necessity led, on December 7, 1942, to the surprise attack on the U.S. naval base at Pearl Harbor, Hawaii. Much of the American fleet was destroyed in this attack, but two essential components survived – the U.S. aircraft carriers, which happened to be out at sea when the attack occurred, and the large military oil supplies that had been brought slowly to Hawaii by U.S. tankers. Because they had the oil and a carrier force able to use it, the U.S. fleet was quickly able to counterattack, and soon

American submarines were patrolling the sea lanes between Japan and Indonesia, and American surface vessels and airplanes had begun to destroy as much of the Japanese fleet as they could locate. By the end of the war, no oil was reaching Japan from Indonesia, and those ships and planes that still survived had so little fuel that they could undertake only limited action.

It is interesting to speculate on what might have happened if Germany had reached the Caspian Sea and Japan had destroyed the oil supplies in Hawaii. Would the Allies have won? Probably they would, but most military experts feel that the war would have dragged on much longer and many more people would have died. Fortunately, without the oil the world was spared that additional agony.

FURTHER READING: Yergin (1991).

With this background in the geology of oil and gas, we now turn our attention to their worldwide distribution and production.

Distribution and reserves of oil and gas

The locations of sedimentary basins and other factors that lead to the formation of hydrocarbons give oil and gas a very uneven distribution around the earth (Fig. 5.17). Consequently, some countries are exporters, some are importers, some areas within countries export to other areas, and there is a continual international and national tension between those who have petroleum and those who do not have it but want it. In order to understand this tension we need to describe reserves, production, consumption, and transportation. We concentrate on oil and somewhat ignore gas for two reasons. One is that the distribution of gas is extremely similar to that of oil, and a discussion of gas production would not add much to our understanding of petroleum distribution. The second reason is that oil is transported both in pipelines and by a worldwide fleet of tankers, whereas gas is generally transported only by pipelines and, consequently, is available only on a more local basis than oil. Table 5.3 summarizes worldwide reserves and production of oil and gas. All figures are reported in barrels, a standard unit of measurement equal to 42 gallons.

The total worldwide reserve of oil is estimated to be about 1 trillion (1,000 billion) barrels. About two-thirds

of this known supply of oil is in the Arabian Gulf and surrounding areas. Some of the figures for individual countries (including those not shown in Table 5.3) are staggering. Saudi Arabia has the largest reserves, nearly 260 billion barrels or about 35,000 barrels per person. Other countries have even higher ratios of reserves to people, the highest apparently being Abu Dhabi, with 92 billion barrels and only 30,000 people, for a ratio of 3 million barrels per person. To put these figures in perspective, the ratio of number of barrels of reserve per person is approximately 80 for the United States and United Kingdom, 5 for Germany, and 0.5 for Japan.

Outside of the Middle East, oil reserves are not particularly restricted to any one geographic area but are concentrated in a few countries. Countries with the most oil include Russia, where poor technology has greatly reduced productivity; Venezuela, whose oil wealth made the country one of the richest in Latin America; Mexico, which has pledged oil revenues as collateral against international loans taken out to stabilize its currency; and Libya and Nigeria, whose reserves overshadow those of any other African countries. These five countries and the Middle East account for nearly 90% of the world's known oil reserve.

Several factors make the figure of 1 trillion barrels of reserve very uncertain. One is that some of the largest reserves are held in countries such as Iran and Iraq, where little modern geologic work has been done and many knowledgeable geologists cannot gain access for political

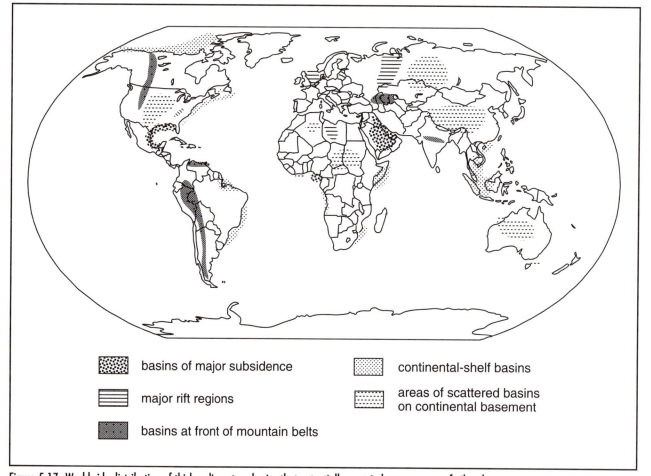

Figure 5.17 Worldwide distribution of thick sedimentary basins that potentially contain large reserves of oil and gas.

reasons. Another factor is that we do not know how much oil will be discovered in the future. Although some areas, such as the United States and North Sea, have been well explored, other areas have received so little attention that future discoveries may be quite large. Whether large or small, we can be sure that frontier exploration will raise international complications. For example, some geologists feel that the largest reserves may be found in the western region of China, a country that presently has difficult relations with the rest of the world, and we use Box 5.3 to illustrate how China's claim to even small reserves can cause diplomatic furor.

In addition to finding new oil, purely economic factors might increase reserves in countries with market economies. The major reason is that the cost of producing oil varies from field to field and well to well, related to such factors as the amount of oil in the reservoir being pumped, the depth and cost of drilling of the well, the cost of methods that might be used to enhance production, and the difficulty of transporting oil from the well

to a pipeline. Consequently, the amount of oil that can be pumped is critically dependent on the price that an owner can sell it for. If the price rises, wells that are uneconomic at a previous price may become profitable, and new wells may be drilled and old ones reopened. A second reason for an increase in future reserves in market economies is that these countries commonly place a tax on known reserves, leading private oil companies not to announce reserves until they have finished exploration and are very confident of their figures.

What, then, is the total amount of oil that the earth can ultimately produce? It will be more than 1 trillion barrels, but how much more? Geologic constraints prevent the ultimate reserves from being more than 2 trillion barrels, but between 1 and 2 trillion the ultimate production is critically related to price. In the 1990s, a barrel of oil costs about $18 (U.S.), fluctuating with small changes in consumption (demand) and production (supply). We would produce more at $50 per barrel, even more at $100, but how much more at any price? We de-

Table 5.3 *Oil and gas reserves and production in 1995*

Country	Oil			Natural gas	
	Reserves (in billion barrels)	Production (in billion barrels per year)	Lifetime (in years)	Reserves (in trillion cu. ft.)	Production (in trillion cu. ft. per year)
North America					
Canada	4.9	0.66	7	67	5.6
Mexico	49.8	1.0	49	68	1.3
United States	22.5	2.4	9	164	22
South America					
Brazil	4.2	0.25	17	5	0.28
Venezuela	64.5	0.94	69	140	1.6
Europe					
Netherlands				65	3.0
Norway	8.4	0.43	20	48	1.5
United Kingdom	4.3	0.92	5	23	1.9
Russia	57	2.5	23	200	23
Middle East					
Iran	86	1.4	62	742	2.1
Iraq	100				
Kuwait	94	0.66	110	52	
Oman	6.1	0.31	20	25	
Saudi Arabia	264	3.3	80	185	2.3
U.A.E.[a]	99	0.8	124	204	1.3
Africa					
Algeria	9.2	0.28	33	128	4.5
Libya	29.5	0.50	59	46	0.5
Nigeria	20.9	0.69	30	110	1.2
Asia					
China	24.0	1.1	22	59	0.5
India	5.8	0.25	23	2.5	0.6
Indonesia	5.2	0.49	11	69	2.6

[a]United Arab Emirates (U.A.E.) include Abu Dhabi, Dubai, and Sharjah.
Sources: American Petroleum Institute (annual); United States Energy Information Administration (annual).

cline to guess, preferring to leave this task to economic forecasters and astrologers.

Production, consumption, and transportation of oil

In contrast to reserves, world production and consumption (which are nearly equal) are very accurately known (Table 5.3). In 1995, total production was 22.4 billion barrels per year, approximately one-third from the Arabian Gulf, 10% each from the North Sea, the United States, and Russia, and 5% each from Mexico, Venezuela, and China. That leaves somewhat less than one-fourth of the world's oil production from other countries. The pattern of consumption, however, was very different. About 30% of world oil production was consumed by the

BOX 5.3 WAR IN THE SPRATLY ISLANDS?

In the southeastern part of the South China Sea is a ridge. Its geologic basement may be simply a thickening of the oceanic crust, or it may be a fragment of continent whose crust is thin enough to be submerged, or it may be an old island arc that never grew thick enough to reach the surface. Whatever it is, this basement was close enough to sea level that reef limestones and similar sediments could grow upward and remain at sea level as the basement subsided. As they grew, more sediment was washed off the sides of the ridge and developed an apron of rapidly accumulating debris. The top of the ridge is barely awash, forming small scattered islets known as the Spratlys. People can (barely) live on one small island, but the rest are completely covered by storm waves and consequently uninhabitable.

Why fight a war over the Spratly Islands? Well, it turns out that the limestones and flanking sediments may contain significant amounts of oil (or they may not — we don't know yet). If they do, then the country that owns the islands owns the oil on the islands, which is trivial, and also the oil in the 200-mile Exclusive Economic Zone around them (see Box 2.3), which may be worth a fortune. The problem is that ownership of the Spratly Islands is claimed by six countries, in alphabetical order

Brunei, China, Indonesia, Malaysia, Philippines, Vietnam. The Philippines have long claimed ownership, and since the 200-mile EEZ would reach almost the coastlines of Brunei and Malaysia on the island of Kalimantan (Borneo), their EEZ overlaps the one claimed by the Chinese. There are two ways of solving this type of problem. The claimants can meet, commonly with neutral nations to monitor the proceedings and advise, and come to a friendly agreement about sharing the profits. The second method is to invade and dare anyone to dislodge you.

China invaded one of the Spratly Islands, appropriately named Mischief Reef, in early 1995 (the exact date of arrival is not known because the occupation was discovered by the Philippines some time later). In order to occupy the island, the Chinese had to build platforms on stilts so that the few dozen men of the invading force would not be drowned by each passing storm. Once on the islands, the Chinese announced ownership and essentially dared anyone to fight for them. Since the Chinese have the largest army, largest navy, and largest air force of all of the claimants, nobody has accepted the challenge. The oil, if there is any, now belongs to China.

FURTHER READING: Chang (1996).

United States, with 5% of the world's population, 20% by western Europe, with 7% of the population, and the remainder by 90% of the people who live in the rest of the world. Thus, with the exception of the United States and the North Sea region (the United Kingdom and Norway), the industrial countries that use most of the oil do not produce it.

Using Table 5.3, we divide reserves by production to estimate the period of time during which productivity can be expected, commonly known as life expectancy, the same term used in Section 1.3 for the predicted lifetime of people. Using 1,000 billion barrels as world reserves and 22 billion barrels as world production, we calculate 45 years of expected production (1,000 divided by 22). This figure varies greatly, however, from country to country. In the United States for example, estimated reserves of 22.5 billion barrels and production of 2.4 billion per year yield a life expectancy of about 9 years (22.5 divided by 2.4). By contrast, the life expectancy is nearly 80 years for Saudi Arabia and more than 125 years for

Abu Dhabi and other members of the United Arab Emirates. Clearly the United States will remain the world's largest importer and the Middle East the largest exporter for the foreseeable future unless the world's total pattern of energy production is reorganized, a topic that we discuss in more detail in Section 5.5.

The difference between places of production and places of consumption requires that oil be moved around the world at a rapid rate. With the exception of oil that enters the United States from Canada and Mexico by pipeline, most of this transportation is provided by the world's fleet of tankers. They pick up crude oil at the coastal terminals of pipelines in producing countries and take the crude to refineries to be processed into gasoline and other products. Then if the refineries are located overseas from the country that wants the gasoline, the tankers load up the refined product and take it to the coastal terminals of pipelines in the consuming country. The transfer process in the 1990s is summarized diagrammatically in Figure 5.18. It shows that about 30% of

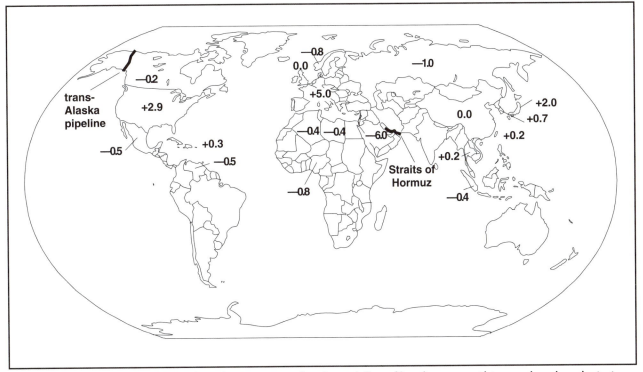

Figure 5.18 Major patterns of worldwide oil transfer. Amounts transferred are in billions of barrels per year, with imports shown by a plus (+) symbol and exports with a minus (−) sign. Two bottlenecks discussed in the text are the Straits of Hormuz and the trans-Alaska pipeline.

world production is transported out of the Middle East to a variety of consumers and about 15% is transported into the United States from many different sources.

Fig. 5.18 also shows present and future bottlenecks where interruption of the flow of oil would have major consequences for the world's economy. We discuss a possible future one in Box 5.4 and mention only two present ones here. The major present bottleneck is the Straits of Hormuz, which connects the Arabian Gulf to the Indian Ocean along a curved and island-dotted waterway about 25 miles wide between Iran and the Arabian peninsula. Tankers passing through these straits carry about one-fourth of the world's oil production, and because their passage could be easily prevented in time of war (or other Middle Eastern instability), many countries in the area have tried to construct pipelines that will carry their crude oil to terminals outside the Gulf, and industrial importing nations have taken military action and conducted ceaseless diplomatic negotiations to keep the straits open.

A smaller bottleneck is important because of the dominance of the United States in world oil consumption. Approximately one-fourth of U.S. domestic production is from the area of Prudhoe Bay, on the North Slope of Alaska. In order to move this oil to the rest of the United

States, it must first be shipped by pipeline for 600 miles across virtually the entire state of Alaska to the port of Valdez, which is tucked into the head of a tiny inlet. The pipeline is approximately 6 feet in diameter, must be heated so that the oil does not congeal in the winter, must be entirely above ground so that the heat does not disturb the permanently frozen ground (permafrost) over which it travels, and must cross two major mountain ranges. When the oil reaches the loading terminal at Valdez, it is stored in an area that was heavily damaged in 1967 by the Good Friday earthquake of southern Alaska, the strongest earthquake ever recorded in the United States (Section 3.3). Finally, the tankers that load the oil in Valdez must be escorted by tugs out of the inlet because the passage is so narrow that they would get stuck if they turned sideways. In a widely publicized incident, one tanker, the *Exxon Valdez*, struck rocks after passing through the narrows, and we discuss it in Box 5.5.

The transfers shown in Figure 5.18 are very different from the patterns that we would have drawn for any period of time older than about the middle 1970s. Many of the countries that are now major producers did not have an oil industry until shortly after World War II or even more recently. These changes in oil production and consumption have had a major effect on world history and

BOX 5.4 GETTING OIL OUT OF THE CASPIAN SEA AREA

Mountains stretching from Turkey to the Himalayas formed within the past 50 million years as various small continental blocks collided with the southern margin of Asia and the Arabian peninsula began to move north and east (Fig. 5.16 and discussion of Arabian Gulf in text). These mountains became a load on the crust and forced basins to the north to subside. The largest basin is the Black Sea, which subsided so far that it is now below sea level and has only a small outlet to the Mediterranean and thus to the rest of the world's oceans. Just east of the Black Sea

is the Caspian Sea, which is completely landlocked and is kept from overflowing only by evaporating as much water as rivers (principally the Volga) deliver to it each year.

The Caspian Sea basin has subsided far enough and been filled rapidly enough that it now contains a thick sequence of sediments that are potentially oil producing in all of the five countries that border it (Fig. 5.19). Oil production started along the Caspian Sea in the late 1800s with development of the large Baku fields, then part of Russia but now in the newly indepen-

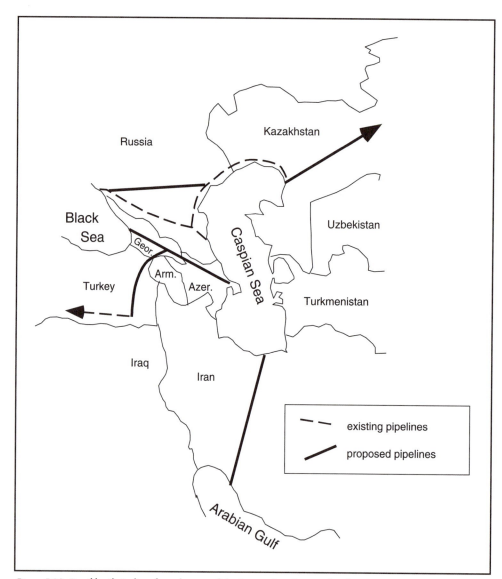

Figure 5.19 Possible oil pipelines from the area of the Caspian Sea. See text for discussion.

dent country of Azerbaijan. The total oil reserve in other parts of the Caspian region is clearly less than in the Arabian Gulf and is also only dimly known, but there is little question that it will be 10 or more billions of barrels. In fact, the Caspian Sea area could contain more oil than the North Sea and the north slope of Alaska, both of which have produced enough oil to help maintain the world's industrial economies. As these and other producing regions begin to lose reserves, the potential oil in the Caspian area looks more and more interesting.

The interest in the oil has spurred numerous oil companies to accept offers from Russia and other countries to conduct extensive explorations. As we write in the middle 1990s, oil is being found – lots of it. The problem is what to do with it. International oil companies can make a profit only if they sell their oil on international markets, not simply to countries around the Caspian Sea, which have little money and whose currencies are generally usable only within their own borders. Consequently, the oil must be moved through pipelines out of the Caspian region, but where. In Figure 5.19 we show five (of many) possibilities:

- Let Iran run a pipeline southward to loading terminals on the Arabian Gulf or Indian Ocean and sell it there. But the United States and several other countries, however, do not want to deal with a government that they feel sponsors international terrorism.

- Run a pipeline westward through the Georgian region of the Caucasus to terminals on the Black Sea, but there are a lot of earthquakes in the Caucasus, not to mention continual political upheavals.
- Let Russia pipe the oil northward along the north slope of the Caucasus. It seems like an ideal route, avoiding mountains and connecting to the existing Russian pipeline system, and then moving on to the Black Sea The route, however, passes through Chechnya, where active rebellion against Russia is in progress, and several other regions with similar separatist feelings.
- Build a new pipeline through Russia north of Chechnya and ultimately reaching the oil terminals on the Black Sea. As with the previous route, this line presumes continued willingness of Russia to permit operations of foreign companies.
- Run a pipeline 3,500 miles eastward, beginning in Kazakhstan and ending on the coast of China. It would, however, cross a major set of mountain ranges in central Asia and would be almost completely under the control of China.

Any ideas?

FURTHER READING: Matzke (1994); Knott (1995).

will probably have an even larger effect in the future. As a dramatic example, our quotation at the start of this chapter is from a KGB memorandum (secret – what else?) warning the government that the colossally poor economy of the Soviet Union was being propped up only by revenue earned by foreign sale of oil and that the government would collapse if those exports stopped and the country had to import oil. The KGB appears to have been right, for the government collapsed in 1991, just after production in the Soviet Union decreased so much that it either had to import oil or close down major sections of its industry.

The other major change from oil exporting to importing was made by the United States. The United States was the world's largest exporter before World War II, and then imported only a small amount of oil until the early 1970s (Fig. 5.20). Partly because of this switch, the United States changed from being the world's largest in-

ternational creditor in 1980, with assets of approximately $500 billion, to the world's largest debtor, in the mid 1990s, with debts to foreigners of approximately $1,500 billion. The country now imports about 3 billion barrels per year at an expense of approximately $60 billion, roughly half of each year's balance of payments deficit. The United States obviously has an economy that can afford oil imports, but the necessity of importing oil has significantly affected its relationships with Mexico, Nigeria, and many countries in the Middle East.

Oil shale and tar sand

Oil and gas occur not only in reservoirs from which they can be pumped but also locked up in sedimentary rocks, commonly referred to as oil shale and tar sands. Oil shale is shale containing preserved organic matter that has undergone some maturation to hydrocarbons from its origi-

BOX 5.5 THE *EXXON VALDEZ*

On March 24, 1989, the tanker *Exxon Valdez* maneuvered away from the loading terminal at the port of Valdez, Alaska, inched through the Valdez narrows into the comparatively open waters of Prince William Sound, and shortly afterward struck a shallow ridge of rock known as Bligh Reef. The rock tore a gash through the thin metal hull that separated the tanker's oil from the icy waters of the sound, and 10.9 million gallons of crude oil floated out. Despite the fact that this accident was the first in more than 8,000 tanker trips in and out of Valdez, and the fact that the amount of oil released was about 15% of the amount emitted from the tanker Amoco Cadiz when it ran aground off the coast of Brittany in 1978, the *Exxon Valdez* became a symbol for people who believed that oil companies exist largely to defile the environment. The accident generated three activities. The first two — we are unsure of which came first — were to start cleaning up the oil and for the involved parties to file lawsuits against each other and anybody else they could think of. The third was to research the environmental effects of the spill and to formulate recommendations to prevent further ones.

Immediate cleanup consisted of placing floating booms around as much of the spill as could be contained and then siphoning up oil that had not already been completely dispersed. These efforts were only partly successful, and oil ultimately floated westward out of Prince William Sound for a distance of several hundred miles along the Alaskan coast. Later cleanup was directed largely at intertidal areas of coasts where the oil had been carried ashore by waves or on high tides. These areas were regarded as particularly important because they are the hatching grounds for herring, an important commercial fish, and are also occupied by mussels and other shellfish that form a major part of the food consumed by birds and sea otters. Nothing could be done about oil already in the mussels, but removal of some of the oil remaining on the shore surfaces was accomplished by blowing hot water and steam on them to wash the oil out to sea.

The legal actions cannot concern us here, and we mention only two aspects. One is that, in searching for long-term consequences of the spill and in defending against suits, investigators discovered that Prince William Sound contains oil from at least two sources in addition to the *Exxon Valdez*. One is normal leakage from the numerous boats that use this very busy area, and the second was natural oil seeps on the floor of the sound. Because the oil from the *Exxon Valdez* and the oil from the seeps contained different proportions of hydrocarbons (see text for types of hydrocarbons in oil), it was possible to determine ap-

proximately how much of the oil was from seeps, but because many of the boats used gasoline refined from oil that passed through the Valdez terminal, their contribution could not be separated. Consequently, the exact percentage of oil in the sound that came from the *Exxon Valdez*, for which Exxon is liable, has not been precisely determined. Our second comment is that the adversarial nature of the legal proceedings prevented investigators working for parties on opposite sides of the issue from cooperating in their investigations of the spill, and in order to maintain complete impartiality, scientists working for the federal government were prohibited from sharing any information with anyone not employed by the government. The consequence of this secrecy has been serious reduction in our ability to study the environmental consequences of the accident.

Research on the accident has tried to separate the immediate consequences from the long-term effects. The immediate result of the oil included the death of approximately 40,000 seabirds, mostly because of coating by oil but partly for other reasons; the death of an estimated 3,000 sea otters; and the cancellation of the 1989 salmon fishing season, scheduled to begin shortly after the accident, because oil on the water would contaminate both fish and nets as the salmon began to crowd into the river mouths for the annual journey upstream. Fishing continued, however, for bottom-dwelling fish, crabs, and lobster and for fish that use deeper waters of the sound.

Continuing effects 5 years after the accident are more problematic than the initial ones. By 1991 the salmon catch reached a record level, possibly because the lack of fishing in 1989 allowed more fish to spawn than in other years. By 1992 the herring catch was large, possibly because only 4% of the shoreline where herring spawn had been affected by oil, and also possibly because only 2% of the original oil remained in intertidal areas. Counts of coastal birds and seabirds are unclear because of large natural variations in their abundance, but by 1993 the closely watched bald eagles appeared to be as numerous as before the accident. Some of the mussels still contain traces of oil, and some of the heavier components of the oil have probably been incorporated with the natural oil in sediments on the bottom of the sound, but their long-term effect is unknown.

The oil spill from the *Exxon Valdez* has taught us several lessons, and we close by listing three and asking some provoking questions:

- The likelihood of oil spilling from a tanker that runs aground would be greatly reduced if tankers had two hulls, an outer one to fend off rocks and an inner one to hold the oil in case the outer one is breached. Are you willing to pay a little more for gasoline in order to require that tankers add an extra hull?

- The elimination of oil from intertidal areas was probably accomplished almost entirely by natural processes of biological and chemical degradation of hydrocarbons. Some investigators have suggested that the steam used to hose off shore-

lines did more damage than the oil itself. Are you willing to wait for natural processes to take care of oil spills?

- The adversarial nature of legal proceedings seriously hampered investigation of the environmental consequences of the accident and possibly also efforts to remedy it. Would you like to restrict the ability of people to sue?

FURTHER READING: Wells, Butler, and Hughes (1995).

nally deposited form. In effect, it could have been a source rock for conventional petroleum accumulations except that the oil and gas did not migrate out of it to reservoirs. Tar sand is, as the name implies, a sandstone in which some of the pore spaces are partly filled by heavy hydrocarbons, commonly known as tar. The origin

of many of these deposits is unclear, but some are probably old oil reservoirs from which the light hydrocarbons escaped, leaving the less-volatile tar behind.

Oil and gas could be extracted from shales and sands by a combination of methods. The one most commonly considered is summarized in Figure 5.21. The rock is

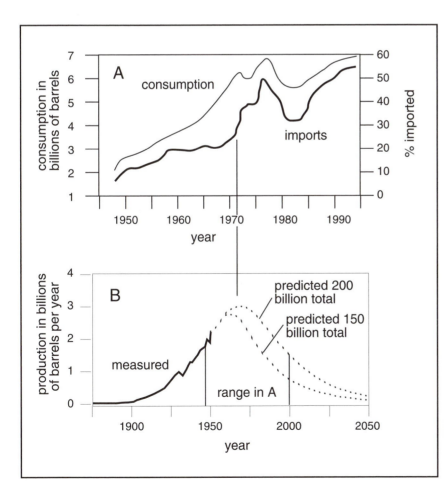

Figure 5.20 (A) Change in U.S. oil consumption and percentage of oil imported from 1945 to 1995. (B) Change in U.S. oil production predicted in 1956 by M. K. Hubbert. The predicted peak production in the early 1970s, regardless of the total reserve, is at the same time as the rapid increase in percentage of imports from approximately 20% of consumption to approximately 40%.

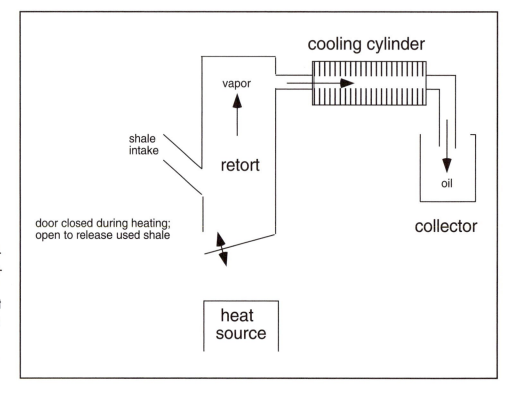

Figure 5.21 Principle of retorting oil shale. The shale is introduced to a retort and heated to drive off vapor. Cooling this vapor causes condensation to oil, which can be easily collected, or to gas, which must be held in a closed container. A retort is usable only if it produces more oil and gas than is needed to heat the shale.

mined, usually by surface stripping, and then cooked (retorted) to drive out the lightest and most volatile fractions. The escaping gas is cooled, condensing the gas to a liquid that is somewhat lighter than ordinary crude oil and needs less refining before it can be used. In order for this process to be useful, it must yield more energy than is used to mine and extract the oil. Consequently, most oil shales and tar sands that are commercially useful require less than one-half gallon of oil per ton of rock (about 1 cubic yard) for mining and extraction and yield about a half gallon to a gallon of oil per ton.

Based on a yield of at least one-half gallon of oil per ton of rock, the world's reserves of hydrocarbon in shale and sand are several times larger than those of conventional oil and gas. Using these reserves, however, encounters three major difficulties. One is that hydrocarbon tied up in shale and sand is more expensive to extract than pumped oil and gas. Thus, the price that an operator would have to receive for sale of the oil or gas from shale and sand must be higher than the normal world prices in the 1990s. In the past, the inability to ensure these high prices has always led to cancellation of projects designed to explore the possibility of large commercial ventures in oil shale and tar sands. In the United States that problem could be overcome if the country, in

an effort to reduce its dependence on imported oil, would either subsidize the production of oil shale or raise the cost of imported oil by placing a tariff on it. Because either action would raise the price of oil, both have always been successfully resisted by United States industries and consumers.

The second difficulty is that the oil shales and tar sands are highly localized. The preponderance of tar sand is in the Athabasca basin of western Canada and almost all of the rest is in Venezuela. Although oil shale is somewhat more widely distributed than tar sand, most of it is in the Green River basin of the western United States (see Box 8.4 for another discussion of the Green River basin). These reserves provide a solution for North America, but other countries could import this oil only if world prices rose significantly above their present level, which would have a serious effect on the economy of the rest of the world.

The third difficulty, particularly applicable to the Green River supplies, concerns the allocation of water resources in the American West. Production of 1 barrel of oil from the Green River shales requires ~200 gallons of water. This water provides the cooling needed to condense the gases retorted from the shale (Fig. 5.21), cools the equipment used in the extraction and retorting process, and washes and cools the shale residue. The Green River shales are near the headwaters of the Colorado River system (Section 4.3), and if the United States decided to use oil from shale for 10% of its 1990s oil consumption, it would have to divert 120 billion gallons of water from the Colorado River in order to produce 600 million barrels of oil. This figure may not seem very large, but it amounts to a high proportion of the total water supply of the Colorado River drainage. We suggest that you look back at Section 4.3 and then try to imagine the outcry from communities, industries, and agricultural interests downstream if water were diverted to the retorting of oil shale.

Geology and distribution of coal

The first land plants evolved about 350 million years ago, and within 100 million years they had established forests over much of the world. Some of the dead plants and their debris were preserved where those forests occupied swamps, particularly coastal swamps (see the previous discussion of conditions under which coal accumulates). The swamps were merely one in a series of continually changing environments in which terrestrial (land-deposited) and marine sediments were laid down during cyclical rise and fall of sea level over a vertical distance of some 50 to 100 m (Figs. 5.22 and 5.23). As sea level

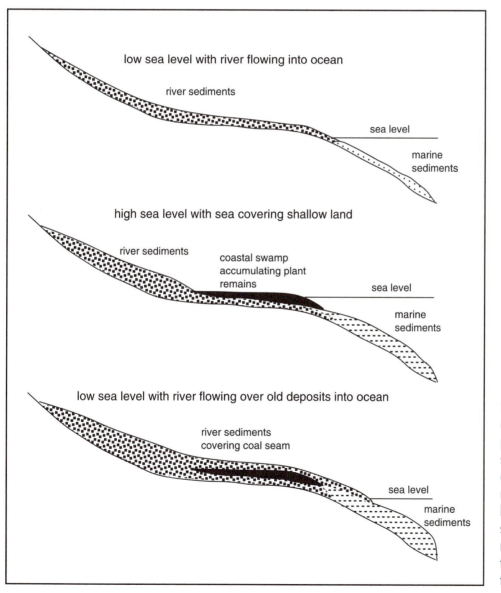

Figure 5.22 Accumulation of coal seams by preservation of plant matter in a coastal swamp and burial beneath other sediments. Repeated raising and lowering of sea level and movements of the shoreline may generate numerous coal seams and interbedded sediments. See further discussion in text.

Figure 5.23 Accumulation of plant remains in Okefenokee swamp of Georgia. Leaves and other fallen vegetation decay in oxygen-deficient water around the stumps of living cypress trees.

fell, the landward side of a coastal swamp was covered by sand and silt from rivers and the seaward side of the swamp moved further toward the coast. Then when sea level rose, the coastal swamp was covered by shallow marine sediments and the swamp advanced landward back over the river deposits. The operation of this process at various times and places generated the present broad distribution of the world's coal reserves.

Once a sequence of sediments containing the swamp deposits was buried, the plant debris was converted to coal, the other sediments were converted to hard sedimentary rocks, and the coal became simply one horizon in a sedimentary section. The thickness and the lateral extent (i.e., the volume) of this horizon depends on the pattern of environments through time as the plant debris was deposited, and the quality of the coal depends on the depth of burial (pressure) and the heat to which the sequence was subjected. Burial caused the original plant remains to progress from peat (semidecayed plants) to lignite (brown coal), bituminous coal (black coal), and finally to a variety known as anthracite, which consists of nearly pure carbon. Because almost all of the energy derived from burning coal comes from its carbon, the progression from peat to anthracite is also a progression in commercial value.

The world's largest coal reserves were formed on broad continental platforms partly covered by shallow seas, and coal reserves in any area are closely related to the size of the platform on which they were formed. The two largest platforms are those of European Russia, with nearly one-third of the world's coal reserves, and North America (mostly United States), with about one-fourth. Smaller platforms occur in China, Australia, parts of western Europe, and South Africa. Together with India, which contains coal in rift basins, these areas contain nearly 90% of the world's coal reserves. The remainder of the world's reserves are scattered through almost all parts of the world, and although there is an enormous amount of coal, most of it is not commercially valuable.

The distribution of reserves only partly controls the production of coal. The value of a coal seam is highly variable and depends on several factors, and the decision of whether to mine it or to buy coal from elsewhere is largely based on economic considerations. One is the cost of mining, which depends heavily on the amount of rock that must be removed in order to get the coal. Thus, coal seams that are thin, extend for only short distances, or are deep in the earth are expensive and commonly not commercially usable, and seams that can be "strip-mined" at the surface are more attractive than those that must be mined underground (Section 6.4). A high cost of mining, however, can be accepted for coal of high quality. For example, anthracite and the high grades of bituminous coal contain high carbon contents and are commonly mined underground, whereas lignite can only be strip-mined. The other factor that determines the value of a coal seam is its closeness to markets, for coal is bulky and expensive to transport.

The decision to mine coal locally or purchase it elsewhere is based both on economics and politics. One example is the decision by Russia (and the former Soviet Union) to subsidize production in the extraordinarily inefficient Soviet coal mines so that the country would not have to spend foreign ("hard") currency to import coal. The opposite decision, based on economics, was made by the United Kingdom, which has very large coal reserves but imports much of its coal because the import price is lower than the domestic one. This decision was highly controversial but was made because the lower coal prices would benefit other parts of the U.K. industry. Based on economics and politics, the widespread occurrence of at least small coal seams permits most countries to use only coal that is mined within the country, and only about 10% of the world's coal production of 4 million tons enters international trade. The pattern of trade is nearly the reverse of the pattern for oil (Fig. 5.24). About one-third of the world's coal exports are from South Africa and western Australia, with another third provided by the United States and Canada, in contrast to the fact that the United States imports about 30% of the world's oil production.

We finish this section by pointing out that we have lived in a "century of oil" for transportation and much of our industry and a century of coal for most of our electricity. During this time the patterns of production and transportation have changed remarkably. Western Europe, which started the industrial revolution largely with its own coal reserves, now is the world's major importer. The United States, which started the oil industry and was once a major producer, now is the world's largest importer. It would be foolhardy to think that these patterns will not change even more rapidly in the future.

Policy questions

If you live in an oil-importing country, what do you think should be done?

- Nothing. Your country's economy is strong enough that you can continue to import.
- Reduce the use of oil by developing alternative energy sources, particularly for automobiles and other forms of transportation.
- Invade a country that has abundant reserves. (We are not seriously suggesting this, but it has been done, and some people are serious.)

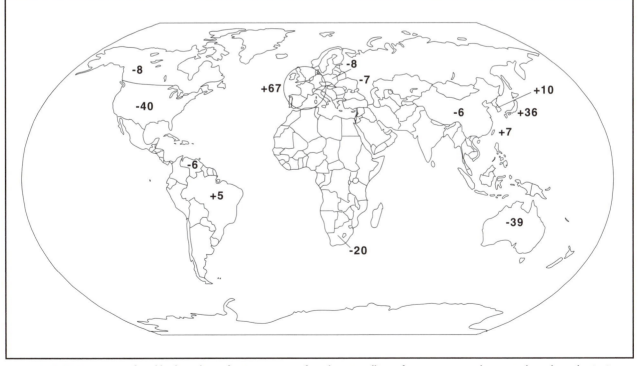

Figure 5.24 Major patterns of worldwide coal transfer. Amounts transferred are in millions of tons per year, with imports shown by a plus (+) symbol and exports with a minus (−) sign.

FURTHER READING: oil and gas – Hunt (1979), Shell Oil Co. (1983); American Petroleum Institute (annual); coal – Gordon (1987).

5.4 NUCLEAR POWER

It might be possible to set up a nuclear chain reaction, liberate energy on an industrial scale, and construct atomic bombs.
– Leo Szilard, September 12, 1933, on the possibility of nuclear fission, in R. Rhodes, The Making of the Atomic Bomb, p. 28

Some of our current sources of energy have been used throughout civilization – putting waterwheels in rivers, warming ourselves in the sun, burning wood and pitch (tar). The entire concept on which much of our current electricity is generated, however, was not even thought of until 1896. Early in that year Professor Antoine Henri Becquerel was studying the effect of sunlight on fluorescence of uranium compounds. During one spell of almost total cloud cover, Becquerel left some completely light-shielded photographic film in a cabinet drawer with some uranium compounds. Development showed that the films had been "exposed" (darkened) by something in the drawer. That something must have been radiation akin to X-rays, which can penetrate light-proof shielding. On the day after Christmas 1898, Marie Curie successfully separated the element radium, which is much more radioactive ("hotter") than uranium, and she and Pierre Curie labeled the "something" radioactivity.

Radioactivity is the result of unstable elements decaying ultimately to stable ones. In the earth, almost all of the natural radioactivity is generated by the heavy elements uranium (isotopes ^{238}U and ^{235}U) and thorium (^{232}Th) and one isotope of the lighter element potassium (^{40}K). These elements contribute significantly (how much is not known) to the earth's internal heat, and the discovery that they were radioactive solved a major problem for 19th-century geologists. They could calculate the temperature that the interior of the earth would have at any time after initial accretion if its only heat source had been the energy generated when the earth formed (Section 3.1). The geologists also knew that the earth is hot enough to produce volcanoes now. In order for the earth to be this hot now, they calculated an age of the earth that was clearly too small to account for all of the geologic processes that they observed, and the discovery of radioactivity showed that the earth could still be hot even if it was much older than originally believed.

In the early 1930s, Leo Szilard (quoted at the beginning of this section) and other physicists realized that radioac-

tivity could be produced in other ways than the decay of naturally occurring unstable isotopes. The first possibility to be investigated was that a mass of uranium of some, as yet unknown, size might become unstable enough to break neutrons and protons out of the uranium nucleus. If these particles collided with other uranium nuclei, thus generating more neutrons and protons, then two much-lighter nuclei could be produced along with a considerable amount of energy. This process was called fission, and the first successful demonstration of large-scale fission was in 1942 in a laboratory near the tennis courts of the University of Chicago. The experiment in Chicago was part of a frenzied program by the Allies in World War II to develop fission of such size and speed that it could be used as a bomb, and to produce that bomb before Nazi Germany did so. The first bomb was tested at Alamogordo, New Mexico, in July 1945, and shortly followed by bombs dropped on Hiroshima and Nagasaki in August.

The third type of radioactivity to be recognized was fusion. Like natural radioactive decay, fusion occurs naturally but not on the earth. The sun derives its energy almost exclusively by the fusion of hydrogen to helium and a few other slightly heavier elements. It occurs on the earth only because we make devices capable of causing it – they are called hydrogen bombs.

Before proceeding to the topic of nuclear power, we discuss in more detail the three types of radioactivity – natural decay, fission, and fusion.

Radioactive decay

With minor exception (potassium 40; ^{40}K), unstable elements decay by the emission of one of two particles – alpha and beta – from their nuclei. An alpha particle is the nucleus of a helium atom, with an atomic weight of 4 and a charge of +2 ($^4He^{+2}$). Its removal reduces the atomic mass by four units and the atomic number by 2. A typical example is the decay of radium 226 ($^{226}Ra_{88}$) to radon 222 ($^{222}Rn_{86}$). Beta rays are electrons with a mass of zero and a charge of −1 (beta^{-1}). A ray's removal consequently does not change the mass but increases the atomic number by one. An example is the decay of $^{87}Rb_{37}$ to $^{87}Sr_{38}$. Uranium (both isotopes) and thorium both have complex decay schemes in which three series of unstable isotopes ultimately produce stable isotopes of lead by

^{235}U—7 alpha —4 beta $\rightarrow$ ^{207}Pb

^{238}U—8 alpha —6 beta $\rightarrow$ ^{206}Pb

^{232}Th—6 alpha —4 beta $\rightarrow$ ^{208}Pb

Each natural decay proceeds at a rate that is constant for that radioactive isotope, and the energies involved in nuclear processes are so high that the rate is unaffected by changes in either temperature or pressure. A constant rate means that a constant percentage of a parent unstable isotope decays in a given period of time to a daughter product. Because the percentage of decay is unchanged, a greater mass of parent produces a proportionally greater mass of daughter in a given time period. Consequently, the amount of daughter produced in a given time period decreases as the amount of parent is reduced, leading to a graph of abundance of parent and daughter through time as shown in Figure 5.25. These graphs are exponential, or logarithmic, and the equation that describes the variation in abundance of parent through time is identical to the equation that describes increase in population through time except that the decay equation has a minus sign in the exponent (Fig. 1.3 and Section 1.2). Thus

$$P_{now}/P_{initial} = e^{-kt}$$

where P is the amount of decaying isotope left at any time t, and k is the decay constant. Furthermore,

$$k = \ln2/t_{half}$$

where ln is the natural logarithm and t_{half} is the half-life. By analogy with the doubling time discussed in Section 1.2, we recognize a half-life in which half of a parent decays to its daughter and can determine those half-lives by laboratory measurements. Table 5.4 shows the half lives of naturally decaying isotopes and also of isotopes formed by fission or other processes significant to us.

It is not quite accurate to say that emission of an alpha particle reduces atomic mass by 4 units and that emission of a beta ray does not change mass. A very small amount of mass is lost in all radioactive processes. The loss of mass results in the emission of a third type of radiation in addition to alpha particles and beta rays. This additional radiation is known as gamma rays, which have higher energy than x-rays (high frequency; see Section 8.1 and Fig. 8.5). Gamma rays result from the destruction of mass according to the equation

$$E = mc^2$$

where E is the energy produced, m is the mass destroyed, and c is the speed of light. This equation was developed by Albert Einstein in 1903, only a few years after the discovery of radioactivity. This energy produced by nuclear fission is the energy (heat) generated by nuclear power reactors that we discuss later.

Fission

As we mentioned briefly, some mass of uranium is needed to generate nuclear fission. If there is not sufficient uranium, the neutrons generated by some initial amount of fission will not strike enough other uranium atoms, be absorbed in them, and cause fissioning to produce enough

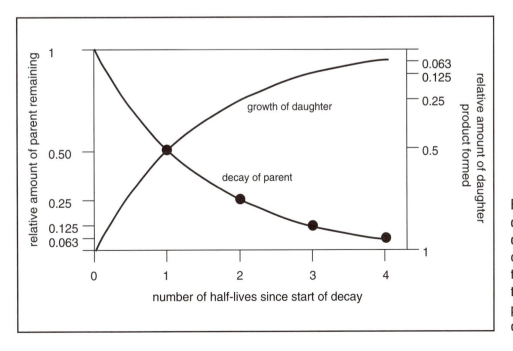

Figure 5.25 Curves showing decrease in abundance of radioactive parent and increase of stable daughter isotope through time. The half-life is the time at which half of the parent has decayed to the daughter product.

Table 5.4 *Half-lives of nuclear fuels*

Isotope	Half-life (in years)
^{232}Th	1.41×10^{10}
^{235}U	7.1×10^{8}
^{238}U	4.51×10^{9}
^{239}Pu	24,400

additional neutrons to strike other uranium nuclei, to be absorbed, and cause fissioning, and the whole process dies down. Given enough mass in one place, however, the number of neutrons generated can be just larger than the number absorbed, and the process cascades into a "chain reaction" until the uranium is used up.

The mass needed for a chain reaction is referred to as a "critical mass" and is dependent on the proportion of ^{235}U and ^{238}U in the uranium. When the earth was formed ~5.45 billion years ago, uranium contained 22% ^{235}U and 78% ^{238}U, but the much shorter half-life of ^{235}U has caused the proportions to change to present values of 0.7% ^{235}U and 99.3% ^{238}U. The greater instability of ^{235}U that leads to more rapid decay also makes it more fissionable, and natural uranium must be enriched in ^{235}U in order to be used for fission reactions. Fissionable material that is used in reactors (see subsequent discussion) must contain a minimum of 3.2% ^{235}U, and material used in nuclear weapons contains much more. These higher percentages of ^{235}U require that natural uranium be separated by very complex processes into an enriched fraction and, consequently, a depleted fraction consisting almost entirely of ^{238}U. Because the half-life of ^{238}U is so long (4.5 billion years), the depleted fraction is essentially inert and may be discarded – some of it can replace lead as radiation shielding, and it is also used by the army for armor-piercing shells.

The fission process, whether in reactors or bombs, breaks down both of the uranium isotopes to produce a complex array of fission products that are also unstable, with half-lives ranging from seconds to a few tens of years, and decay directly to stable isotopes by alpha or beta emission (see discussion of the problem of nuclear waste in Section 7.4). The fission process also produces energy because the total mass of the fission products is slightly less than the mass of the original uranium nucleus. The lost mass is converted into energy and emitted as gamma rays according to the same Einstein equation that we discussed for natural radioactivity. We discuss the

construction of fission reactors, plus a related form of reactor known as a breeder, and also the manufacture of bombs later.

Fusion

Fusion is the combining of light nuclei to make a heavier one. The sun and other stars derive their power from this process, almost exclusively by the combination of hydrogen to helium, the next lightest element (atomic number 2), by the very generalized equation

$$4^{1}H \rightarrow {}^{4}He$$

The fusion process was first recognized in the sun, and no one considered the possibility that it could occur on earth. It does not occur naturally, and it was not until after World War II that the Russian physicist and human-rights advocate Andrei Sakharov suggested that it could be made to occur on earth. Russia and the United States then engaged in a ferocious race to be the first to cause fusion. Both were successful when they generated the first hydrogen bombs.

Fusion reactions derive their energy from the same loss of mass as fission reactions (see the previous discussion of the Einstein equation). The ability of heavy elements to lose energy while breaking into lighter ones and light elements to lose energy when they combine to form heavier ones implies that elements in the middle of the periodic table have the least excess mass and cannot be either fissioned or fused. The most stable element is iron (atomic number 26).

Nuclear reactors and bombs

Assembly of a critical mass of fissionable material can cause either a controlled reaction or a bomb depending on how it is done. A reactor obviously needs to be controlled, and conventional reactors do this in two ways – one increases the efficiency of the fissioning, and the other prevents the reaction from running wild. In order to increase efficiency, the neutrons emitted by the fission reaction must be slowed down so that they do not simply pass through other nuclei without splitting them. This is accomplished partly by material, such as graphite, in control rods and partly by fluids that circulate through the reactor. One of the best materials for slowing neutrons is deuterium (^{2}H), the hydrogen component of heavy water, and its manufacture was essential in efforts to construct the first reactors and bombs.

Two methods are used to prevent the fission reaction from running wild. First, reactors use uranium with 3% to 4% ^{235}U, the lowest possible enrichment to sustain a chain reaction. Second, they intersperse fuel, commonly in "rods" consisting of uranium dioxide (UO_2) or metallic uranium with "control" rods containing material that absorbs neutrons without fissioning (Fig. 5.26). These control rods consist of a variety of materials, and they can be pulled partly out of the reactor to speed up fissioning or pushed farther into the reactor to slow down the process. A controlled reactor, therefore, generates a steady supply of heat for the generation of electrical energy. The conventional reactor uses the heat to boil water and run the generator in the same way that fossil fuels are used. The temperatures produced by reactors, however, are sufficiently high that other methods than boiling water have been devised to turn a generator. The principal, but not the only, method is to melt sodium in the reactor and force it through a web of pipes out of the reactor and through a generator. This serves the dual purpose of running an electrical generator and also cooling the reactor.

A reactor can go out of control for several reasons. Fuel rods with 3% to 4% ^{235}U are not enriched enough to become a bomb, but they can cause a lot of damage. One of the simplest problems is cracking of water pipes in the reactor, thereby releasing steam that has come into contact with fuel rods and has dissolved some of the radioactive fission products. This type of breakage occurred at Three Mile Island, causing the worst accident in U.S. history, but resulting in no radiation damage to people. The most extreme problem is meltdown of the reactor core. This meltdown could occur only if virtually all of the cooling and control systems failed to operate effectively, thus permitting an uncontrolled chain reaction. The molten core can run into the ground, vaporizing the groundwater and sending radioactive steam and entrapped soil particles into the atmosphere. The only reactor in the world to have undergone a core meltdown is the one at Chernobyl, Ukraine, when it was part of the Soviet Union, and we describe it in Box 5.6.

In addition to making reactor designs and operating conditions as safe as possible, two other major safeguards are built into reactors. The primary one is a containment vessel around the entire reactor complex (Fig. 5.27). This containment vessel should hold all of the steam released in case of breakage of water pipes and some of the radioactive steam and particles from a core meltdown. The second safeguard is to prevent a molten core from coming into contact with the ground by placing the core above a series of concrete conduits that would separate the molten uranium into widely spaced, isolated, vessels that could cool harmlessly.

Two types of reactors use fuel that is more enriched than 3.5% ^{235}U and are, consequently, somewhat more dangerous. A few European reactors now use uranium so enriched that it would cause a small nuclear explosion if it became uncontrolled. And some reactors, principally

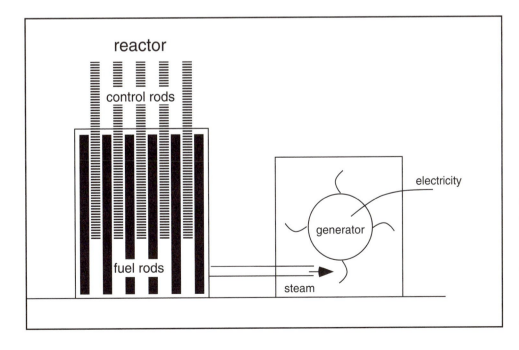

Figure 5.26 Basic features of nuclear reactor. Fuel rods are interspersed with control rods, which can be removed or inserted to maintain the reaction at a proper speed. Steam from the reactor (or a boiler above it) is used to run an electric generator as shown in Figure 5.5.

BOX 5.6 CORE MELTDOWN AT CHERNOBYL

At 1:24 in the morning of April 26, 1984, someone pressed the panic button at the no. 4 reactor of the Soviet power station at Chernobyl in the Ukrainian Republic (now the independent country of Ukraine). It was a tacit admission of the failure of a test designed to determine whether the reactor could shut itself down in the event of a complete loss of electrical power at the station. Whether it could have done so if the test had been conducted properly will never be known. What happened instead of a shutdown was a rapid overheating of the fuel in the reactor core as cooling water could no longer circulate through it (see text). As the fuel began to melt, the reactor became so hot that water still in it vaporized and exploded, blowing out parts of the reactor and damaging the roof of the building that housed it. This explosion was followed about 20 minutes later when hydrogen formed by breakdown of the water also exploded, shattering the reactor building and blowing pieces of building, reactor, and fuel into the air. Then came real trouble. The reactor core had been made largely of graphite with spaces for the fuel rods, and now that the building was completely open, the material spewing upward sucked air into the core, and the graphite caught fire. That produced essentially a minivolcano, erupting burning graphite, steam, melted fuel, fission products, and anything else that the superhot core came in contact with.

Because the Soviet government was still in control of the Ukraine at the time of the meltdown, no reliable information was available immediately. The first real indication that anything was wrong was made in western Europe when winds that were blowing northwestward over Chernobyl at the time of the explosion carried radioactive fallout that was detected by routine radiation monitors. The fallout in western Europe was not particularly severe, although all countries took precautions to measure radiation in food and milk (Finland even imposed a ban on reindeer meat during the summer following Chernobyl). Because of these precautions and because the total fallout was so small, no effects on the health of people in western Europe have been demonstrated in the 10 years since the explosion, although longer-term consequences cannot be ruled out at this time.

The immediate consequences of the explosion in the vicinity of Chernobyl stand in stark contrast to those in western Europe, but the long-term effects are puzzling. In the first 2 weeks after the explosion, 31 people died from high radiation doses and blast effects, and more than 200,000 were evacuated from an "exclusion zone" up to 30 km downwind (northwest) of the reactor because of radioactive fallout (the exclusion zone is now mostly in the newly independent country of Belarus). In the first 2 months, hundreds of thousands of people in and near the exclusion zone were found to have radioactive iodine (^{131}I) in their thyroid glands, and fear of other radiation effects was widespread. Ten years after the explosion, however, the only measurable effect of radiation is an increased incidence of thyroid cancer in children, and the widely expected increase in leukemia rates has not appeared. Whether the radiation will have longer-term effects, including genetic defects in newborn children, will only be known as the years pass.

Effects on animals and vegetation, including crops, may ultimately turn out to be more serious than those on people. Nearly a square mile of pine forest was killed almost immediately, and areas up to 25 miles away received enough fallout that plant growth was stunted. In the first growing season after the accident, vegetables were so contaminated that they had to be discarded, and fallout in the grass made cows produce milk that could not be consumed. Radiation in the soil and water has decreased markedly over the past 10 years, and much of the region is now being farmed again, but as with the effects on people, it will be many years before the ultimate consequences are known.

The situation around Chernobyl will surprise readers from industrial countries. The giant "sarcophagus" made of cement to encase the reactor has begun to leak, periodically emitting more radiation into the air. Regardless of this risk, the two reactors that did not explode, and which had resumed operation within months after the accident, continue to operate simply because Ukraine has no way of replacing the electricity that they supply. People in both Ukraine and Belarus have moved back into much of the affected area because they need land to make a living on, and agricultural land is not available elsewhere.

Poor countries don't have many options.

FURTHER READING: Medvedev (1990); Marples (1995); Stone, Williams, and Balter (1996).

Figure 5.27 Containment vessels at Oconee Nuclear Power Plant, North Carolina. Courtesy Duke Power.

in France and Japan, are referred to as breeder reactors because they are designed to produce fissionable material from the ^{238}U that normally does not take part in ordinary fission reactions. Breeders convert ^{238}U into an isotope of plutonium (^{239}Pu) by a series of reactions

$$^{238}U + n \rightarrow {}^{239}U$$

$$^{239}U \text{—beta} \rightarrow {}^{239}Np$$

$$^{239}Np \text{—beta} \rightarrow {}^{239}Pu$$

where n is a neutron, and Np is the element neptunium. Reactors can be designed to use some of the plutonium as they breed it, but a reactor referred to as a breeder uses fuel rods that have been manufactured partly from plutonium that has been extracted from spent fuel rods. This

process is known as reprocessing, and because it is intimately related to the issue of supplies of nuclear fuel, we discuss it later under that heading.

All reactors eventually use up enough of the fissionable material in their fuel rods that the rods no longer constitute a critical mass. This point is reached when the fission products are sufficiently concentrated that they absorb too many neutrons for the chain reaction to continue. At this time, the fuel rods must be removed from the reactor and replaced with fresh ones. The spent rods are very rich in radioactive fission products and also, to some extent, in plutonium. That is, they are very dangerous, and here is where the major problem with nuclear power raises its controversial head. What do we do with the spent fuel rods? We discuss this issue in Section 7.4.

We finish this discussion of reactors by describing how to make nuclear weapons. The basic principles are not difficult, although the mechanics of assembly are tricky. You need a supply of bomb-grade uranium and/or plutonium – less than a 100 pounds will do for a small bomb. Then you must store this explosive in separated pieces that are too small to explode by themselves and have a way of bringing them together so rapidly that they do not undergo significant fission before they form one mass that fissions explosively – this process generally must be complete in less than millionths of a second. Finally, if you want to make a really big bomb, you enclose a cannister of hydrogen inside the uranium/plutonium ring, and when the fission bomb explodes, the hydrogen will be compressed to form a fusion bomb. Naturally there are a few other technicalities, but basically the process is very simple.

Fuel supplies for reactors

Plutonium does not occur in nature, but uranium is widespread enough that almost all rocks contain tiny amounts. Ordinary granite, for example, has U concentrations of 1–5 parts per million (ppm; $10^{-4}\%$), but economic concentrations of uranium occur in only a few places. Although many of the deposits are hydrothermal veins, which we discuss in Section 6.2, most commercial uranium occurs in sedimentary rocks. Sedimentary deposits form because uranium occurs in nature in two different oxidation states: uranous (U^{+4}), with a valence of $+4$; and uranyl, in which the uranium has a valence of $+6$ and is combined with two oxygens to form the ion UO_2^{+2}. The U^{+4} is quite insoluble in water and precipitates as a combination of oxides and hydroxides, but the UO_2^{+2} is easily soluble in most surface waters, as shown by the fact that ordinary seawater contains a concentration of about 10^{-6}%.

Sedimentary deposits form because of the effect of the earth's oxidizing atmosphere on the oxidation states of uranium. Oxygen-bearing water acts on exposed uranium veins, which contain minerals with both U^{+4} and U^{+6}, and oxidizes all of the uranium to UO_2^{+2} and dissolves it in river water and groundwater. If this uranium is washed into a reducing environment, such as a swamp that forms coal (Section 5.3), then the soluble UO_2^{+2} can be reduced to the insoluble $U+4$ and precipitated as an oxide or hydroxide. These organic-rich deposits, however, must undergo a further enrichment before they reach commercial grade. A typical process is for oxygen-bearing water to dissolve the uranium out of the organic sediment, transport it into sands and shales, and simply deposit a wide variety of uranyl salts by evaporation.

Exploration for uranium is conducted by large companies using sophisticated techniques, such as multispectral radiation detectors, but has also been avidly pursued by amateur prospectors who felt that simple geiger counters, which measure total radiation, should enable them to locate ore. One of the most intense periods was shortly after World War II, when the U.S. government attempted to develop domestic uranium reserves and offered prizes to anyone who could locate one. The deserts of the Southwest were virtually alive with prospectors, many of whom had never been outside of a major city and had to be given water and have their cars pulled out of sand so that they wouldn't die. One of us (JR) remembers being approached by a person, obviously new to the desert, who asked one simple question – "Do you know where I can find a uranium mine that hasn't been discovered yet?" The answer was no. The current major producers of uranium are Australia, Canada, the United States, Brazil, South Africa, and Niger (Section 6.3).

Policy questions

Are you opposed to nuclear energy because reactors are dangerous? (If you are opposed because they produce dangerous waste, see section 7.4 before reaching a conclusion.)

Are you in favor of nuclear energy because (1) it releases us from dependence on oil and coal, and (2) much of the industrial world already depends on it?

FURTHER READING: Rhodes (1986); Mounfield (1991).

5.5 PRESENT AND FUTURE ENERGY USE

When America runs out of oil, we will raise the price. – The vice-emir of a district near the Arabian Gulf

Now that we know where and how we get energy, we must understand why we need it, how we use it, and what might be available in the future.

Energy and the economy

To a first approximation, people who have more energy live better. "Better" is generally defined in terms of the gross domestic product (GDP), which we used in Section 1.4 to compare the economic health of countries and their birthrates. With this definition, plotting the per capita total energy use against per capita GDP yields the graph shown in Figure 5.28, with the same categories of countries as in Figure 1.6. The upward slope to the right does not prove that rich people are happier than poorer ones, but it does at least show that rich people have more things to play with.

The general relationship shown in Figure 5.28 is amplified in Table 5.5. Some of the figures are quite startling to people who have never thought much about energy consumption. The per capita energy use for the world is 50 million Btus per year. The United States consumes energy at 5 times the world average, Western Europe 3 times, and sub-Saharan Africa only one-tenth as much. These disparities correspond to vast differences in other indices of economic health, such as life expectancy, average years of schooling, and frequency of debilitating disease. Many of these differences result from the increased food production and distribution that energy makes possible (Section 2.2), and others reflect the ability of energy-rich countries to reduce the risk of infectious diseases by purifying water supplies and generally cleaning up the environment (Chapter 7).

Although the general trend shown by Figure 5.28 is very clear, close inspection reveals significant differences in the efficiency with which various countries use energy. Thus, countries above and to the left of the general trend (high energy use) are relatively inefficient, whereas countries below and to the right of the trend are efficiently producing a healthy economy with comparatively low energy consumption. Major countries that are efficient (shall we call them frugal?) include Switzerland and Japan. By contrast, the less-efficient countries (profligate?) are mostly major oil exporters, and the appearance of waste results at least partly from the necessity to use energy to produce the oil.

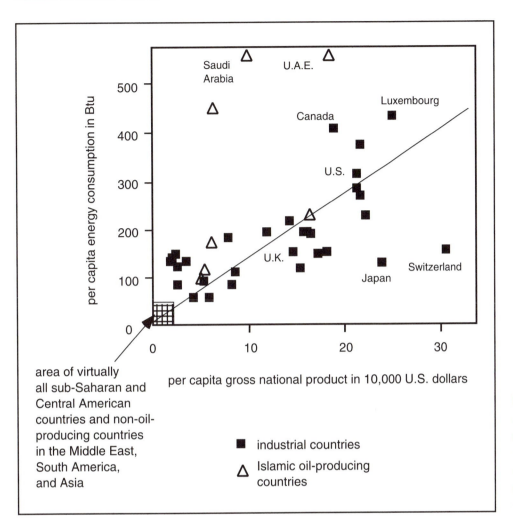

area of virtually all sub-Saharan and Central American countries and non-oil-producing countries in the Middle East, South America, and Asia

per capita gross national product in 10,000 U.S. dollars

■ industrial countries

△ Islamic oil-producing countries

Figure 5.28 Plot of energy consumption versus gross domestic product, with different symbols for different country groups (identical to the groups used in Figure 1.8). See discussion in text.

Present energy use

We can understand some of the differences in energy consumption by taking a closer look at the ways in which countries obtain their energy. In addition to the variations in total energy consumption, countries differ in the proportions of the energy produced by different sources (Table 5.5). This table shows that:

- Rich, industrial (OECD) countries generally import energy. The extreme example is Japan, which has no oil and only a little coal and imports virtually all of its energy. Another extreme is Singapore, a city with a small countryside that is not a member of OECD but has a major commercial economy. Singapore has no reserves of energy or any other natural resources and must import everything to maintain its high standard of living.

- At the other end of the economic spectrum is sub-Saharan Africa, which uses very little commercially produced energy and has the poorest economy of any part of the world. Furthermore, the dependence of sub-Saharan Africa on wood and other plant material for heating and cooking may foretell a decline in an already low standard of living as deforestation reduces the limited energy supplies now available.

- Many major exporting countries do not have a particularly high rate of domestic energy consumption. These countries include Mexico, Iran, Nigeria, South Africa, China, and Indonesia, plus numerous smaller countries not listed in Table 5.5. The wealth generated by exports of oil and coal has generally not improved the lives of most of the people in these countries.

- Industrial countries have a reasonable mix of energy production that includes oil, gas, coal, hydropower, and nuclear energy. Most other countries depend almost exclusively on one source, such as oil in the Middle East and coal in South Africa and China. This de-

Table 5.5 *Annual energy consumption and production*

Country	Per capita consumption (in 10⁶ Btu)	% energy consumption imported	% energy production exported	% energy consumption from wood	% commercial production supplied by				
					Oil	Gas	Coal	Hydro	Nuclear
North America									
Canada	400		22	0	35	37	16	10	3
Mexico	53		42	1	82	13	3	5	0
United States	308	18		2	30	36	29	2	3
Central America									
Average	25	84		44	17	0	0	83	0
South America									
Argentina	62		5	5	54	41	1	3	1
Brazil	47	33		30	55	6	5	33	1
Venezuela	92		70	1	83	13	1	2	0
Average	47			20					
Western Europe									
Average	139	38		1	22	21	45	4	7
Russia (and other FSU)ᵃ	194		20	1	37	39	22	1	1
Middle East									
Iran	44		65	1	87	12	1	0	0
Israel	128	100		0	0	0	0	0	0
Kuwait	222		86	0	94	6	0	0	0
Oman	93		85	0	94	6	0	0	0
Saudi Arabia	167		79	0	91	9	0	0	0
U.A.E.ᵇ	552		81	0					
North Africa									
Algeria	27		83	3	61	38	0	0	0
Egypt	23		48	4	87	12	0	1	0
Libya	113		80	1	89	11	0	0	0
Sub-Saharan Africa									
Nigeriaᶜ	13		82	62	96	4	0	0	0
South Africa	10		20	81			100	0	0
Average	19	97		80	Very little commercial production				
Asia									
China	25		11	6	20	2	76	22	0
India	12	13		25	21	5	71	3	1
Indonesia	15		73	47	65	30	3	1	0
Singapore	139	100		0	No production				
Rapidly developing countriesᵈ	52	Variable		10	18	43	27	2	0
Australia	214		38	3	18	11	70	1	0
New Zealand	191	18		0	19	46	15	19	0
Islands	11	100		49	No production				

pendence on one energy source clearly poses dangers if any event interrupts the supply and production.

These comments based on Table 5.5 show that countries that export energy (mostly oil and coal) do not necessarily have high standards of living, and countries that import energy do not normally see much decrease in their standards of living. It all depends on whether the imported energy is used to support industrial production. For example, in terms of foreign assets Japan is the world's wealthiest country despite importing all of its energy. The Japanese have acquired those assets by directing most of their energy to industry and maintaining a life-style that many people in the United States and Europe would find uncomfortable (a prime example is small houses that are cold in the winter and hot in the summer). By contrast, the use of imported oil in the United States largely to allow people to drive cars at high speeds is counterproductive and has led to an increase in the nation's foreign debt, but thus far (middle 1990s) the only indication of economic problems is a growing unease among many people that they are not living as well as they expected.

In countries with poor economies, the necessity to import energy can be serious. For example, because Bangladesh is on the delta of the Ganges River system (Section 3.7), it contains no rock or gravel that can be mined for the aggregate in concrete, and concrete can be manufactured only by using a coal or gas oven to heat river mud to a sufficiently high temperature that it partly fuses to a solid rock, which can be broken into gravel-sized pieces (Section 6.2). Imports of fuel in Bangladesh and other poor countries must be paid for in currency that the country earns abroad by sales of its own products, thus effectively converting the country's work force into (very low-paid) employees of foreign interests.

This leads us to the issue of future energy use.

Future energy use

How will we use energy in the future? Just as we do now or in a different way? Differently. Make no mistake about

that. At some time in the 21st century, very likely during the lifetimes of many people reading this book, changes in the sources of energy and patterns of their consumption will be forced on us by at least two developments and possibly two more.

The first problem is in the world's poorer countries, where so much energy is derived from burning wood. As we discuss in more detail in Section 8.3, most of these countries are stripping forests for lumber and firewood at rates that cannot be sustained. Much of this wood, plus twigs, brush, and anything else that is burnable is used to cook meals because people cannot digest raw grain (Section 2.1). Consequently, people who depend on wood will either find other sources of energy or they will endure widespread starvation. Which of these two possibilities happens probably depends on the ability of poor countries to compete economically for the world's energy resources. If they cannot compete internationally, then competition within each country will become even more fierce than it is now, with a smaller proportion of people successfully buying energy and a larger proportion unable to do so and dying. If they can compete, then their increased demand for energy will raise the worldwide price and make energy more expensive even for those countries that use most of it now. Along with the competition for current energy sources such as oil and coal, those countries that can afford it are looking largely to nuclear power as the energy source of the future. Naturally the technology and nuclear fuel are being provided by industrial countries, and they generally try to exercise control over reactor safety and, particularly, over reprocessing that might provide a country with bomb-grade uranium and/or plutonium.

The second inevitable change is a decrease in the availability of oil. Within a century, probably sooner, the supplies of oil that can be pumped from the ground will dwindle to the point that even the present worldwide demand for oil will drive the price to very high levels, and if poor countries become competitive the price will be even higher. The price increase will probably lead to some combination of a switch to alternative energy for

Notes to Table 5.5

Notes: Wood includes all recently growing plant matter. Exports and imports are almost exclusively oil and coal.
[a]Data for Russia and other countries of the former Soviet Union are highly variable and of uncertain accuracy.
[b]United Arab Emirates (U.A.E.) include Abu Dhabi, Dubai, and Sharjah.
[c]The import/export data for sub-Saharan Africa omit Nigeria, Zaire, Angola, and Gabon, which export energy.
[d]Rapidly developing countries in eastern Asia include South Korea, Malaysia, Taiwan, and Thailand; Malaysia exports and the others import energy.
Sources: United States Energy Information Administration (annual) and World Resources (1992, 1994).

Table 5.6 *Comparative costs of electricity production in United States*

Energy type	Producer cost (in $/kwh)	Consumer cost (in $/kwh)	Start-up cost and related issues
Commercial production			
Coal	0.032	0.06 to 0.12	Hundreds of millions mostly to expand existing facilities
Gas	0.065	0.07 to 0.15	Not planned
Oil	0.073	0.09 to 0.18	Not planned
Hydro	0.007	0.05 to 0.08	Not planned
Nuclear burner	0.022	0.06 to 0.12	Several billion for new plant
Solar	0.12 to 0.20	0.30 to 1	Tens of billions for large-scale commercial production
Wind	0.04 to 0.07	0.07 to 0.15	Probably not large-scale operation
Geothermal	~0.05	0.07 to 0.10	Limited resource
Tidal			Possible in future
Nuclear fusion			Possible in future
Private production			
Solar electric			2,500 to 7,500 for 50% of house
Wind			700 to 5000 for 50% of house

transportation and greater efficiency in engines that use oil, although improvements in efficiency will only postpone the price rise. Another likelihood of a rise in the price of oil is the same type of competition among individuals that we projected for poor countries – people with money will be able to drive cars, and those without it won't.

A third, slightly less definite, reason for change in the patterns of energy consumption is the inability of societies to decide on some way to dispose of the waste from nuclear power reactors. We discuss this problem more in Section 7.4 and here simply say that at present (middle 1990s), spent fuel from nuclear reactors is in "temporary" storage around the reactors. We put "temporary" in quotes because nearly everybody agrees that the spent fuel should be stored by the reactors only until a permanent place is found for it, but since no one knows when this will happen, temporary may mean a very long time. The result could well be public pressure to reduce or eliminate nuclear power, although if that causes the price of electricity to rise, then people may demand their reopening, and if that happens, then – but we are entering an endless loop, and the only way out is to cut it off.

The fourth possible reason for a change in future patterns of energy consumption is the effect of burning coal on the world's atmosphere. As we discuss in Section 8.2,

global temperatures have been increasing during the past century, and most investigators attribute the rise primarily to carbon dioxide (CO_2) released by burning coal (see Section 5.3 for a comparison of burning coal, oil, and gas). If the temperature rise is demonstrated to have serious effects on the earth's ability to support people, then we may be forced to eliminate or at least curtail the use of coal.

A potential solution to all of these energy problems is the development of energy sources not based on fossil fuel or nuclear energy. Both of those sources are nonrenewable in the sense that the earth contains only a finite supply of fossil fuel and minable uranium, but alternative sources offer a possibility of boundless energy because they are either continual (e.g., wind and solar energy), renewable (e.g., tree farming or growing corn for alcohol), or in almost limitless supply (e.g., oil shale). At present, however, none of these sources are important because they are more expensive than energy based on fossil fuel, and the likelihood that they will become more significant in the future depends on their cost relative to the cost of present sources. The important costs include both the continuing cost of producing energy and, for new sources, the start-up (capital) cost of establishing new plants and equipment. We offer Table 5.6 as a summary of very approximate operating and start-up costs for energy supplies (it is based on costs in the United States because they are so highly variable around the world that we cannot hope to summarize them for more than one country).

Several of the alternative sources shown in Table 5.6 will probably not become important. Oil shale is abundant in the United States but not in most countries, and even in the United States it is more expensive than pumped oil, and its production has severe environmental consequences (Section 5.3). Hydropower is almost completely developed in most industrial countries, and the capital costs of opening major power dams in the nonindustrial world may well be beyond reach (see Box 5.1). Tidal power shows great potential along coastlines with large tidal ranges, such as the Bay of Fundy in Canada and long estuaries in the United Kingdom and Scandinavia, but it is limited along other coastlines and obviously nonexistent in landlocked countries. Wind power has similar limitations, and geothermal power can be utilized only in a few special places with high heat flow. Tree farming and growth of other crops, such as corn for alcohol, compete for increasingly scarce land that could be used for food production.

Consequently, the most likely new source of energy is active solar power using photovoltaic cells (technology discussed in Section 5.2). Three developments could make it more important. One is an increase in the cost of energy from oil and gas because of a decrease in their abundance and from coal and nuclear power because of environmental problems. The second development is technological improvements that reduce the cost of producing solar energy, and if any combination of decrease in the cost of solar energy and increase in the cost of conventional energy makes solar energy cheaper, then it could suddenly become a principal energy source. The third development is technological improvements in our ability to store energy. If we have better batteries, learn to use hydrogen as a storage method, or discover some wholly new technology, then it may be possible to switch much of the world to a solar-powered economy.

Any major change in energy supplies will require large capital expenses (Table 5.6). Imagine, for example, replacing the world's gas stations with stations where we can recharge batteries that we have used to drive our cars. Consider the necessity of covering thousands of square miles with photovoltaic cells and constructing new electric distribution systems to bring energy from these grids instead of from the present locations of coal or nuclear plants. In order to make these changes, societies will have to choose to allocate a large part, possibly most, of the economy to building this new power system. That is a lot to ask of most people.

Finally, what would a science-fiction writer dream of? Fusion based on hydrogen in the world's oceans? Satellites orbiting the earth, collecting solar radiation, and sending it to the surface by microwaves? Processes so sensitive that they could generate electricity by cycling fluids through the very small temperature changes in the oceans or in land areas now regarded as only poor candidates for geothermal power? All of these sources have already been considered, and how many more haven't even been thought of yet?

Policy questions

How much energy, and what type of energy sources, do you expect to use 25 years from now?

FURTHER READING: Miller and Miller (1993); World Energy Council (1993); United States Energy Information Administration (annual).

PROBLEMS

1 The heat of combustion of propane (C_3H_8) is 530.6 kcal/mol. Using the procedures explained on p. 180, show how propane compares with other heat sources in terms of the ratio of heat produced to CO_2 released.

2 Most of the oil consumed in the United States is used for cars, trucks, buses, and trains (data in Table 5.5). If these oil users were replaced by electric cars and by trains running on electrified tracks, and if this electricity were generated by coal-fired plants, what would be the approximate percentage increase in CO_2 released in the United States?

3 A radioactive isotope with an atomic number of 61 and an atomic weight of 145 decays through a series of steps that involve the emission of two alpha particles and two beta rays. What is the atomic number and weight of the daughter product?

4 An ornamental reflecting pool is 10 m long, 5 m wide, and 5 cm deep. It receives sunlight for 12 hours

a day at a rate of 1.35×10^4 kcal/m² (Section 2.2). If 10% of this radiation is absorbed by (not reflected from) the water in the pool and the water is heated uniformly during this time, how much hotter is the water at the end of the day than at the beginning?

5 One set of energy units applied to the Einstein equation for the conversion of mass to energy yields the formula

$$\text{energy (in Btu)} = 10^{-10} \times \text{mass (in grams)} \times c^2$$

where c is 3×10^{10} cm per second. If a barrel of oil yields 5.8×10^6 Btu, how many barrels of oil are equivalent to the loss of 1 g in a nuclear reactor?

6 If you wanted to serve a dinner consisting of cold ice tea and a hot casserole, which of the following should you do? (a) take the casserole out of the oven, put it on the table, and then pour the ice tea. (b) put the ice tea on the table and serve the casserole immediately after taking it out of the oven.

7 Using data in this chapter and Chapter 1, calculate the ratio of (per capita oil consumption in the United States)/(per capita consumption in the entire world).

8 Using data in Sections 1.1, 5.1, and 5.4, calculate the percentage of the earth's land surface that would have to be covered by solar converters in order to generate all of the world's electricity by solar power.

Assume 10% efficiency in converting solar radiation to electricity.

9 The table here shows the following information for three countries (A, B, and C): population in millions; per capita oil consumption in barrels per year; and oil production in millions of barrels per year.

Country	Population	Per capita consumption	Production
A	20	20	200
B	50	4	300
C	10	8	100

Compare the countries in terms of their present per capita gross domestic products. How might these comparisons change in the next 25 years?

10 In the middle 1990s, estimated U.S. reserves of crude oil are 25 billion barrels, and reserves of natural gas are 170 trillion cubic feet. The United States uses about 6 billion barrels of oil and 25 trillion cubic feet of gas each year. Based solely on these numbers, how long will supplies of oil and gas last? What additional information is needed to determine when supplies of oil and gas will be used up?

11 Use data in this chapter and Chapter 1 to estimate the amount of world oil production that is imported by OECD countries.

REFERENCES

American Petroleum Institute (annual). *Basic Petroleum Data Book*. Washington, D.C.: American Petroleum Institute.

Chang, F. K. (1996). Beijing's reach in the South China Sea. *ORBIS* 40(3): 353–74.

Gordon, R. L. (1987). *World Coal*. Cambridge: Cambridge University Press.

Howes, R., and Fainberg, A. eds. (1991). *The Energy Sourcebook – A Guide to Technology, Resources, and Policy*. New York: American Institute of Physics.

Hunt, J. M. (1979). *Petroleum Geochemistry and Geology*. San Francisco: W. H. Freeman.

Johansson, T. B., Kelly, H., Reddy, A. K. N., and Williams, R., eds. (1993). *Renewable Energy – Sources for Fuels and Electricity*. Washington, D.C.: Island Press.

Knott, D. (1995). Caspian Sea activity picking up off former Soviet Union republics. *Oil and Gas Journal* 93 (January 30): 31–5.

Marples, D. R. (1995). Belarus' ten years after Chernobyl. *Post-Soviet Geography* 36: 323–50.

Matzke, R. H. (1994). Challenges of Tengiz oil field and other FSU joint ventures. *Oil and Gas Journal* 92 (July 4): 62–5.

Medvedev, Z. A. (1990). *The Legacy of Chernobyl*. New York: W. W. Norton.

Miller, E. W., and Miller, R. M. (1993). *Energy and American Society*. Santa Barbara, Calif.: ABC–CLIO.

Mounfield, P. R. (1991). *World Nuclear Power*. New York: Routledge.

Pearce, F. (1995). The biggest dam in the world. *New Scientist* (January): 145 (1962): 25–9.

Remnick, D. (1993). *Lenin's Tomb – The Last Days of the Soviet Empire*. New York: Vintage.

Rhodes, R. (1986). *The Making of the Atomic Bomb*. New York: Simon and Schuster.

Schipper, L., and Meyers, S. (1992). *Energy Efficiency and Human Activity: Past Trends, Future Prospects.* Cambridge: Cambridge University Press.

Scientific American (1991). *Energy for Planet Earth – Readings from the Scientific American Magazine.* New York: W. H. Freeman.

Shell Oil Co. (1983). *The Petroleum Handbook.* Amsterdam: Elsevier.

Stone, R., Williams, N., and Balter, M. (1996). The explosions that shook the world. *Science* 272: 352–60.

Sullivan, L. R. (1995). The Three Gorges Project: Dammed If They Do? *Current History* 94: 266–9.

United States Energy Information Administration (annual). *International Energy Annual.* Washington, D.C.: U.S. Department of Energy.

Wells, P. G., Butler, J. N., and Hughes, J. S., eds. (1995). *Exxon Valdez Oil Spill: Fate and Effects in Alaskan Waters.* Philadelphia: American Society for Testing Materials.

World Energy Council (1993). *Energy for Tomorrow's World – The Realities, the Real Options and the Agenda for Achievement.* London: Kogan Page.

World Resources Institute (1992, 1994). *World Resources.* Oxford: Oxford University Press.

Yergin, D. (1991). *The Prize: The Epic Quest for Oil, Money, and Power.* New York: Simon and Schuster.

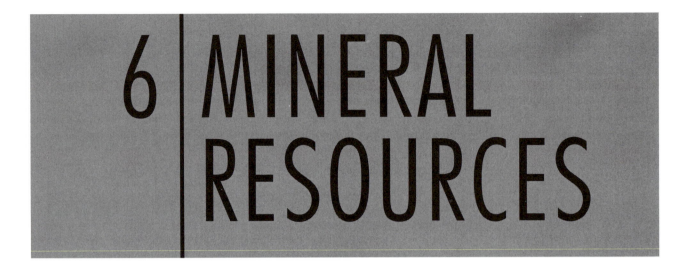

6 | MINERAL RESOURCES

6.0 INTRODUCTION

If you can't grow or catch it, you have to mine it. — Bumper sticker in the western United States in the 1980s

Now that we have discussed water and energy resources, we move on to the remaining mineral resources, metals and the so-called industrial rocks and minerals. Before we do so, take a piece of paper and a pencil, sit back in your chair, and look around you. Of all the things in view – your desk, books, perhaps the remnants of a meal, the walls, the floor, the window, a computer – jot down those things that you believe required nonfuel minerals in their manufacture or production.

Did you find many? Perhaps some were obvious. If metal coins or steel paper clips are scattered about your desk, you undoubtedly identified them as iron, copper, or nickel mined from the earth. You may also have realized that a brick wall or a china coffee mug was made from fired clay and that the circuits of your computer, your phone, and your stereo contain metals that allow electrical currents to flow or data to be stored. Did you know that the glass in your windows was once sand that was mined, melted, and extruded to cool into thin sheets? Even less apparent, however, is the paper you wrote on. It began its career as wood pulp, a nonmineral natural resource, and then was bleached with sulfuric acid and mixed with clay or chalk as a filling agent. Sulfuric acid is made from mined sulfur or extracted from petroleum and natural gas; clay and chalk must also be mined. Your pencil contains graphite perhaps from mines in Sri Lanka, South Korea, or China. Your desk top was smoothed and polished with sandpaper undoubtedly containing garnet or quartz sand, and if you overindulged last night, you may have taken an antacid with sodium bicarbonate – all

mined. If you had a milkshake at your favorite fast-food joint, it may well have contained inert mineral fillers to give it a smooth texture. A leftover slice of pickle from your lunch was pickled with salt that might have been mined in New York, Michigan, Utah, or Louisiana, and the cucumber that was harvested to make that pickle was grown with fertilizers from phosphate rock quarried in Florida, North Carolina, or West Africa and potassium salts from Saskatchewan. Your walls are likely to be plaster or wallboard made from mined gypsum, and the paint on the walls uses titanium oxide pigments that are a refined product of the mining of two common minerals, rutile and ilmenite. In short, it is very hard to find anything produced without minerals.

It wasn't always so. Primitive people had few minerals at their disposal and even fewer needs for them. Preoccupied as they were with food, shelter, and clothing, their mineral use was no more sophisticated than picking up a rock and using it as a tool for crushing seeds or as a projectile for bringing down game. A pretty stone might become an amulet or fetish, and red oxides of iron may have been used as pigments in paints or for body adornment. Eventually, some distant ancestors realized that simply grinding the rough edges off of a grindstone or flaking off a chip from a spear head would make a better grinder or projectile. As this realization grew, *Homo sapiens* moved from being simply a tool-using to a tool-making animal. Many of these tools (especially the ones we find tens of thousands of years later) were made from rocks and minerals, and we commemorate these discoveries by saying that people had entered the Neolithic ("new stone") Age. From this time on we mark people's development by their progressive and increasingly sophisticated use of mineral resources (Table 6.1). From flint

Table 6.1 *Eight ages of people*

Age	Approximate date of start
Paleolithic (Old Stone)[a]	500,000 B.C.
Neolithic (New Stone)	8000 B.C.
Copper	5000 B.C.
Bronze	3000 B.C.
Iron	1400 B.C.
Coal	A.D. 1600
Petroleum	A.D. 1850
Nuclear	A.D. 1950

[a]Use of stone tools; mining begins about 100,000 B.C.

spear point or basalt axhead, to clay pots, to copper daggers, to bronze helmets, to iron chains, and so on, civilization has progressed.

So, among our many other sobriquets and notable achievements, we are the mineral-using animal. In addition to using minerals for tools and other purposes, our ancestors also began to search for minerals in out-of-the-way places and began to modify and alter the minerals they found. We became prospectors, miners, and metallurgists as well as tool users. We now use a diverse collection of materials that mineral economists lump together as the "nonfuel" minerals. Some, like gems, are quite valuable but essentially useless, whereas others are basic, mundane materials like sand and gravel. Some are the stuff of legends (Box 6.1); others are so common that most us are wholly unaware of our dependence on them.

We investigate our use of minerals by defining them and describing their characteristics (6.1), their geologic features (6.2), their location (6.3), the process of extracting them from the ground (6.4), legal and property issues connected with mineral use (6.5), and finally the economics of their present and future use (6.6).

6.1 DEFINITION AND CHARACTERISTICS OF MINERAL RESOURCES

Think of a mineral deposit as some money that you accidentally dropped into a hole. If it is a lot of money, you will probably try to recover it. If it is only a few pennies, you might forget about it. — *S. E. Kesler, Our Finite Mineral Resources, p. 5*

A resource is anything held in reserve for future use. In this sense, a savings account is a resource, as is food in a freezer or a pile of wood stacked as fuel for next winter. Natural resources, in contrast to cultural and human resources, are obtained from the earth, and the inorganic resources are commonly referred to as "minerals." To a geologist, any material is a mineral if it is a naturally occurring inorganic, crystalline, solid with a composition that varies only within a limited range. Thus, minerals are the physically distinct, but irreducible, building blocks of the earth, and geologists use the term "rock" for assemblages of one or more minerals. Basically, minerals are to rocks as cells are to organs in biology. In a resource context, however, a mineral is quite different. A mineral resource is any rock or mineral that (1) someone is willing to pay for and (2) can be found in sufficient quantity that an enterprising group of people can remove it from the earth, make necessary physical or chemical modifications to produce a marketable product, and deliver it to consumers at a profit.

Thus, economics defines what is and is not a mineral resource. For example, a limestone quarry on the shores of Lake Michigan is a mineral resource of significant proportions because limestone satisfies the huge demand for cement, the need for agricultural lime, and the requirements of the iron and steel industry of the Great Lakes region (Section 6.3). Proximity to one of the great inland waterways of the world provides an inexpensive means of conveying a bulk product, such as limestone, to consumers, but the same limestone rock in Nebraska, Ecuador, or Zaire is not currently a mineral resource. As another example, 25 years ago the vast mercury deposits of the Almaden region of Spain were an important mineral resource, as they had been for nearly 1,000 years. With increased environmental awareness, however, the worldwide demand for mercury in herbicides, insecticides, fungicides, and amalgams has dropped by more than half, and this once vast and important resource has lost much of its value.

Based on this economic definition, we discuss five characteristics of mineral resources in this section: (1) sustainable versus nonsustainable resources; (2) scarcity of mineral resources; (3) the concept of resource inventories; (4) the problem of the fixed location of resources; and (5) the risk assumed in exploration for resources.

Sustainable versus nonsustainable resources

Imagine yourself stranded on an island. Your survival depends on your ability to sustain life. At the very least, you will need fresh water, food, shelter, and (depending on where the island is) clothing. If the island is big enough,

BOX 6.1 JASON AND THE ARGONAUTS

Most myths and legends are found to have some basis in fact. King Solomon's mines were probably rich copper mines located at Timna, a tiny town along the west banks of the Jordan River, 12 miles north of Aqaba. The gold that overlaid the cedarwood columns of Solomon's Temple likely came from mines belonging to the queen of Sheba in modern-day Yemen, or perhaps from the Arabian peninsula about halfway between Mecca and Medina. One of the more enduring legends of mineral wealth is that of Jason and the Golden Fleece. Can there be any truth in this story?

The legend of Jason is long and complex, a soap-opera in classical guise. Jason was a Greek prince deprived of his rightful inheritance. To win back his throne, he had to perform the heroic act of sailing east across the known world to wrest a golden fleece from Aietes, king of Colchis. The fleece itself was from a ram that had been ridden by two other legendary figures in some other storyline. So Jason assembled a group of adventurers including such luminaries as Hercules, Castor and Pollux (the twin sons of Zeus), and Argus, the shipbuilder, after whom the ship (the Argo) and the travelers take their names. After many adventures, with the requisite shipwrecks, fire-breathing bulls, interventions of gods and goddesses, an elopement with a princess, Jason wrested the Golden Fleece from its rightful place hanging from a tree (guarded by your basic dragon) and returned to Greece. Despite this success, Jason didn't live happily ever after because his betrothed turned out to be Medea, who had an unusual predilection for cutting her relatives into small bits. She

never got Jason, but did serious damage to their children — but that is another story told better by Euripides.

Is the story pure fantasy, or did ancient Greek poets base their tales on glimmers of fact and history? Why a golden sheepskin? It turns out that Colchis, the kingdom to which Jason is sent on his heroic task, is an area along the eastern shore of the Black Sea at the foot of the Caucasus Mountains in modern Georgia. Rivers and streams that drain the Caucasus carry fine gold that settles out in river gravels just as once happened on the western slopes of the Sierra Nevada in California. At some time in the second millennium B.C. — the dates vary from 2400 to 1200 B.C. — a thriving gold industry was located here. To extract the fine gold that was mixed with sand and gravel, the miners shoveled the river sediments into flumes lined with sheep skins. As water rushed down the flumes, the denser gold sank to the floor and was trapped in the thick, oily wool of the sheepskins. Mud, sand, and gravel were carried off. Whenever a sheepskin was charged with gold, it was hung on trees to dry, and the gold was then simply shaken off and, presumably, melted down to a more marketable form.

So there *were* golden fleeces 4,000 years ago. A 1,000-mile voyage across unknown and uncharted waters, through the Dardenelles and Bosporus, and along the coast of an inland sea inhabited by strangers, would have constituted a heroic task in anyone's mythology. The Argonauts were probably just a bunch of prospectors and gold speculators who, much later, got trapped inside a really good epic.

the climate hospitable, and rainfall plentiful, the island may be able to sustain you indefinitely. However, if any single resource is limited — we might say "finite" — then the odds of sustaining life decline. If there is only one palm tree (the classic cartoon version of being stranded on a desert island), then the competing needs for palm fronds for shelter from storms, coconuts for food, and wood for fuel to burn in winter may present you with a set of competing resource requirements that make long-term sustainability impossible.

With sufficiency, careful planning, and wise use, many of our natural resources are sustainable — water, food and fiber crops, timber, and fisheries can be managed to last indefinitely. (That they often are not is, as we have seen in earlier chapters, commonly a result of ignorance,

greed, or just bad planning.) Most minerals, however, are a nonsustainable resource no matter how carefully we manage them. There is no second harvest, to say nothing of a third or fourth. As we will see, it takes millions of years for nature to make and preserve most deposits, and people cannot wait that long for a second crop. If we have removed all of the ore from a deposit, we have just three choices. First, we can recycle the commodity that we have already mined, thus converting it into a sustainable resource. Second, we can recognize that the ore is a nonsustainable resource but look for new resources that can supply us for some additional period of time. This search simply delays the time at which the ore is exhausted. Third, we can find a substitute for the commodity. If we are wise (which happens rarely), we do all three at once – recycle as much as possible and look for new ores to last until our scientific research finds a substitute.

The problem of recycling can be explained best in terms of energy consumption. Over millions of years, natural earth processes using vast amounts of heat energy may have concentrated a mineral resource in a fortuitous location where it is economically exploitable. People extract the mineral, use it, and then, in most cases, discard it. Vast expenditures of additional energy would be required to reassemble all of the dispersed pieces of iron or tin, all the lumps of clay or chalk. For most minerals, it is cheaper to seek out a new, natural, occurrence than to attempt to recycle minerals scattered about the earth's surface. For example, salt cannot be recycled since it is highly soluble and quickly dispersed into the world's oceans after it is used. Zinc is hard to recycle because most uses involve mixing it with other metals, such as copper to make brass; this dispersion makes the recovery and repurification of zinc difficult and, consequently, expensive. Also, some recycling is simply not possible because the mineral product is so altered that it no longer has any value; examples are the common clay used in bricks and limestone in cement.

Consequently, recycling is important only for a small subset of mineral resources. In the United States, for example, about one-third of the annual aluminum demand is met by recycling. This recycling is possible because aluminum is: (1) chemically inert, so that it does not dissolve in rainwater or corrode; (2) light, so that transportation of waste aluminum is inexpensive; (3) rarely used in combination with other metals, so that it can be collected for recycling in its pure form, and the scrap tossed into an electric furnace, melted, and recast at a saving of about 95% of the energy required to extract aluminum from rocks or minerals. Gold is another example of a metal that can be, and always has been, recycled. An estimated 90% of the gold ever mined is still in circulation as currency, in ornamental uses such as jewelry and art, in technological settings as wires, foils, and coatings, or simply as gold bars in bank vaults, government storage, and hiding places of all descriptions all over the world.

Even mineral products that can be recycled are not completely sustainable because they require some addition of new supplies each year, either because the recycling isn't perfect or simply because the world's growing population requires more than the previous generation. Consequently, all mineral deposits are exhaustible and depletable (nonsustainable), and resources mined today must be treated as if they have been removed from the earth's resource base forever. The immediate issues, however, are questions such as, When will we run out? and Will it make any difference if we do? These are much harder to answer because knowing when a finite resource is depleted or the impact of depletion requires knowing the answer to other, difficult questions. For example: What is the total inventory of any mineral resource in the earth? What fraction of that inventory is accessible? What is the future demand for any mineral resource? What fraction of any resource can be recycled, reused? What new technologies might make new resources economic? What new technologies might make this resource less essential?

Clearly, we can say that the ultimate limit on any mineral resource's availability is a combination of physical and economic factors. There is *both* a physical limit to the ultimate availability of a mineral, since the earth's crust is finite, and an economic limit since people will only pay so much for a resource. A person dying of thirst may be willing to exchange all worldly possessions for a drink of water, but in the midst of the Namib Desert, this material sacrifice would be in vain.

Scarcity of resources

Many mineral commodities are scarce. Few of us have ever seen elemental copper or gold, and no one has ever seen elemental aluminum in nature. Valuable minerals generally are present in only fractions of a percent in common rocks. A typical ton of granite might contain 4 mg of gold (about 1/10,000 ounce). Even when present at abundant levels of a few percent or so, the resource we need – for example, aluminum or iron – is normally just one chemical constituent of a mineral. Removing metallic aluminum or iron from common minerals requires significant amounts of energy to break chemical bonds and to separate out unwanted elements like silicon or calcium. To get a sense of the magnitude of the problem, Table 6.2 shows the estimated crustal abundance of 14 of the more common metals we mine. For example, the average crustal rock has a zinc concentration of 70 parts per million (70 g of zinc per metric ton of rock). At typical zinc prices of $0.75/kg, all the zinc in a ton of average crustal rock would be worth about $0.05. Considering that mining zinc involves unearthing, removing, transporting, crushing, and dissolving rocks, it is clear that we cannot mine a ton of ordinary rock to extract five cents' worth of zinc.

Thus, to mine scarce zinc economically, some set of natural processes must concentrate the zinc to greater

Table 6.2 *Crustal abundances and economic cutoff values of common metals*

Metal	Crustal abundance (in ppm)	Cutoff value (in ppm)	Concentration factor
Aluminum	82,000	400,000	5
Iron	56,000	250,000	4.5
Manganese	950	250,000	260
Chromium	190	400,000	2,100
Nickel	75	10,000	130
Zinc	70	25,000	360
Copper	55	50,000	900
Lead	12.5	30,000	2,400
Uranium	2.7	100	40
Tin	1.7	5,000	2,950
Tungsten	1.5	3,000	2,000
Silver	0.08	50	625
Platinum	0.005	2	2,500
Gold	0.004	1	250

than crustal average values. Table 6.2 shows the abundance of each metal that is currently needed to allow that metal to be economically mined, commonly called the "cutoff value" or minimum grade. For zinc this cutoff is 25,000 ppm (2.5%), a 360-fold natural concentration above crustal averages. With the zinc price used earlier, this means that the cost of mining a ton of zinc-rich rock must be slightly less than $18.00 per ton ($360 \times \$0.05$) in order for such an operation to be economic. The realization that the inherent scarcity of zinc requires natural concentrating processes to occur before we can begin to mine zinc at a profit brings us to the heart of mineral resource geology. Any rock that can be mined at a profit is called "ore," and an ore deposit is an accumulation of valuable minerals whose extraction is economically feasible. Thus there is a geological aspect to an ore deposit – natural processes that concentrated mineral resources to the minimum grade level – and an economic aspect: relating to market conditions, price, locational advantages, and mining feasibility and efficiency.

Some people claim that there is no limit to resource availability because, after all, the amount of metal contained in the crust or in the oceans or on some nearby planetary body is enormous. That is true, but the net market value of *all* of the scarce metals in a ton of average granite is less than $10.00, and it is simply impossible with current or foreseeable technology to mine and process granite and extract its constituent metals for $10 a

ton. Similarly, the oceans contain so little dissolved metal that we would have to process about 1 million gallons of seawater to get an ounce of gold. As for lunar or Martian mining, that is best left to visionaries who foresee space colonies in our future.

Resource and reserve inventories

Now that we have a sense of the economic implications of scarcity, we must consider how mineral economists make an inventory of the world's mineral resources. Inventories for nonmineral resources are not too difficult to prepare. If we want to know how much wheat or sugar or cotton is available for consumption, we add up all the fields under cultivation for that crop, look at predicted yields, factor in weather, determine the amount of each commodity in storage and generate a figure for, say, how much cotton is likely to be available 6 months hence. Since these crops are grown anew each year, we probably find that the amount of wheat, sugar, or cotton produced each year closely balances consumption; that is, these crops are "sustainable" resources.

An inventory of mineral resources is similar to that for crops but with two major differences. One is that mineral production is nonsustainable. Once mined, a mineral resource is forever gone, so a mineral economist wants to know how much of each mineral resource is in the ground, available for future extraction. The second dif-

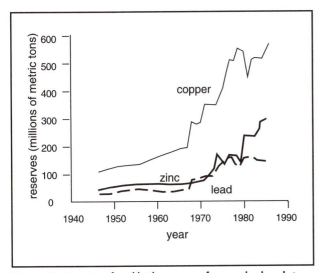

Figure 6.1 Estimates of worldwide reserves of copper, lead, and zinc between 1940 and 1990.

ference is that we can rarely determine the total resource of a mineral, which would be enormous if we could mine average rock, but can only tabulate a quantity known as "reserves." Reserves are quantities of ore, the amounts that we are confident can be extracted profitably from the ground in the foreseeable future, and because we have added an economic qualification to our definition, reserves are always much smaller than resources.

Having defined the concept of reserve, we use Figure 6.1 as an example of an inventory for three common metals over an arbitrary 50-year period. The vertical axis shows mineral reserves, the amounts of copper, lead, or zinc known to be available, in the ground, for mining. Notice that reserves generally stay stable or increase as mining occurs. How can that be? If the mineral resource base is finite and we mine minerals continually, should we not run out? Even if we mined copper in 1950 and subsequent years at half the 1989 production rate (3.55 million metric tons per year), we should have run out of copper in 1983. Instead, in spite of continuous production over that 33 year period, copper reserves increased fivefold. Is someone growing copper? Hardly.

The apparent constancy of reserves results from our definition of the term reserve, and we expand the consequences of that definition in Figure 6.2. The vertical axis estimates the feasibility of economic recovery – that is, it summarizes those factors that can make a mineral deposit more or less desirable for exploitation. How far is the mineral deposit from markets, from a sufficient skilled labor force, from power? How strong is the market for that mineral? Are other large mines about to open that can

produce the same mineral cheaper? How deep in the earth's crust is the ore? In some instances, such as fine clay for china or monumental stone, it is critical that the rock be free of minor contaminants or unattractive fractures or stains. These and other factors can make the resource recoverable (a reserve) or marginal, and if it is neither, it isn't even a resource.

The horizontal axis of Figure 6.2 shows the degree of confidence that we have that the amount of a resource is, in fact, present. To continue our comparison with crop production, a farmer can estimate yield in the week before harvest by walking the fields and seeing how healthy the crop is. By contrast, a farmer who tries to estimate crop yield in December, before spring planting, cannot know the weather conditions 6 months in advance or the market price for corn at harvesttime. A geologist trying to estimate the amount of a mineral resource yet to be mined is like a farmer in December. Because most of an ore deposit is buried in the earth and no one can see all of it until after it has been mined out, a geologist may have only the most indirect indications of its size and average grade.

The method of estimating reserves is shown in Figure 6.3, a diagram of a vertical cross section through an imaginary underground mine. The diagram shows a vertical access way, called a shaft, two horizontal tunnels, called drifts, and a single drill hole, called a "diamond drill hole" because it is drilled with a pipe that has diamond chips imbedded in its cutting edge and produces a continuous core sample of the rock that the drill penetrates.

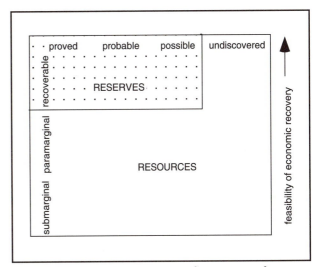

Figure 6.2 Diagrammatic representation of various types of reserves and resources (the "McKelvey Box," originally prepared by V. S. McKelvey).

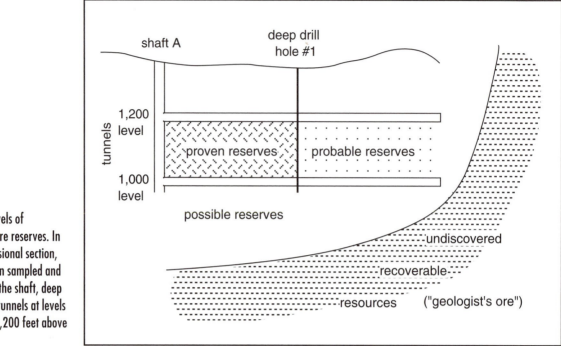

Figure 6.3 Levels of confidence in ore reserves. In this two-dimensional section, rocks have been sampled and assayed along the shaft, deep drill hole, and tunnels at levels of 1,000 and 1,200 feet above sea level.

The geologist assays the drill core and samples taken along the shaft and each drift. If all of the rock samples in the area of Figure 6.3 bounded by the two tunnels, the shaft, and the deep drill hole assay at or above ore grade, we can confidently identify a block of rock that contains "proven" reserves. We might consider the rock in the block to the east as "probable" reserves because we do not have information along the eastern margin of this block. Similarly, rocks in other blocks are "possible" reserves – our confidence is dwindling as our data get skimpier. More drilling is called for. Outside of a line beyond current exploration might be the area of "undiscovered, recoverable" resources. Mining engineers, who often see geologists as somewhat overoptimistic, like to call this latter "geologist's ore," which is not a compliment. Of course, we have only looked at two dimensions and a full determination of the volume of reserves requires a similar analysis into the third dimension.

The procedure just discussed allows us to establish degrees of confidence in the extent of the ore deposit and translate them into identified or undiscovered minerals with a subclassification of identified minerals into proven, probable, or possible. Identified ore, for which economic recovery is presumed to be feasible, becomes ore "reserves." A mine operator typically only bothers to classify as reserves that portion of the potential mineral resource as reserves that is needed to assure that investment in the operation will be returned with sufficient return on capital. New discoveries, extensions of old ore deposits, even changes in technologies that make new reserves out of what were previously just resources or submarginal resources, occur over the lifetime of the mine as old reserves are mined out and new reserves are needed to keep the mine open or to justify new capital investment.

The consequence of further exploration may be to transfer probable, possible, and unidentified reserves into the proven category, and when they are proven they are listed as "reserves" in tabulations of mining statistics and in diagrams such as Figure 6.1. This continual upgrading of reserve categories explains the increase of reserves as mining progresses and, on a broader scale, the decline in copper prices as this finite resource is mined (Fig. 6.4). It is also highly important to mine operators because a mineowner can probably borrow money from the bank using proven reserves as collateral. Probable and possible reserves are the kind of resource one takes to investors who are more comfortable with high risk in return for high yield. There are stock markets, like the Vancouver Exchange, that specialize in such "penny stocks" for those who like to speculate in companies with holdings that are best described as possible and unidentified reserves. Investment in undiscovered resources is for neither the faint of heart nor the light of wallet (we describe one such venture in Box 6.2).

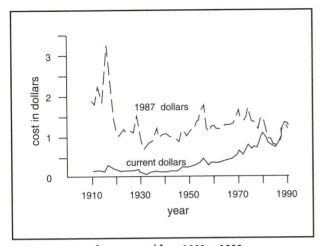

Figure 6.4 Prices of copper metal from 1900 to 1990.

Fixed location of mineral resources

Unlike most other natural resources, mineral resources are firmly and eternally attached to the earth. A farmer can chose among a dozen farm sites as long as the soil, water, and climate are amenable to the crops to be grown. A fisherman can sit outside the territorial waters of an unfriendly nation and wait for the fish to come out or chase the schools of fish as they seek food. A logger chooses among stands of pines or hardwoods and even decides which areas to reseed for future crops. A miner finds a tungsten deposit that is economic, and there it sits. The miner must negotiate with whoever owns the property because there may not be another such deposit for 1,000 miles, or at least not one known to the miner, and the miner will have to meet whatever requirements the local governing agencies place on mining. Also, if the mine is an ocean away from tungsten markets, the miner will have to factor the cost of transportation into any economic analysis of future profitability.

Actually, distance from civilization can be an advantage if the mineral is valuable enough. Mines seem to rank right behind landfills as undesirable neighbors in the minds of many. Often, there are competing uses for any parcel of land that happens to contain valuable minerals. Politically based decisions on land use – Should we have a molybdenum mine or ski resort, a quarry or subdivision? – are often made with an eye to counting votes rather than to the long-term strength of the local economy. This is a problem for the miner. A subdivision can be built anywhere, desirable ski mountains are fewer, but hardly rare. There are only two economic molybdenum mines in the United States.

Discovery risk

All economic ventures carry risk of failure. A farmer can see a lifetime's investment vanish in a hailstorm, and a manufacturer can be bankrupted by changing tastes, fashions, or regulations. Mineral production has its risks from bad engineering, poor investment decisions, losses of markets, and competition, among other factors, but there are some added economic risks in mining, many of them resulting from the fact that mineral deposits generally are very hard to find. They are commonly buried beneath soil, sediments, or other rocks that contain no valuable minerals (called waste or barren rock by prospectors). There is, as often as not, no hint on the surface of the mineral wealth beneath.

A social get-together of prospectors (or "exploration geologists" as they call themselves in polite company) is mostly a succession of stories, many even true, of how luck, ingenuity, accident, or fate allowed some lucky soul to find a hidden ore body. We have heard tales of diamond drill holes in the wrong direction that accidentally intersected a mineralized vein that created a mining camp. Every prospector with more than one season's experience can tell of a final unauthorized set of samples taken from a very unpromising-looking rock on the last day of the field season that assayed at ore grade. There are tales of exploration geologists who ran out of targets to drill and, rather than bring home unspent money from a drilling campaign for fear of seeing a budget cut the next year, simply drilled a random, blind hole into what became a bonanza ore body. A story is even told of a prospector in Canada who slipped and fell on wet moss as it shifted beneath his feet and landed with his hand on a vein rich in silver minerals. Not only that, but the soft silver minerals contained the imprints of hobnailed boots, which meant that other prospectors had come that way, walked over the very same moss, left their "footprint," and passed on – as he would have if he had not slipped. And finally, the discovery outcrop for the giant Hemlo gold district in Ontario was along the Trans-Canada Highway, a place that thousands of geologists must have seen and hundreds visited.

In spite of such legends and stories, however, most ore bodies are discovered by lengthy campaigns of mapping, sampling, drilling, assaying, remote sensing, and a good deal of creative geologic reasoning. Good ore-finders have visited hundreds of economic and subeconomic deposits, read shelves of dry, too-often poorly written reports, spent long days in the sun, wind, and rain followed by nights in cinder-block motels, and still count them-

BOX 6.2 OK TEDI

The center of New Guinea is a mountainous region of dense rainforest that gets almost 25 feet of rain a year with an average of only 26 days of sunshine. Along the crest of the Star Mountains, at an elevation of more than 1 mile, the Tedi River (Ok Tedi in the native Papuan language) drains the south slope of Mt. Fubilan, about 10 miles of the border between Papua New Guinea (PNG), an independent country, and Irian Jaya, a province of Indonesia. The indigenous peoples met their first European in 1963, and less than 30 years later, a huge gold and copper mine containing over 400 million tons of ore was in production on the flanks of Mt. Fubilan.

This deposit is one of the best examples of miners going where the minerals are in spite of the physical obstacles created by remote settings. Five years after the first arrival of government patrols into the area, a geologist with Kennecott Copper Corporation discovered the mineralization. Kennecott explored the region but decided, in 1975, that the logistical problems were insurmountable. In 1976 an Australian company, Broken Hill Proprietary (BHP), signed a concession agreement with the PNG government to try to mine this deposit. They sought international partners to help capitalize the project and, by 1981, began construction of the mine that is now known as Ok Tedi. BHP mined a gold-rich cap on the deposit for the first 4 years (1984–7) because it required far less investment in infrastructure than mining the deeper copper. Fifty-five tons of gold worth about $650 million (at a gold price of $350 per ounce) were flown out of the mine to be sold on the international market. From investment capital combined with the profits from the gold mining, BHP was able to expand its mine and begin extraction of the 150,000 tons of copper in the ore body in 1987.

To accomplish the Herculean task of building an open pit gold-copper mine in the remote wilderness of New Guinea, BHP ultimately invested $1.4 billion. About $375 million went for roads and airports, railways and barge shipping facilities; $160 million was invested in hydroelectric power generating plants to provide energy to the mines, the mills, and the local town that grew up to house the workers; almost $200 million was invested in houses, schools, hospitals, and other "social" amenities required to attract skilled miners and workers to a remote place where it rains all the time. The balance of the investment was spent on the mines and mills. Add this up and you find that 50% of the start-up costs of such a remote mine involved non-mine-related expenses. Consequently, BHP needed both a rich deposit and a bold set of investors to get into this business. They included, at various times, German, American, Japanese, and World Bank partners, in addition to the government of PNG as a silent partner.

BHP needed this help for the simple reason that when development of the gold-rich cap began in 1981, gold prices were at a historical annual high of nearly $600 per ounce, and BHP projected its earnings based on such a high, stable price. But three years later, when gold production began, gold had fallen below $350 an ounce and the Ok Tedi project was in serious financial trouble. In fact, the government of PNG was so nervous about the losses that BHP and its partners were projecting that, in early 1985, the government would refuse to renew the mineral concession for fear that the company would mine out the gold and abandon the property before investing the vast sums needed to develop the much larger copper mine — a practice known as "high-grading," where a miner removes all the high-grade ore and leaves the rest behind. This loss of the concession shut down gold mining for several months. Ultimately, BHP and its various partners were able to assure the government of PNG that they would stay the course, and the concession was renewed later in 1985. By 1987, copper production began and continues to this day (1996).

Even with the financing, the project has been overwhelmed with problems. Indonesian insurgents have crossed the border several times and the mining community has been evacuated more than once. The incessant rains often cause landslides that block roads and railways, and the mine has had to shut down for periods of several weeks while landslide debris is cleared. Ironically, for one of the wettest regions in the world, droughts downstream on the Fly River, which is used to transport barges laden with copper concentrate, have required the mine to shut down because there was no way to get concentrate to market. Strikes and worker riots over the living conditions have caused shutdowns and damages to mine property in excess of $10 million. And, even in remotest New Guinea, environmental problems have been severe. Because of the heavy rainfall and threats of torrential floods, BHP was allowed to discharge its finely ground rock waste and tailings directly into the Ok Tedi rather than storing them behind a tailings dam. The initial rationale for this was that if a tailings dam failed during a heavy rain, catastrophic flooding would occur downstream, endangering thousands of villagers, but by the early 1990s this practice was no longer tolerable, and 7,500 members of the local Miripiki clan sued BHP in Australian courts for damages arising from destruction of their historical habitats along the Ok Tedi by unrestricted dumping of mine tailings. This suit eventually had more than 30,000 claimants seeking damages approaching $4 billion but eventually willing to settle for $500 million.

In short, this has been a monumental project. Billions of dollars have been invested in a remote jungle, and billions more in mineral wealth have been removed. Jobs have been created, and taxes and royalties paid to one of the poorest nations in the world. Western investors have reaped huge profits, and PNG natives have received wages, education, and health care previously unknown and unavailable. Pristine environments have been destroyed, and indigenous peoples and their cultures forever changed. Is this good? We won't answer that. We don't think it is that simple.

FURTHER READING: *Mining Magazine* (1988); Shipes (1995).

selves fortunate to have found one ore body in a career. Buried mineral deposits are illusory targets and a good deal of money is spent in their discovery. In 1990, a fairly typical year, U.S. and Canadian mining companies reported expenditures of just under $325 million on mineral exploration. George Kohls, the vice-president of Gold Fields Mining Corporation, once reviewed a 15-year period of his company's exploration activity. During that time, Gold Fields acquired 53 projects, found 256 individual exploration targets on those properties, and ended up with three operating gold mines. In that same period, it spent $110 million to find 6 million ounces of gold ore, for a discovery cost of $18 per ounce. This effort turned out to have been worthwhile because gold has been worth anywhere from $350 to $400 an ounce over the past decade. No wonder explorationists call mineral prospecting "the world's biggest and best gambling business."

Once the difficult task of discovery has been accomplished, the work of turning a mineral find into a mine has only just begun. It typically takes from 3 to 10 years to bring a mine into production (we describe one example in Box 6.2). Here the discovery risk changes from hard work and disappointment to true economic risk since the development period may coincide with changes in worldwide mineral demand and commodity price (Fig. 6.4) or in the political climate of the country where the deposit was found. Some students of the subject have even argued that investors in minerals so commonly misunderstand the cyclic nature of mineral commodity prices that companies routinely and perversely wait until prices are high to invest in new mines with an almost certain guarantee that production will not begin until market prices have cycled back to their lowest level. Instead of following the time-honored tradition of "buying low and selling high," they buy high and sell low.

Policy questions

Do you think your local, state (province), or federal government should devise some way to reduce the risk of searching for mineral resources? If so, what?

FURTHER READING: Wolfe (1984); Craig, Vaughan, and Skinner (1988).

6.2 CLASSIFICATION AND FORMATION OF MINERAL DEPOSITS

Minerals and their compositions are the memories of formative processes. – *Guilbert and Park,* The Geology of Ore Deposits, *p. xii*

In this section we provide a classification of nonfuel mineral resources and describe their diverse geologic origins.

Classification

Nonfuel mineral resources include a highly diverse set of commodities, and we must classify them somehow. One classification is shown in Table 6.3, which puts together minerals that have similar uses and, therefore, experience similar market conditions. A primary distinction is between the metals and everything else, these latter often called industrial rocks and minerals or nonmetals.

Metal is a chemist's term that refers to a varied group of chemical elements with similar properties. In general, they are electron donors in chemical reactions (they form positively charged cations). They tend to be strong, ductile, good conductors of heat and electricity, opaque to visible light, and capable of being polished to make them reflective. Commodity specialists further subdivide

Table 6.3 *Classification of nonfuel mineral resources*

Metals	Industrial rocks and minerals (nonmetals)
Precious (noble) metals	**Fertilizer minerals**
Gold, silver, platinum, palladium, rhodium, iridium, osmium	Phosphate rock, guano, potash
Ferrous metals	**Chemical minerals**
Iron, manganese, nickel, cobalt, molybdenum, chromium, vanadium, tungsten	Salts (halite, sylvite, borates), trona
Nonferrous (base) metals	**Building (construction) materials**
Copper, lead, zinc, tin, mercury, cadmium	Crushed rock, sand, gravel, building stone, common clay, perlite, gypsum
Light metals	**Glass raw materials**
Aluminum, lithium, magnesium, titanium	Silica, feldspar
Rare (specialty) metals	**Insulation materials**
Beryllium, bismuth, cesium, gallium, germanium, indium, tantalum, zirconium	Asbestos, vermiculite
	Abrasives
	Garnet, diamonds, corundum
	Refractories, ceramics, and fluxes
	Limestone, fluorite, olivine, clay
	Fillers and pigments
	Kaolin, hematite, sericite, titanium oxides, diatomite
	Drilling materials
	Expanding clays, barite
	Filtration materials
	Zeolites
	Gems and semiprecious stones
	Inert gases
	Helium, carbon dioxide

the metals into groups with common properties that make them valuable (Table 6.3). Precious metals are used primarily in supporting currencies, manufacture of jewelry, or in measuring wealth, and most of them have little industrial value. Ferrous metals include iron and others capable of alloying with iron to manufacture different varieties of steel. Conversely, the nonferrous (base) metals have many different uses but are classified together because there is a high demand for them in industrial societies – all were known and mined to some extent in the twilight of prehistory. Light metals are relative newcomers to the markets because they are prized primarily for their high ratio of strength to weight. This ratio makes them important for airframes, automobile bodies, and railroad cars since the less the vehicle weighs, the more paying freight can be carried for a fixed expenditure of energy. The specialty metals are simply everything else that has minor and highly specialized uses. Most of them were isolated by chemists only in the past 150 years and

find uses in the manufacture of electronic components and other high-technology products.

In general, metallic mineral resources are more valuable than nonmetals, although this is clearly not the case for gems and some other less-familiar commodities like high-quality kaolin or some zeolites. Economists would say that metals have a high "unit value." Gold is worth $350 to $400 an ounce, copper $1.00 to $1.25 a pound. This high unit price means that, in many cases, metals can be mined almost anywhere. There are metal mines north of the Arctic Circle, beneath the ocean, on the flanks of some of the highest peaks in the Andes, in the middle of the Namib Desert. There would undoubtedly be mines in Antarctica if they were not banned by international treaty, and minerals would be mined from the deepest ocean floors were there not doubts about who owns them. Another feature of metals that separates them from most nonmetals is that, following mining, they must be converted to their elemental form before being marketed. There is no market for sphalerite, ZnS, the common ore mineral of zinc, but there is most assuredly a market for elemental zinc. This means that the metals industry is more than simply an extractive industry. Multiple stages of mineral concentration and purification are required to produce an end product that commonly is an ingot or bar of pure metal.

By contrast, most industrial rocks and minerals are valuable in their own right or, at most, involve minor amounts of size classification or washing to remove impurities of one kind or another. Many nonmetallic resources are simply the rock itself. Most have lesser unit value than equivalent amounts of metals, but they are very valuable if located in the right place; that is, they have a high place value. If you own a crushed rock quarry near a major, growing metropolis you are likely to become a comfortably wealthy person. Although crushed rock may be worth only tens of dollars a ton, the average residential home contains several tons of crushed rock, and every highway, commercial building, airfield, parking lot – you name it – requires many tons of crushed rock as ballast. There are more than 6,000 rock quarries in the United States that produce more than 2 billion tons of rock every year.

Location alone does not guarantee the value of non-metal deposits because most of them must meet significant quality standards before any customer will purchase them. Unlike metals, whose high unit value makes copper from a mine in Indonesia no different from one in the Urals, industrial rocks and minerals must satisfy pickier buyers. Kaolin clay for the china trade, for example, cannot have more than a very small amount of iron because firing an iron-rich kaolin will discolor the porcelain product. Or, to consider a more mundane product, the crushed rock ("road metal") that is layered beneath asphalt roadways to allow them to drain and resist the punishing forces of trucks and cars must meet stringent requirements for strength and chemical inertness to assure that the road bed will stay stable and firm.

The result is that each mineral commodity has its own specifications and unique markets. Consequently, mineral-producing companies themselves tend to specialize in a range of related commodities. Some produce only precious metals; others produce only nickel, cobalt, and copper, which are commonly found together in nature. Some aggregate companies only produce sand, gravel, crushed rock, and large rough-cut stone for riprap (rock piles such as seawalls) and ballast; while others specialize in cut stone for building facings and headstones. There are fertilizer companies that produce potash and phosphate, companies that specialize in construction prod-

ucts like plaster and gypsum wallboard, diamond producers who control these precious stones from the time they are mined until the cut stone is sold in Amsterdam, Hong Kong, or New York.

Geological ore-forming processes

In the space available, we cannot come close to describing the variety of geological processes that can produce an economic mineral deposit. Consider copper, for example. About 9 million tons of copper is mined each year in dozens of countries, from rocks as old as 3 to 4 billion years to as young as a few tens of millions of years, in geological settings as varied as young oceanic island arcs, ancient continental shields, and active continental tectonic belts. It has been mined as copper carbonate coatings and growths on coaly material and as oxide cements in sandstones (Nacimiento mine, New Mexico), from copper sulfides that settled to the bottom of $1,000°-C$ magmas in the rubble left behind from a meteorite impact (Sudbury, Ontario), from ancient accumulations on the sea floor at midoceanic spreading centers (Kidd Creek in Quebec, Cyprus), from native copper nuggets that appear to have formed as thermal waters moved through an ancient pile of basaltic flows and interbedded conglomerates (Keweenaw Peninsula, Michigan), from the tops of cooling intrusive bodies that were feeding magmas to volcanoes several kilometers above (Bingham Canyon, Utah), from hot aqueous emanations from the flanks of undersea volcanoes in island arcs (Kuroko, Japan), and from shales that were once fetid, organic-rich marine muds on the subsiding margin of an ancient continent (Megen, Germany). And those are the geologic settings of just some of the major deposits.

In general, ore deposits are rare because they require a sequence of low-probability events, essentially a set of geological coincidences, that must occur in the right order to form, preserve, and expose valuable resources in an appropriate setting. We classify ore-forming processes into six categories and summarize the environments in which they occur in Figure 6.5.

MAGMATIC PROCESSES   Many economic minerals are concentrated in magma chambers during the cooling of an intrusive body. Magmas are mixtures of molten silicate rock and crystals that rise in the crust and cool to form igneous rock bodies known as plutons (more in Section 3.4). As magmas cool, some minerals begin to crystallize while the remainder of the magma remains liquid. As cooling progresses, more crystals form and less liquid remains until the body becomes entirely solid. If the ear-

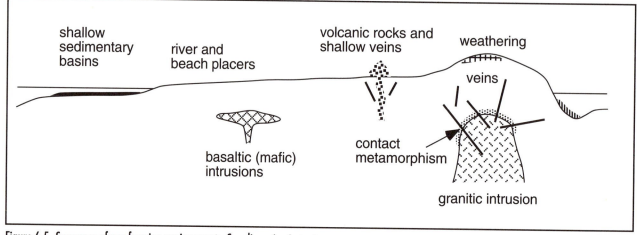

Figure 6.5 Summary of ore-forming environments. See discussion in text.

liest minerals to crystallize are denser than the surrounding liquid, they sink to the bottom of the chamber; if lighter, they float to the top. This gravitational settling or buoyant floating segregates these minerals from the remainder of the magma and concentrates them. If they are valuable, this concentration may form an ore deposit provided that: (1) there is enough magma to yield a sizable source of valuable minerals; (2) the concentration process continues long enough to make thick and rich accumulations; and (3) no disruption by later intrusives or other processes disturbs the rich layers of minerals. Where this happens, mafic (basaltic) magmas form deposits of nickel, platinum, and chromium. Examples include the Duluth gabbro body in Minnesota, the Stillwater complex in the Beartooth Mountains of Montana, and the world-renowned platinum and chromite deposits of the Bushveld Complex in South Africa.

As early-crystallizing minerals form, some cooling magmas separate into two liquids that are immiscible (cannot mix to form one liquid). For example, as a magma rich in sulfur cools, silicate minerals that contain no sulfur crystallize, and the remaining magma becomes richer in sulfur. Because most magmas have temperatures of 900°C or more, well above the melting point of sulfur, droplets of liquid sulfur separate from the magma if it becomes saturated with sulfur as cooling continues. Many metals, including iron, nickel, cobalt, platinum, and copper, are preferentially enriched in the sulfur-rich liquid, forming a dense, metal-rich liquid sulfide that may settle to the bottom of the magma chamber and cool to form rich mineral deposits. The world's largest nickel deposit at Sudbury, Ontario, and the large deposits at Norils'k, Siberia, formed by such a process.

Near the end stages of crystallization, when perhaps 90% or more of the silicate liquid has crystallized, the water content of the remaining liquid commonly exceeds the solubility of water in the magma. This water separates from the magma just like sulfide liquid, but unlike the dense liquid formed by sulfur, the water forms a fluid that is much less dense than the molten magma. This exsolution (out-of-solution) water commonly strips the magma of those chemical elements with higher solubility in water and may cause a catastrophic (for the magma chamber) volume expansion (see discussion of volcanic explosions in Section 3.4). The water-rich fluid that escapes from the magma chamber ultimately cools and deposits its dissolved minerals. Some deposition occurs partly within the magma chamber, but more commonly the fluid escapes and deposits its minerals in cooler rocks around the chamber (see the subsequent discussion of hydrothermal deposits).

In addition to producing metals, the igneous rocks themselves can be a valuable mineral resource. Granite, for example, forms blocks that are hard, tough, and durable. They are quarried and shipped, especially by barge, long distances to be used in breakwaters and seawalls. Cut, dressed, and often polished, granite and many other igneous rocks are found as building stones, ornamental facings, monuments, and headstones. Another example is an igneous rock whose sudden increase in value is causing concern in some parts of the southwestern United States. It is a relatively fragile volcanic rock known as pumice, a frothy volcanic glass that looks like mineral cotton candy and is preserved in only the youngest and often most scenic volcanic fields. Pumice is light enough to float and, being a bubbly mass of silicate glass, is hard and abrasive. It is these latter characteristics that makes pumice so desirable and is causing stripping of some of the best scenery in the southwestern United States. Why? Think of stone-washed jeans. Newly minted blue jeans are tumbled in vats of water with pumice to abrade the material and give them their currently fashionable (mid 1990s) appearance.

HYDROTHERMAL PROCESSES   Also associated with magma bodies, but in a less direct fashion, are accumulations of resources loosely called hydrothermal (hot-water) ore deposits. Everyday experience tells us that hot water can dissolve almost anything better than cold (that is one reason why we use boiling water to make coffee or tea). In addition, chemical reactions occur faster at high temperatures. Thus, natural hot waters, like those at Yellowstone or at Rotorua, New Zealand, contain significant amounts of dissolved metals along with other elements. The "rotten-egg" smell so characteristic of geyser basins is produced by sulfur gases escaping from the cooling thermal waters. This increased efficacy of water as a solvent at high temperatures is enhanced by the increase in pressure, which raises the boiling point of water by many hundreds of degrees Celsius at depths of only a few kilometers.

Some areas of the crust are hotter than others and consequently produce more hydrothermal activity. They include the margins of cooling magma bodies, the vicinity of volcanoes and their associated accumulations of hot ash and lava, and the crust at oceanic spreading centers. In these areas, water is heated to many hundreds of degrees Celsius, has a low density, and ascends in the crust as cool water moves down to replace it. The hot ascending water can mix with magmatic waters rich in minerals extracted from the cooling pluton and will probably dissolve minerals from the rocks they come in contact with as they pass through. Eventually, near the surface in the fractures and conduits that channel the rising, hot, mineral-laden waters, the fluid mixes with colder surface waters, reacts with new, cold rocks, or boils as the pressure drops. Each of these processes reduces the capacity of the water to carry metals, and the result is deposition of metal-rich minerals in veins, in subcylindrical pipes, and in throughgoing disseminations (Fig. 6.6). If enough metal is present in a small enough area (and we can find it!), we have a hydrothermal ore deposit.

Hydrothermal deposits form at all depths in the crust. Deposits in the midcrustal region are commonly called postmagmatic and include some of the largest copper deposits in the world, for example, at Bingham, Utah, and El Salvador, Chile; all of the world's molybdenum, pri-

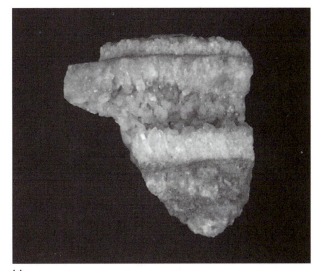

(a)

(b)

Figure 6.6 Types of deposition from hydrothermal solutions: (a) quartz crystals growing into cavity from both sides of vein; (b) manganese oxide precipitated from solutions occupying thin fracture surface (only one surface of the fracture is shown).

marily in Colorado and British Columbia; famous deposits of silver, like the Comstock Lode in Nevada and Potosi and Pachuca in Mexico; and much of the world's tungsten and tin, largely in Bolivia and eastern Asia. Other hydrothermal deposits occur in and near volcanoes, including much of the gold from Alaska, Nevada, Mexico, and Brazil, and all of the world's mercury, antimony, and sulfur. Eventually, volcanic fluids can vent to the surface as hotsprings or geysers (Fig. 5.4) or to the sea floor. At the surface, or in mixing with seawater, the metal-rich fluids boil, cool, and undergo chemical changes that cause rapid deposition of metallic minerals, generally sulfides. These deposits on the sea floor have the imposing name of volcanogenic polymetallic massive sulfide deposits and commonly consist of 40% or more (thus, *massive*) copper, zinc, lead, silver, iron, gold, and barium (thus, *polymetallic*) in sulfur-rich minerals (mostly *sulfides*) adjacent to or on the flanks of volcanoes (*volcanogenic*). These deposits are important resources in Japan and other Pacific Islands and in Canada.

SEDIMENTARY PROCESSES   The sedimentary environment is a rich source of ore deposits. As with igneous rocks, sedimentary rocks themselves are resources. Limestone is used as a building stone (Fig. 6.7), in making cement, in the smelting process for iron (Section 6.3), for agricultural lime, as an additive to chicken feed so that laying hens will have enough calcium to make durable eggs. In its principal use for cement, the limestone is fired at high temperature with quartz sand to drive off the carbon dioxide (Section 8.2) and form a hydrous calcium silicate, which "sets" to hardened cement shortly after it is mixed with water. One formation in the American Midwest, the Salem Limestone, is a very uniform, well-cemented, fossiliferous rock that is easily quarried and cut into handsome, durable blocks of all sizes and shapes. Most colleges, universities, county courthouses, and federal buildings in the United States have Salem Limestone pillars, steps, lintels, or pilasters. Sand and gravel from stream, river, or glacial deposits are valuable as aggregate for road surfacing and concrete (a mixture of gravel and

cement). In some places, especially where natural gas is cheap and abundant, shale is mined, crushed, and fed into rotary kilns, where the water in the impermeable clay of the rock instantly boils. The shale fragments explode like tiny popcorn, are fired by the heat of the kilns, and come out as lightweight but remarkably tough and durable expanded aggregate. It is used to make ultralight concrete that can save significant amounts of money in construction.

Detrital sedimentary processes can produce metal deposits as well as nonmetal ores. Figure 6.8 is a schematic diagram of a gold vein that is eroding. Gold is insoluble and dense, many times denser than grains like quartz, feldspar, or small rock fragments. Consequently, when each fragment of gold is released from the grasp of its surrounding minerals and storm waters wash the unconsolidated detritus of the eroding vein into nearby streams, the gold quickly collects on the stream bottom, especially in riffles and rills where fast-moving and turbulent waters carry everything else away. The resulting accumulations of gold are called alluvial placer deposits, and virtually every gold district in the world was discovered by an intentional or unintentional prospector who found a gold nugget on a stream bottom. Many an eager prospector worked the local stream beds in the aftermath of this discovery, and smarter ones realized that abandoned stream beds and old stream terraces in the local alluvial valleys would also contain placer gold. The really smart ones,

Figure 6.7 Marble (metamorphosed limestone) quarry near Carrara, Italy. This area has supplied stone for building and sculptures (including those of Michelangelo) for more than 2,000 years.

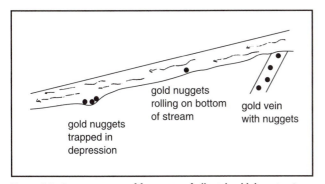

Figure 6.8 Representation of formation of alluvial gold deposits. See discussion in text.

however, figured out that the gold had to be eroding out of the earth upstream and set out to find the so-called mother lode.

The California gold rush of the 1850s followed the discovery of placer gold in 1848 at Sutter's Mill. While the "Forty-Niners" and their successors panned and washed every stream bed on the west flank of the Sierra Nevada, others built high-pressure hoses to wash the terrace gravels for gold. This process, called hydraulic mining, is an environmental disaster since it washes otherwise-stable floodplain and terrace deposits into stream beds. (A report written in the early 1900s by G. K. Gilbert describing the wholesale destruction of the marshes at the northern end of San Francisco Bay by the influx of mud and sand from hydraulic mining on the Sacramento River is considered to be the first environmental geology paper ever written.) As hydraulic mining in California was proceeding, other people sought the upstream outcrops of the eroding gold veins, and the hard-rock mining camps of the Sierras were born. So were fortunes and misfortunes associated with such famous figures of the last half of the nineteenth century as Fremont, Crocker, Hearst, Strauss (as in Levi Strauss of blue jean fame), and Stanford.

Gold from stream placers created fortunes in countries besides the United States. One of the remarkable 19th-century fortunes was that of an Englishman, Cecil Rhodes, who made most of his money mining alluvial placer gold in southern Africa. This placer gold, however, is distinctly different from the placer deposits of California. In the Witwatersrand Basin of South Africa, the alluvial gold was deposited in stream gravels more than 2 billion years ago. The placers were then buried and metamorphosed into a hard, quartz-cemented meta-conglomerate peppered with gold. These ancient placers are still mined, even to depths approaching 12,000 feet

below the surface, and make South Africa the world's largest gold producer.

Not all placer deposits form in streams, and most do not contain gold. All minerals that are insoluble, dense, and durable can be concentrated both by stream flow and by waves and currents along beaches. Much of the tin from Malaysia is mined in stream placers that contain tin oxide (SnO_2; cassiterite), and diamonds, zircon, ilmenite, rutile, and magnetite are mined from beach placers in numerous countries.

Another important sedimentary environment for mineral resources occurs in arid regions where isolated ocean basins or intermontane lakes can evaporate and form saline brines. Eventually, these brines become saturated with salts and minerals begin to precipitate. Typical marine evaporite basins produce thick and valuable accumulations of halite (NaCl), sylvite (KCl), and gypsum ($CaSO_4 \cdot 2H_2O$). Some of the best-known deposits are from basins that developed in the Permian (about 200 million years ago) over vast portions of northern Europe, in the Devonian (400 million years ago) in the American Midwest, and in the Jurassic (150 million years ago) in the coastal region of the Gulf of Mexico. The salts of marine evaporites are basic raw materials for the manufacture of many chemicals, and it is no accident that the American upper midwest and Lower Mississippi valley, along with the Ruhr basin of Germany, became major centers for the modern chemical industry. Nonmarine evaporites, deposited mostly in temporary lakes in desert basins (Fig. 6.9), contain many important salts that differ significantly from marine evaporites. They include minerals such as borax ($Na_2B_2O_4$) and trona (Na_2CO_3), which are used in soaps, detergents, glass, and fire-retardant chemicals.

DEPOSITION OF METALS IN CONTINENTAL BASINS Petroleum forms where thick accumulations of mostly marine sediments are deposited in large sedimentary basins (Fig. 6.10; see also Section 5.3). Deposition in these basins also develops one of the most important classes of metal deposits, the carbonate-hosted lead-zinc-fluorite ores, commonly referred to as Missouri Valley Type ores because of their abundance in the central United States. They are formed by upward migration of waters that were originally trapped in the muds and silts of the subsiding basin. As the weight of the thickening pile of sediments compacts the deeper beds, they become warmer and their contained water is pressed into more porous sandy layers. This water removes metals that were adsorbed on the marine clays and carried them up and out of the basins, commonly into carbonate rocks at the basin margins.

Figure 6.9 Salt accumulating by evaporation of water on the floor of Death Valley, California: (a) white salt deposits left as water trickles toward Badwater (lower right), which is 280 feet below sealevel and the lowest point in North America – Courtesy Allen Glazner; (b) expanse of salt at Badwater – Courtesy Catherine Flack; (c) close-up view of salt precipitating at Badwater – Courtesy Allen Glazner.

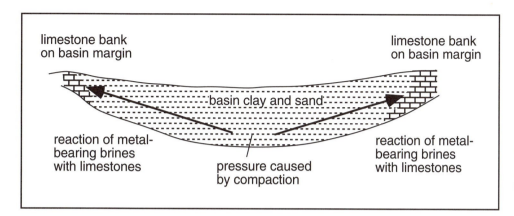

Figure 6.10 Representation of formation of Missouri Valley Type (MVT) lead–zinc deposit. See discussion in text.

Limestone and dolomite are highly reactive, and apparently the metal-bearing brines react with the carbonate rocks and deposit lead and zinc minerals as well as fluorite (CaF_2) and barite ($BaSO_4$) in pores in the carbonates. Deposits of this kind have formed in the North American midcontinent, in Australia, and northern Europe and provide most of the lead and zinc mined in the world.

A similar process appears to operate even in the absence of limestone and dolomite. In basins containing only clastic sediments, it is possible that metal-rich, basinal brines may seep into pores of sandstones and siltstones and somehow deposit metals in them. However they get there, deposits in clastic sediments produce some of the largest sediment-hosted mineral deposits in the world at places like Sullivan, British Columbia, and Mt. Isa and Broken Hill, New South Wales, Australia.

SURFICIAL PROCESSES Surface processes like weathering and groundwater migration can concentrate minerals and make valuable ore deposits by removing mineral substances that are worthless and concentrating those that are valuable. In most soils the SiO_2 in the quartz, as well as that liberated from the breakdown of feldspar, becomes a relatively insoluble acid (H_4SiO_4), and in temperate climates with moderate rainfall this SiO_2 remains in the soils. In the tropics, however, where rainfalls are heavy and the water in soils is typically basic, SiO_2 and most other chemical species are leached, and the only component that remains is aluminum hydroxide. Where aluminum-rich rocks, like some granites, are extensively leached by continuous and deep weathering, the resulting accumulation of aluminum hydroxide forms bauxite, the world's only significant aluminum ore. Similar ore-forming processes occur in the formation of kaolin for fine china, in concentrating nickel in some residual soils in the South Pacific, and even in the formation of a mineral known as vermiculite, whose absorbent properties make it very attractive to owners of indoor cats.

As groundwater slowly moves through porous rocks, it can also concentrate dispersed minerals, especially in places where oxygen-rich groundwater can take a metal into solution and then reprecipitate it where the water encounters sediments or sedimentary rocks rich in organic material. The oxygen in the water oxidizes the reduced carbon in the organic matter, and if the metal that was soluble in its oxidized state is insoluble in its reduced state, minerals form at this so-called redox (reduction-oxidation) front. This process occurs most importantly in a class of uranium deposits often called "roll-front" (Section 5.4) deposits and has also led to the development of some copper and vanadium deposits.

BIOLOGICALLY MEDIATED PROCESSES Deposition of some ores apparently requires the activity of bacteria or other microbes. Three important ores formed by this process are phosphate, banded iron formations, and manganese nodules.

Phosphate is a critical ingredient in fertilizer (Section 2.2). Particularly in the midlatitudes of the North Atlantic Ocean basin in the late Tertiary (past 30 million years), deep, cold, phosphate-rich waters welled up onto continental margins. This phosphate interacted with nutrients from coastal rivers and the abundant oxygen of the near-surface environments to create a rich bloom in algae and many other marine plankton. As the organisms died, their phosphate was deposited in organic-rich sediments on the shallow sea floor. If the rates of clastic sedimentation had been high, this organic material would have been dispersed and buried, but off the coasts of Florida, the Carolinas, Morocco, and Western Sahara there was very little clastic material being delivered to the continental shelf at this time. Consequently the phosphate-rich organic materials accumulated in the thin muds on the sea floor, where continuing degradation by bacteria, worms, sea cucumbers, and other bottom-dwelling organisms concentrated the phosphate into the valuable mineral apatite, with the formula $(Ca_5(PO_4)_3(OH,F)$, thus forming a rock known as phosphorite.

Another time-restricted mineral resource is BIF, an acronym for banded iron formation. BIF consists of thinly layered sediments that contain primarily the two iron oxides (hematite, Fe_2O_3; magnetite, Fe_3O_4) plus quartz, iron carbonates, and iron hydroxides (Fig. 6.11). The layers differ only in the ratios of magnetite/hematite plus iron silicates to silica plus carbonates, and individual layers less than 1 centimeter thick can be traced for miles. Huge, billion-ton reserves of BIF occur in the Lake Superior region of the United States, Labrador, the Carajas and Minas Gerais regions of Brazil, Cerro Bolivar in Venezuela, the Hammersley Basin of Western Australia, the Transvaal of South Africa, and the Kursk region in Russia. In each location the rocks formed between 3,800 and 1,800 million years ago. Why? We know from our studies of black smokers that iron is continually introduced to the oceans in the relatively soluble, reduced (Fe^{+2}) form at the midocean ridges. Today, that Fe^{+2} iron is quickly oxidized by deep, oxygenated waters and the resulting Fe^{+3} iron is so insoluble that it is immediately removed from the oceans. It is likely that the deep oceans before 1,800 million years ago lacked oxygen because photosynthesis was probably limited and the earth's atmosphere itself was oxygen-deficient. Thus, the deep oceans were iron-rich, and they may have welled up onto continental margins and deposited BIF just as phosphate-rich waters today deposit rich phosphorites in areas where clastic sedimentation is restricted. Because of the lack of significant fossil remains, however, it is also possible that the BIF formed by purely inorganic processes.

Manganese (ferromanganese) nodules occur across some of the deepest portions of the Pacific Ocean basin, where sedimentation rates are on the order of 1 millimeter or less per thousand years, and the material collecting on the abyssal sea floor consists mostly of terrestrial dust and includes such bizarre components as micrometeorites, fish teeth, and whale earbones. In these areas we can dredge up nodules about the size of an orange that consist of delicate, concentric, layers of iron and manganese oxides. Photographs of the sea floor show areas littered with millions of nodules like golf balls on a driving range after a long summer weekend. Their onionlike structure suggests that they have grown from the precipitation of iron and manganese oxides from seawater, and

Figure 6.11 Core of banded iron formation. The white bands are mostly quartz, and the dark bands are mostly magnetite and some hematite.

it seems unlikely that this is a wholly inorganic process. Many investigators have suggested that some bacterial agent removes metals derived from terrestrial weathering and from black smokers and deposits them on the surface of these nodules.

Whatever is forming the nodules, the U.S. Bureau of Mines once estimated that just one of these fields of nodules contains 20 times the known reserves of manganese – a 2,000-year supply at current rates of consumption. In addition, the nodules contain significant amounts of other metals, including copper, nickel, and cobalt. The nodules, however, are a good example of the difference between a resource and a reserve. Most lie at depths greater than 4,000 m, and no one yet has figured out how to sweep them from the ocean floor. Furthermore, even if we could mine them, who owns them? Is access to the sea floor away from any nation's exclusive economic zone (200 miles from the coast; Section 2.4) on a first-come, first-served basis, as the ocean surface is? Is the deep seafloor a commons like Antarctica or, perhaps, the moon? A Law of the Sea Treaty drafted in the 1970s tried to resolve this issue, but there were significant differences of opinion on who owned the sea floor between a few wealthy nations like the United States and Japan, which had the technological capability to exploit this resource, other coastal nations who feared the unknown environmental consequences of deep-sea mining, and poor, landlocked nations who wondered how they might benefit from exploitation of a common resource. Keep tuned – no one has a clue how to mine these riches or to whom they belong.

FURTHER READING: Guilbert and Park (1986); Peters (1987); Harben (1990); Evans (1993).

6.3 LOCATION OF NONFUEL MINERAL RESOURCES

I know it's down there. I just have to dig deep enough.
— Prospector in the California desert standing at the bottom of a 4-foot hole that he had begun digging with shovel and pickax 6 weeks earlier

Where do we find resources? Almost anywhere. That might seem a flippant answer to an important question, but we phrase it this way to reinforce once more the difference between resources and minable, exploitable minerals. Resources are relatively abundant; reserves are not. The locations of minable ore deposits are controlled by the sequence of geological events in a region, and because no two regions of the earth's crust have experienced the same events in the same sequence at the same time, no two areas have the same assemblages of mineral deposits. Some areas, like oceanic islands or the sea floor, are geologically young and have been subjected to only a few geological processes. Other areas, particularly the cores of large continents (referred to as Precambrian shields), are billions of years old and have undergone volcanism, sedimentation, weathering, magmatic intrusion, metamorphism, uplift, and erosion in a complex, nearly indecipherable, sequence that includes almost every mineral-producing environment described in Section 6.2. Consequently, the mineral-resource inventory of oceanic crust is much smaller than that of Precambrian shields. When we gain ready access to shield areas – where they are not buried beneath desert sands, glacial ice, younger sediments, water, or snake-infested swamps and frozen tundra – we often find rich deposits of metals.

The uneven distribution of mineral resources would not be an intractable problem were it not for the human propensity to draw arbitrary political boundaries across the landscape with little or no attention to geological considerations. Most such boundaries are drawn at easily located or defensible limits. The Rhine River is a convenient boundary between France and Germany, but the geological reality is that similar geology and mineral resources, particularly iron and coal, on both sides of the river have made this boundary contentious at various times in recent history. In the United States, government surveyors certainly made their jobs easier when they chose 109° west as the north–south boundary between Colorado and Utah (north) and Arizona and New Mexico (south), and 37° north as the east–west boundary between Utah and Arizona (west) and Colorado and New Mexico (east). In doing so, they neatly created the only place in the United States where four states meet, but for

exploration geologists they bequeathed an administrative headache since this "Four Corners" region is one of the richest uranium districts in the world. Each state has its own laws and regulations concerning land acquisition, land use, environmental practices, and mine permitting. Add to that a dozen or so Native American reservations and land grants, and the lawyers begin to salivate.

Because of the lack of correspondence between political and geological boundaries and the significant geographical variations in geological history, there are significant inequities in the distribution of nonfuel minerals, just as there are of energy resources and water, that must be taken into account in understanding world politics. Some nations are rich in minerals, the most important being Canada, Chile, South Africa, Australia, Russia, and China. In every case, this mineral largesse is the legacy of geological accident; each nation includes within its borders major geological provinces rich in mineral wealth. In all of these examples except Chile, the geological province that provides most of the national mineral wealth is a sizable tract of Precambrian shield. Chile has a young mountain range, the Andes, where the balance of uplift and erosion just happens to expose shallow- to midcrustal levels of continental crust, a geological environment rich in minerals (Section 6.2). If we wanted to predict other nations with potential for future mineral wealth, we would be tempted, and rewarded, by looking for other areas with accessible Precambrian shields, such as Brazil, Zimbabwe, Zambia, and India, or with young mountain belts, for example, Chile, Peru, Bolivia, Colombia, and Mexico.

Conflicts resulting from the inequitable distribution of mineral resources have occurred throughout most of human history (Box 6.3). The view that most international behavior is resource driven is called geopolitics, and although many conflicts (e.g., Northern Ireland) are not discernibly affected by resources, many struggles are closely related. For example, during the period of Western colonialism from the 1700s to middle 1900s, the prevailing economic theory was mercantilism, which held that those countries that prosper must have access to, and production of, all vital resources. As a result, poor nations with rich mineral resources were subjected to economic and political domination by industrial nations. Resources also partly determine the course of wars. For example, we cannot understand the British and German battles over northern Norway in World War II until we realize that they were fighting for control of railroads connecting the great iron district of Kiruna, in neutral Sweden, with ice-free ports on the Atlantic. On the

BOX 6.3 THE CHILE NITRATE WAR

Wars are fought for mineral resources. Ancient conflicts over Cyprus between Egyptians, Greeks, Persians, Phoenicians, and Romans were originally battles to control rich copper deposits — the "strategic oil fields" of the Bronze Age. The French and the Germans for many years traded the rich iron-ore districts of Alsace and Lorraine as spoils of war. In 1917, on declaration of war with Germany, the United States eagerly expropriated the molybdenum mine at Climax, Colorado, from German interests that controlled it, but not before the Germans had exported enough molybdenum to make the tough and durable steel for the gun-barrel of Big Bertha, the cannon that fired shells 75 miles into Paris.

This idea that competition for mineral resources is a cause of international conflicts is far too simple a picture of a complex social, economic, and political decision that also has profound nationalistic and racial overtones. Once in a while, however, there are good examples of wars fought over minerals. The "War of the Pacific" between Chile, Peru, and Bolivia is one. After achieving their respective independence from Spain in the 1820s, these countries had disagreed quietly over their adjoining borderlands in the Atacama Desert region along the coast of the Pacific Ocean. It was a dry, desolate, region with some sporadic mining of sodium nitrate in low, coastal plains. Because the nitrate was only marginally valuable in the manufacture of explosives and fertilizers in America and Europe, at first no one cared who owned what.

In the 1870s, this nonchalance changed as mechanized farming in the United States, Canada, and Europe created a huge demand for nitrate fertilizers. Suddenly the economic significance of the nitrates in the Antofagasta District, which nominally belonged to Bolivia, and Tarapaca District, which was considered part of Peru, became apparent. Peru decided to expropriate the mines from Chilean owners in Tarapaca and Bolivia decided to heavily tax those in Antofagasta. Chile declared war in 1879. The war went Chile's way and resulted in the Treaty of Ancon in 1883, which ceded both districts to Chile. Bolivia ended up land-locked; Peru's economy was left a shambles; and Chile became a dominant turn-of-the-century Pacific power on the basis of $1.4 billion in nitrate exports between 1883 and 1915 when the Haber process made recovery of atmospheric nitrogen economic. The nitrates of the Atacama desert dictated the economic and social well-being of three major Latin American countries for almost 40 years.

FURTHER READING: O'Brien (1982).

---

diplomatic, rather than military, level, geopolitics also explains the U.S. policy toward South Africa prior to the end of apartheid because of U.S. dependence on South African chrome and manganese.

The uneven distribution of mineral resources is particularly well shown by patterns of trade. Most OECD nations import large quantities of nonfuel minerals, and Figure 6.12 shows the major nations that export them. We use exports, rather than production, to construct this figure because exports show the economic winners in the global exchange of minerals. The United States, for example, produces vast amounts of nonfuel minerals, but its consumption levels are so high that it is a net importer of nearly every mineral commodity, including ones that it produces like iron, copper, and zinc. This importing adds considerably to the deficit in the U.S. trade balance, but it is very small compared to the money that the United States sends abroad each year to purchase petroleum (Section 5.5).

Although major exporting countries such as Canada, South Africa, and Australia export several minerals, Figure 6.13 illustrates that other countries are essentially mineral "monocultures" in the same sense as some countries are agricultural monocultures (Section 2.5). These are nations where the geological legacy provides the country only one economic mineral resource (Box 6.4). They include particularly a number of tropical nations (e.g., Jamaica, Guyana, Surinam, Guinea, and Sierra Leone) where aluminum is the only significant export. Because tropical soils are poorly suited for agricultural production (Section 2.2), these countries tend to be net importers of food and, thus, wholly dependent on the world aluminum market. Jamaica, for example, depends on mining of bauxite for 60% of its exports. A drop in aluminum demand due to an economic recession in the industrial world or to increased recycling of aluminum cans, or a decline in reserves because of depletion, may have a disastrous impact on nations already among the poorest in the world.

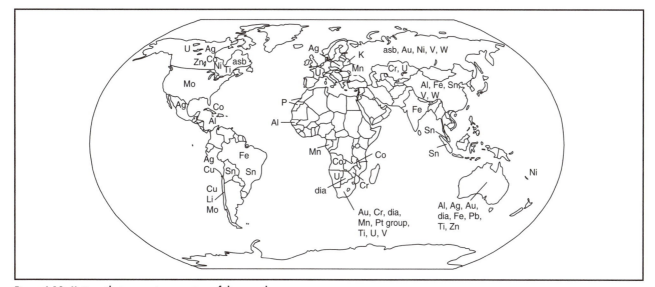

Figure 6.12 Nations that are major exporters of the ores shown.

Policy questions

Are you concerned about a world situation in which industrial countries are rich enough to import whatever they need and lesser-developed countries must export raw materials in order to maintain even weak economies?

FURTHER READING: Crowson (1994).

6.4 METHODS OF MINING ORE AND PREPARING IT FOR MARKET

There have been many speculations on the manner in which the metallurgical arts were begun . . . an Egyptian lady dropping some malachite [green copper ore] into a charcoal brazier, and copper ores dropped into campfires.
— P. J. Flawn, Mineral Resources, p. 75

In simplest terms we have this material, rock, that needs to be moved from where we find it, in the earth, to where we can use it. Then, depending on the material, we may have to process it further so that it can be used. The removal from the earth, we call mining, and the processing is beneficiation. In this section we discuss various modern methods of mining and beneficiation and then give a very brief history of mining and the credit that society gives miners for their efforts.

Mining can be as simple as shoveling sand into a wheelbarrow or as complex as drilling tunnels, blasting rock, and lifting ore 10,000 feet to the surface of a refrigerated gold mine in South Africa. We can mine by going underground in vertical or inclined shafts and in horizontal tunnels that all connect with huge underground galleries, called stopes, where we can extract rock and transport it to the surface (Fig. 6.14). This process requires expensive and specially designed lifts and drills, haulage vehicles, and ventilation systems to allow the safe and efficient movement of men and materials in a warren of underground passages. Increasingly, we prefer to mine at the surface. Presently, more than six out of seven active metal mines in the United States are surface mines, and worldwide 90% of all mines and 99% of all nonmetallic mineral mines are surface mines.

There are two basic surface-mine designs, a strip mine and an open mine, or quarry (Fig. 6.14). The difference is in the configuration of the ore body. If we are mining a horizontal bed of rock near the surface that is not so hard or consolidated that it requires large amounts of blasting, we strip mine it by removing the overburden that obscures the ore and then mining the layer of ore with giant power shovels. We can (although we don't always) return the overburden to the area from which the ore has been stripped as mining progresses and regrade, replant, and return it to a nearly natural state. A strip mine, then, sort of marches across the countryside with only a small fraction of the ore being exposed and mined at any given time. An open pit or quarry (Fig. 6.15) is more appropriate for blob-shaped, near-surface ore bodies, especially in hard rock that requires explosives to turn the rock to chunks that can be loaded onto trucks or conveyor belts. The problem with open pits and quarries is that as long as mining proceeds, the pit gets deeper and wider. All waste

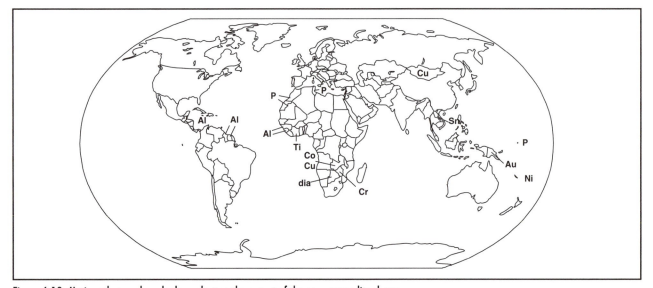

Figure 6.13 Nations that are largely dependent on the export of the one commodity shown.

rock must be stored somewhere else until mining ceases, and we leave behind a huge hole in the ground and massive unsightly spoil piles. Generally we do not refill the hole because, if it took twenty years to make the hole, it is going to take several years to fill it back in, and no one makes money while the hole is being filled.

Why do we allow open pit mining? There are two compelling reasons. First, surface mining is much cheaper than underground mining. We can, for example, mine ore as lean as 0.6% copper in an open pit, but the same ore mined underground would likely require a minimum grade of 2% to 3% to be economic. Second, an open-pit mine is very safe. It is essentially a giant construction project with huge earth-moving equipment operating under a canopy of blue sky, and training, vigilance, and safe practices can make it as safe as a highway construction project. Statistics kept by the U.S. government routinely demonstrate that a miner is safer at work than a farmer or a truck driver, and this safety record is undoubtedly a measure of the predominance of surface mining in the United States. By contrast with open pits, underground tunnels are inherently dangerous. The very processes that make the ore deposit – fracturing of rock to increase permeability, and the reactions of fluid with rocks that cause the precipitation of metals – make the rock weak and unstable. Drilling and blasting openings into such rock creates an environment where the most vigilant and careful miner is at risk.

Most underground mine accidents occur in coal mines, where natural gas accumulates and the smallest spark can ignite an inferno or cause an explosion. Coal-mine explo-

sions occur every year, killing dozens. Until the 1960s, not a decade passed in the United States without a coal-mine explosion killing 100 men or more, and as recently as 1984, 27 miners died in a single explosion in West Virginia. Because nonfuel, underground mines do not release explosive gases and the rock itself is noncombustible, they are safer than coal mines. Most "hard-rock" mine accidents involve collapsing mine workings, rock falls, or heavy machinery. As recently as 1972, however, 91 miners died in a silver mine in Coeur d'Alene, Idaho, as a result of fires in hot, dry, abandoned mine workings that consumed the available oxygen and filled the underground workings with toxic or suffocating fumes.

Once mined, the materials must generally be modified ("beneficiated") depending on what they are and who wants to buy them. Some combination of crushing, grinding, and milling to reduce size, washing to remove impurities, classifying to sort the material into size fractions that can be processed, and physical separation to isolate one rock or mineral fraction is typically performed near the mine. All of these processes reduce transportation costs because the waste product of the beneficiation process can be left on the mine site.

A common problem with mining and beneficiation is the generation of useless rock (waste) that must be moved and dumped to gain access to the ore plus all of the residue from beneficiation. Discarding this waste in unsightly, sometimes unstable, piles of rock upon which nothing can grow is a problem in its own right, but when it is exposed to the atmosphere and rain, the waste commonly reacts to produce acid runoff or other dangerous

BOX 6.4 EXPORTING BIRD DROPPINGS, STAMPS, AND WEIGHT LIFTERS

An example of a mineral monoculture that seems almost allegorical, but is no less true, is that of the tiny, island nation of Nauru. Almost no one outside of philatelists and residents of the southwestern Pacific had ever heard of Nauru before the summer Olympics of 1996. At that media-saturated event, much was made of the weight lifters from Nauru, an 8- by 16-mile dot of rock located 600 miles northeast of the Solomon Islands (the P, for phosphate rock, in Figure 6.13). Nauru is the smallest independent nation in the world. Its economy is based on bird droppings. To be more specific, Nauru survives by the mining of natural fertilizers that were produced over many millennia by the interaction of bird droppings (guano) with marine sediments exposed at the surface. Essentially depopulated during Japanese occupation in World War II, about 8,000 Nauruans now live on 1,100 of the 5,236 acres that are not mined to produce fertilizer for markets in Japan, New Zealand, and Australia. Once the phosphate rock is mined, the remaining land consists only of what has been called "inhospitable pinnacles of rock." To inhabit a small island made entirely of phosphate ore is to give the term "finite resource" a very graphic meaning. When the resource is gone and mining ceases, currently estimated to occur in the year 2000, there will be no exportable product for Nauru except stamps and weight lifters. Even with reclamation, Nauru is unlikely to become a major tourist stop since there will be little of the island left that has not been stripped of every pound of exportable phosphate rock.

Fortunately for Nauru, it seems to have been served by wise leaders who, seeing an end to the export of their only resource, have tried over many years to invest the profits of the Nauru Phosphate Corporation in assets that would continue to provide for the nation. They own a 52-story office complex, Nauru House, in Melbourne, Australia; an airline; a small shipping line; an insurance company; and interests in the Queen Victoria Hospital and a brewery in Melbourne. They even had an interest in the Fitzroy Football Club of Australia, before it went bankrupt in 1995. How successful they have been in assuring a stable future for the residents who remain when mining ceases remains to be seen. Nauru, like many others in the 1980s, invested heavily in real estate in the so-called Pacific Rim nations, including Indonesia, Malaysia, the Philippines, New Zealand, Australia, and the United States, and many of these real-estate ventures have soured in the mid-1990s. In addition, Nauru has taken an understandable interest in making development loans to other island nations, such as the Cook Islands, in the southwestern Pacific, but in several cases, the loans have been defaulted upon.

Will Nauru survive the final depletion of its mineral resources? We should know in a few years. Is Nauru an allegory for other mineral monocultures or, even, for other mineral-exporting nations? The answer to that question will take a little longer to answer.

FURTHER READING: Viviani (1970).

---

effluents (Section 7.3). As an example, consider an ore mineral dispersed throughout a rock with a concentration of 2.5% by weight. If the miner can only sell a concentrate that is at least 90% by weight, the miner will have to discard at least 35 tons of silt-sized waste or tailings for each ton of ore sold.

For many nonmetals, beneficiation is the final process. The washed and classified crushed rock is sold directly to the consumer as gypsum, clay, cut and polished stone, or other materials. For most metals, and some nonmetals, however, significantly more modification of the concentrate is required. Phosphate rock, for example, must be treated chemically to convert it from the insoluble mineral apatite to a water-soluble fertilizer called "superphosphate" (Sections 2.2 and 7.5). In order to form copper metal, a concentrate of copper sulfide minerals must be melted and the sulfur removed, commonly as sulfur oxide gases that go up the smokestack and into the atmosphere to return as acid rain (Section 7.6). Even native elemental gold must be purified to meet market standards. Smelting and refining, as these operations are often called, are notorious for the air and water pollution they produce. Since they may or may not occur near the mining site and most require abundant, cheap energy, it is increasingly common for smelting and refining operations to be located in lesser-developed nations where environmental concerns often take a secondary interest to jobs and investment in heavy industry.

Some of the most extensive beneficiation consists of converting iron ore to steel. The historic Bessemer pro-

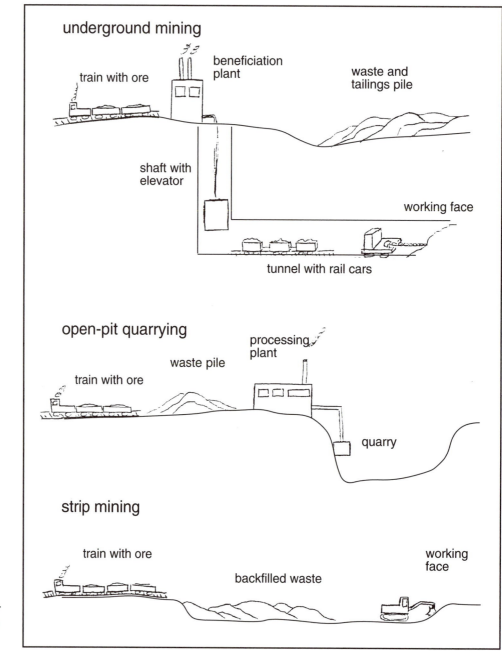

underground mining

train with ore

beneficiation plant

waste and tailings pile

shaft with elevator

working face

tunnel with rail cars

open-pit quarrying

processing plant

train with ore

waste pile

quarry

strip mining

train with ore

backfilled waste

working face

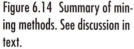

Figure 6.14 Summary of mining methods. See discussion in text.

cess consists of mixing iron ore with coke (from high-quality coal) and limestone at a temperature high enough to melt native iron. The coke causes reduction of the hematite (Fe_2O_3) in the ore to elemental iron, which filters down to the bottom of the blast furnace and leaves a "slag" of impurities floating on the top. This molten iron can then be drained off the bottom and further purified by passing blasts of air through it, resulting in oxidation of remaining impurities that float to the top as a

new slag. Once the iron has been thoroughly purified, it can then be remelted and mixed with other ("ferroalloy") elements to make different types of steel, the most familiar of which is a high-chromium variety known as stainless steel. Because steel plants require delivery of iron ore, coal, and limestone, they are located where transportation of these commodities is least expensive. As industry in the United States surged in the late 1800s, this location was western Pennsylvania, part of West Vir-

Figure 6.15 Open-pit mine.
Photo by P. G. Feiss.

ginia, and eastern Ohio because: (1) iron ore could be shipped by water through the Great Lakes from mines in Minnesota and Wisconsin; (2) coal could be shipped by rail from mines in the Appalachians; and (3) limestone was available locally.

Finally, we should note that the nature of the mining profession has undergone major changes through time. For millennia before the feudal era, mining was the work of prisoners and slaves. The Greeks and Romans operated their mines with captive armies or local indigenous peoples who came under their control, and the life of a miner was short and brutish. Even to this day we refer to undesirable work environments as "salt mines" and think of breaking rocks as fit duty for convicted felons. Conditions in the Middle Ages were much better, and mining was a noble profession. Mining guilds in Germany and England were powerful organizations because they held the key to production of metals needed for coinage and arms. They were often wooed by the nobility and royalty, given privileges no others in society had, and allowed to operate under laws that assured them of a steady livelihood. Well into the twentieth century in the United States, the United Mine Workers union was a powerful presence in the political life of America. Some of that power undoubtedly obtained from the fact that during the turbulent years of labor unrest from about 1880 to 1930, miners had access to, and knowledge of, one commodity that striking workers wanted – dynamite.

Now, in the 20th century, mining has again become a relatively invisible activity. Mines are messy, and some are dangerous. Few people want them or the plants that beneficiate the ores in their backyard, and most people prefer jobs that are physically easier to do. Consequently, most mines are located in out-of-the-way places, and mechanization that increases efficiency and productivity leaves only a very small percentage of the work force in most industrialized nations involved in nonfuel mining. There are fewer workers in metal mines in the United States today than in any time since data began to be kept in the late 19th century. There are now more teachers, farmers, construction workers, and retail workers than miners. No wonder we take for granted the mineral necessities of life, including the ones that you counted at the start of this chapter and give little credit to the people who supply them.

Policy questions

Assume that you are a member of the governing board of a town when a mining company asks to open a limestone quarry within the town limits. If the quarry will *both* (1) reduce the unemployment rate in the town by 50% and (2) leave an unsightly pit behind, would you approve the company's request?

FURTHER READING: Peters (1987).

6.5 LEGAL AND PROPERTY ISSUES

The "greater fool" theory of mining investment . . . states that "there will always be a greater fool who will pay more for a [mineral] property, notwithstanding how much one has paid for that same property." – Harris and Kesler, "BLASH vs. GF Theory"

At first glance, the ownership of nonfuel mineral resources seems to be straightforward. After all, they are not fugitive resources like oil or water. We can't drill a hole and pump the gold or aluminum from our neighbor's property the way we could the groundwater. We can lease a building to a renter or sell the timber to a paper company, but the land and all the access rights continue to be ours. Should it not be the same with minerals? Maybe it should, but it isn't. There are several reasons why mineral ownership rights, or simply mineral rights, are not always neatly conveyed with the transfer of land from one owner to the other (see Box 6.5 for an extreme example of the problem).

Regalian versus accessory rights to minerals

The first difficulty lies far back in the mists of time. Early metal mining was almost always conducted either to obtain iron or base metals for weapons or to acquire precious metals for making the coin of the realm. In the days before representative government had been invented, the ruler of any city or state or a tribal chieftain would need to be assured that access to raw materials for arms manufacture and the minting of currency was secure. Consequently, this ruler might allow the people who lived on the land to control the land's surface but would be reluctant to give up control or ownership of the subterranean minerals – we call this concept the "regalian tradition" of mineral rights.

The original regalian concept worked well as long as mining was an arcane profession carried out by slaves and contract workers who produced raw materials that only a king could use anyway. Most people had all they could do to produce food and fiber to stay alive and had no real use for copper, gold, or silver. Through time, however, the regalian tradition came to have interesting and significant consequences. First, as kings gave way to bands of nobles and eventually to various democratic styles of government, the mineral rights were often transferred to the governing body that replaced the ruling family. (We will see below how this has interesting consequences in the United States.) A second consequence is more philosophical. Clearly, if one subscribes to the regalian doc-

trine, then mineral rights *can be separated* from surface rights. You can own your farmland and do whatever you wish with it, but someone else can own the coal or tin or diamonds that lie beneath.

The oddity of separating mineral and surface rights can have interesting consequences. In countries that subscribe to the regalian tradition, the government can lease these mineral rights to any mining venture – state or private. Can you keep the miners off your land? Well, maybe not if they could tunnel under your land to gain access to their minerals. But what if the minerals are almost at the surface? Whose right is better, yours or theirs? These issues are most often resolved in civil law courts. As an example, in Cornwall, England, in the midcenturies of this millennium, Cornish miners wanted to exercise their right to mine the king's tin under land owned by minor nobles. The king wanted his share of the profits on the sale of this tin. The nobles wanted to keep the miners out of their game preserves, off their peasant's croplands. To resolve the difficulties, the king set up the stannery courts (from *stannous*, Latin for tin) to resolve these conflicts. The kings, nobles, and free miners are gone, but the stannery courts exist to this day in England to resolve property and land disputes.

In some places – for example, ancient Rome – the codification of law altered the regalian tradition. The Justinian Code declared that the rights to minerals beneath the land were an accessory to surface ownership, thus the name "accessory rights." They allow the mineral and surface rights to be separated, but unless this is formally done, the minerals under your land are yours and certainly not the government's. This isn't going to help if your great grandfather sold the mineral rights, but at least you are otherwise protected. Well, most of the time. In the United States, the separation of coal rights from surface rights led to serious abuses in the poverty-stricken hills of Appalachia when mining companies who had purchased the right to strip-mine coal 50 years earlier showed up one day to claim their rights, and many a poor farmer was deprived of land and livelihood. To prevent this, many states now have laws that require owners of separated mineral rights to exercise that right within 5 to 10 years of purchase or lose that right ("use it or lose it"). Even in ancient Rome, the Justinian Code only applied to Roman citizens and was not applicable in conquered lands. In fact, Roman emperors got into the habit of paying off victorious generals with the mineral rights to gold and silver in conquered lands in return for one-seventh of the gold and silver mined (this practice of retaining a portion of the minerals for the ruler is the origin of the term "royalty"). In Rome, giving distant mineral concessions to military heroes had several advantages. It kept gold and silver flowing from the mines in remote lands and it kept victorious generals and their idle legions far away from Rome in case they felt any spontaneous urges to be in charge themselves.

The two traditions of regalian and accessory rights, much modified, continue today in different parts of the world. In many countries, minerals are viewed as part of the commonwealth of the nation – the regalian tradition modified to accommodate whatever the current form of government happens to be. Rights to mine the nation's minerals are leased or rented as mining concessions to the highest bidder (or to the person offering the most generous bribe). Often a lease block or concession is opened to bid, and mining companies are given 6 months to a year to review all public geological data and to conduct whatever investigations they wish. At the end of this period, sealed bids are submitted according to terms published in advance, a lease or concession is granted that provides access to some or all minerals in a region for a period of years, and the government is guaranteed to receive some percentage of any minerals or profits from the concessionaire. In other countries, where the accessory model prevails, mining companies deal with whomever the owner of the mineral rights may be and reach an agreement just as they would on any land transaction.

The problem of the public lands in the United States

There is a special set of problems that occurs with public lands. For example, in the United States, the legal tradition applied to private land is generally consistent with the accessory model of mineral rights. But what happens on the public lands? Who owns these minerals and how is access assured for every citizen? With the exception of the eastern seaboard of the United States, essentially all of the lands west of the Appalachian Mountains were purchased or acquired by treaty agreements, and the federal government thus became the owner of record. Even today, more than 85% of the land in Nevada and half of the land in California belongs to the federal government, and since the government in a democratic country is the people, the lands belong to everybody. So do the minerals, with exceptions made for those areas where people already lived and thought they owned the land by a previous agreement with Spain or France or Russia or Mexico or whomever the United States obtained the land from.

The exceptions for previous ownership in the United States were very minor, and at the time of acquisition, more than 95% of western lands were unoccupied in the traditional sense of someone owning a legal, registered deed of trust. (Native Americans who had used these lands for millennia as hunting and grazing ranges might not have characterized these lands the same way.) So, by default, the U.S. government became the biggest landowner of all. Lest you think this was some nefarious plot of Big Government, the truth is that the first thing the U.S. government did every time it acquired such a tract of open land was to try to unload it. Vacant land pays no taxes, but settlers demand services – mail, protection from marauding bands of cattle rustlers, roads and canals, land surveys, and guarantees of civil and property rights. Thus, the federal government, eager to pay for the acquisition of the land and needing revenue to provide the new services demanded of it, began in the first years after independence to encourage settlement and the transfer of land title to private land-owners. Land was sold, leased, given away – all in hopes of generating revenues via tariffs and taxes of one kind or another.

For the first 50 years of the American republic, this land policy progressed in a relatively systematic, if not smooth, manner. Because most of the lands of the Louisiana Purchase were fertile farmlands with sufficient water to support subsistence farms, people wanted to settle on them. Many properties were purchased with small down payments at federal land offices (which did a "land-office business" in a country of land-hungry immigrants). Few of these settlers ever paid off the balance because, as soon as enough new land owners moved in and before they were forced to pay up or be evicted, they petitioned for statehood and elected a senator who promptly introduced legislation to forgive all residents of the new state their debts to the federal government for land purchases.

Some mineral-rich areas were acquired as well as farmland, but the expense involved in mining made ownership of these mineral resources an arcane and technical subject. It was a long way from the early mining camps of the Lake Superior area and Mississippi Valley to Washington, D.C., most of the mining corporations included influential politicians who could keep the federal authorities at bay, and no one much cared anyway. Also, most of the land acquired in the Mexican War was inhospitable desert – no one wanted it or worried about who owned the minerals on the lands that were public. So it remained until 1848 when gold was discovered on public lands in California, and everything changed. "Who owned this gold?" became an important issue. A prospec-

BOX 6.5 THE BUTTE UNDERGROUND WAR

The Mining Law of 1872 (see text) was created to bring order to the process whereby prospectors could lay claim to minerals on public lands. Just about the time the first transcontinental railroad crossed North America, the only economically viable mineral resource in the wild and remote regions of the western United States was the rich bonanza vein deposit. Geometrically, most such bonanza veins are steep to shallowly inclined, sheet-like masses from a few tens of centimeters to meters thick. The prospector who stumbles on the outcropping of the vein has no idea of how deep the vein goes or how far it extends to the side. But since the boundaries of property and thus ownership rights are presumed to extend like vertical walls to the center of the earth, what is a prospector to do to protect a discovery of a rich vein if it isn't vertical and strays beneath a neighboring claim?

To protect against the eventuality of straying veins, the clever legislators in the 43rd U.S. Congress created something called "extralateral rights." The idea is simple. The highest point on a vein was called its "apex," and the law says that whoever discovers and lays claim to land on which the apex of a vein is found owns that vein to the "center of the earth" regardless of whose land it may pass under en route. This is strange enough, but what happens if the vein apex lies beneath your land but,

since it does not reach the surface, other people find it first in their mine? Well, then, the law says it is theirs if they can prove they found it first. What happens if a fault cuts the vein off at depth and then the vein reappears on someone else's land (Fig. 6.16)? What if that other person, knowing nothing about this fault since it is covered by river sediments, has already mined it. Can you sue? What happens if the vein splits in two or joins another vein. Who owns what?

That this law creates problems is not simply speculation on our part. By 1886, 14 years after the Mining Law went into effect, the great copper and silver mines of Butte, Montana, the "richest hill on earth," were annually producing more than $13 million in ore. Far greater amounts of money were being invested by East Coast interests in mines, smelters, railroads, and other commercial interests. The problem was that the ore at Butte was not contained in a single vein, but in hundreds of criss-crossing veins that intersected, merged, and offset one another. The hill itself was a patchwork of claims made over a period of years. Some overlapped each other due to surveying errors, there were unclaimed gaps between active properties, some claims lapsed due to failure of the claimant to perform development work or to pay for his patent. Into this complex tangle of competing interests stepped one of the more remarkable

Figure 6.16 Diagrammatic representation of the questions involved in determining ownership of underground veins. Vein A belongs to owners 1 because its apex is in their property. Vein B also belongs to owners 1, but how do they know this if owners 2 discover it first? Who owns veins A1 and A2? Is vein D really vein C offset along a fault? Can owners 3 mine vein D because its apex is on their side of the river?

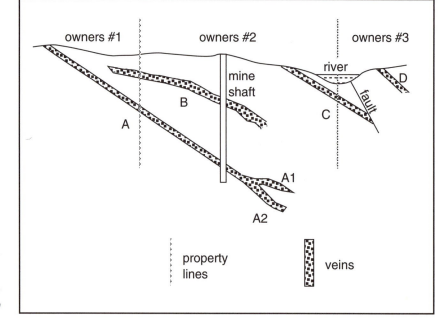

figures in mining history, F. Augustus Heinze, a poor but well-educated mining engineer from New York.

Heinze went to work at the standard wage for a mine surveyor of $5 a day and learned all he could about the underground geology at Butte. He learned quickly that, with dozens of rich veins going every which way, no one really knew what vein apexed on whose property. He learned, too, that there were tiny unclaimed gaps between some of the richest mines on the hill. One of these, which was less than 40 square yards in size, he obtained and called the Copper Trust. It had the immense advantage of lying next to one of the richest mines in Butte, which belonged to the Amalgamated Copper Company. From his worthless mines, he connected underground tunnels with those of adjacent rich mines and stole ore from their underground workings. When told to cease, he claimed that these veins apexed on one of his various properties. The choice for the aggrieved party was to go to court at great expense or to surrender to Heinze for a handsome cash settlement. This latter was often advantageous since, in a jury trial, the complexities of underground surveying, the descriptions of veins and ore minerals, and the jargon of mining often led to the winner being the one with the best attorneys or the most friends on the jury — Heinze always made sure he had both. In fact, he also owned a federal judge, which made many people reluctant to take him on in court.

Heinze's ultimate test came, however, over his battles with Amalgamated Copper Company, which came under the control of John D. Rockefeller's Standard Oil group about the turn of the century. By this time, Heinze was stealing ore from Amalgamated mines and bringing it up out of the ground from his shafts at a prodigious rate. Rockefeller, however, was not a man to surrender, and he could hire more and better lawyers than Heinze. But, while battles over ownership were waged in courts across Montana and the west, Heinze's miners and Amalgamated's miners waged a more physical war underground over who would mine what ore from where. One author has described it wonderfully:

Imagine the Empire State Building . . . completely buried in the earth. Assume half of its rooms and corridors, some large suites, some whole floors, are filled solidly. Many of its corridors, instead of running straight, are curved and twisted, turning abruptly at odd angles or making wide arcs. Here and there at unexpected points are open shafts with ladders dropping from the corridors to levels perhaps a hundred feet lower. Here and there a great hole in the ceiling leads up to higher levels, accessible only by ladders. Narrow railroad tracks lead along some corridors. Mud and rock pave others. Plank chutes come down at many points to carry material, tumbling and roaring into bins for transfer to dump cars and hoisting to the surface. No one knows what lies below. No one is certain what is contained in the vast areas filled with solid rock. . . .

Multiply it by one hundred to include scores of mines under a square mile of surface of Butte hill and you may have some idea of the physical environment in which the underground battles between Heinze's miners and the Amalgamated miners were fought. (Glasscock [1935], pp. 262–3)

And fight they did. As miners for each side discovered tunnels from their opponent's mines entering their property, they blasted them shut, often injuring miners in the debris avalanches. Eventually, opposing miners encountered one another in the damp darkness of the mines and they directed high pressure water hoses, steam, and even clouds of lime driven by blowers against one another. They burned rags in one another's workings to create suffocating clouds of smoke. And, of course, since dynamite was always handy, there was a good deal of tossing of sticks of dynamite at one another.

In spite of all the excitement, only two men are known to have been killed. The miners reported years later that the dynamite exchanged had fuses long enough for those on the receiving end to get well away. Even when railway tracks were booby-trapped by connecting them to electrical generators, they never charged them more than "sufficient to knock [a man] down." One observer of the events said, "More powder was burned than in the Russo-Japanese War," but that was likely hyperbole by an overstimulated miner.

Eventually, the courts prevailed. In 1903 a series of rulings fined Heinze $22,000 for jumping the Amalgamated claims, though local wisdom always said Heinze won anyway since he stole at least $1.6 million worth of ore before being caught and fined. Amalgamated Copper soon acquired and consolidated all claims on Butte hill and became, after the breakup of Standard Oil, the Anaconda Copper Company, which controlled Butte well into the 1980s.

FURTHER READING: Glasscock (1935).

tor with a mule, a shovel, and a gold pan could be a millionaire by working hard and keeping the gold. With the money made in the California gold fields, the answer to the question, Who owned all the silver and lead and zinc and copper that were found on western public lands? became a pressing issue to more and more prospectors and investors. As it turned out, no one could answer this question. In the mining camps of the West, they had just sort of made things up as they went along. The prospector who found the gold was entitled to it. In some camps, an individual could only hold two or three claims; in others, it was first come, first served and help yourself to as many as you could hold. Some camps required forfeiture of unworked claims, in others you could reserve a claim simply by leaving a shovel or pick stuck in the ground (shovels and picks were neither cheap nor abundant at first). All camps, however, agreed that "finders were keepers." If there was an argument over who found it, there were nonjudicial, often terminal ways of settling rival claims.

Only after the American Civil War (1865) did Congress turn its attention to the issue of who, in fact, owned the minerals on the public lands. And when Congress did so, it was carefully guided in its law making by a bevy of western governors and senators who had made fortunes, or had rich friends who made fortunes, by extracting minerals from the public lands for personal profit. Thus was born a remarkable piece of legislation that still applies to the public lands of the western U.S. – the Mining Law of 1872. Here are some highlights:

- The law allows any U.S. citizen to stake a claim to any nonfuel minerals on those federal lands open to mineral entry. This open area is enormous, although it constitutes less than one-third of the 809 million acres of public lands because the courts have closed national parks, national monuments, wilderness areas, wildlife refuges, military bases, any national forest lands purchased from or donated to the federal government by private citizens or corporations, and many other exceptional land categories.

- The law guarantees the discoverer a clear title to all minerals *and* surface rights on the claim for a nominal charge (about $2.50 to $5.00 an acre for up to 20.66 acres per claim). All the claimant must do is perform a small amount of exploratory work during a several-year probationary period to show sincerity and then file for a deed.

- Once title to the land is acquired (the technical term is to "patent" the land), the lucky person has to pay no royalties on minerals mined, no lease fees, nothing other than normal property and income taxes that derive from owning any parcel of land. Despite this favorable situation, only 0.4% of federal lands have been patented.

- Once a patent is obtained, strangely enough, the owner of the mining claim is under no obligation to mine. Ski resorts, hotels, condominiums, and, in one well-publicized case, a brothel have been located on mining claims acquired from the public lands for nominal cost.

In some ways the Mining Law of 1872 has worked very well. Because it is a self-initiated and equal-opportunity process that rewards the individual who does the hard work of exploration and discovery, it resulted in the discovery of hundreds of mines that provided livelihoods to hundreds of thousands of people and thus helped to jump-start American economic development in the century following the Civil War. Great fortunes were made, and there are many philanthropic monuments to these riches, such as the Guggenheim Museum of Art in New York City and dozens of geology buildings on colleges and universities across the country. In some ways, however, the Mining Law has worked poorly. Because the law applies only to metals, there are completely different bodies of law for energy minerals like coal and petroleum and for the nonmetallic minerals like limestone, crushed rock, and stone. The law also contains a number of arcane rules that result from the fact that, at the time it was devised, the only hardrock mining envisioned was an underground mine that chased a mineral-rich vein into the earth (Box 6.5). These same rules are wholly inapplicable to large disseminated mineral bodies mined by open-pit methods.

The Mining Law's major flaw, however, is that it gives public lands away at essentially no cost. This has caused spectacular abuses. A 1974 study found that only 7 of a randomly selected 93 patents granted over a 22-year period were ever mined. The remaining 86 were acquired at a cost of $12,411 and had, in 1972, a fair market value to the owners of more than $1 million. One famous case involved 80 acres of land in Arizona acquired under a mining patent in 1955 for $200 and sold in 1972 to housing developers for $368,000.

Will the 1872 Mining Law be changed? How should it be changed? In 1996, we can answer neither question. Ideas as diverse as leasing federal lands, charging royalties, denying the right to permanent title to federal lands until mining commences, and charging mining companies fair market value for public lands have been discussed. Many opponents claim that such practices will end mining in the United States, with predictable economic costs. Is this likely? Probably not. Most countries charge royalties and demand fair market price for land. So do owners of private lands. As we have seen before, prospectors have no choice but to go where the minerals are. They will explore on federal lands with or without the Mining Law of 1872. One mining expert, responding to the premise that changing the law would drive mining companies overseas, said he doubts it and told the story of the famous bank robber, Willie Sutton, who, when asked why he robbed banks for a living, responded, "Cause that's where the money is." Prospectors will go where the minerals are.

Policy questions

How do you think that ownership of mineral resources should be determined?

FURTHER READING: Flawn (1966).

6.6 THOUGHTS ON MINERALS AND THE FUTURE

In many quarters deterioration of the mineral position and industrial strength of the United States is accepted as inevitable – E. M. Cameron, At the Crossroads, p. ix

Strong economic systems are diversified. They are independent of any one crop or any one source of goods for their economic vitality. They should be dynamic, certainly growing as fast as the population, and active in all economic sectors: raw materials production, manufacturing, service. It remains to be seen in this postindustrial age how strong a national economy can be if it loses or forfeits its basic extractive industries like mining and mineral processing. Many gurus of the 21st century like to say the developed world is passing on to a new kind of economic system – the knowledge-based or "postindustrial" economy. They talk with the glazed eyes of zealots about a world huddled around computer screens, moving products, ideas, Eurodollars, services, and who-knows-what-else from place to place without ever leaving sight of their hot tub. We suspect not. That computer and the circuits that connect it to other computers require copper and gold, aluminum and iron, sulfuric acid and silica to operate. The food, even the junk food, that our computer user consumes needs phosphate and nitrates, limestone and gypsum before it ever arrives at the market. The foundation beneath and the roof above, as well as the hot tub, need crushed rock and gravel, clay, and iron to be stable and weather-proof. If we don't mine these things, someone, somewhere will have to do it for us. It is not possible to foresee an increasing population of the earth without an increasing requirement for nonfuel minerals. More people, more mines.

Clearly, then, we will produce nonfuel minerals ourselves or buy them elsewhere. We will despoil some fraction of the earth's environment to extract and produce minerals, and we may or may not clean that mess up depending on our political will and sense of environmental ethics. We will run out of some resources at current prices and either (1) substitute some other material that will, in all probability, still have to be mined, (2) pay higher prices for that resource, thus creating new mines, or (3) do without and incur a predictable impact on our standard of living. And we would be naive to think that we can ship our mines and mining-generated environmental problems overseas without political consequences. A nation dependent on foreign raw materials is a nation with foreign entanglements, responsibilities, concerns. Global economies require global engagement. Finally, to understand the role of the nonfuel minerals in a nation's economy, we cannot conclude that minerals are unimportant because they constitute only a few percent of the gross national product.

FURTHER READING: Cameron (1986); Youngquist (1990).

PROBLEMS

1 A company is extracting uranium from a strip mine. The ore contains 2% uranium, and after the rock is removed from the mine, it is crushed and the uranium dissolved (leached) from it by acids. Assume that the original rock has a pore space of 5% but that the fragments of crushed rock fit together so poorly that the

crushed material has a pore space of 25%. After leaching the uranium, what percentage of the hole can be refilled with the crushed rock?

2 On the accompanying map, where do you think it would be most profitable to locate a plant that makes concrete? (To answer this question, you must take into account the costs of the various processes that are required for the manufacture of concrete.)

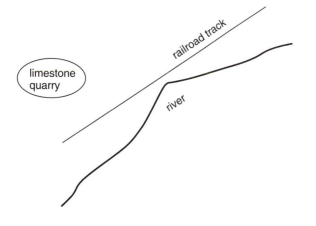

3 A thick banded iron formation (BIF) is at the surface (in outcrop) over an area of 40 square miles. Over most of this area, the BIF contains 50% quartz (SiO_2) and 50% magnetite (Fe_3O_4), but one small area was altered by dissolving (leaching) one-fourth of the quartz and oxidizing all of the magnetite to hematite (Fe_2O_3). What percentage increase in the iron content of the BIF has been caused by this alteration? (Note – the Fe content of magnetite is 72.4% and of hematite is 70.0%.)

4 A company is mining gypsum at a quarry 50 miles from the coast and wants to ship it to a city on the coast 500 miles away. Assume that a railroad line 500 miles long makes a direct connection between the quarry and the city, and a line 50 miles long connects the quarry with a port on the adjacent coast. If the cost of rail transportation is three times as much as the cost of transportation by ocean freighters, would it be cheaper to ship the gypsum directly by rail or to send it to the coast and load it on freighters for delivery to the city?

5 Titanium (Ti) is obtained mostly from two minerals – rutile (TiO_2) and ilmenite ($FeTiO_3$). Assuming that mining and smelting costs are the same for rocks with either rutile or ilmenite, would you derive more profit from mining a rock containing 2% rutile or one containing 4% ilmenite? (Note – some chemistry is required for this problem; atomic weights are: Ti, 48; Fe, 56; O, 16.)

6 Country A exports 20% of the aluminum in international trade, country B 30% of the phosphate, and country C 2% of the copper, 4% of the uranium, and 3% of the iron. Based solely on these data, which country do you feel has the strongest economy and which the worst?

REFERENCES

Cameron, E. M. (1986). *At the Crossroads: The Mineral Problems of the United States*. New York: John Wiley.

Craig, J. R., Vaughan, D. J., and Skinner, B. J. (1988). *Resources of the Earth*. New York: Prentice-Hall.

Crowson, P. (1994). *Minerals Handbook, 1994–95*. London: Macmillan.

Evans, A. M. (1993). *Ore Geology and Industrial Minerals*. 3rd ed. Boston: Blackwell Scientific.

Flawn, P. T. (1966). *Mineral Resources*. Chicago: Rand McNally.

Glasscock, C. B. (1935). *The War of the Copper Kings*. New York: Grossett and Dunlap.

Guilbert, J. M., and Park, C. F. (1986). *The Geology of Ore Deposits*. San Francisco: W. H. Freeman.

Harben, P. W. (1990). *Industrial Minerals: Geology and World Deposits*. London: Metal Bulletin.

Harris, J. H., and Kesler, S. E. (1996). BLASH vs. GF theory in Mining Investment. *Society of Economic Geologists Newsletter*, no. 26 (July): pp. 23–4.

Kesler, S. E. (1976). *Our Finite Mineral Resources*. New York: McGraw-Hill.

Mining Magazine (1988). Ok Tedi targets 70,000 t/d production level. 159 (September): 150.

O'Brien, T. F. (1982). *The Nitrate Industry and Chile's Crucial Transition, 1870–1891*. New York: New York University Press.

Peters, W. C. (1987). *Exploration and Mining Geology*, 2nd ed. New York: John Wiley and Sons.

Shipes, H. R. (1995). Profile of a mine in the wilderness. *American Metal Market* 93: 16.

Viviani, N. (1970). *Nauru, Phosphate, and Political Progress*. Canberra: Australian National University Press.

Wolfe, J. A. (1984). *Mineral Resources: A World View*. New York: Chapman and Hall.

Youngquist, W. (1990). *Mineral Resources and the Destinies of Nations*. Portland, Ore.: National Book.

7 | WASTE AND POLLUTION

7.0 INTRODUCTION

When irreplaceable benefits are claimed for the continued production of a toxicant, the people at risk do not benefit, and the people who benefit are not at risk. – *B. Magnus Francis*, Toxic Substances in the Environment, *p. 296*

The plastic cups, candy wrappers, and pages of old newspapers that blow through the streets of London testify to its wealth. It is a society that can throw things away. In many unrealized ways, we characterize other places by the human effluent that they produce: the tidy streets of Kyoto where the ever-scrupulous Japanese pick up everything; the blanket of stinging smog that hangs over Mexico City, produced by the unchecked emissions of the factories and cars packed into the former lake basin in which it is built; the water of Bombay, which you cannot drink because of the untreated sewage that filters through its soil; the city of Cracow, Poland, where the abundant, but dead, waters of the Vistula River are so polluted by hazardous chemicals from upstream factories that the residents must pipe their water in from 25 miles away; and the town of Hanford, Washington, home to the largest nuclear waste dump in the United States.

This chapter ponders the source and fate of our waste products. First we discuss how waste accumulates in the soil and how it is transported by surface and groundwater and by air (7.1). We then divide waste products into five arbitrary and somewhat overlapping categories. The first is bulk, including household garbage, junk cars, and debris from demolished buildings (7.2). Hazardous chemicals are some of the compounds manufactured by our industrial society (7.3). Because they do not occur natu-

rally on the earth, people, other animals, and plants have never adjusted to their presence, and many are toxic to one or more types of life. A second type of manufactured hazard is radioactive waste (7.4). Biological waste is natural but, because of the increasing population density of people and animals, has reached concentrations that are dangerous to the ecosystem (7.5). Because it is manufactured for the purpose of enhancing biologic activity, we include fertilizer in the discussion of biological waste. The final section is on air pollution, principally smog and acid rain (7.6). Each section contains suggestions for the disposal and management of the waste discussed in it.

All of the topics in this chapter refer to local, rather than globally distributed, waste. We defer to Chapter 8 the discussion of the global effects of human activity including climate warming, ozone depletion, decrease in biodiversity, and worldwide lead pollution. This separation of topics leaves this chapter with fairly small controversies, although some of them are intense, and places the most contentious ones in the next chapter.

The problems discussed in this chapter also, to some extent, have solutions. We can reduce waste, we can render it harmless before it is discharged, and if we cannot destroy it, we can put it in a safe place where it will not bother us. Where is that safe place? Well, somewhere else. Some place where we can't see it. Certainly NIMBY – not in my back yard. Probably not in yours either. The solution: we'll put it in *their* back yard. And, as we will see, that is exactly what we do as long as *they* do not have the political, economic, and sometimes the military power to stop us.

7.1 DISPERSAL OF WASTE IN SOIL, SURFACE WATER, GROUNDWATER, AND AIR

Waste that is not properly managed, perhaps sewage or the smoke from a factory, can enter the atmosphere, surface water, and groundwater (Fig. 7.1). Some waste is deliberately discharged in the mistaken impression that the amounts are so small that they will have no ecological consequences, and in many cases that is true. The water released by burning gas and oil has no effect on the atmosphere, and the only importance of the CO_2 from fossil fuels is to make a contribution to global warming, which we discuss in Section 8.2. In rivers, waste that is highly soluble becomes so dilute that it simply washes downstream, perhaps ultimately into the oceans, without bothering anything. Water from sewers is commonly so purified in sewage plants that it can be discharged and reused further downstream.

Other emissions, however, range from annoying to deadly. Burning coal releases ash, which, if not properly contained, settles as a white powder over the surrounding countryside. Coal and oil also release unburned hydro-carbons, CO, SO_2 and various oxides of nitrogen that are heavier than air and either collect in place or move along the ground downwind from their source. Waste that is insoluble or only slightly soluble in water may float on top of a river if it is lighter than water or sink to the bottom and accumulate as a sludge if it is denser. Dense sludge also accumulates in oceans where contaminated rivers enter them or where sewers from coastal cities have their outlets on the shallow sea floor.

The most common method for the escape of waste is by seepage from above-ground storage areas or from improperly designed shallow landfills, which are almost invariably above the water table (Section 4.4). Some of the waste from these pits percolates only short distances into the surrounding soil and is absorbed into clays or onto the surfaces of other soil particles. Molecules that are soluble in water, however, are usually dissolved by water moving downward and, once below the water table, are dispersed widely in the predominant direction of flow. This groundwater may ultimately discharge into a river and contribute to contamination in both the river and its area of discharge into the ocean.

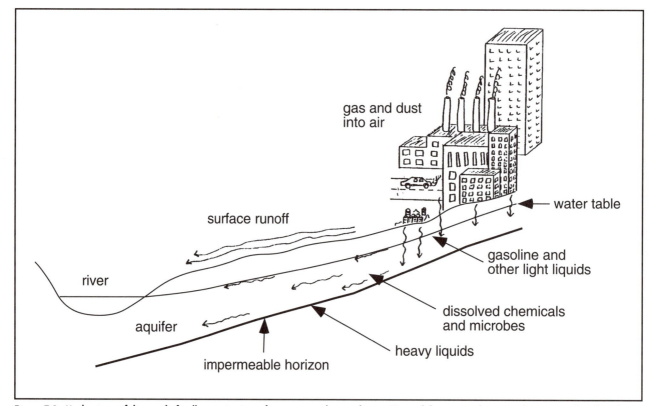

Figure 7.1 Mechanisms of dispersal of pollutants in air, surface water, and groundwater. Gas and dust enters the atmosphere. Some liquids and solids enter streams and rivers directly. Liquids discharged into the ground either float on the water table if they are insoluble and have a lower density than water, dissolve in the groundwater and are dispersed, or sink to the bottom of the aquifer if they are insoluble and have a high density.

Before the age of modern environmental concern, movement of waste products was the principal way of getting rid of them. Bacteria seeping from outhouses and pig farms would move through the ground, where they would be largely destroyed by soil microbes, or would be oxidized as they flowed downstream in surface water. One rule of thumb is that water is safe to drink if it is more than 10 miles from a source of bacterial contamination. In our industrial society, however, this rule is no longer valid. Some of our manufactured chemicals cannot be broken down by microbes or oxygen, and they are toxic at concentrations far smaller than any waste of a preindustrialization society. We can no longer rely on the maxim "The solution to pollution is dilution." It is in this spirit that we investigate specific waste products and possible methods of handling them.

FURTHER READING: Blackman (1993); OECD (annual a, annual b).

7.2 BULK WASTE

No one needs more than one camel can carry. – Geologist from North Africa comparing the simple life led by his ancestors with his life led amid a collection of "stuff" (his term)

Archaeologists digging into the floors of houses in ancient Troy find successive layers of clay containing animal bones, small artifacts, and other trash. Apparently the occupants of the houses let this junk accumulate on their floors until finally they couldn't stand it any longer and solved their waste problem by bringing in a layer of fresh clay and spreading it thickly enough to form a nice clean floor. After doing this a few times, they had to crawl through their doorways and finally resolved the problem by raising the roof and cutting the door entrance higher. This process repeated throughout the city caused the mound on which Troy was built to grow upward about 3 to 5 feet per century, approximately the estimated rate of accumulation in other ancient cities.

The waste generated by the Trojans and other ancient people was confined largely to their cities and immediate surroundings. The countryside was so unpopulated that the little trash produced by people who wandered through it could be discarded with no significant effect. This does not mean that ancient people were sensitive to environmental problems – in Section 8.3 we show that they brought numerous animal species to extinction – but merely that their trash is noticeable only where they congregated together in cities.

The difference in waste accumulation between cities and rural areas continued as productivity and population began to grow. A farm family in the Middle Ages produced virtually no waste. On those rare occasions when animals were killed, the meat was eaten, the fat was processed into soap and candle wax, the hide was used for clothing, and the skin was occasionally converted into parchment for writing. Inedible parts of vegetables were composted, and any fibers woven into clothing were used until the garments fell apart. Most families owned almost no metal, which was reserved for weapons and jewelry for the rich, and when wooden furniture could no longer be repaired it was simply broken up for kindling.

Urban families had a somewhat greater problem. They tended to discard the inedible parts of carcasses and vegetables and to throw out rusted or decayed metal and hopelessly broken furniture. As wealth increased, old clothing could not only be discarded but the fibers in it could be converted into paper. This production of paper encouraged the invention of the printing press by J. Gutenberg about 1440, which in turn encouraged the production of more paper, and then the mostly urban citizens who could read began throwing away paper. The situation was serious enough that in the late Middle Ages cities began sending carts around to collect trash and haul it out to the formerly verdant countryside.

When the population was still relatively small before the industrial revolution, the small amount of waste could be spread into the countryside without serious effects. When productivity increased and the population began to grow rapidly in the late 1700s, however, the problem of trash quickly became a crisis. Wood, metal, and glass containers began to accumulate. More horse-drawn wagons appeared, and although the 20 pounds of manure that each horse produced per day could be used as fertilizer in the countryside, it first had to be carted out of the cities. The ash that accumulated when coal became a more prominent energy source in the 1800s – approximately 1,200 pounds per U.S. household each year by the end of the century – had to be disposed of, commonly by dumping in the poor sections of the city. Food that decayed because of lack of refrigeration had to be dumped somewhere.

In an age of electricity generated by power plants outside of the cities, of automobiles instead of horse carts, of household items that are cheaper to replace than to fix, we have other problems. We use fewer glass bottles and very little wood for packaging, but we produce more than 100 metal cans for each OECD person per year. And plastic seems to be everywhere! Each year the United

States produces about 200 pounds of plastic per person – bottles, fibers for clothing, wrappings, insulation, even little bubbles that have to be punched open before we can take a medicine pill. And paper! The great cyberspace revolution that was supposed to replace paper with electronic communication and storage actually produced more than 100 million printing presses as everyone with a computer turned out "hard copy." We add that paper to the cardboard boxes, junk mail, newspapers, and magazines.

This waste is generated by populations that are growing and in which the percentage that live in cities is increasing even more rapidly (Section 1.3). Each year OECD countries now produce about 2 tons of industrial waste for each person in them. This waste includes not only the residues of factories but also the factories' products that have outlived their usability. Worldwide about 2 million cars are junked each year, and although some of them can be stripped for replacement parts for other cars, many of them have to be converted into scrap metal and a few waste plastics. Automobile tires are another large waste item, roughly 250 million in the United States each year, a figure roughly equal to the population. Some of this junk can be recycled within the OECD, but much is now sent abroad for recycling (more in Section 7.3).

In addition to industrial waste, OECD countries annually produce about 0.5 ton of municipal waste per person. About three-fourths of the municipal waste consists of relatively small items, including 40% paper and 10% each of plastic, metal, and glass, and a small amount of food waste and other organic material. The other fourth consists of larger debris from construction and demolition of buildings, such as wood, bricks, and concrete; most of this debris is harmless, but some people regard part of it as potentially hazardous (see the discussion of asbestos in Box 7.1). If we do not want our cities to turn into mounds growing upward like Troy, we have to put this waste somewhere. Basically, there are three options: (1) put it in a landfill and leave it there; (2) burn it at a high enough temperature so that most of the organic and hazardous material is broken down and then put the ash residue in a landfill; (3) use some mixture of the first two options while reducing the volume of waste by recycling and/or modifying society so that we produce less.

Landfills built prior to about 50 years ago consisted mostly of junk that was thrown on a low spot in the ground, which in many cases consisted of a swamp (Box 7.2). Many of these fills have been leaking since they were first used, putting chemicals (some of them toxic) into groundwater and in a swamp almost directly into surface water. That hazard plus the fact that many of these old landfills are getting full requires construction of new landfills, and in most OECD countries they are better designed. Figure 7.2 shows a modern landfill with a presumably impervious plastic liner covered by a bed of absorbent clay. After trash is dumped into the fill, it is covered by dirt, and then dirt and trash are interlayered until the landfill can hold no more (Fig. 7.3).

The sample landfill shown in Figure 7.2 looks simple but encounters two problems. One is the sheer volume of material that must be stored. Already states in the crowded northeastern United States are sending half or more of their waste to dumps further west, in some cases more than two-thirds of the way across the country. The second problem is that most landfills are unsightly nuisances, and some are hazardous. Any organic material, such as waste food, in the lower levels of a dump undergoes anaerobic (without oxygen) decay by microbes. Some of the products are acids that attack the plastic lining of the landfill and cause leakage into the groundwater. The decay is sufficiently intense in large dumps that temperatures reach 60°C, and the dump generates a large amount of methane. The methane has no odor, but it carries associated decay products that smell bad enough that the neighbors want the dump put somewhere else. For this reason, truly modern garbage dumps are now being designed with caps that funnel the methane into pipes where it can be burned, either simply to get rid of it or, in the larger dumps, to be used as a fuel.

Burning trash has both benefits and drawbacks. The good news is that burning gets rid of virtually all of paper and other wood products and the more than 75% of food waste that is combustible. The ash that must be stored, therefore, has only about half of the volume of the unburned waste and virtually none of the odor. The bad news is that sometimes we try to burn tires, plastics, and other things that produce both a terrible smell and also small amounts of dioxins and other toxic chemicals. The solution to this problem is the same as it is with landfills. Put the incinerator upwind from someone else, preferably someone who is poor and powerless.

Faced with these more-or-less unsolvable problems, people are turning to a third option – reduce the amount of trash by recycling. Recycling policies are almost always set by local governments and, consequently, vary enormously throughout the OECD. For this reason, we simply list the policies in effect in 1996 in Orange County,

BOX 7.1 THE GREAT ASBESTOS PANIC _____

Asbestos is a commercial term, not a mineralogical one. It refers to several hydrous (water-bearing) silicates that occur in nature as long fibers, the most important of which are chrysotile, a type of serpentine, and several varieties of amphibole. All of them are nonflammable, chemically inert, resistant to heat, easy to manipulate, and comparatively inexpensive. When these properties were fully realized in the 1930s, asbestos became the miracle industrial fiber. It was used to reduce the risk of fire in construction materials, including roof shingles, wallboards, and the much-advertised stage curtains in theaters. It was sprayed on surfaces to render them fireproof and to act as an insulator. Asbestos was incorporated in cement and synthetic tiles to make them tougher. It was great.

By the 1960s, when industrial societies were beginning to be more safety conscious, studies showed that people needed to take better care of their lungs. Lung cancer and respiratory diseases were clearly induced by inhaling large amounts of dust while working in coal mines, textile mills, cement factories, or hundreds of other occupations. These investigations led to very necessary efforts to reduce airborne particulates in workplaces and to require masks and breathing apparatus for people in particularly hazardous situations. During the investigations, it was discovered that asbestos miners and millworkers had as high an incidence of lung problems as workers who inhaled other types of dust, and consequently asbestos was placed on lists of dangerous inhalants.

Unfortunately, instead of simply requiring safety precautions for asbestos workers, people began to worry about the asbestos that was already in the buildings they inhabited. At this point panic set in. It apparently began in Europe and spread rapidly to the United States, where Congress passed the Asbestos Hazard Emergency Response Act (AHERA) in 1986. This act was stringently interpreted by a public whose fears were fanned by groups of people including politicians, the media, "environmentalists," and school boards. Additional fanning was provided by entrepreneurs who expanded existing companies or established new ones that promised to rid the world of asbestos

(naturally at a modest cost). In the 10 years that have elapsed since the AHERA, somewhat more than $100 billion have been spent to remove asbestos from public buildings.

Unfortunately, few people seem to have bothered to collect scientific information before spending all this money. That information would have showed:

- Chrysotile, which accounts for more than 95% of all asbestos used, has never been demonstrated to have any biological effect unless it is inhaled in very large amounts by asbestos workers. One study that demonstrated this fact showed that residents of the town of Asbestos, Quebec, where asbestos is mined, have the same rate of lung cancer as people living elsewhere in Quebec unless they work in the mines and mills.
- Amphibole asbestos, which is a small amount of asbestos used commercially, is about as dangerous as other airborne dusts, and there is no reason to apply more strict safeguards to it than to other particles.
- Asbestos already in buildings rarely makes its way into the air that people breathe. Some studies showed that the concentration of asbestos in air in schoolrooms with asbestos in the insulation was actually lower than in normal outside air.
- Some of the materials that we have used to replace asbestos may, in the long run, prove to be more dangerous than the asbestos.

Why have we told this little story? To point out that industrial societies did need to take steps to protect some people from asbestos but that they have spent more than $100 billion, perhaps several times that amount, removing asbestos that did not need to be removed. That's a lot of money. Could we have used it for something useful, like education, research, and environmental cleanup. Nahh. It's more fun to panic.

FURTHER READING: Mossman et al. (1990); Guthrie (1992); Ross (1995).

BOX 7.2 FRESH KILLS ISN'T SO FRESH ANYMORE

When the Dutch colonized Nieuw Amsterdam, they gave the word "kill" to small streams in the surrounding area, particularly where they entered the sea. When New Amsterdam became New York, the new owners continued to refer to streams as kills, and ultimately they said that wonderful, sparkling streams and a bay on nearby Staten Island were so pure that they should be called Fresh Kills. All of this was some 300 years before New York, which now included the borough of Staten Island, began reaching toward its present population of 7 million, not to mention millions more in the adjacent parts of what is now one of the world's largest metropolitan districts.

About 50 years ago, the city of New York was looking for a solution to its mounting problem of trash and spotted Fresh Kills. As a result, it began dumping garbage in the marsh and bay, feeling relieved that it could put its trash in almost the only unoccupied land in the city limits. A fleet of 103 barges now reach Staten Island from other parts of the city, load the garbage onto waiting trucks, and instantly return for more. New York City produces 13,000 tons of garbage each day, requiring round-the-clock filling of Fresh Kills. The dump now covers nearly 5 square miles and is approaching a height of 500 feet, making it the highest point along the entire eastern seaboard of the United States and a navigational hazard for planes landing and taking off at nearby Newark Airport.

Fresh Kills has become so large and rancid that the neighbors are beginning to complain. They point out that decay of garbage in Fresh Kills produces 5.7% of all the methane released in the entire United States, that the methane is a fire hazard (landfills have caught fire elsewhere), and that the methane carries with it volatile organic compounds that both smell and make people sick. They feel that the policy of spraying pine oil deodorant on the dump is not a long-term solution. They are also concerned about the leakage of toxic chemicals from the landfill and also the sheer volume of garbage. Will a landslide suddenly bury some of the neighbors? They want something done, and they are following the standard American policy for getting it done. They are suing the city of New York to stop using Fresh Kills.

There is good news! The city has promised to stop. Other than a mandatory recycling program that will reduce some of the volume of garbage, however, the city hasn't decided what to do with the 13,000 tons of daily garbage. Perhaps they will . . .

AIEE! Heads up!

FURTHER READING: Rathje and Murphy (1992).

North Carolina. The following items can be placed at the curb to be picked up by trucks:

Aluminum cans, but not foil and cooking trays
Metal cans except for those that contained flammable or toxic chemicals
Glass bottles, but not light bulbs, window glass, or mirrors
Most plastic bottles, but not plastic bags (which can be recycled at groceries); the types of plastics that are recyclable are very specific (Table 7.1)
Newspapers
Glossy magazines provided they are separated from other paper

In addition to the curbside pickup, recycling centers at strategic places will accept all of the items just listed plus:

Corrugated cardboard, provided it is not coated with wax and is separated from other trash
Milk and juice cartons and boxes
Aluminum foil and pans

In an effort to reduce the volume of waste and keep hazardous material out of storage, Orange County operates two exchange programs. On the first Saturday of each month the landfill has a special collection of such household items as pesticides, fluorescent light bulbs, batteries, motor oil and fuels, solvents, paints, cleaners, and aerosol cans. The paint collection works partly on an exchange basis, with some that is still usable picked up by other people who may need only a small amount to finish a job. The second exchange program is a salvage shed for such items as old furniture, toys, and books that someone may want.

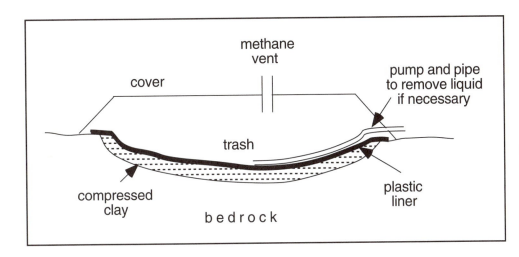

Figure 7.2 Diagram of sanitary landfill. The clay and plastic form a double barrier to leakage, and the cap controls the release of methane (CH_4) and other gases.

The most effective method for further reduction of waste volume is for people to buy less that has to be thrown away. Orange County refers to this as "precycling" and encourages the following policies (from a pamphlet distributed in 1996 by the Orange County Landfill Authority, Hillsborough, North Carolina):

- Buy products that have the least packaging – for example, avoid large plastic bubble packages that contain only a small item.
- Buy only products packed in recyclable materials.
- Buy reusable items rather than ones that must be thrown away; examples are sponges instead of paper towels, rechargeable batteries, and towels made of cloth instead of paper.
- Take your own bags and containers to grocery stores instead of getting new paper or plastic bags to carry food home in (a normal practice in all countries except the United States).
- Repair items such as clocks and radios instead of throwing them away and buying new ones.
- Put vegetable and fruit waste in a compost pile to be used as fertilizer in your garden.

Figure 7.3 Sanitary landfill. The truck at the right is bringing new trash, and the grader at the left is smoothing it into layers with dirt. Courtesy Chapel Hill, North Carolina, Department of Solid Waste Management.

Table 7.1 Types of household plastic

Number	Chemical name	Examples of use	Use after recycling
1	Polyethylene terephthalate	Soda and drink bottles	Carpeting, fiberfill, clothing, new bottles
2	High-density polyethylene	Bottles for milk, detergent, bleach	Plastic lumber, drain pipes, toys, new bottles
3	Polyvinyl chloride	Bottles for vegetable oil	Floor mats, pipes, hoses
4	Low-density polyethylene	Plastic bags	New plastic bags
5	Polypropylene	Bottles for syrup, ketchup, sauces	Paint buckets, fast food trays, cases
6	Polystyrene	Food containers, packing materials	Toys, trash cans, flower pots
7	Various	Detergents, squeezable bottles	Plastic lumber

Policy questions

What can you, your business, and your community do to reduce the amount of waste that must be taken to a landfill, and are any of these efforts being made?

FURTHER READING: Rathje and Murphy (1992).

7.3 HAZARDOUS CHEMICALS

"When I use a word," Humpty Dumpty said . . . , "it means just what I choose it to mean, neither more nor less."

"The question is," said Alice, "whether you can make words mean so many different things."

"The question is," said Humpty-Dumpty, "which is to be master – that's all."
– *Lewis Carroll.* Through the Looking Glass

The alarm goes off. You turn on the lights. Push back the covers. Get out of bed. Shower. Walk through the house to the kitchen. Fix juice, coffee, toast and jelly, possibly a bowl of cereal with milk. Wash the dishes (and maybe last night's?). Read the newspaper. Leave the house and lock the door behind you. Settle into the car, being sure to fasten your seat belt. Start the ignition. Put your foot on the gas. The morning has just begun, and you haven't even gotten out of the driveway, but already you have sent a stream of toxic chemicals into the earth's water, air, and soil. Let's start this section by counting some of them.

The electricity in your house reaches it through wires and transformers that are insulated by synthetic covers on the wires and commonly by organic liquids in the transformer boxes. The mattress in the bed is probably filled with some light synthetic material that required or-

ganic solvents when it was manufactured, solvents when it is cleaned, and an organic gas (foaming agent) to put air holes in it and make it soft. The sheets and blanket required similar solvents when they were manufactured and detergents to keep them clean. You used other detergents in the shower, walked over a synthetic rug (more solvents) laid down on a wood floor covered by shellac or other preservative, and looked at walls that had been painted (dyes and more solvents). The bread and cereal were probably produced from land that had been fertilized and sprayed with pesticides. The bread is in a plastic wrapping, and juice and milk are in plastic containers or possibly coated cardboard (paper products that required bleaches and solvents during their manufacture). The newspaper required the same chemicals and the ink even more. You probably used different detergents for the dishes than for your shower. The metal in your door lock was manufactured by processes that required acids, and is probably coated by a light organic film to prevent discoloration. The metals in your car and the upholstery inside all used more chemicals, and the battery that ignites the motor contains lead and acid, both of which frequently become part of our waste dumps. Had enough?

We have reviewed this use of chemicals to demonstrate that industrial society functions by using an enormous array of chemical compounds that exist on the earth only because people produce them. Most of this product is chemically inert, such as cement and steel, and causes problems only when we have to throw it away (Section 7.1). Some of the other compounds, such as medicines, are not dangerous unless they are used incorrectly, but many chemicals are dangerous both while they are used and after they are discharged. These chemicals differ in their stability when they enter the air, water, and soil. Some decompose rapidly and are dangerous only immediately after they are released, such as the poisonous gas at Bhopal, India (discussed later), whereas other chemicals are stable and persist for years, even decades, such as most of the pesticides that have now been banned in industrial societies (also discussed later).

Regardless of its stability or method of discharge, any dangerous chemical is potentially a "hazard" to some or all of the earth's ecosystem. The problem with a simple definition of hazards is that some chemicals affect all species, some affect only a few, some are dangerous only in high concentration or after prolonged exposure, some cause minor irritation such as itching, and some may kill. For some chemicals the "toxicity" – exposure necessary to cause an effect – depends largely on the concentration of that chemical during a brief exposure. An example is sulfuric acid (H_2SO_4), which is highly dangerous in concentrated form but causes little or no damage to people when they are exposed to acid rain containing up to 0.01% sulfuric acid (the acid rain has serious environmental consequences, however, and we discuss it in Section 7.6). Conversely, serious illness and/or death may result from long-term exposure to concentrations of dissolved lead that are far too low to have an instantaneous effect.

Because of these complications, we leave the precise effects of various hazardous chemicals to other books and simply provide examples of chemicals that have significantly affected people and the environment with which they most closely interact. We use two categories: (1) industrial chemicals that are discharged accidentally into the environment or as designated waste; and (2) pesticides and herbicides that are deliberately distributed into the environment. At the end of the section, we summarize methods for safe handling of the waste.

Industrial chemicals

Industrial chemicals include the half-filled cans of dried paint that sit in your garage, the stream of liquids discharged as residue from the manufacture of plastics, the oil drained from your car during an oil change, the waste solvents from the paper industry, and on and on. Almost all investigations of discharged chemicals use different classifications, many of them very elaborate. In order to give you a break, we use only five very general categories, many of which overlap.

ORGANIC SOLVENTS This category includes volatile hydrocarbons (Section 4.5) and accounts for a high percentage, and in some classifications the majority, of all of the toxic waste generated by industry. They are used in the manufacture of plastics, textiles, wood and paper

products, pesticides, pharmaceuticals, paint – in short, almost everything that we use. In addition to manufacturing, organic solvents are also necessary for cleaning most of the finished products before they are sold and for keeping them clean afterward ("dry" cleaning, for example, is dry only in the sense that clothes are cleaned with an organic solvent instead of water). Some of the solvents, like alcohol and acetone, are soluble in water and are dispersed without significant environmental effect. Others, such as benzene, are virtually insoluble, and have effects ranging from giving a bad taste to water in very small concentrations to causing serious health effects at higher concentrations.

WASTE OIL In most classifications, waste oil (which includes some volatile hydrocarbons) is second only to organic solvents in total amount of discharge. Some of the oil is discharged from industrial operations where it has been used as a lubricant or a fuel. Perhaps more is gasoline that leaks from the storage tanks of filling stations, although data on this problem are hopelessly inadequate even in closely monitored countries (see Box 7.3 for an extreme example of discharged gasoline). In addition, both fuel and lubricating oil leak onto roads from cars, into waterways from boats, and seep from junked cars and motor-driven machinery.

Because all of this oil and fuel is hydrocarbon, none of it is soluble in water, and all of it is less dense than water. Therefore, it concentrates in, or on the top of, specific horizons of groundwater and may accumulate on the surface of lakes and rivers. In small concentrations it causes the same bad taste and, ultimately, health effects as waste solvents. On water surfaces it poses another danger – the possibility of catching on fire. (Yes, river and lake surfaces have been known to burn!) Because of their toxicity and pervasiveness, oils are included among the chemicals measured at almost all stations monitoring water quality, even though other chemicals may be ignored.

POLYCHLORINATED BIPHENYLS (PCBS) PCBs are a special group of chlorinated hydrocarbons that have a high boiling point and are stable at high temperatures (Fig. 7.4). They were originally developed as heat-transfer liquids for electrical transformers and heat exchangers but were later used as lubricants, hydraulic fluids, and in numerous other applications. Because PCBs are stable and insoluble in water, those that leak from equipment and landfills or are accidentally discharged during manufacturing can accumulate in soil and water to dangerous concentrations. There they pass into the food chain, where they are concentrated in the fatty tissues of fish and animals and ultimately may be eaten by people.

BOX 7.3 COFFEE, TEA, A LITTLE RUSSIAN GASOLINE?

They do not speak publicly about it in eastern Europe. Virtually nothing is put in writing. But if you talk to people, you can find out a little about what the Russian armies did with their gasoline when they pulled out of the camps that they had occupied since the end of World War II.

The disintegration of the Soviet empire began in the mid 1980s, was most memorably marked when the Berlin Wall was torn down on November 9, 1989, and finally led to independence both for countries within the Soviet Union and also for those nominally independent countries in eastern Europe that had been run by puppet governments backed up by the Russian army. Even after independence, the Russian armies stayed on in the countries that they used to control. They stayed because there was no place for them to go. In the chaos that was Russia, there was no housing, little food, and nothing for the soldiers to do. So they waited and only slowly began to move back to Russia itself. When they moved, they took trucks and other equipment that could still move and abandoned much that couldn't.

Because some of the Russian vehicles were too decrepit to move, and because the Russian army had been storing gasoline in their camps when they controlled the occupied countries, the Russians did not need all of the gasoline and could not carry the remainder back to Russia. That left them with two choices. They could leave the gasoline for the new governments of the independent countries or they could get rid of it. In all too many places — no one is sure how many — they got rid of it. They opened the valves on their storage tanks and poured the gasoline on the ground.

That may not seem too serious, but consider what happens when gasoline is poured on the ground. It filters down through the open spaces in the soil, ultimately reaching the water table (Section 4.4). Then because gasoline is less dense than water, it spreads out on the top of the water table and begins to disperse in whatever direction the water is flowing. When water is withdrawn from the well, the gasoline at the top comes out first and keeps filtering into the well as water continues to be withdrawn. How serious is this?

We can make an approximate estimate. The many different hydrocarbons in gasoline (Section 5.1) have different acceptable concentrations in water for human use, but generally they should not exceed about 5 parts per billion (ppb; 5×10^{-9}). This means that 1 gallon of gasoline can render 100,000 (10^5) gallons of groundwater unusable. Using the fact that 1 cubic foot = 7.5 gallons and assuming that the gasoline, because of its low density, will contaminate only the top foot of water below the water table, we then calculate that 1 gallon of gasoline will contaminate an area of slightly more than 250,000 square feet. Thus, a spill of 10,000 gallons (only 250 barrels) will contaminate 100 square miles. Depending on the amount of gasoline dumped, serious contamination may extend through a large part of the area formerly occupied by Russia.

Regardless of the exact numbers, which are still unknown, the gasoline is a serious threat to a region that has already suffered from environmental degradation under the Russian regime. It probably can't be cleaned up either. Just about the only feasible method for removing gasoline from groundwater is to pump the water out, spray it in the air, and let the gasoline evaporate. Estimates of cost for cleanup of gasoline-contaminated soils around leaking tanks in U.S. filling stations run about $100,000 per station, which occupies about one-eighth of an acre. We leave you to estimate the cost of cleaning up 100 square miles and the likelihood that the newly independent countries of eastern Europe can afford it. No, we don't think they can either.

Few deaths have been attributed to PCB poisoning, but continued exposure to even low levels of contamination causes liver damage, stiffening of joints, and possibly birth defects. The more serious problems arise from the tendency of PCBs to break down if held at high temperatures for long periods of time or particularly if they are involved in a fire. Some of the breakdown products are dioxins (Fig. 7.5), which are so dangerous that they have never been produced for any commercial purpose. As unintended contaminants of PCBs and other oils, however, they have killed people downwind from fires, and in the widely publicized case of the small town of Times Beach, Missouri, dioxin contamination of oil used to settle dust on dirt roads forced the abandonment of a town of nearly 1,000 people.

PAINTS, PIGMENTS, GLUES, AND PRESERVATIVES In the early part of the 20th century, paint consisted largely of pigments dissolved in oil and was used on most outdoor and

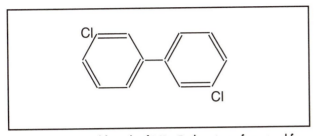

Figure 7.4 Structural formula of PCBs. Explanations of structural formulas of organic compounds are in caption for Figure 2.1.

indoor surfaces that could not be covered with "whitewash," an inexpensive suspension of a white pigment in a base consisting almost entirely of water. Gradually, paint was developed in which pigments were dissolved in various synthetic (essentially plastic) compounds that were soluble in water and could be used on "flat" surfaces that did not need the shine imparted by oil-based paints. These compounds reduced the amount of oil that was left over from painting operations, and were sufficiently harmless that any excess could simply be washed down the drain. Recently, synthetic compounds have been developed that give surfaces the same shine as oil paints, thus further reducing the amount of oil that must be disposed of.

Paint pigments have changed as much as the media in which they are suspended. For example, some lead-based pigments were used to give a high shine to painted surfaces, and some were used as a preservative, such as the "red lead" painted on exposed steel in order to prevent rusting. Because of its severe health effects (Section 8.4), however, a major effort has been made to eliminate pigments that contain lead, and no lead-based paint is now available to the general public anywhere in the industrialized world. Other metals were generally not as dangerous as lead, but virtually all of them have now been replaced with presumably harmless synthetic pigments.

Despite this success with paints, the amount of relatively dangerous organic chemicals used in the wood and metal industry continues to increase. A large part of that increase comes from the wood pulp industry, which uses solvents, glues, and cleaning agents to produce a growing stream of paper, cardboard, and synthetic wood such as plywood. Even ordinary lumber is treated with a variety of chemicals that include natural derivatives such as creosote that are pressed into wood to prevent decay, synthetic products used to coat wood to prevent water damage, and even more to repair it when it cracks. Metals also are protected from decay by being dipped or sprayed with thin plastic coatings. None of these chemicals are particularly harmful after they are applied and hardened, but they are toxic when they are produced and applied.

RESIDUE FROM THE METALS INDUSTRY The metals industry generates hazardous waste at all stages, from mining to installation of the finished product. Separation of metals from their ores (smelting; see Section 6.4) commonly generates waste acids. Some of these acids are residues from those added deliberately to break down ore minerals. Even larger amounts of waste acid are formed when ore sulfides are brought to the surface by mining, left in waste dumps at the mine site, and then oxidized to sulfuric acid that leaks into the surroundings.

More waste is generated when the separated metals are fabricated for further use. Metals in everyday use, such as brass (an alloy of copper and zinc) have commonly been cleaned with acids and organic solvents, polished with abrasives, and sprayed with more organic chemicals to give them a surface that does not corrode. All of these chemicals must usually be discarded, and finally, more waste is generated when metals have been used and thrown away. In waste dumps, the metals corrode, forming more acid and releasing organic chemicals attached to their surfaces, all of which are added to the ever-widening pool of toxic chemicals from other sources.

Pesticides and herbicides

The best-known pesticide, which is no longer used, is dichloro-diphenyl-trichloroethane, more commonly known as DDT (Fig. 7.6). DDT is a white powder that was first synthesized in the middle 1800s, but its ability to kill insects was not fully realized until the 1930s. That recognition turned out to be a stroke of extraordinary good fortune for the Allied armies in World War II. Because it is not absorbed through the skin, DDT is safe to sprinkle on soldiers to kill lice, which transmit typhus when they bite people. It also kills mosquitoes and could be dusted on ponds and swamps to reduce the incidence

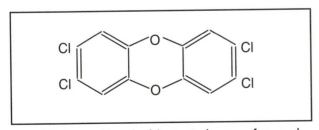

Figure 7.5 Structural formula of dioxins. Explanations of structural formulas of organic compounds are in caption for Figure 2.1.

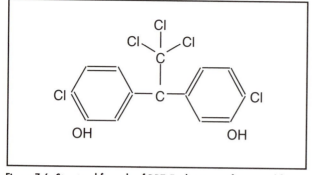

Figure 7.6 Structural formula of DDT. Explanations of structural formulas of organic compounds are in caption for Figure 2.1.

of malaria. Without DDT, tens of thousands of Allied soldiers might have died or become permanently disabled by insect-borne illnesses.

Because of the success of DDT, much of the world saw pesticides as a solution to the problem of maintaining an adequate food supply for a population that began to increase rapidly at the end of the war. It was a logical conclusion based on observations that insect pests destroyed approximately 30% of the agricultural harvest in the United States in 1950, somewhat less in Europe, and considerably more in most of the world. During the succeeding half century, we developed new and better pesticides and distributed them widely over the earth. When one was banned in one or more countries, we developed another pesticide and used it until it was banned. We tell this strange story in slightly more detail here and ask the reader to measure its success by the statistics that, in the 1990s, insect pests destroyed approximately 30% of the agricultural harvest in the United States, somewhat less in Europe and other now-industrialized areas, and considerably more in most of the world.

The first major pesticides to be used were chlorinated hydrocarbons similar to DDT. By interfering with the metabolic and reproductive processes, chlorinated hydrocarbons kill not only insects dangerous to humans but also locusts, aphids, and other species that eat food crops. Unfortunately, shortly after World War II, it became apparent that these insecticides were too effective, not only killing the targeted insects but also wiping out whole insect populations that formed the base of the food chain for fish, birds, and the larger animals that lived on them. Also, because they do not break down after they are distributed, they washed out of soils and plant leaves, were carried downstream by rivers, and finally accumulated in high concentrations along river floodplains and in swamps. There they began to affect the metabolism and

reproduction of animals other than insects. Although they had virtually no effect on humans and most other mammals, which protect their embryos inside a womb as they develop, the unprotected eggs of birds, fish, reptiles, and frogs were seriously affected by the chlorine-bearing chemicals. Eggshells became thin, and offspring were born deformed or not born at all.

In 1962 this assault on the earth's ecosystem was brought powerfully to the world's attention by Rachel Carson's masterpiece *Silent Spring*, which takes its title from the proposition that the world will be so polluted that there will be no birds to sing when spring arrives. This book and the concerns of many other scientists led rapidly to the banning of most chlorinated hydrocarbons and to efforts to find alternatives. This ideal insecticide would affect only a targeted insect (which would not be eaten by birds or other animals), would decompose almost immediately after use, and would have no effect on people who accidentally spilled it on themselves, breathed it, or even swallowed it. Several possibilities have been tried, almost all of them after extensive testing that showed that they were safe for use.

One alternative was based on organic phosphates (organophosphates). Some varieties of organic phosphates are essential to all organisms. They include DNA, the genetic template that controls all life, and adenosine triphosphate (ATP), which is responsible for the transfer of energy through both plants and animals. Other varieties, however, are poisonous to one or more types of plants and animals. This poisonous ability had been known since early in the 1900s, and organic phosphates had been investigated for possible military use in both world wars. One of the chemicals studied at this time was sarin (Fig. 7.7), a particularly potent variety of nerve gas, which can painfully destroy the functioning of the ner-

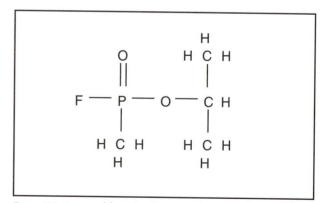

Figure 7.7 Structural formula of sarin. Explanations of structural formulas of organic compounds are in caption for Figure 2.1.

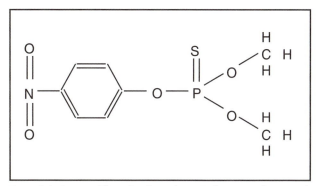

Figure 7.8 Structural formula of parathion. Explanations of structural formulas of organic compounds are in caption for Figure 2.1.

vous systems of people and other animals. Some measure of the dangers of all organophosphates is the shaking and paralysis that affected nearly 10,000 people exposed in the United States in the 1930s, and similar effects in nearly as many people who used accidentally contaminated cooking oil in Morocco in 1959.

No one (we hope) ever considered using nerve gas as a pesticide, but when it became apparent that virtually all organic phosphates as well as most chlorinated compounds were too dangerous to use, attention shifted to other chemicals. Organic phosphates with sulfur (thiophosphates) were tried in an attempt to restrict the effects of the pesticide to very specific species. The first major compound was parathion (Fig. 7.8), whose use resulted in several thousand deaths worldwide, and it is now banned in almost all industrial countries.

At this point, generally in the 1960s, several other groups of chemicals were tried. One class of compounds is cyclodienes (Fig. 7.9), used mostly to control soil insects such as termites. A compound closely related to cyclodienes is Mirex, which was hailed a solution to the problem of encroachment of fire ants into the southern

United States but was used mostly as a flame-retardant chemical rather than as an insecticide. Unfortunately, the cyclodienes are so stable that they seeped into the basements of houses sprayed for termites, persisted in soil and water, accumulated in animals and fish, and ultimately began to make people ill. Virtually all cyclodienes were banned in the 1970s in most of the industrial world. A second class of compounds is carbamates, derived broadly from carbamic acid (H_2NCOOH; Fig. 7.10), that has been used for insects, snails (mollusks), and other agricultural pests. Because carbamates are very soluble in water, they do not accumulate to high concentrations in swamps and pools, but their high toxicity to humans and all other mammals caused them to be banned by the 1980s in most industrial nations.

Another type of pesticide was based on the long-known observation that insects tend to avoid pyrethrum bushes, which are related to chrysanthemums. They do so because pyrethrums contain a class of chemicals known as "pyrethrins" (Fig. 7.11), which are toxic to most insects. Perhaps, it was reasoned, this naturally occurring insecticide could be used without adverse effects on people or other organisms, and pyrethrins were then mixed with organic liquids and sprayed around the countryside. Unfortunately, the concentration of pyrethrins needed for insecticides also began to kill a variety of animals and fish and to give people skin diseases and other illnesses. Pyrethrins were banned in the 1980s.

Because much of the damage caused by pesticides resulted from their chemical stability, which caused them to remain in water and soil for many years, efforts were soon made to find chemicals that would kill the pests and then break down (degrade) quickly into harmless chemicals. Some of these chemicals are simply unstable in the presence of water and air, but others break down only when they are processed by soil microbes and, hence, are referred to as "biodegradable." One of the most commonly used degradable pesticides is carbaryl and various derivatives (Fig. 7.12), which are sprayed either as a powder or in solution in oil. Carbaryl kills chewing and sucking insects on vegetables and then disappears within two weeks of application. Sounds ideal, but unfortunately there are a few side effects. One is that carbaryl kills not only pests but also beneficial insects such as bees. Another is that most of the birds that happen to be around when carbaryl is applied also die. And, finally, a principal method of manufacturing carbaryl uses the gas methyl isocyanate, whose escape from a plant in Bhopal, India, killed several thousand people and injured nearly 200,000 (more in Box 7.4).

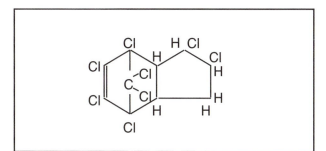

Figure 7.9 Structural formula of chlordane, a type of cyclodiene. Explanations of structural formulas of organic compounds are in caption for Figure 2.1.

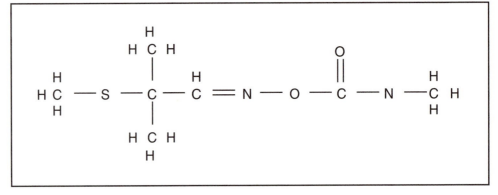

Figure 7.10 Structural formula of aldicarb, a type of carbamate. Explanations of structural formulas of organic compounds are in caption for Figure 2.1.

At the same time that some scientists were searching for better and safer pesticides, others were attempting to find suitable herbicides. Both in agricultural areas that are naturally fertile and in those whose nutrients have been enhanced by artificial fertilizer, plants that some people regard as undesirable (weeds) grow at least as rapidly as food crops. Thus, people looked for a herbicide that kills weeds without affecting the food plant and without having undesirable side effects on humans or other animal life. Some varieties currently in use are chlorinated hydrocarbons, which should be applied very carefully, but others consisting solely of carbon, hydrogen, and oxygen have been developed. The advantage of the latter compounds is that they can be designed to kill plants with which they come in contact but can be broken down immediately when the plant dies and its leaves fall onto the ground. These biodegradable herbicides are decomposed into carbon dioxide and water by microbes in the soil and, therefore, are safe to use as long as you do not spray them onto yourself or anyone else.

Finally, the entire problem of pesticides and herbicides might be avoided by a blend of modern and traditional agricultural techniques. The modern techniques are based on increased understanding of genetics and pest behavior. We now know enough about the actions of individual genes that we can genetically alter plants to be at least somewhat resistant to pests. We know enough about insects that we can release sterile males into an insect population, thus reducing reproduction rates because females that mate with the sterile males produce no offspring. And we can simultaneously release chemicals (pheromones) that enhance the attraction of females to the sterile males.

We also understand some of the traditional farming techniques better. We now know that native plants are more resistant to pests than the commercial varieties of corn, wheat, and other grains that are distributed so widely around the world (Section 2.2). Furthermore, if we do not use pesticides and kill all of the insects, we can fight pest insects with other insects that are harmless to

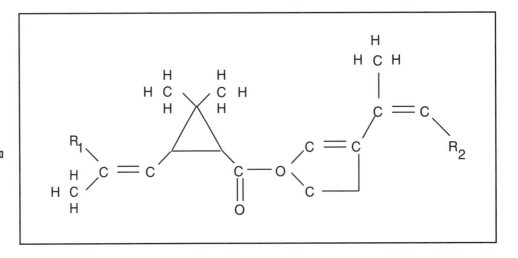

Figure 7.11 Structural formula of pyrethrins. R_1 and R_2 represent various chemical groups. Explanations of structural formulas of organic compounds are in caption for Figure 2.1.

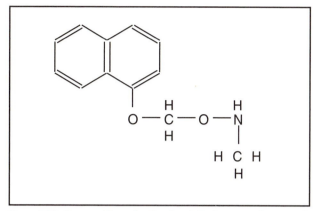

Figure 7.12 Structural formula of carbaryl. Explanations of structural formulas of organic compounds are in caption for Figure 2.1.

plants and people, possibly even importing "friendly" insects for this purpose. The praying mantis, for example, effectively stands guard over tomato plants by eating many of the insects that attack them, ladybugs eat aphids of all types, and certain varieties of wasps that are too small to bother people devour ground-inhabiting insects such as caterpillars and bollworms. We also are more sophisticated in our use of plants to repel insects. By planting strongly aromatic herbs and other plants such as mint and nasturtium among rows of grain and vegetables we generate odors that are very pleasing to people, attractive to insects that pray on other insects, and either repel many of the pest insects or confuse their sense of smell so that they cannot locate the food plants.

Even these modern varieties of traditional agriculture might be avoided if we returned further to the hand labor that dominated agricultural methods of the past. We would not need herbicides if we removed weeds by hoeing, and we would need fewer or no pesticides if we picked leaves off of plants by hand as soon as they showed signs of infestation. The problem with this suggestion is that it runs counter to the basic premise of modern industrial, wealthy societies – that mechanization increases productivity and increases the standard of living for everyone. For example, in order to keep food prices as low as they are, people hired to hoe weeds and pick infested leaves probably could not be paid as much as they could earn working in a factory or a service occupation in a modern city. We could, of course, pay them more, let food prices rise, and justify the extra cost as a necessary expenditure to prevent environmental degradation. Are you willing to do this? We ask the question again at the end of the chapter.

Disposal of hazardous waste

What are we going to do with all this waste that supports a comfortable life-style? The answer to that question depends partly on knowing how much there is, and quite frankly we don't. Numerous local, national, and international organizations attempt the frustrating task of estimating the amount of toxic waste that is discharged each year. In the United States, the Environmental Protection Agency (EPA) publishes an annual Toxic Release Inventory (TRI), which is reasonably accurate except for companies that attempt to minimize, or completely hide, their discharges. The OECD (Section 1.4) estimates that discharges by the world's principal industrial economies amount to several hundred pounds of toxic waste annually for each of its roughly 1 billion inhabitants. The estimate is only an approximation, however, because some countries do not keep adequate records, different countries report their waste in different categories, and chemicals that are classified as toxic in some countries are regarded as harmless in others. And finally, efforts by the United Nations Environmental Programme to estimate the discharges for the entire world are almost hopeless because industries and municipalities in so many poor countries simply discharge their untreated waste wherever they can without bothering to report the types or amounts.

People have used a variety of methods for dealing with this largely unknown quantity of waste (Fig. 7.13).

The simplest method is to ignore it. That is all that we can do for pesticides that have already been sprayed, and environmental cleanup simply means deciding to use less of them in the future. Unfortunately, we also ignore much of the waste that we could do something about. Factories put barrels of dioxin-contaminated PCBs outside the back door, and homeowners let half-empty spray cans of insecticide stay on the cupboard shelf.

Another very simple method is to throw the waste out when no one is looking. The term "midnight dumping" refers to the practice of sending trucks through the countryside in the middle of the night and either throwing trash or spraying toxic chemicals onto the side of the road. Perhaps on a larger scale this is what happened to 16,000 tons of ash from a Philadelphia incinerator that were loaded on a freighter in 1986. Two years later, after being denied entry to numerous ports around the world, it arrived in Singapore. Empty. Well, the ocean is a big place, and not too many people are watching when you throw something overboard at midnight.

One of the most commonly used methods for disposing of toxic waste is putting it in landfills. About 1% of

BOX 7.4 POISON GAS AT BHOPAL, INDIA

Alarm sirens rang at 12:30 A.M. on the morning of December 3, 1984, in Bhopal, India, at a plant owned by Union Carbide. The internal siren was to warn workers at the plant that a toxic gas leak had occurred, and the public siren was to warn people living around the plant. Not realizing the extent of the emergency, the public siren was almost immediately shut off, and it was not until 45 minutes later that city authorities learned that people were fleeing the area and not until 3:00 A.M. that they learned that some people were already dead. Ultimately, it was to be recognized as the most deadly industrial accident in history, claiming thousands of lives, permanently disabling tens of thousands, and leaving perhaps as many as 100,000 people with some aftereffects. Why do we not use more precise numbers? Well, perhaps some of the deaths that night or later were partly caused by other accidents or illness, possibly some of those disabled were really not all that badly injured and some of the disabled were not reported or found, and in a large industrial city in a nonindustrial country people are hard to count anyway.

What happened is slightly more clear than the effects. After operating in India for many years, in 1969 Union Carbide built a plant in Bhopal to manufacture pesticides and herbicides based on carbaryl (see text). The plant was 50.9% owned by Union Carbide in the United States and 49.1% by Indian investors and was placed in Bhopal partly because the Indian government and state and local authorities wanted to provide jobs and stimulate economic growth in the region. For the first 10 years of operation Union Carbide imported the major ingredients for pesticide manufacture and simply combined them at the Bhopal plant, but in 1979 the Indian government overruled objections of authorities in Bhopal and gave Union Carbide permission to manufacture all of the necessary ingredients in Bhopal. By 1984 the Union Carbide plant was storing large amounts of the two key ingredients, alpha naphthol and methyl isocyanate gas (MIC), and making carbaryl by the reaction

$$\text{alpha naphthol} + \text{MIC} \rightarrow \text{carbaryl}$$

which, using formulas, is

$$C_{10}H_9OH + CH_3NCO \rightarrow C_{10}H_7OCH_2ONHCH_3$$

Unfortunately, for reasons that are still somewhat in dispute, in the early morning of December 3, 1984, water leaked into one of the tanks holding the MIC. This caused an immediate reaction that broke down some of the MIC to various other organic compounds and generated large amounts of CO_2 and heat (the reaction was exothermic; see Section 5.2). This combination of hot gases ruptured the control system and poured out into the streets of Bhopal. People who inhaled large quantities of the gas died immediately; their internal organs almost dissolved. People receiving lower doses had serious corrosion of their lungs (causing many later deaths), and many were blinded.

The ensuing financial and legal complications are as sickening in their own way as the gases themselves. They centered around two questions: Who was responsible for the accident, and who should be compensated for it? Union Carbide in America offered an immediate $200 million as a humanitarian gesture but disclaimed responsibility because the plant was operated by an Indian subsidiary and not under direct control of the parent company. The Indian government rejected the offer as insufficient and spent considerably less money to take care of the families of those who had died and those who were too ill to continue working. Hordes of U.S. lawyers descended on Bhopal, attempting to get victims to let them bring suit in U.S. courts, where any compensation awarded would presumably be larger than in an Indian court and, not incidentally, would pay the lawyers for the efforts. As we write 12 years later, neither of the two questions has been satisfactorily answered.

Nor have most of the victims received adequate compensation.

FURTHER READING: Morehouse and Subramaniam (1986).

---

the volume of a municipal garbage dump consists of materials that range from mildly toxic (can of old paint) to very toxic (batteries). If properly designed, these landfills (Section 7.1) can hold toxic chemicals with no or minimal leakage for at least some period of time. Many of the chemicals will decompose into harmless compounds, but

they may require tens or hundreds of years in order to do so. Unfortunately, most good landfills are younger, and most people who have studied them do not expect them to be effective for more than about a century.

Some facilities exist for burning toxic waste. The idea of incinerating ordinary garbage (Section 7.1) can be ap-

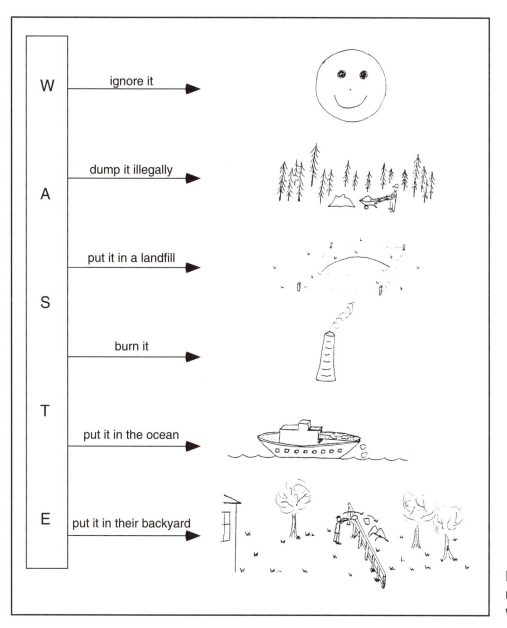

W   ignore it

A   dump it illegally

S   put it in a landfill

T   burn it

   put it in the ocean

E   put it in their backyard

Figure 7.13 Summary of methods for dealing with toxic waste. See discussion in text.

plied to toxic chemicals. Ideally, the waste should be decomposed to carbon dioxide, water, possibly hydrogen chloride if the chemicals contained chlorine, and only a minimum amount of unburnable ash. This much decomposition, however, requires very high temperatures and is consequently very expensive. For this reason most organizations that operate incinerators are content to break down the toxic chemicals just enough that they are no longer regarded as dangerous. That sounds like a good compromise until we remember that dioxins can be formed by burning PCBs. The other problem is that the ash may contain some of the original toxic chemicals or

new ones formed during burning, and then some method for disposing of the ash must be found.

A more recent method for dealing with toxic waste is to put it down deep wells. Wells that penetrate more than 1,000 feet into the ground offer an apparently easy solution to the waste problem. Both liquid waste and slurries of solids can be pumped under pressure to the bottom of a well and hopefully "injected" into the surrounding rock. If the rock is reasonably porous and also sufficiently impermeable to groundwater, then the waste will stay near the bottom of the well. And if it doesn't, well at least it's out of sight.

Some people advocate putting toxic waste in the ocean. Tens of thousands of metal ships now lie on the bottom of the ocean, and we must consider the possibility of adding more metal (Box 7.5) and also toxic waste and maybe radioactive waste (Section 7.4). Burial in drums inserted into sea floor sediments theoretically puts three barriers between the waste and people or even the abundant marine life of the upper part of the ocean. The first barrier is a drum made of materials that will not corrode in seawater or the water in the pore spaces of sea floor muds. It isn't much of a barrier because no one has invented a material that can do this. The second barrier is the surrounding mud, and it is useful because some sea floor sediment really is capable of absorbing both metal ions and toxic organic chemicals. The third barrier is the ocean itself. In some areas of the ocean floor the bottom currents are so slow that waste escaping from the drum and mud has little chance of migrating long distances. Thus, the ocean may be an effective repository for toxic waste, but we are not certain, and the political problem of putting it there will be enormous.

Finally we should discuss the most popular disposal method – send it somewhere else. We described the transport of ordinary garbage from crowded to relatively unpopulated areas in Section 7.2, and because some of that garbage already contains hazardous waste, we might as well consider getting all of the hazardous chemicals out of sight in the same way. As industries in OECD countries became more and more productive following World War II, this option became more attractive. In

---

BOX 7.5 SINKING THE BRENT SPAR

Oil can be produced from the oceans only if we are willing to build platforms to drill the wells, to tend the wells, and, unless each well is attached to a pipeline to shore, to build storage buoys where the recovered oil can be kept until it is picked up by tankers. In order to operate all of this equipment, the platforms must provide space for the technical equipment plus housing, meals, and recreation for the workers. Most of them have helicopter pads as well as facilities for docking boats in order to transport workers to and from the mainland, a frequent occurrence because work shifts for the platform crews usually require several days or a week of intense work alternated with frequent time spent at home with their families.

Like everything else that people build, platforms wear out, and when they do they must be discarded. When the Brent Spar, one of Shell Oil Company's major storage buoys in the British-controlled part of the North Sea, reached this point, the problem was what to do with a structure that has an area slightly larger than an acre, a height of 450 feet, and consists largely of a submerged oil tank weighted down by concrete and moored to the sea floor. In March 1995, the company said that it intended to tow the Brent Spar out into the Atlantic and sink it. The announcement provoked an immediate reaction from Greenpeace, an international organization interested in environmental issues. On April 30, Greenpeace sent boats out to the Brent Spar, where protesters climbed onto the buoy and occupied it until May 23, when they were removed by British police.

Having failed at this effort to stop removal of the buoy, Greenpeace announced that it would organize an international boycott of Shell products if the Brent Spar was dumped in the Atlantic. Greenpeace wanted the platform towed to land and dismantled, and whatever could not be recycled stored there. From this point on, arguments flew back and forth among Shell, Greenpeace, and numerous other organizations and governments. The platform would make an insignificant contribution to the pile of rotting metal from sunken ships already resting on the sea floor – no, the platform contained oil residues and other chemicals that would pollute the oceans. It would be much cheaper to sink the Brent Spar than to dismantle it – fine, Shell is rich. Dismantling the platform was as dangerous as any other construction/demolition activity, and someone might be injured or killed in the process – no, the dangers were insignificant. Storing the platform debris on land would pollute the land – no, the metal could be recycled and all of the rest disposed of properly. While all of this was going on, since something had to be done and the water off the British coast was too shallow for the buoy, Shell had the Brent Spar towed into a temporary storage position in deep water off the coast of Norway.

As we write about 1 year after the original announcement to dump the Brent Spar in the ocean, the upshot of all of these arguments is – but you have already guessed it – the Brent Spar is still off the coast of Norway waiting for something to be done.

FURTHER READING: Knott (1995).

particular, crowded Europe sent waste to Africa, where it was improperly stored, and people living near the waste dumps began getting sick and dying. This placed governments that accepted the waste in the position of deciding between the health of their people and the money that industries were paying them for storage. Because the government, usually a military dictator, almost always took the money for its (or his) own pockets, international organizations began to make feeble attempts to police this "transnational" movement of waste.

The first formalization of this protection was the Basel Convention, signed in Switzerland by the OECD and many poor countries in 1993. The Basel Convention attempted to regulate the dumping of many types of toxic waste but was undercut immediately because there were virtually no penalties for violating it and also because, just before it was signed, the European Union (EU) reclassified much of its hazardous waste as nonhazardous. The EU, closely followed by other OECD countries, now has three categories of waste:

- Red waste includes PCBs, dioxins, and other chemicals that everybody agrees are dangerous, and they are reasonably well controlled.
- Amber waste includes such materials as Pb-acid batteries and most solvents. It can be exported 30 days after the importing country or organization signs a letter of consent, which can probably be obtained if the exporter is willing to pay for it.
- Green waste is material that can technically be recycled. It includes most metals (including Pb), plastics, tires, and some types of ash from power stations. Provided this material is labeled "for recycling," it can be exported without consulting the EU or any other international organization.

When people and organizations objected to export of waste that caused health problems, they met an unexpected opposition. Governments and many entrepreneurs in importing countries had become accustomed to the money they were paid to take the waste, and instead of trying to stop the trade, they wanted more. Now the world's industrial countries are legally pouring scrap iron, lead (including acid batteries), other metals, and discarded plastics all over the world. The main importing countries are in eastern Asia, where some of their poorest people are put to work smelting the metals without adequate safeguards, separating Pb from batteries without protection from either the Pb or the acid, and sorting plastics that may still be contaminated with hazardous chemicals or germs.

The United States sends somewhat less waste abroad than Europe, largely because there are more places to put it within the country. This does not mean, however, that all Americans share the problem of storing the waste. Much of it goes to areas occupied either by African Americans or Native Americans (Indians), a process sometimes referred to as "environmental racism." The African American communities generally have had little choice in the matter and received no benefit from the storage, a situation that has led them to take political and legal steps to stop the process. Indian reservations, on the other hand, have legal title to their land, and the waste can be stored there only if the tribal government that runs the reservation agrees to accept it. Some have done so – at a price. Where the money is used for the betterment of everyone on the reservation, this storage policy may be acceptable. Unfortunately, in some cases the money apparently goes to only a few people, which is not satisfactory.

We conclude that we can clean up much of our toxic waste (without reclassifying it!), but we need the will to do so. We know how to design industrial processes that produce less-hazardous waste, to store or dispose of waste that is hazardous, and to control some of the chemicals that already are in our soil, water, and air. When these policies are rigorously applied, we have had great success (Box 7.6). Unfortunately, we usually take the path of least resistance when faced with a toxic-waste problem.

Policy questions

If the industrial countries can maintain their present standard of living only by producing toxic industrial waste and/or by distributing hazardous pesticides, then which of the following attitudes is closest to yours?

- We should continue our present industrial economy because pollution from hazardous chemicals is not high enough to be serious.
- We should continue our present industrial economy because new technologies will soon enable us to reduce emission of toxic chemicals without affecting economic productivity.
- We should reduce emissions as much as possible even if it means a decline in our standard of living.

FURTHER READING: Carson (1962); Blackman (1993); MacKenzie (1993); Woodside (1993); Broadus and Vartanov (1994); Francis (1994); United States Environmental Protection Agency (annual).

BOX 7.6 CLEANING UP THE RHINE RIVER _____

In November 1995, a group of French biologists found salmon and sea trout at a dam on the Rhine River near Strasbourg. So what? So this.

It had been more than 10 years since sea trout, salmon, or any other ocean fish had ventured into the Rhine River (Fig. 7.14), and for much of that time even freshwater fish couldn't live in the river. The Swiss pharmaceutical and other industries put toxic chemicals into the Rhine, partly as a normal process of waste disposal and partly as a result of industrial accidents. As these chemicals made their way downstream, they acquired municipal pollutants from French and German towns along the river, salt from mining operations in northeastern France, heavy metals and other industrial pollution from the German industrial heartland along the Ruhr, and finally gathered industrial chemicals, municipal wastes, and fertilizers in the Netherlands, one of the world's most crowded and heavily fertilized countries. By the time the Rhine reached its mouth in Rotterdam, only very hardy or suicidal fish would venture near it.

In the 1970s, five countries in the Rhine drainage basin embarked on a massive campaign to clean it up. They placed strict controls on industrial emissions, leading companies either to dispose of their residues better or, in many cases, to switch to processes that used chemicals that were less hazardous. They also required better processing of municipal waste and at-tempted to place limits on surface runoff and groundwater seepage from agricultural areas. In a special effort to reduce salt concentration, they required the French mining operations to store unused salt for disposal elsewhere rather than simply let it wash into the Rhine.

The results of these efforts proved to be remarkable. The salt content of water as it entered the Netherlands, which had been too high for Dutch farmers even to use for irrigation, was reduced from nearly 600 ppm to about 200 ppm. The concentration of heavy metals that came from industrial outflow and corrosion of pipes and equipment in both homes and industries was reduced to 10% of its former value. Biochemical oxygen demand (BOD; see text) was reduced to about 20% of its original value largely by requiring better sewage treatment, and chlorinated organic compounds were reduced below 10% by banning or restricting the use of chlorine-bearing pesticides and chemicals used to bleach paper.

The problems of the Rhine have not disappeared. It still has high nitrate concentrations because the surrounding areas are still heavily fertilized, but the river certainly is remarkably cleaner. Even the salmon and sea trout are happy.

FURTHER READING: Malle (1996).

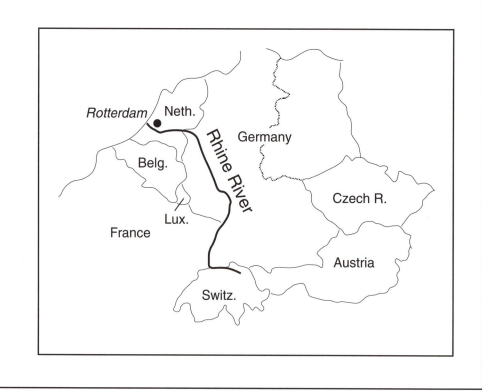

Figure 7.14 Rhine River.

7.4 RADIOACTIVE WASTE

We of this generation are enjoying the benefits of cheap electricity and are thereby creating a hazard for generations to come. – K. Krauskopf, Radioactive Waste Disposal and Geology, p. 134

All of the waste problems that we have discussed thus far have affected people throughout history. We have always generated some kind of solid waste, sewage, polluted air, even certain types of toxic chemicals such as lead. The problem of radioactive waste, however, simply did not exist until the first nuclear reactor was built in the 1940s (Section 5.3) and fission products began to accumulate. At first, no one gave much thought to what to do with the waste, but recently the seriousness of the problem has become evident. As a start on finding a solution we divide radioactive waste roughly into two categories – high-level waste (HLW) produced in the cores of nuclear reactors and by the atomic weapons industry, and low-level waste (LLW) discarded by hospitals, research laboratories, and industries. These two different types of waste create very different storage problems, and we investigate each one separately after a short discussion of the dangers posed by radioactivity. The remainder of this section assumes that readers are at least somewhat familiar with the concepts of radioactivity and the operation of nuclear reactors discussed in Section 5.4.

Dangers of radioactive waste

Radioactivity is dangerous because it causes ionization of biological tissue. When stable, nonionic, organic molecules are struck by radiation of sufficient intensity, some break into two or more ionized parts and are no longer able to perform their biological functions. Consequently, exposure to very large amounts of radiation can cause death because of complete disruption of bodily processes. Exposure to lesser amounts commonly leads to various types of cancer, particularly the cancer of the bone marrow that produces the fatal disease leukemia (the relationship between radiation and cancer was first demonstrated, but not understood, when Mme. Curie died in 1934 of leukemia – presumably induced by her work with radium). Laboratory experiments have also demonstrated severe modification of genes in reproductive organs, leading to mutations and birth defects.

Radiation damage can be caused by X-rays and by all three types of nuclear radiation (alpha, beta, and gamma) discussed in Section 5.4. Alpha particles, however, are so large and sluggish that they can be absorbed by almost anything (a page of newspaper works fine), and they pose little danger except to someone who eats a large quantity of alpha-emitting nuclei. Beta rays (electrons) are somewhat more of a problem, although they can be absorbed by a thin plate of glass or a few meters of air and have little effect unless they are ingested. The major danger from beta rays in nuclear waste comes from the fact that one of the principal fission products in nuclear waste is strontium 90 (^{90}Sr), which replaces calcium (Ca) in bones. Consequently, anyone ingesting ^{90}Sr from a leaking waste dump could potentially suffer serious bone damage because of the beta ray bombardment. The major problem is gamma rays, which account for more than 99% of the radiation emanating from nuclear waste, and we concentrate on them in this section.

Radiation is measured both in terms of number of nuclear *disintegrations* and in terms of the amount of ionization caused by the *radiation*. The *curie* was named for Mme. Curie, the discoverer of radium, and one curie (Ci) is defined as 3.7×10^{10} nuclear disintegrations per second, approximately the number of disintegrations that take place in 1 gram of radium. An alternative unit is a *becquerel*, named for the discoverer of radiation, with one becquerel (Bq) defined as one disintegration per second. The fundamental unit of radiation is the *roentgen*, named for the discoverer of X-rays. One roentgen of radiation forms one electrostatic unit of charge (1.6×10^{12} ion pairs) in 1 cubic centimeter of air at normal temperature and pressure. Radiation of biological tissue produces about 80% to 90% of the number of ion pairs as an equal amount of radiation absorbed in air, making the *radiation absorbed dose* (rad) slightly smaller than the roentgen. A further refinement of the rad is the *radiation equivalent man* (rem), which adjusts the number of rads for the type of biological tissue absorbing the radiation. Use of these different terms would be confusing except that, for all practical purposes, one roentgen = one rad = one rem, and radiation scientists have mercifully decided that the symbol *r* can be used for all three units. The most commonly used unit in radiation measurements is the milliroentgen (mr), equal to one-thousandth of a roentgen.

Using these units to measure the amount of radiation, we now want to quantify the relationship between the amount of radiation absorbed and the amount of damage done to the body, known as a "dose–response" curve, and Figure 7.15 is a very rough effort to construct one. It shows death within a short time after a dose of 1,000 or more rads, death after a few weeks at doses down to ~100 rads, and longer-term effects, not necessarily fatal, at lower doses. Within those lower doses we have plotted a

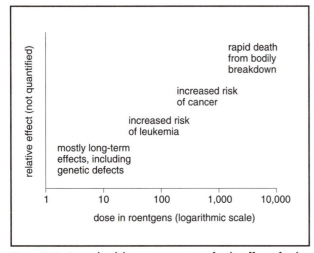

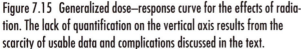

Figure 7.15 Generalized dose–response curve for the effects of radiation. The lack of quantification on the vertical axis results from the scarcity of usable data and complications discussed in the text.

few studies that appear to demonstrate that the incidence of various types of cancers increases linearly with dose.

Figure 7.15 contains at least four inadequacies (maybe more). The first is the lack of data. Much of our information has been provided by studies of victims of the atomic bombs in Hiroshima and Nagasaki, Japan, and even 50 years later, the study continues on people who were children and far enough from the blasts to have received small enough doses to survive until now. These data are augmented by studies of the much smaller number of victims of nuclear accidents (see Box 5.7 for a discussion of Chernobyl) and people exposed to X-rays and radioactive drugs given during medical treatments. These studies are very fragmented, however, and become increasingly difficult to interpret as the radiation dose decreases.

The second problem with Figure 7.15 is that it plots information on amounts of radiation received over relatively short periods of time, perhaps only a few hours at Hiroshima and Nagasaki. The same dose, however, received in smaller increments over a longer period of time generally has far less effect, and, except in the case of nuclear war, it is that long-term effect that we are most concerned with. Nuclear waste would be a danger only if it leaks from a waste dump and continually exposes people to small amounts of radiation. Unfortunately, we do not know how to convert the meager data on instantaneous exposure to the effects of this long-term exposure and must take qualitative data that are even less clear.

The third difficulty with Figure 7.15 is determining any effects at all in the range of low doses. All of the ill-

nesses resulting from excessive radiation, including cancer and birth defects, also have other causes, most of which are unknown. Consequently, proving that some particular cancer or defect has been caused by radiation has thus far proved to be almost impossible. The problem is accentuated by the fact that everyone receives some radiation all of the time. This "background" comes from a variety of sources shown in Table 7.2, the most important ones being cosmic rays and radioactive isotopes in the soil and water. Cosmic rays are charged particles reaching the earth from the solar system, and because they are absorbed by air, their intensity is controlled largely by elevation, with typical exposure values being a few mr at sea level and up to about 50 mr at 15,000 feet. The intensity of radiation from rocks, soils, and water varies within the ranges shown in Table 7.2, with high values generally derived from areas underlain by granite and low values from areas of exposed sedimentary rocks. Taking these variations into account, the average person receives 200 to 300 mr per year, perhaps slightly more in industrial countries where they have periodic X-rays of their teeth or emergency ones to check for broken bones or illness.

The fourth problem with Figure 7.15 is the concept of threshold, which is easy to define but far more difficult to analyze. Threshold is a dose below which there are no consequences of exposure to radiation. Proponents of this concept argue that any threshold must be greater than the background radiation because people have obviously adjusted to this level since they first evolved. That is, how could the human race have survived if everyone was constantly bombarded by radiation above the level at which they would be seriously affected? Opponents of the threshold concept argue that the human race has been affected through all of its history because the background radiation is responsible for at least some of the cancers and genetic defects that are our common lot.

So why don't we just measure the threshold and find out? That question takes us back to the other defects in Figure 7.15. We have reasonably good dose–response data at high levels of radiation, poorer data at moderate levels, and none at low levels. Unfortunately, the threshold can be determined directly only by measurements at those low levels, and without them we have to extrapolate the dose–response line from high levels back to zero response (Fig. 7.16). If we pass that line through zero response only at zero dose, then we conclude that there is no acceptable level of radiation to which anyone can be exposed. But if we pass the line through zero response at some dose equal to or greater than the threshold, then we

Table 7.2 *Sources of natural background radiation*

Source	Dose (in milliroentgens)	
	External exposure from air and rock	Internal exposure (mostly in bones)
Cosmic rays	36	
Uranium and its decay products	10	124
Thorium and its decay products	16	18
Potassium 40	15	18
Total	~80	~160
Total average exposure ~240 mr, but highly variable from ~150 to ~400		

Sources: Granier and Gambini (1990).

conclude that there is some level of radiation below which there is no effect. How would you extrapolate the line in Figure 7.15? Would it, or would it not, show a threshold?

The preceding argument is not simply an academic one. If there is no threshold – that is, if any amount of radiation causes damage to people – then we are probably obligated to close down our nuclear industry as fast as possible. That decision would come as a relief to people who oppose nuclear power and to those who feel that we must dismantle our nuclear weapons systems. If, conversely, there is some acceptable level of radiation to which people can be exposed, then we can try to establish some standard and design our nuclear programs to adhere to it. That decision would please people who believe that nuclear energy is an acceptable alternative to coal for producing electricity (Section 5.3), to those who receive medical treatment from isotopes produced in reactors, and to people who feel that a nuclear deterrent is still necessary in an age of uncertainty about the spread of nuclear weapons from the now dismantled Soviet Union. Despite these uncertainties, our society has clearly decided that the benefits of using radiation outweigh the risks. So we had better decide what to do with the wastes.

High-level nuclear waste

High-level waste is produced in the cores of nuclear reactors. As we discussed in Section 5.4, neutron bombardment of fresh fuel rods consisting almost entirely of ura-

nium gradually converts the U into fission products and into a small amount of plutonium 239 (^{239}Pu) and other elements heavier than uranium (transuranic elements; TRU). When the concentration of uranium 235 (^{235}U) has dropped too low and the concentrations of fission products have become too high to sustain a chain reaction, the fuel rods must be removed. Because most of the fission products are highly radioactive, these spent rods are extremely dangerous and shortly after they are removed from the reactor they are commonly transferred to pools of water (holding ponds) near the reactor and allowed to cool. Because all of the radioactive material is wrapped inside a sheath of metal, the water will not become contaminated until the metal reacts with it, and there is no known limit to the amount of time that a spent rod can sit in a tank at the reactor site. Fortunately, more than 90% of the radioactivity is from isotopes with half-lives ranging from seconds to days, and in a few weeks the spent rods can be transported to some place of more permanent storage.

After cooling for 10 years, the spent fuel rods contain a mixture of radioactive isotopes in the range of concentrations shown in Table 7.3 and Figure 7.17. The fission products have atomic masses ranging around 90 and 140 (90 + 140 = 230, the approximate mass of uranium). The transuranic elements formed when neutrons and other particles are added to the uranium nuclei (Section 5.3) consist largely of ^{239}Pu and very small amounts of other plutonium isotopes plus americium (Am) and neptunium (Np). Also, as the residual uranium and the TRU decay, the concentrations of radium 226 (^{226}Ra) and lead

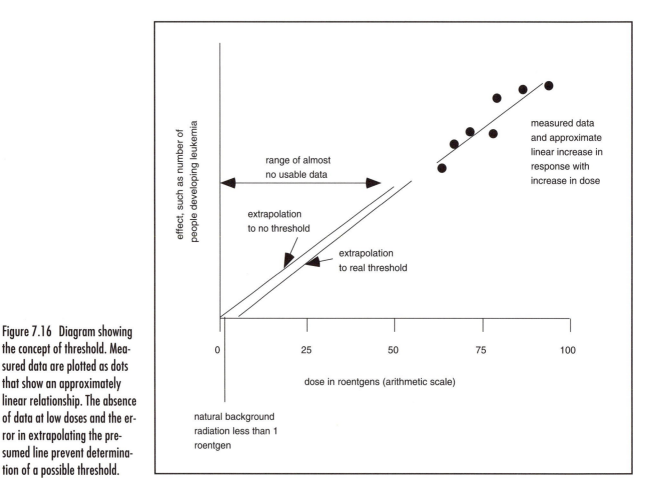

Figure 7.16 Diagram showing the concept of threshold. Measured data are plotted as dots that show an approximately linear relationship. The absence of data at low doses and the error in extrapolating the presumed line prevent determination of a possible threshold.

210 (^{210}Pb) become important components after a few thousand years and remain at moderate concentrations for several million years because of the long half-lives of their parent elements.

The half-lives in Table 7.3 indicate the length of time for which the waste must be stored. The general rule of thumb is that waste is dangerous for 10 half-lives of each element, at which time only 0.1% of the element remains ($1/2^{10} = 1/1,024$). On this basis, most of the fission products will decay to insignificant amounts in a few hundred years, the technetium 99 (^{99}Tc) has a long enough half-life that it will remain active for about 2 million years, although it is not particularly dangerous, and most of the TRU will be gone in a few tens of thousands of years. The plutonium 239 (^{239}Pu), however, has a 24,400-year half-life that makes it a dangerous isotope and also requires it to be stored for about 250,000 years. That is the principal problem in designing methods of disposing of nuclear waste.

Using this information, there are two principal ways to handle waste from spent fuel rods. One is to leave it

more or less intact and to search for a storage place of the bulk waste, possibly mixed with glass or other synthetic material to dilute and stabilize it. The OECD has now accumulated about 100,000 tons of spent fuel rods in this form, approximately half of them in the United States and most of the rest in the United Kingdom, France, Japan, and Germany. Russia has accumulated probably about the same amount, although the secrecy and lack of records make accurate information impossible to obtain. Although the figure of 100,000 tons sounds enormous, it represents only a small volume of material, approximately the amount that would cover a football field (American football) to a depth of about 50 feet, and it should be possible to find some storage place or combination of places.

The second principal method of handling the waste is to "reprocess" it in order to extract plutonium and uranium. The principal method of reprocessing consists of dissolving the spent fuel and then separating the plutonium and uranium in organic chemicals. This process separates the waste into two parts, drums containing liq-

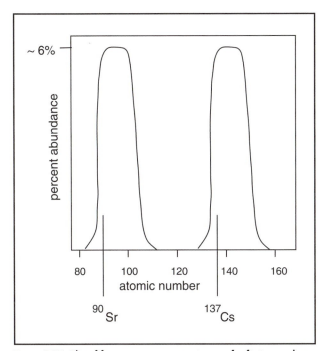

Figure 7.17 Plot of frequency versus atomic mass for fission products. The two broad peaks are roughly centered at atomic masses of 90 and 140. The two elements that pose the greatest danger are ^{90}Sr and ^{137}Cs (Table 7.3).

uid waste with the composition shown in Table 7.3 and pieces of plutonium and uranium metal that are commonly shaped to the size of hockey pucks. This liquid waste is combined with the liquids left from the production of Pu for nuclear weapons, and OECD countries are currently storing about 400,000 tons of liquid waste and a few thousand tons of plutonium, mostly in the United States. Russia has reprocessed a higher percentage of its nuclear waste than OECD countries and now either has more liquid waste than the OECD, or has less because it simply pours much of into rivers and lets it wash downstream into the Arctic Ocean or, well, who knows?

In the 1990s the problem of nuclear waste underwent a major reorganization when the presumed end of the cold war between Russia and Western allies led to partial dismantling of the nuclear weapons programs. At the height of tension, both Russia and the United States had about 20,000 nuclear warheads, each containing at least the minimum of ~40 pounds of ^{239}Pu or ^{235}U in order to produce an explosion, but both sides had dismantled part of their weapons by 1995. In the United States, the Department of Energy, which is responsible for nuclear weapons production, placed more emphasis on cleanup of its facilities than on construction of new weapons and

began to look for places to store the explosives from the decommissioned weapons. In Russia, nuclear weapons production also stopped, but because their military and power industries had been closely integrated, there appears to have been little reduction in waste generation. The most serious problem resulting from dismantling of Russian weapons is probably the storage of their old explosives, some of which appear to be making their way into international trade (a cheering prospect).

Several options are available for storing unprocessed fuel, reprocessing liquids, and plutonium. The one currently in use is to leave the unprocessed waste from power reactors in holding ponds on the reactor site even after it has cooled down to the extent that it could be moved. The advantage of this method is that the waste does not have to be transported and is readily available if

Table 7.3 *Principal radioactive isotopes that must be stored in waste sites*

Isotope	Half-life (in years unless otherwise specified)
Fission products	
^{90}Sr	28
^{99}Tc	2.1×10^5
^{137}Cs	30
^{129}I	1.6×10^7
Isotopes generated in the waste	
^{210}Pb	
^{226}Ra	
^{234}U	2.5×10^5
^{237}Np	2.1×10^6
^{242}Pu	3.8×10^5
^{243}Am	7,400
Other transuranic isotopes	
^{239}Pu	2.4×10^4
Medical and industrial waste	
^{24}Na	15 hours
^{32}P	14 days
^{60}Co	5.3 years
^{82}Br	36 hours
^{85}Kr	10.7 years
^{125}I	60 days
^{192}Ir	74 days
^{198}Au	2.7 days

a good disposal method is found. People opposed to nuclear energy also believe that accumulation of waste around the reactors will encourage companies to stop using nuclear reactors. The disadvantages of on-site storage is that the holding ponds may leak, spreading contaminated water into the surroundings, and that terrorists and other vandals have comparatively easy access to the waste. This type of storage is used only in the United States, largely because the search for a permanent storage site has been completely unsuccessful.

A second option is to seal the unprocessed waste into cannisters and put it into underground storage for about 250,000 years. In order to be acceptable, a storage area of this type must leave space between the cannisters so that the heat that they generate will not become excessive and should also provide "multiple barriers" between the waste and the surrounding environment (Fig. 7.18). The first barrier might be fusing the waste into an inert material, such as glass, concrete, or possibly a synthetic material known as "synroc," which consists largely of minerals whose lattices easily include the principal radionuclides. A second barrier is a metal casing, or possibly more than one casing, plus a clay or synthetic material that can absorb any leaked waste. The third barrier should be the rock itself. It must be stable so that the cannisters will not move around, and ideally it also would consist of minerals that could absorb leaks; the ability of volcanic

ash to absorb metal ions is one of the reasons that the United States has proposed a high-level waste facility at Yucca Mountain, Nevada. And finally, the storage area should be above the water table so that the waste will not dissolve and be dispersed in groundwater if all of the barriers fail.

The underground storage could either be "retrievable," left open, or made "permanent" by closing the opening to it with explosives or a concrete fill after the waste is deposited. Permanent storage is based on the assumption that our technological civilization may not last – perhaps our descendants will not know how to detect and handle dangerous radiation and be prevented from accidentally entering a nuclear storage site. Retrievable storage, however, assumes that future generations will either maintain or continually improve our present technology. They might, for example, want to dig up the waste if technology improves to the point that they have learned to deactivate nuclear processes. Another advantage of retrievable storage is that the waste can be moved if, for example, the climate changes to such extent that an underground site in an arid region originally above the water table is in danger of becoming below the water table because of increased rainfall. We discuss the problem of climate change in Section 8.2 but should mention here that the history of the past million years shows us that we can expect at least two major

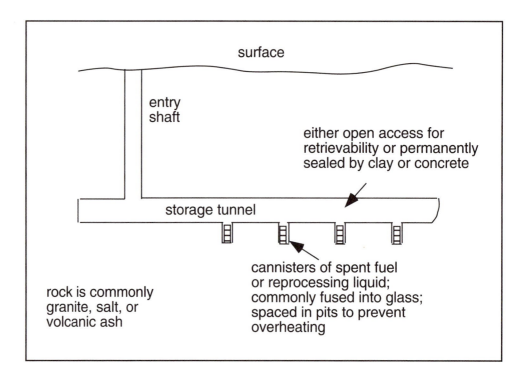

Figure 7.18 Diagram of site for disposal of high-level nuclear waste. See discussion in text.

glacial advances and retreats during the 250,000-year history of a waste dump, with each glacial advance bringing cool wet climates to areas of the Northern hemisphere that are now arid (including the site of the Yucca Mountain dump in the United States).

A third option for storage of high-level waste is to put the unprocessed waste into weighted metal cannisters and drop them into clay sediments on the deep ocean floor. There are two advantages to this method of disposal. One is that dropping the cannisters into deep water and burying them in mud puts the waste almost permanently out of reach of any but the most technologically advanced civilizations. The second advantage is that the seafloor mud can act as a second barrier around the waste by absorbing metal ions in the same way that volcanic ash would act on land. The disadvantage of sea floor burial is that some waste might leak over the long lifetime of the deposit, and it could become a danger to marine life if it were ever circulated into shallow water where most marine organisms live.

And fourth, we could consider burying only the reprocessed waste (mostly fission products) after removal of Pu and U and dispose of the reprocessing liquid separately. The advantage of this method is that virtually all of the radioactivity in the reprocessed waste will be gone in a few hundred years (Table 7.3), a period of time in which we can more confidently predict the future of civilization and climate than the hundreds of thousands of years required for decay of plutonium. Then the reprocessing liquid containing the plutonium and uranium can be handled in three ways. One is to fuse it into glass or synroc and bury it in the same way as unprocessed waste is buried. The advantage of burying the plutonium and uranium separately is that its volume is so small that waste dumps are easier to design. Another disposal method is to inject the liquid into wells 1,000 or more feet below the surface and let it mingle with groundwater on the assumption that any of the waste that ever reaches the surface will be so dilute that it will not pose a danger (we hope). The third disposal method for the reprocessing liquid is to separate metallic plutonium and uranium as described earlier and then reuse them in reactors. The uranium would have to be further enriched to concentrate the ^{235}U to reactor grade, and the plutonium would become more fuel for breeder reactors (Section 5.4). The advantage of this method is that it permanently destroys plutonium and ^{235}U by nuclear reaction, leaving only an accumulation of ^{238}U, which is not dangerous and can be disposed of easily. The disadvantage is that it increases the world's supply of plutonium.

By now you can see, that to solve the problem of disposing of high-level waste, all we have to do is find the answers to a few questions. How much radiation can people tolerate without ill effect? Will there be a nuclear war in the next few hundred years? Will anything leak from a waste dump in the next 500 years (a period equal to the time since Columbus reached North America)? What will civilization be like 10,000 years from now (a period of time covering virtually all of human history to the present)? What will happen to the earth's climate in the next 100,000 years? Simple, isn't it?

Low-level waste

Dozens of artificially radioactive elements produced either by irradiation in reactors or by bombardment in accelerators are used in industrial, medical, and research operations (Table 7.3). Their uses include destruction of tumors and other cancers by implanting radioactive material in or near the tumor or by bombardment with gamma rays produced by isotopes outside of the body; medical diagnosis and tracing the effects of drugs by following radioactive isotopes as they pass through the body; tracing industrial reactions, measuring rates of flow of materials, and finding leaks by inserting small amounts of radioactive isotopes whose locations and movement can be determined by radiation detectors.

All of this radioactive material is classified as low-level waste along with discarded laboratory debris (gloves, contaminated equipment, etc.) and the irradiated metal tubes of fuel rods after the fuel has been removed. As the half-lives in Table 7.3 show, finding a place to put it is far less difficult than finding a place for high-level waste. Most of the low-level waste has half-lives of hours or days, and some decays so rapidly that it must be taken from a reactor and used immediately, thus leaving virtually nothing to be discarded. Only three of the isotopes shown in the table have half-lives measured in years, and even the longest one (^{137}Cs, 30 years) needs to be stored for only 2 to 3 centuries. Of course, 2 to 3 centuries is a very long time, but it is shorter than 250,000 years.

Because of these short half-lives, finding a storage place for low-level waste poses no more technical difficulties than finding a place for toxic waste. Both types of waste need to be buried in some secure landfill, covered so that no one is accidentally exposed to them, and then monitored for leaks. In many situations, the radioactive isotopes are far less hazardous than many of the chemicals discussed in Section 7.3. Unfortunately, the word "radioactive" arouses fear and opposition among people even tens of miles away from a low-level dump and outright hostility among people who are closer. For this reason, much of the low-level material that could be buried with comparative safety is still stored in temporary, insecure, sites around hospitals, laboratories, and factories.

Policy questions

Artificial radioactivity has become a part of the lives of everyone in an industrial society, but we must make choices about how to use it in the future. Which of the following possibilities would you prefer?

- Close all nuclear reactors, so that we do not produce more high-level waste, and leave accumulated waste on the reactor site until it is safe to store elsewhere. If you choose this option, please indicate what changes you would make in our electric supply system (see Section 5.4).
- Continue reactor operation and require federal and local governments to designate a high-level waste site or sites. If you choose this option, please indicate how you plan to enforce this requirement.
- Permit production of high-level and low-level waste only by organizations that can prove that they can dispose of the waste safely. If you choose this option, please indicate how you plan to enforce this requirement.

FURTHER READING: Roxburgh (1987); Krauskopf (1988); Granier and Gambini (1990); United Nations Environmental Programme (1991).

7.5 BIOLOGICALLY ACTIVE WASTE

[People] and animals are really the passage and the conduit of food. – Leonardo da Vinci commenting on the origin of biological waste

Biologically active waste is produced in two principal ways. One is the waste products of animals, which are part of the natural cycle of life, in which microbes discharged from the insides of animals are normally killed by exposure to air, and nutrients promote the growth of plants that nourish more animals. The second method is by spreading fertilizers, which provide even more nutrients to enhance the load from animal waste. These processes concern us only because the earth now con-

tains more total people, more people crowded into cities, more domesticated animals, and more fertilized fields than ever before. This section is divided into two parts – the management of sewage and the problem of excess nutrients, commonly referred to as eutrophication. Before discussing them, however, we should mention one more possible source of biologically active waste; in Box 7.7 we discuss the highly controversial possibility that plastics and other industrial chemicals have affected reproductive processes of people and other animals.

Sewage

People living in industrial countries at the end of the 20th century may want to live only where sewer systems have been established, but that is not the case in most of the world today nor was it true of anyone living before the industrial revolution. The royalty of Europe, the emperors of Asia, and the priest-kings of Central America lived in magnificent palaces that are now visited by tourists who think enviously of the luxurious lives that their owners must have led. Perhaps these tourists should note that all of the toilets in these palaces are recent additions. We discuss the history of the development of sewer systems before describing the operation of modern disposal facilities.

By the late 1700s, when the industrial revolution was just beginning to flower, the statistical differences between the health of people in cities and those in rural areas became too large to ignore. The life expectancies of people living in cities was about 10 to 20 years less than for rural people, and many investigators attributed this to the difference in sanitary conditions. In the countryside, both people and animals were so sparsely distributed that people could use outhouses, and their sewage and that of their animals would be oxidatively destroyed before it accumulated in sufficient concentration to cause problems. As cities expanded, however, the old practice of dumping sewage into the nearest river, or simply throwing it into the street, simply had to stop.

The first efforts to reduce contamination by sewage consisted of sending carts (sometimes called "honey wagons") through the cities to collect pots full of it. This primitive system was then replaced by sewers, whose chief mission was to get the sewage out of town by whatever means possible (put it "somewhere else"). In many cities that meant continually running water through pipes or drainage ditches that discharged into the local river or even into fields beyond the town limits. Not only did this system waste water, but the absence of treatment

BOX 7.7 WHAT'S WRONG WITH MEN? _____

"Every man in this room is half the man his grandfather was," said Louis Guillette, a biologist at the University of Florida, addressing a U.S. congressional hearing. It was a remark that attracted attention (to put it mildly). The purpose was to highlight a series of studies that some people interpreted as indicating a 50% decline in sperm production by men during the past 50 years. Similar studies, apparently consistent with the human sperm data, indicate deterioration of male reproductive organs in birds, alligators, otters, whales, and other wildlife. Although the effects seem to be most severe in industrial countries, limited data suggest comparable changes throughout the world.

These data spawned three different reactions. One reaction is from scientists who examined the data and found them inconclusive. Some of them believe that the data do not indicate sufficient sexual problems among men and males of other species to warrant more than limited continuing investigation. We pay attention to these evaluations and doubts. A second reaction is from nonscientists who feel that there is absolutely nothing wrong with the world, except possibly that it contains too many intellectuals who want to study it, and consequently feel that further scientific investigation is a waste of time and money. We ignore them. The third reaction is from scientists who find the data sufficiently interesting to warrant further study and who make preliminary efforts to explain them. We follow this reasoning in the rest of this box.

The culprits, if there are any, in the decline of male potency are chemicals that resemble the various female hormones known as estrogens (Section 2.1; Fig. 2.5). Many cells, both in women and men, have locations known as "estrogen receptors," where estrogens can bind to components of the cell. Men and women, however, react very differently to estrogen. Estrogen receptors are essential in women because some attached estrogen is required to regulate the menstrual cycle and reproductive activity, but numerous studies have shown that too much estrogen increases the risk of breast cancer. Men can tolerate some estrogen without ill effect, but too much exposure, particularly when they are fetuses in their mother's wombs, apparently reduces their ultimate sexual development and increases the risk of testicular cancer. This effect occurs because sexual activity in men requires the male hormone testosterone (the principal androgen; Section 2.1; Fig. 2.5). High estrogen levels reduce the effects of testosterone, and if androgen receptors are filled by chemicals that behave like estrogens, the testosterone cannot reach them.

The extensive data on the increase in breast cancer among women and the more controversial data showing decrease in sexual potency among men and other male animals suggest, but do not prove, that industrial activity is doing something that has the effect of increasing exposure of people and animals to estrogens. But what? Although artificial estrogens are easy to manufacture, we obviously are not pouring them indiscriminantly into the air and water. Recently, however, scientists have discovered that estrogen receptors, and possibly androgen receptors, are not particularly discriminating. They can be fooled by certain other chemicals and will bind with those chemicals just as if they were real estrogens — these chemicals are given the general term "estrogen-mimicking." Unfortunately, estrogen-mimicking chemicals include chlorinated hydrocarbon pesticides, such as DDT (Fig. 7.6), PCBs (Fig. 7.4), and phenolic chemicals (Fig. 2.1) that are common in some types of plastics. That is, unwittingly, we appear to have polluted the air and water with chemicals that have a similar effect as increasing the actual estrogen concentrations.

Now the issue is whether this pollution has really had an effect. Because the pesticides are soluble in fats, their concentrations are particularly high in fatty tissue, and high levels of chlorinated pesticides and related chemicals have been found in wildlife and domesticated animals such as cows and chickens. High levels have been proved in the fatty tissue of women with breast cancer and may have been found — although the data are not yet adequate — in men with inadequately developed testes or with testicular cancer. Presumably these concentrations result from eating pesticide-bearing meat and drinking water with undetected small amounts of pesticides.

So industrial, including agricultural, pollution has increased the concentration of estrogen-mimicking chemicals in some people, perhaps in all of us to some small extent. In order to demonstrate that this increase has had any effect, however, we must show that the cancers and other illnesses are caused by the chemicals. This has proved to be extremely difficult. Experiments on laboratory animals have generally shown that injections of estrogen-mimicking chemicals at concentrations much higher than those found in people and wildlife had no significant effect. Then, in June 1996, new experiments showed that simultaneous doses of low concentrations of two or more estrogen-mimicking chemicals could have major effects on both male and female reproductive systems in laboratory animals.

BOX 7.7 WHAT'S WRONG WITH MEN? *(continued)*

Here is the situation as we write this book. We have polluted the environment with estrogen-mimicking chemicals. No chemical by itself has been shown to have an effect on animals, but two or more chemicals together apparently do. If there has been significant increase in breast and testicular cancer and reduction in

sperm counts, then *possibly* the estrogen-mimicking chemicals are responsible. Until we know something more certain, we recommend that both women and men should avoid eating pesticides.

FURTHER READING: Dold (1996).

of the discharge simply shifted the burden of disease downstream. Both defects had to be solved. Saving water was first accomplished in the late 1800s when the flush toilet was invented by Mr. Thomas Crapper, a British plumber living in Chelsea (now a borough of London). Rendering the discharge both safe and usable has now become a highly refined technology.

The discharge of wastewater in OECD countries is about 100 to 200 gallons per person per day. As it enters sewage treatment plants this water contains dangerous microbes that must be killed, solids that must be removed, and a variety of chemicals whose concentrations must be reduced (Section 4.5). The three major chemicals are dissolved nitrogen (a total of approximately 35 ppm; $35 \times 10^{-4}\%$), phosphorus (approximately 7 ppm),

and an equivalent of about 200 ppm of dissolved and suspended organic material that reacts with oxygen, creating a "biochemical oxygen demand" (BOD) and reducing the dissolved oxygen content of the water.

Figure 7.19 schematically shows the processes that take place in a modern sewage treatment plant. The first task is screening and initial settling to remove coarse material and grit. This material is either shredded and returned to the liquid for further processing or removed and stored. The next process is sedimentation (clarifying) of the sewage in tanks of quiet water that allow sludge to settle to the bottom and the clearer water to be removed from the top of the tank. The separated sludge is then commonly compressed ("dewatered") and is either placed in landfills, burned, or sold to agricultural op-

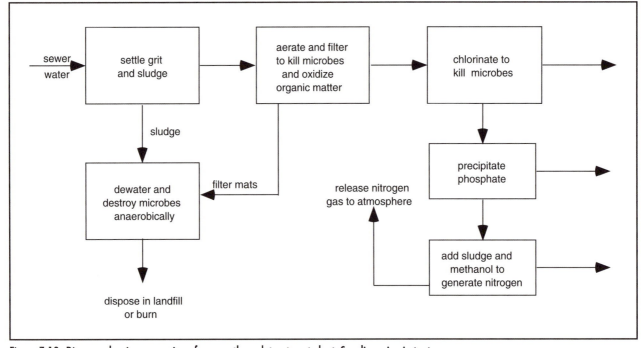

Figure 7.19 Diagram showing processing of sewage through treatment plant. See discussion in text.

erations as a soil conditioner. The last principal stage is biological filtration, in which the clarified liquid is aerated and filtered past mats of bacteria, protozoa, and/or fungi. The nutrients in the liquid promote the growth of these organisms, which consequently remove more organic matter from the liquid and form a continually thickening layer of anaerobic microbes at the bottom of the aeration tank. Ultimately the interiors and bases of these mats die and slough off into the liquid, where they can be removed and added to the solid organic waste for further processing.

The liquid that flows out of the last process contains much less organic material than the incoming wastewater and a much smaller BOD (Table 7.4). This effluent water is commonly pure enough to be used for agricultural irrigation, but water that will be used only short distances downstream for drinking water can easily be treated with chlorine (Section 4.5) to kill all remaining microbes and reduce the amount of chlorination that will have to be done in the next water supply plant. The principal problem with the effluent is that normal sewage treatment has very little effect on phosphorus and nitrogen (Table 7.4). Because of the problem of excess nutrients downstream, it is sometimes necessary to reduce their concentrations by a series of complicated and unfortunately expensive methods.

Phosphorus concentrations can be reduced by limiting the amount in sewage and by precipitation as the waste water leaves the sewage plant. About half of the phosphorus that enters a treatment plant is organic phosphorus from animal waste and half is inorganic phosphate (PO_4) from phosphate in fertilizers and detergents. The amount of inorganic phosphate can be reduced by switching to natural, rather than artificial, fertilizers (Section 2.2) and by using nonphosphatic varieties of detergents and other chemicals. Because this inorganic fraction is only 50% of the total, however, even a complete elimination of the inorganic input (which is impossible) still leaves a high total phosphorus concentration in the wastewater. By the time the processed water leaves the sewage plant, all of the phosphorus is phosphate (PO_4^{-3}) because of oxidation of the organic P, and the only satisfactory way to remove it is by adding either aluminum sulfate ($Al_2(SO_4)_3$) or ferric chloride ($FeCl_3.6H_2O$) and precipitating aluminum or iron phosphate ($AlPO_4$ or $FePO_4$).

Removing nitrogen from wastewater is more difficult than removing phosphorus. Animal waste contains organic N, much of it in the chemical urea ($CO(NH_2)_2$) in urine, and bacteria in the sewage convert it to ammonia

Table 7.4 *Partial compositions (in ppm) of typical raw sewage and effluents from treatment plants*

Type	Sewage	Effluent
Organic matter as suspended solids	24	3
Biochemical oxygen demand	20	3
Inorganic nitrogen	2.2	2.4
Organic nitrogen	1.3	0.2
Inorganic phosphorus	0.4	0.4
Organic phosphorus	0.3	0.3

Sources: Hammer and Hammer (1996).

(NH_3). This ammonia augments the ammonia put into wastewater by fertilizers, and all of it is oxidized in sewage treatment plants to nitrate (NO_3^-). Because nitrate is highly soluble, it is emitted from the sewage plant in nearly the same concentrations as in the influent water and cannot be precipitated. The only procedure for reducing nitrogen in effluent water, therefore, is to convert it to nitrogen gas and let the gas escape into the atmosphere. This requires a complex biochemical process in which some of the sludge previously removed from the liquid is mixed with methanol and added back to the effluent from the sewage plant. The denitrifying bacteria in this sludge then use the methanol and the nitrate in the effluent to synthesize some type of oxidized organic matter and release N_2 gas.

Excess nutrients (eutrophication)

A normal supply of nitrogen and phosphorus into fresh and salt water establishes the base of a balanced food chain in which plants (green algae) are eaten by animal plankton, plankton are eaten by small fish, small fish are eaten by large fish, and all of the animals ultimately die to provide further nitrogen and phosphorus (Fig. 7.20). Where the effluents from sewage plants, high concentrations of animals, and fertilized fields provide more nitrogen and phosphorus than is normal, the normal food chain becomes unbalanced. The extra nutrients cause excessive growth of aquatic plants and blue-green algae (which are bacteria that photosynthesize like plants and are generally inedible by plankton). The aquatic plants

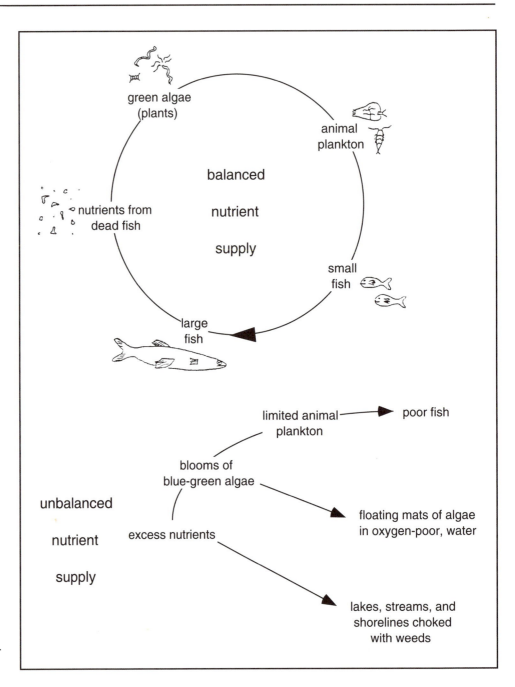

Figure 7.20 Life cycles in healthy environments compared with those with excess nutrients (undergoing eutrophication). See discussion in text.

choke lakes and marshes, and the blue-green algae turn the water murky and may form floating mats of decaying vegetation. The process is referred to as "eutrophication," and where it is carried to extreme the oxygen and nutrient content of the water may become so low that fish can no longer live in it, and the water is effectively dead.

Dead water is not only unsightly and smelly, it also has a serious effect on the food chain. An environment that cannot support freshwater fish also cannot support the young and larval stages of saltwater small fish and crus-taceans such as shrimp. As we discussed in Section 2.4, many of these organisms either give birth in coastal marshes or migrate into them for protection while they are still young and small. Eutrophication of their hatching and development areas is contributing to the world-wide reduction of fish and shrimp stocks. We discuss this problem further in Section 2.4.

To summarize this section, from the standpoint of general sanitation, we are certainly in better condition than we were 100 years ago, and some advances have been made in poor countries. The main problem now is the escape of nutrients into the ecosystem and resulting eutrophication. Those nutrients that come from sewage plants can be reduced or virtually eliminated by one or more complex technologies applied to the effluent from the plants. Nutrients that enter groundwater and surface water from fertilizers and high densities of domestic animals cannot be removed in any practical manner, and the only path to their reduction is to use less fertilizer and spread our food animals out over larger areas of ground.

All of these solutions cost money, although firm figures are almost impossible to obtain. Complete removal of nitrogen and phosphorus from sewage might add about 25% to the average water and sewer bill. Using organic methods of growing crops rather than artificial fertilizers might add 10% to the typical food bill. And spreading out animals will require more land for the same amount of production, perhaps adding another 10% to the average food bill. In the long run, of course, eutrophication of the world's waterways may cost more in lost food production than the costs of present cleanup.

Policy questions

Would you be willing to do either of the following?

- Pay more for food by restricting the use of fertilizer and spreading out farm animals (see also Section 2.2)?
- Pay more for water and sewage in order to remove nutrients from sewage?

FURTHER READING: Reyburn (1969); *Rodale's Organic Gardening* (serial); Eden and Haigh (1994); Hammer and Hammer (1996).

7.6 AIR POLLUTION

Britain's coal-smoke, foul and black,
Sinking o'er the land is seen;
Clokes and smirches on its track
Every freshening shoot of green.
— H. Ibsen, Brand (a dramatic poem set in Norway in the late 1800s), translated by F. E. Garrett, Lyrics and Poems from Ibsen

Throughout history city dwellers have always contended with smells produced by the accumulation of garbage, sewage, and animal waste and have extolled the virtues of the clean air of the countryside. Rich people tried to live on hills or facing the coast, where wind could blow clean air into their houses and, if they could afford them, they built additional houses outside of the cities. Poor people, by contrast, lived along the banks of the garbage-laden rivers and breathed the foul air of the lowlands.

As the industrial revolution arrived, many of the pollution problems of cities began to be solved, only to be replaced by others. Until about 1950 the principal culprit was coal. When it was burned in open hearths or later to produce electricity, coal spewed out ash, unburned tar, and sulfur and nitrogen oxides. The ash drifted through streets and into houses and water systems, and water droplets that coalesced around the particles formed the "pea soup" fogs of London and other European cities. The coal tars stuck to buildings, putting a black to brown grime on the formerly white cathedrals of Europe and the office towers of the United States, coatings that are now being scraped off at great expense in order to restore the cities to some resemblance of their former beauty. The sulfur and nitrogen oxides reacted with water in the air, forming dilute sulfuric and nitric acids that dissolved and discolored exposed building surfaces.

The increased use of coal also began to affect the formerly pure air of the countryside. People living near railroads welcomed the new ease of transportation but wished that the engines did not shower them so continually with smoke and soot. As industries used more coal, more sulfuric and nitric acids made their way into the atmosphere. The problems caused by this acidity were recognized as early as 1661 in England, 100 years before the traditional start of the industrial revolution, and we discuss them in this section under the general term "acid rain."

The days when coal was the principal problem were replaced in the middle 20th century by the age of the automobile. As cities tried to repair the ravages caused by coal, they began to be filled by the exhaust from cars and trucks. Part of the exhaust consisted of lead compounds, but leaded gasoline was banned in most of the industrial world starting about 1980 and today is used primarily in poor countries (because of its significance as a global pollutant, we discuss lead in Section 8.4). The major pollution produced by gasoline now consists of organic compounds that oxidize in the atmosphere and consequently are not as widely distributed as lead. When auto exhaust mixed with SO_2 from burning coal, it gave us the term smog, originally used in London for a mixture of smoke and fog. With reduction in the sulfur emissions from coal, the term smog has now been largely taken to mean the pollution caused by gasoline, and we discuss it under that heading below.

Acid rain

All rainwater is mildly acidic because ionization of atmospheric carbon dioxide (CO_2) that dissolves in it produces a solution with a pH of approximately 6.5 (compared with a neutral pH = 7.0; more in Section 8.1). The acidity produced by CO_2 can be significantly increased by solution of sulfur and nitrogen oxides (SO_2 and NO_2). Not enough sulfur and nitrogen are emitted into the atmosphere by natural earth processes to have a significant effect on rainwater acidity, but when their concentrations are increased by industrial activity, they can produce rain with a pH less than 5. We describe the reactions of sulfur and nitrogen in the atmosphere before briefly discussing the consequences of acid rain.

Sulfur is naturally released into the atmosphere in three main forms: (1) organic sulfides from decomposition of oceanic algae and from soils and marshes; (2) hydrogen sulfide (H_2S) generated by bacteria in soils and wetlands; and (3) sulfur dioxide (SO_2) from volcanoes and by oxidation of ore sulfides. When they reach the atmosphere, all of these forms are initially oxidized to SO_2, and ultimately the SO_2 from all sources is further oxidized to sulfate (SO_3), at which time it reacts with water to form sulfuric acid (H_2SO_4). Unlike H_2CO_3, sulfuric acid ionizes completely to H⁺ions (it is a strong acid), and the only reason that rainwater has the pH produced by H_2CO_3 is that there is too little sulfur in the atmosphere to have a significant effect.

Nitrogen is naturally released to the atmosphere by somewhat different mechanisms than sulfur. Ammonia (NH_3) is released as a waste product from animals and from some of the bacterial activity in soils, and various types of nitrogen oxides are generated by soils and by oceanic activity. In the atmosphere all nitrogen compounds are ultimately oxidized to nitrate (NO_3^-), which dissolves in rainwater to produce nitric acid (HNO_3), which is a strong acid that ionizes readily to H⁺. Like sulfuric acid, the nitric acid would greatly increase the acidity of rainwater if there were enough of it in the atmosphere, but the concentrations are generally too small to have an effect.

Industrial activity has increased the concentrations of both S and N in the atmosphere to the extent that they now have effects on pH greater than the effect of CO_2. The SO_2 emitted by industry, virtually all from coal burning, is now approximately twice the amount of S emitted by all natural processes together. A similar increase in N comes from burning gasoline and wood (including other biomass). The net effect is to reduce the pH of rainwater below 6 in much of the world and below 5 in areas particularly affected. In addition to being precipitated in rainwater, both sulfuric and nitric acids settle onto the earth's surface as "dry" molecules and enhance the acidity caused by the acid rain.

Acidity caused by rain and other precipitation has several effects. It can reduce the pH of lake water below the level that can be tolerated by various types of fish, and numerous lakes in the most severely affected regions are now effectively dead. It also weakens soil by leaching nutrients and by releasing Al, which is poisonous to many plants, from clays. This reduction in fertility may not be serious where soils are very rich and for plants that can tolerate broad ranges of nutrient availability. Plants that are easily affected and growing on poor soils, however, may grow more slowly and even die (pines turn yellow – see the Ibsen quotation). When the vegetation begins to die, soil erosion is enhanced and lakes and streams are filled with sediment that makes them even less receptive to fish.

The effects of acid rain and other precipitation vary enormously from place to place. The most affected large region is the eastern United States and eastern Canada, with broad areas of Europe in about the same category. Both areas are east of (downwind in the belt of westerlies) the largest concentrations of industry anywhere on the earth. The worst region is probably an area at the southeastern corner of Germany, the Czech Republic, and Poland, which is commonly referred to as "the black triangle" (Box 7.8) because of the widespread destruction caused by inadequate pollution controls on heavy industries established by the Soviet Union in their former satellite states.

Smog

The ideal car engine would convert all gasoline into CO_2 and H_2O. Unfortunately, neither the engine nor the gasoline is ideal, and the exhaust pipe also produces some unburned gasoline, a small amount of carbon monoxide (CO) formed instead of CO_2 by incomplete burning, and a variety of nitrogen oxides, which act as catalysts to form ozone (O_3) from ions produced by the unburned gasoline. The mix of unburned gasoline, CO, nitrogen oxides, and O_3 forms the basis of smog (smoke and fog) in most cities of the industrial world (where 20% of the world's population owns 80% of the cars). In poor countries where emissions from coal-burning plants are not closely controlled, however, car exhausts mix with SO_2 to form an even more potent set of gases. Once the SO_2

BOX 7.8 HOW TO BLACKEN A TRIANGLE

A British doctor (K. Chopin) visiting Poland in 1991 reports this comment from his Polish host, the deputy minister for health: "When Poland faces a tricky international football [match], the odds are that it will be played in Katowice; if the surroundings do not incapacitate the opposition, the oxygen-free atmosphere usually takes a few yards off their pace." Katowice is the principal city of the Silesian region of southwestern Poland, home of Polish coal production and, consequently, of its steel and most of its other heavy industry. Silesia has always been rich in resources but poor in quality of life for its inhabitants. It was occupied by Prussians in the 1800s, seized by Germany in World War II, and then taken by the Soviets at the end of the war. Instead of using its resources to improve the life of the people who lived there, all of these conquerors, and many others, forced them to produce goods for the benefit of the occupiers.

During Soviet control, Polish Silesia joined with eastern parts of the Czech Republic (then Czechoslovakia) and southeastern parts of East Germany to form the major site of heavy industry in Russia's satellites in eastern Europe. Little, if any, thought was given to pollution control. Dr. Chopin writes, now in his own words, "The air is sulphurous here. The snow is acid snow. The lakes and rivers still look alive. Not so the forests; whole tracts of once glorious birch and pine stand naked, stripped of their foliage, or stunted, resembling the skeletal survivors of a nuclear storm." Because of this desolation, the area has been referred to as the "black triangle" (Fig. 7.21), and all three countries that share it have been trying to develop strategies to clean it up.

The cause of this devastation is the high sulfur content of the poor, but abundant, Silesian coal. When burned without adequate emission controls, of which there are virtually none anywhere in the triangle, the released SO_2 forms acid precipitation more destructive than that of anywhere in the rest of the world. Strangely, this extreme atmospheric pollution does not extend very far. Forests only a few tens of miles outside of the triangle show the telltale signs of stress, gray leaves and yellow pine needles, but not wholesale killing of trees. Water-borne pollution, however, renders groundwater unusable for tens of miles around the triangle and washes down the major rivers for hundreds of miles (Fig. 7.22).

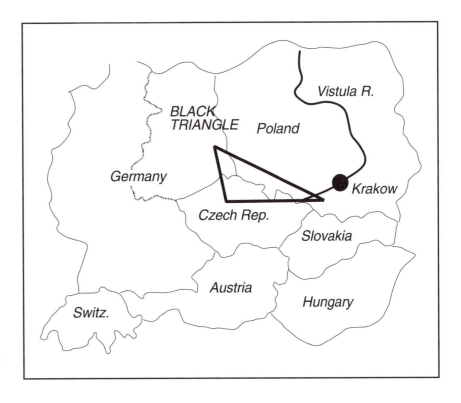

Figure 7.21 Location of "black triangle" in central Europe.

BOX 7.8 HOW TO BLACKEN A TRIANGLE *(continued)*

The effect on people's lives is more dramatic than on the forests. The inhabitants are described as tired and prematurely aged, suffering from asthma and other respiratory problems, and having a life expectancy approximately 10 years less than those elsewhere in eastern Europe. When they try to raise vegetables in small gardens, the groundwater is so polluted that one-fourth of the product is inedible.

Surely something can be done, but what. If a polluting factory is closed, people lose jobs. If money is used to clean up the triangle, it is not available for use elsewhere. Right now, not much is being done.

FURTHER READING: Chopin (1992).

Figure 7.22 Vistula River at Krakow, Poland. The river is lifeless, without even birds landing on it, because of toxic chemicals entering it from heavily industrialized areas in the Black Triangle. See text for further discussion.

and exhaust gases are in the air, they combine with water droplets and any solid particles that may be present to form a blanket of polluted air that is too dense to rise into the upper atmosphere and commonly hangs over cities like the proverbial black cloud.

This cloud, which we call smog, is a serious health risk for many people. SO_2 and solid particles constrict lung passages and are particularly dangerous for people with asthma. The oxidizing power of ozone also adds to lung damage, promotes asthma in people who are susceptible, and causes almost everyone's eyes to sting. CO has the opposite effect of ozone, removing oxygen from the blood by interfering with the action of hemoglobin. And finally, the hanging cloud of stale air simply smells bad.

Table 7.5 provides qualitative estimates of the smog in the world's twenty largest cities, and we offer a few examples of individual problems.

- Mexico City is built on a former lake (Section 3.3), and SO_2 produced by industries in the city joins car exhaust gases in a cloud that cannot escape the basin. The government has tried to reduce pollution by restricting the driving of private cars, including setting aside some days on which they cannot be driven at all.
- Sao Paulo, rapidly becoming the world's second largest city, has smog generated both by expanding industrial activity and rapid purchase of cars, giving the city one of the highest ozone levels in the world.
- Tokyo is rapidly cleaning up its smog problem by a series of steps begun in the 1960s. They include removing sulfur from coal before burning, trapping SO_2 produced by burning, using only unleaded gasoline, and placing catalytic converters on cars.
- Cairo has the double problem of increasing industrial activity and its position in the Sahara Desert. The

Table 7.5 *Types of air pollution in 20 large cities*

City	Sulfur dioxide	Lead	Carbon monoxide	Nitrogen dioxide	Ozone
Bangkok		Medium			
Beijing	High				Medium
Bombay					
Buenos Aires					
Cairo		High	Medium		
Calcutta					
Delhi					
Jakarta		Medium	Medium		Medium
Karachi		High			
London			Medium		
Los Angeles			Medium	Medium	High
Manila		Medium			
Mexico City	High	Medium	High	Medium	High
Moscow			Medium	Medium	
New York			Medium		Medium
Rio de Janeiro	Medium				
São Paulo				Medium	High
Seoul	High				
Shanghai	Medium				
Tokyo					High

Notes: All cities except London, New York, and Tokyo are regarded as having high levels of suspended particulates (dust). Data was unavailable for many categories of pollution in different cities, including four in Buenos Aires, three in Karachi, three in Manila, and four in Shanghai.

Sources: World Health Organization as cited by *Environment* (1994).

dust generated by cement and other factories is added to the dust periodically blown in from the desert, sometimes in choking storms so dark that people have trouble finding their way.

- Los Angeles is in a basin surrounded by mountains that trap wind from the Pacific Ocean and prevent them from blowing the smog away. The city now has very stringent controls on both industrial and car emissions, but occasionally a brown cloud of NO_2 will hang over it.

- Dacca, Bangladesh, is generally regarded as having one of the worst air-pollution problems in the world. We do not show it in Table 7.5, however, because the government does not keep usable records.

We have several ways to limit or eliminate the effects of smog. The first is to reduce the amount of SO_2 in the atmosphere by moving coal-burning plants away from cities and using scrubbers that can catch the soot and SO_2, or at least move them out of cities and send the effluent elsewhere. The second remedy is to limit car emissions as much as possible to CO_2 and H_2O. The easiest way to do this is to reduce the use of private cars by encouraging better development of public transportation, which is far less used in the United States than anywhere else in the world. We can also reduce emissions by using a more oxygen-rich fuel or inserting "catalytic converters" in a car's exhaust system to force the gasoline to be burned more effectively and thus reduce the CO and the unburned organic compounds that are the precursors to ozone. Neither process has much effect on the nitrogen oxides, but in the absence of the other gases they generally disperse fairly easily into the atmosphere.

The cost of these remedies is as difficult to determine as most of the other costs that we have discussed. Scrubbers on coal-burning factories add a few percent to the cost of fuel, possibly raising the price of goods produced in them, thus making them noncompetitive with goods produced elsewhere and putting their employees out of work. Both oxidized fuel and catalytic converters add a few percent to the cost of operating a car, which might require less purchase of other items and possible loss of jobs among people who produce them. Ideally these costs should be compared quantitatively with the costs of medical treatment of people who breathe the smog, but unfortunately none of the costs are sufficiently well known for this comparison to be done accurately.

Policy questions

Here is another set of unanswerable questions.

- Should industrial societies pay more for manufactured goods in order to reduce the effects of acid rain on soils and waterways?
- Should poor countries close some of their factories and accept present unemployment as a consequence of better health for their citizens in the future?
- Should people in industrial countries pay a little more to run their cars in order to reduce health risks for people who are particularly susceptible to smog?

FURTHER READING: Bridgman (1990); Mason (1992); Radojevic and Harrison (1992); Boubel et al. (1994).

7.7 SUMMARY AND CONCLUSIONS

So how is the earth's environment doing? Well, the news is both good and bad. We summarize the topics of this chapter, considering first the good news, to highlight our accomplishments, and then the bad news, to show where problems still need resolution.

Bulk waste

There is good news for both lesser-developed and industrial countries. Most of the lesser-developed countries generally do not produce enough waste to exceed the capacity of their countryside to store it, and industrial (OECD) countries that produce most of the world's solid waste have begun to recognize the problem. By using materials that are biodegradable and by recycling everything possible, they are making efforts to reduce the total volume of material that must be stored. These countries are also using innovative methods, such as high-temperature

incineration, to reduce the quantity that must be placed in landfills. We probably will not cause our cities to rise as mounds to rival ancient cities that did not throw away their garbage.

The bad news is that industrial countries still produce more waste than they know what to do with. Many of them export waste to other (generally poor) countries, and particularly in the United States, populous areas export waste to other areas that still have some space left.

Hazardous chemicals

The good news is that we now recognize that many chemicals that we have used indiscriminantly are toxic both to people and to other organisms that a healthy environment needs. For this reason, industrial countries now ban many dangerous pesticides and are continually looking for ways to reduce discharges of hazardous chemicals used by industry. As a result, those of us who live in industrial countries are probably safer today than we were 10 years ago.

The bad news is twofold. One is that lesser-developed countries still use many of these dangerous chemicals in both industry and agriculture. Much of this use is prompted by industrial societies that depend partly on food produced abroad and items manufactured in low-wage countries that have few (or no) environmental safeguards. The second piece of bad news is that much of the cleansing of industrial countries results from their ability to export dangerous chemicals that they do not want around and the willingness of governments of poor countries to be paid for importing them.

Radioactive waste

The good news is that we now understand the dangers of both high-level and low-level waste, we know how to store low-level waste safely, and we have some promising ideas for disposal of high-level waste.

The bad news is that we aren't doing anything. Political problems have prevented construction of low-level dumps in most places and high-level dumps anywhere. The low-level waste may not be a major danger, but the temporary storage of high-level waste could be catastrophic.

Biologically active waste

The good news is that we know the sources of the excess nutrients that are causing eutrophication and other problems, and we know what to do about them. Much of the problem can be solved by using less fertilizer, and we have developed technological processes to make sewage water nonpolluting. Where these methods have been used, particularly in western Europe, they have remarkable results.

The bad news is that the apparent cost means that nothing much is being done in most of the world. Many people feel that reduced use of fertilizer will make agricultural products more expensive, and many more do not want to bear the cost of extra sewage treatment. The situation is understandable in poor countries, but rich countries like the United States have little excuse for not reducing their output of excess nutrients.

Air pollution

The good news is that that we understand most of the reasons for air pollution and what to do about them. Consequently, by reducing the polluting emissions from automobiles and factories, we have made the air cleaner in many of the world's major cities. By rigorously enforcing these procedures, there is no reason why we should not enjoy further improvement.

The bad news is that the air in the cities of lesser-developed countries continues to deteriorate. Without major improvement in the economies of these countries, this problem will probably become worse.

PROBLEMS

1 An area of approximately 150 square miles in and around New York City generates 12,000 tons of garbage per day. If that garbage was not disposed of and simply accumulated in the city, compare the rate at which the ground of New York would rise with the rate of accumulation in ancient Troy (Section 7.2). Assume a density of about 0.33 ton per cubic yard for the accumulating garbage.

2 A nuclear waste dump receives two elements: A, with a half-life of 30 years, and B, with a half-life of 5 years. The amount of element B delivered originally is 100 times the amount of element A. Estimate how long it will be before the amount of B is less than the amount of A.

$$
\begin{array}{ll}
\ln 1 = 0 & \ln 6 = 1.79 \\
\ln 2 = 0.69 & \ln 7 = 1.95 \\
\ln 3 = 1.10 & \ln 8 = 2.08 \\
\ln 4 = 1.39 & \ln 9 = 2.20 \\
\ln 5 = 1.61 & \ln 10 = 2.30
\end{array}
$$

3 A radioactive waste product has a half-life of 30 years. How long must it be stored until only 1% of the original amount remains?

4 A manufacturing company accidentally discharges 500 kg of a soluble lead compound into a small lake that is 10 m deep and has a surface area of 25,000 square meters. If the upper limit for lead in drinking water is 50 ppb (50×10^{-7}%):
   a. Is the water in the lake drinkable?

   b. If inflow and outflow to the lake make the residence time of water in the lake 10 years, how long will it be before the water in the lake is drinkable?

5 Two pesticides (A and B) are discharged in equal amounts into a drainage basin. A is 100 times more toxic than B. A is biodegradable (degraded by organisms in water and soil) at a rate such that one-half has disappeared within 2 months after discharge. B is not biodegradable but is washed out of the drainage basin at a rate such that only one-half remains after 2 years. After 5 years, which pesticide is more dangerous?

6 Compare two lakes (A and B). Lake A is undergoing eutrophication, and its water contains 8 ppm (8×10^{-4}%) phosphate and 19 ppm dissolved nitrogen. Lake B has a normal biota, and its water contains 9 ppm phosphate and 7 ppm dissolved nitrogen. In order to prevent eutrophication of Lake B, how necessary is it to reduce the use of phosphate fertilizers in its watershed?

7 From the standpoint of air pollution, is it better to burn wood or bituminous coal? (Note – refer to discussion in Section 5.3.)

8 Iodine is more effective than chlorine in killing microbes in drinking water, with an iodine concentration of less than 1% needed for purification. For this reason, backpackers sometimes carry iodine when they are going to be drinking from streams and springs of unknown purity (for data on water requirements, see Section 4.2). How much iodine should a person carry for a 2-week trip?

9 A city produces waste that contains approximately 30% paper and cardboard, 5% to 10% each of plastic, glass, and metals (including aluminum cans), and approximately 50% noncombustible or recyclable material. If it costs the city $20 million per year to put all of the waste in a landfill, how much should the city pay to operate an incinerator, and how much should it pay to recycle as much as possible?

REFERENCES

Blackman, W. C., Jr. (1993). *Basic Hazardous Waste Management*. Boca Raton, Fl.: Lewis.

Boubel, R. W., Fox, D.L., Turner, D. B., and Stern, A. C. (1994). *Fundamentals of Air Pollution*. San Diego: Academic Press.

Bridgman, H. A. (1990). *Global Air Pollution: Problems for the 1990s*. London: Belhaven Press.

Broadus, J. M., and Vartanov, R. V. (1994). *The Oceans and Environmental Security: Shared US and Russian Perspectives*. Washington, D.C.: Island Press.

Carson, R. (1962). *Silent Spring*. Greenwich, Conn.: Fawcett.

Chopin, K. (1992). Pollution most foul: Effect of pollution in Poland on citizens' health. *British Medical Journal* 304: 1495–7.

Dold, C. (1996). Hormone hell. *Discover* (September): 53–9.

Eden, G. E., and Haigh, M. D. F., eds. (1994). *Water and Environmental Management in Europe and North America – A Comparison of Methods and Practices*. New York: Ellis Horwood.

Environment (1994). Air pollution in the world's megacities. *Environment* 36(2): 4–26.

Francis, B. Magnus (1994). *Toxic Substances in the Environment*. New York: Wiley-Interscience.

Garrett, F. E. (1912). *Lyrics and Poems from Ibsen*. London: E. P. Dutton.

Granier, R., and Gambini, D.-J. (1990). *Applied Radiation Biology and Protection*. New York: Ellis Horwood.

Guthrie, G. D., Jr. (1992). Biological effects of inhaled materials. *American Mineralogist* 77: 225–43.

Hammer, M. J., and Hammer, M. J., Jr. (1996). *Water and Wastewater Technology*. Englewood Cliffs, N.J.: Prentice Hall.

Knott, D. (1995). Protest over Brent Spar disposal claims spotlight off NW Europe. *Oil and Gas Journal* 93(48): 23–8.

Krauskopf, K. B. (1988). *Radioactive Waste Disposal and Geology*. London: Chapman and Hall.

MacKenzie, D. (1993). Europe's toxic waste (Basel Convention seeks to regulate exporting of waste to poorer nations). *World Health* 46(5): 6–8.

Malle, K.-G. (1996). Cleaning up the Rhine River. *Scientific American* 274 (January): 70–5.

Mason, B. J. (1992). *Acid Rain: Its Causes and Its Effects on Inland Waters*. Oxford: Oxford University Press.

Morehouse, W., and Subrmaniam, M. A. (1986). *The Bhopal Tragedy*. New York: Council on International and Public Affairs.

Mossman, B. T., Bignon, J., Corn, M., Seaton, A., & Gee, J. B. L. (1990). Asbestos: Scientific developments and implications for public policy. *Science* 247: 194–301.

OECD (annual a). *Environmental Data – Compendium*. Paris: Organization for Economic Cooperation and Development.

OECD (annual b). *State of the Environment*. Paris: Organization for Economic Cooperation and Development.

Radojevic, M., and Harrison, R. M., eds. (1992). *Atmospheric Acidity – Sources, Consequences, and Abatement*. London: Elsevier.

Rathje, W., and Murphy, C. (1992). *Rubbish!: The Archaeology of Garbage*. New York: HarperCollins.

Reyburn, W. (1969). *Flushed with Pride – The Story of Thomas Crapper*. London: Redwood Burn.

Rodale's Organic Gardening (serial). Emmaus, Pa.: Rodale Press.

Ross, M. (1995). The schoolroom asbestos abatement program: A public policy debacle. *Environmental Geology* 26: 182–8.

Roxburgh, I. S. (1987). *Geology of High-Level Nuclear Waste Disposal – An Introduction*. London: Chapman and Hall.

United Nations Environmental Programme (1991). *Radiation: Doses, Effects, Risks*. Oxford: Blackwell.

United States Environmental Protection Agency (annual). *Toxic Release Inventory*. Washington, D.C.

Woodside, G. (1993). *Hazardous Materials and Hazardous Waste Management: A Technical Guide*. New York: Wiley-Interscience.

World Resources Institute and United Nations Environmental and Development Programmes (1994). *World Resources*. Oxford: Oxford University Press.

8 | GLOBAL CHANGE

8.0 INTRODUCTION

If the public believes something to be true, even if science has shown it to be false, that belief remains an important fact for policymakers. — Litfin, Ozone Discourses, p. 33

We have discussed numerous problems facing people as they attempt to use the earth for their benefit. Too many people. Not enough food. Imminent disasters such as earthquakes and floods. Not enough fresh water. Shortages of energy and various raw materials. Too much pollution of groundwater and soil. Do any of these issues really matter? If some country is overpopulated, does that affect the whole earth? Since we already produce as much food as the world needs, will the entire population ever starve to death? Do dwindling supplies of oil and steel mean that we will face permanent shortages of energy and construction materials? If 50,000 people are killed by a hurricane, will that destroy the human race? If one city is evacuated because of water pollution, is there any reason that its population can't move elsewhere?

No. None of these events change the surface environment of the whole earth, the ecosystem to which we are adjusted. Some of our activities, however, may irreversibly modify the whole atmosphere, waters, and land surface. We have saved contemplation of this marvelous eventuality until the end of the book, dividing the issue into three interrelated processes that are partly, but probably not wholly, caused by human activity.

MODFICATION OF THE ATMOSPHERE The emission of carbon dioxide (CO_2), and other gases that absorb radiation from the earth's surface may cause global warming, com-

monly termed the "greenhouse effect." We discuss the history of the atmosphere and the processes that control it in Section 8.1, and in Section 8.2 we evaluate the evidence for recent temperature increase and attempt to distinguish between natural changes and those induced by people.

DESTRUCTION OF ORGANISMS The ability of people to control their environment enables us to make limited choices about the success of both plant and animal species. For example, we domesticate cows and wheat and eliminate predators and large swaths of forest. Our investigation in Section 8.3 concentrates on species extinctions and reduction in the amount of forest cover.

PROLIFERATION OF DAMAGING CHEMICALS People send a number of chemicals that do not exist naturally on the earth into its surface and atmosphere, and we discuss them in Section 8.4. On a global scale they include lead, which causes neurological and metabolic damage to people, and industrial gases that reduce the ability of the atmosphere to shield us from incoming ultraviolet radiation.

All of these issues are the subject of denial. Some people deny that the earth is undergoing any significant changes, or they conclude that any changes are not our fault and may even be beneficial. For example, in 1995 a member of the U.S. House of Representatives stated that the possibility of global warning might be "liberal claptrap." Other people virtually deny that the human race can make any use of the earth without irreparably harming it. We will try to find a balanced approach to these problems by outlining the choices faced by all of us as we try to preserve a world that we can inhabit.

8.1 HISTORY AND CONTROLS OF CLIMATE AND ATMOSPHERE

It is quite right to think of the present ice cap as a relic of a much greater ice mass, which existed here when other parts of the world were also in the grip of the so-called great Ice Age. – L. M. Gould, Cold, p. 233

The earth's canopy of air is responsible for life as we know it. We think initially of the oxygen that we breathe, without which all life except bacteria would die. The atmosphere, however, is essential in other ways. It provides the CO_2 from which plants synthesize themselves. It is the source of rainwater, which replenishes surface water and groundwater and makes life on land possible. It absorbs solar energy "reflected" (see subsequent discussion for a better word) back from the surface of the earth. Without this absorption, the mean temperature of the earth's surface would be about 30°C colder than it is now, freezing all water and killing virtually all life. Furthermore, the upper atmosphere contains ozone that absorbs incoming ultraviolet radiation which, if intensified, would cause a variety of afflictions from skin cancer to increased mutation rates. Before we discuss how the atmosphere performs all of these functions, we need to discuss its history, both with respect to composition and climate.

The composition of the earth's initial atmosphere, about 4.5 billion years ago, was probably similar to the gases that constitute the gas-giant planets, such as Jupiter. Major components probably included ammonia, methane, water, carbon dioxide, and hydrogen. The H_2 escaped rapidly from the atmosphere, leading to the breakdown of NH_3 to N_2 and reducing the methane (CH_4) contents to low levels through a complex of little-understood reactions. By the time the first life began to develop on the earth, about 3.5 to 4 billion years ago, the atmosphere was likely to have been largely a mixture of N_2 and CO_2, plus some water.

A major change in the atmosphere occurred about 2 billion years ago, when free O_2 first appeared. The evidence is that iron in soils and sediments formed since that time is mostly the oxidized form (Fe^{+3}, ferric), whereas older rocks show that the more soluble reduced form (Fe^{+2}, ferrous) was freely transported in surface water. The supply of oxygen (O_2) gradually increased, reaching about 5% when the first plants evolved approximately 350 million years ago. That figure is derived from the fact that 5% O_2 is the optimal concentration for plant growth. Since that time, by a variety of poorly understood processes, the O_2 concentration of the atmo-

Table 8.1 *Composition of the atmosphere*

Component	Concentration (in %)
N_2	78.1
O_2	20.9
H_2O	0.3
CO_2	0.9
Ar	0.01
N_2O	3×10^{-5}
CH_4	1.7×10^{-4}
Total CFC	5×10^{-7}

Notes: Concentration of water vapor is highly variable and is shown only as a worldwide and annual average. Total CFC is the sum of the percentages of all chlorofluorocarbons (Section 8.4).

sphere has increased to its present 18%. The present composition of the atmosphere is shown in Table 8.1 and will be discussed further.

Much of the earth's climate is determined by the distribution of surface temperatures, and temperature is controlled largely by the abundance of CO_2 in the atmosphere and the intensity of the sun's radiation (see the subsequent discussion of "greenhouse gases"). Solar radiation apparently has increased about 25% since the early stages of earth history, thus increasing surface temperature. This increase, however, has been matched by reduction of atmospheric CO_2, which kept the earth cooler. The earth's surface is probably cooler now than it was in its early history, but temperatures below the boiling point of water and above its freezing point have been maintained for nearly 4 billion years. Within that broad range, the climate shows gradual changes over intervals of several million to tens of millions of years and, in the recent past, periodic fluctuations over tens of thousands of years.

All major climatic fluctuations are closely tied to the advance and retreat of continental glaciers. Glaciation occurs when the earth is sufficiently cool and when drifting continents are positioned so that some large land mass is in a polar region, and glacial periods ("icehouses") have been identified at various times during approximately the past 1 billion years of earth history. The last major icehouse developed about 25 million years ago when the Antarctic continent was isolated from other continents by circumpolar currents, thus terminating a warm "greenhouse" that had apparently reached its peak about 100 million years ago (Fig. 8.1). Glaciation in the Northern Hemisphere began about 3 million years ago,

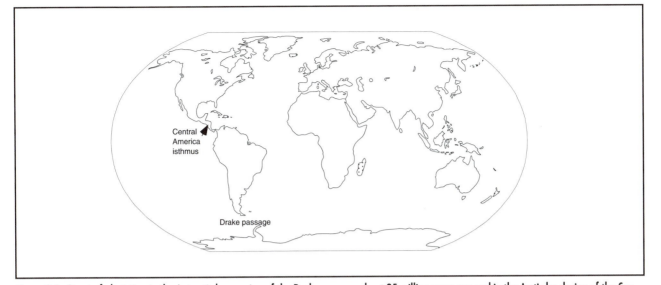

Figure 8.1 Onset of glaciation in the Antarctic by opening of the Drake passage about 25 million years ago and in the Arctic by closing of the Central American isthmus about 3 million years ago.

caused by reorientation of oceanic currents when volcanic processes and plate movements closed the seaway between North and South America by construction of Central America.

About 15 to 20 major periods of glacial advance and retreat have occurred in the Northern Hemisphere in the past 2 million years. They have a periodicity of approximately 100,000 years, consisting of a gradual cooling that lasts about 80,000 to 90,000 years followed by rapid warming for 15,000 to 20,000 years (Figs. 8.2 and 8.3). The reasons for this periodicity are highly controversial (readers know by now that geologists can never agree on anything), but the basic cause is related to variations in the earth's path of movement about the sun. The last

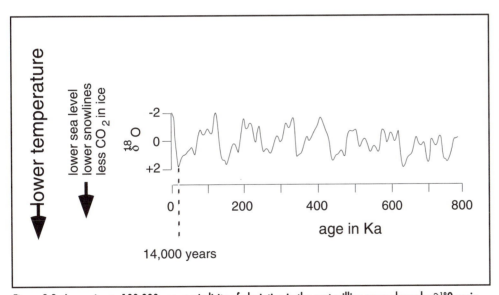

Figure 8.2 Approximate 100,000-year periodicity of glaciation in the past million years shown by $\delta^{18}O$ variations in marine calcareous fossils. ($\delta^{18}O$ is [($^{18}O/^{16}O$)sample $-$ ($^{18}O/^{16}O$)standard]/($^{18}O/^{16}O$)standard] $\times$ 1,000.) Increase in ^{18}O content of seawater indicates increase in volume of ice sheets, which are impoverished in the heavy ^{18}O isotope. Correlative variations that indicate lower temperatures in polar regions include lower sea level, lower snowlines and less CO_2 preserved in gas entrapped in ice. From Rogers (1993), figure 8.15

(a)

(b)

Figure 8.3 Retreat of glaciers: (a) Worthington glacier, Alaska; (b) east coast of Greenland.

warming to an "interglacial" phase began about 15,000 years ago. At this time both plants and animals moved northward as temperate climates replaced Arctic ones, and spring meltwater from North American glaciers left erosion scars showing that the Mississippi River was nearly 50 miles wide during the peak of discharge of glacial ice into the Gulf of Mexico. As meltwater flowed into the oceans, sea level rose $\sim$100 m to its present position over an interval of a few thousand years. Within this postglacial (or interglacial) period are several smaller episodes of warming and cooling, although none of them reached the low temperatures of the last glacial period. Because they are closely related to human activity, we discuss them in Section 8.2.

Control of atmospheric composition

The atmosphere contains the diverse components shown in Table 8.1. We do not discuss N_2 because its concentration is so large that exchanges with the surface are too small to be relevant, or O_2 because there is more than an adequate supply for the world's animal life and the metabolic needs of plants (Table 8.1; discussion of plant metabolism in Section 2.2). We concentrate, therefore, on CO_2 because of its importance as a "greenhouse gas," and on H_2O and CH_4, both of which show important variations from time to time.

CARBON DIOXIDE   Distribution of carbon calculated as CO_2 between the atmosphere and other parts of the surface is summarized in Figure 8.4. The figure shows various

"reservoirs" that contain most of the surface CO_2 and the rate of transfer of CO_2 among them. Where carbon exists as organic carbon, as in plants and animals, we show the amount of CO_2 that would be formed if all of the organic carbon were burned. By far the largest reservoir of potential CO_2 is the amount stored in carbonate rocks, principally limestones. As limestones are deposited and buried, some of them are removed from contact with the atmosphere, and their CO_2 is retained until the limestone is weathered. The precipitation and weathering of limestone is immeasurably slow on a human time scale (hundreds to thousands of years), however, and we do not include it in Figure 8.4. The next largest reservoir is the amount of CO_2 that would be formed by burning fossil fuel (coal, oil, gas, etc.; see Chapter 5). This carbon is

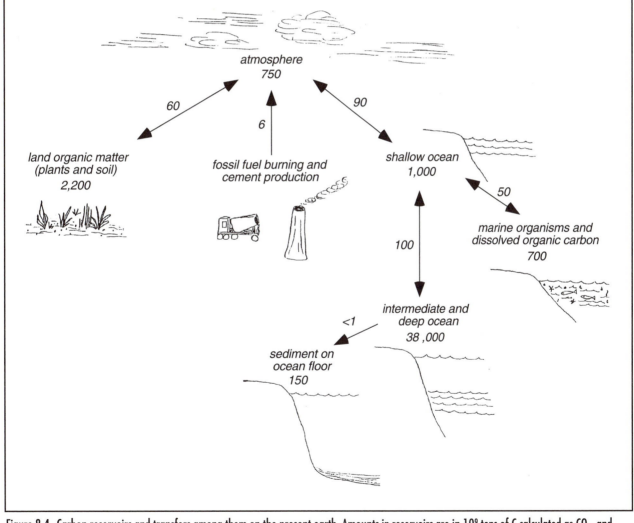

Figure 8.4 Carbon reservoirs and transfers among them on the present earth. Amounts in reservoirs are in 10^9 tons of C calculated as CO_2, and rates of transfer are in 10^9 tons of C calculated as CO_2 per year. Adapted from Houghton et al. (1994).

the remains of plants and animals that died and were pre-served from decay by burial away from the oxidizing ef-fects of the atmosphere (Section 5.2). Organic remains are circulated back into the atmosphere only by natural weathering of exposed rocks, such as coal seams, and by human burning as a fuel. The amount that we burn as fuel is shown in Figure 8.4, and we discuss the current rapid increase in this amount in Section 8.2.

The amount of carbon currently contained in living organisms is much less than the amount accumulated as fossil fuel, and most of it is continually cycled through the atmosphere as CO_2. The cycling results from the ex-traction of CO_2 from the atmosphere by plants during photosynthesis and its emission is a product of animal metabolism, processes that we discuss more in Section 2.1. Unless other processes affect the equilibrium, if the concentration of CO_2 in the atmosphere remains con-stant, then the amount extracted by photosynthesis must equal the amount emitted by animals.

The CO_2 in the atmosphere is in equilibrium with CO_2 dissolved in water, principally the oceans. When CO_2 dissolves, it first reacts with water to form carbonic acid (H_2CO_3) and then ionizes to bicarbonate, carbon-ate, and hydrogen ions by the following reactions:

$$CO_2 + H_2O \rightarrow H_2CO_3$$

$$H_2CO_3 \rightarrow H^+ + HCO_3^-$$

$$HCO_3^- \rightarrow H^+ = CO_3^{-2}$$

The H^+ ion undergoes the further reaction

$$H^+ + OH^- \rightarrow H_2O$$

where

$$[H^+] \times [OH^-] = 10^{-14}$$

This reaction illustrates the concept of pH, which is defined as $-\log_{10}(H^+)$ and is a convenient way to de-scribe whether a solution is acidic, neutral, or basic. At pH = 7, the concentrations of both H^+ and $OH^- = 10^{-7}$, and the solution is neutral. At pH < 7, the concentra-tion of H^+ is > 10^{-7}, the concentration of OH^- is < 10^{-7}, and the water is acidic. Conversely, at pH > 7, the OH^- is more concentrated than the H^+ and the water is basic. Because of the formation of H^+ when CO_2 dis-solves, all freshwater is acidic, with a pH between 5.5 and 7. Seawater, however, contains enough carbonate and bi-carbonate ions washed into the ocean by weathering of limestones on land that they offset the tendency of CO_2 to form an acid, and seawater consequently remains at a pH of 8.2 (basic). This high pH permits CO_3^{-2} ion to combine with Ca^{+2} in seawater to form shells and ulti-mately be preserved as limestone.

The ability of water to dissolve CO_2 is determined by three factors. Increasing temperature drives any gas out of solution (e.g., a cold can of carbonated drink fizzes less than a warm can when they are opened). Increasing pres-sure drives gases into solution, a fact utilized by drink manufacturers when they put pressurized CO_2 into the can. Also, higher pH (more basic) promotes greater solu-bility because a lower H^+ concentration permits more ionization of carbonic acid. These factors combine to generate two competing effects on solubility:

- An increase in the pressure (partial pressure) of CO_2 will increase the concentration of dissolved carbonate in the oceans. Although an increase in dissolved CO_2 would lower the pH of pure water (make it more acidic), it will have only a slight effect on the pH of the seawater because the pH is held at a mildly basic value of ~8.2 by dissolved HCO_3^- and CO_3^{-2} (we say that the bicarbonate and carbonate "buffer" the pH).

- If an increase in atmospheric CO_2 causes atmospheric, and hence oceanic, heating, then the higher tempera-ture will reduce the solubility of CO_2. In this situa-tion, the CO_2 will equilibrate back into the atmo-sphere, which will further increase the partial pressure of CO_2 in the atmosphere. An increase in atmo-spheric temperature of 3°C will cause a 10% decrease in the solubility of CO_2 in the surface layers of the ocean but will not affect the deep ocean water, which contains virtually all of the dissolved CO_2 (Fig. 8.4), unless the temperature remains high for a long enough time to heat the deep ocean (about 1,000 years) or un-less some other event causes heating throughout the ocean (Section 8.2).

In addition to considering the amounts of CO_2, or po-tential CO_2, in the various reservoirs, we also must con-sider the rate of transfer between the reservoirs. Figure 8.4 shows these rates as balanced between atmosphere and ocean and between atmosphere and living organ-isms, but the emission of CO_2 by burning of fossil fuel is a one-way process on a human time scale and may be dis-turbing the equilibrium (more in Section 8.2). Further-more, because the equilibrium between atmosphere and ocean requires virtually instantaneous transfer of an equal amount of CO_2 into and out of solution, these rates cannot readily be measured. The important rate in Figure 8.4 is the amount used in photosynthesis and the amount emitted by animals. This rate shows that approximately one-sixth of the amount of CO_2 in the atmosphere is removed by plants each year and returned by animals. We say that CO_2 has a 6-year "residence time" in the atmosphere (see further discussion in Section 4.1).

WATER The transfer of H_2O into and out of the atmosphere is, in principle, similar to the distribution of CO_2 (see Section 4.1 for further discussion). The oceans contain 97% of the earth's water, with most of the remainder in icecaps and groundwater. The atmosphere contains 13 trillion (13×10^{12}) metric tons, an amount that would cover the earth to a depth of 1 inch if it were all "squeezed" out of the atmosphere. This volume of atmospheric water is maintained essentially constant by the evaporation of 500×10^{12} metric tons from surface water each year and an equal amount of rainfall. This precipitation corresponds to a world average rainfall of approximately 30 inches (75 cm) per year. The residence time (see above) of water in the atmosphere is calculated as

$$\frac{13 \times 10^{12} \text{ mt}}{500 \times 10^{12} \text{ mt/yr}} = 0.026 \text{ years (approximately 9 days)}$$

METHANE The methane (CH_4) content of the atmosphere is low because it oxidizes in air to H_2O and CO_2 in the same way that natural gas burns (Section 5.3). Because of its instability, CH_4 is present in the atmosphere only because the slow oxidation rate gives it a residence time of ~10 years. At present approximately 500 million (500×10^6) metric tons of methane are emitted into the atmosphere per year, roughly equally divided among four sources: (1) as a gaseous waste product by ruminant animals (cows, goats, and sheep; see Section 2.3); (2) as a decay product from natural marshes and other swampy ground where lack of O_2 prevents complete oxidation of organic material; (3) as a decay product from rice paddies; and (4) as a product of the burning of fossil fuels (Section 5.3). The marshes, largely including subarctic tundra, and some of the ruminants are natural emitters, but emissions from domesticated ruminants, rice paddies, and fossil fuels are created by human activity (Section 8.2). The natural concentrations and emissions are essentially balanced and yield atmospheric CH_4 concentrations of approximately 0.7 ppm (0.7×10^{-4}%), but human activity now adds approximately 6% more CH_4 to the atmosphere per year than can be oxidized, and pres-

ent CH_4 levels have reached approximately 1.7 ppm. Some recent periods appear to have been times of excessively high emission of CH_4, and we discuss one of these episodes in Box 8.1

Control of climate

The earth's climate is controlled largely by three factors: (1) the earth's surface and its rotation; (2) the composition of the atmosphere; and (3) the oceanic "conveyor belt" that helps distribute heat around the world.

EARTH'S SURFACE AND ROTATION The decrease in amount of solar radiation from the equator to the poles combines with the earth's rotation to yield patterns of precipitation and air movement, which we also discuss briefly in Sections 2.2 and 4.1. Precipitation is generally heaviest in equatorial regions and in a band centered around latitude 60° north and south. Conversely, the driest air and the world's deserts are centered in belts from about 20° to 40° north and south latitude. Latitudes 30° and 60° also separate the earth's surface into three bands of contrasting wind direction in each hemisphere. From the equator to 30°, the dominant wind is trade winds that blow from the east toward the west (they are easterlies). Between latitudes 30° and 60°, the dominant winds are westerlies that blow from the west toward the east. From latitudes 60° to both poles, the prevailing winds shift back toward the west.

In addition to latitudinal variations in solar radiation and wind, the earth's surface exerts an important control over climates by the distribution and topography of continents. The principal effect of their distribution results from the fact that ocean water cools and heats more slowly than land surfaces. Consequently, cool air forms over oceans and, because of its density, moves over continents as "onshore" winds. Also, islands and other land areas adjacent to oceans show lower climatic extremes than landlocked areas; that is, lands with a "maritime" climate are commonly warmer in the winter and cooler in the summer than continental interiors. Furthermore, the fact that the Northern Hemisphere is primarily land and the Southern Hemisphere primarily water means that more solar radiation is absorbed in the Southern Hemisphere than in the northern one.

The topography of continents exerts only minor influence over global climates but important control on a local scale. Flat continents permit air masses to move freely, and because all continental margins around the Arctic Ocean are inactive (Section 3.1), cold Arctic air blows unimpeded across much of the Northern Hemi-

BOX 8.1 LEAKING SOVIET PIPELINES _____

"Damn the torpedoes, full speed ahead!" said Admiral David Farragut, commanding his Union fleet to attack the Confederates in Mobile Bay on August 5, 1864. Farragut was referring to mines, a known danger that stood between his fleet and its objective. A similar attitude based more on ignorance and carelessness was displayed by much of the Soviet hierarchy when its objective was sufficient industrial development to enable the leaders to stay in power.

In the spirit of carefree development, during the 1970s (the Brezhnev era) the Soviet Union began a rapid, almost helter-skelter, development of its natural gas resources. The gas was necessary because of its profligate use, and waste, by industries and consumers in the Soviet Union and eastern Europe. Much of the production was in the West Siberian basin, which holds approximately one-third of the world's resources of natural gas (methane; CH_4). From there, pipelines fanned out over much of the Soviet Union, satellite countries in eastern Europe, and then onward to provide much of the gas supply for the industrial economies of western Europe.

All of this development made good economic sense as measured in short-term profits. The problem was the long-term environmental costs. In their haste and carelessness, the Soviets paid little or no attention to the quality of their production methods and of their pipeline system. To put it bluntly, the pipelines leaked like sieves. Although gas leaks from modern pipelines in industrial countries are commonly about 2% of production, loss from the Soviet system may have been as much as 10%. Despite this inefficiency, gas production increased throughout the 1980s, reaching a peak of about 500 million tons of methane in 1990. At this time, the Soviets could no longer ignore the dangers – a pipeline explosion in 1989 killed more than 400 people – and they called on Western experts to overhaul the system.

Repair of the pipelines in 1991 roughly coincided with a change in the rate of increase in CH_4 in the atmosphere. During the 1980s, the CH_4 concentration had been increasing by approximately 0.1 ppm (0.1×10^{-4}%) per year. This rate of increase was caused by the emission of approximately 500 million tons per year (see text) and the ability of the atmosphere to oxidize only about 95% of that amount. By comparison, as we discuss in Section 8.2, atmospheric methane has increased from approximately 0.7 ppm in preindustrial times to approximately 1.7 ppm at present, a rate of increase of approximately 0.05 ppm per year over the past 200 years. In 1992, the rate of global atmospheric methane increase returned to approximately 0.05 ppm per year, with the reduction in emissions particularly noticed in the Northern Hemisphere.

Is it a coincidence that repair of the Soviet pipelines occurred at the same time that methane emissions returned to a more normal level for an industrial world? Other explanations are possible. The most likely is the effects of the eruption of Mt. Pinatubo in the Philippines in 1991. The aerosols emitted by this eruption could have increased the oxidizing capacity of the atmosphere because of enhanced penetration of ultraviolet radiation through an ozone-depleted stratosphere that had been slightly depleted by the eruption (Section 8.4). Also, the slight global cooling caused by Pinatubo's atmospheric dust might have caused slight reduction in CH_4 production from wetlands and rice paddies. None of these effects seem quantitatively significant, however, and the coincidence of pipeline repair and reduction in methane buildup, plus the absence of any other reasonable explanation, leave us with the likelihood, but not proof, that Soviet pipelines were supplying nearly half of the excess methane emitted to the atmosphere in the 1980s. We cannot be sure, but we do hope that the successor countries to the Soviet Union can keep their pipelines in better repair.

FURTHER READING: Dlugokencky et al. (1994).

sphere during the winter (perhaps responsible for the fact that Santa Claus is thought to ride a sleigh). In contrast to flat countryside, mountain ranges force air to move up and over them. This rise cools the air, causes high rainfall on the slopes of the mountains toward which the wind is blowing (the "windward" side), and leaves deserts on the opposite ("lee") side. Deserts formed in this fashion are called "rain shadow" deserts.

ATMOSPHERIC COMPOSITION The atmosphere modifies the climatic controls just described because of its ability to absorb some parts of the spectrum of electromagnetic (EM) radiation (Fig. 8.5), and we describe the spectrum before discussing atmospheric effects. EM radiation moves with the speed of light (with the symbol c; 186,272 miles per second) and is described in terms of its wavelength and frequency. Wavelength (l) is the dis-

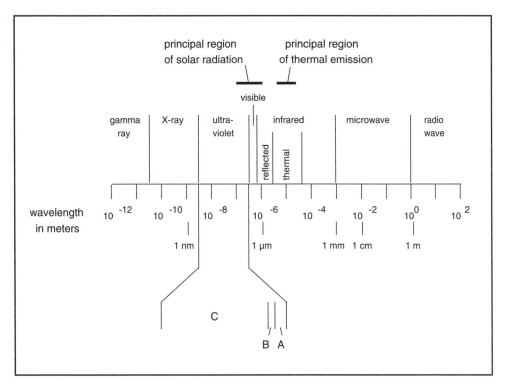

Figure 8.5 The spectrum of electromagnetic (EM) radiation.

tance between peaks, or troughs, of the cyclical wave; that is, it has the same meaning for EM radiation as for ocean waves. Frequency (*f*) is the number of cycles that pass a point in one second. Thus,

$$l \times f = c.$$

The energy in a packet (photon) of radiation is proportional to its frequency (inversely proportional to its wavelength). This relationship is represented by the equation

$$E = h \times f,$$

where *h* is Planck's constant of radiation. Thus, high-frequency waves contain more energy per packet than low-frequency waves, and the high-energy waves are capable of penetrating farther into solids and liquids than low-energy waves. That is why gamma rays and X-rays are particularly dangerous and radio waves are not.

Absorption of radiation is partly dependent on the radiation frequency. Solids and liquids, such as the land and ocean surfaces of the earth, absorb radiation largely "in bulk," but gases consist of individual molecules that selectively absorb radiation at characteristic frequencies, and they are generally unaffected by radiation at other

frequencies. Frequencies absorbed by a molecule provide energy that causes the molecule to "resonate" at the absorbed frequency. Resonant frequency has an identical meaning to the frequency (pitch) of a tuning fork, except that the tuning fork has only one frequency at which it both emits and absorbs sound waves. Molecules, however, have several absorption frequencies resulting from their ability to rotate, stretch, and bend in various directions.

Based on these principles, Figure 8.6 summarizes the fate of solar radiation as it reaches the earth's atmosphere. The frequencies of radiation emitted by the sun are determined by its high surface temperature of approximately 6,000°C (its "black body temperature") and are largely in the visual part of the spectrum but extend some distance into both the infrared (IR) and ultraviolet (UV) ranges. About one-third of solar radiation is reflected directly back into space from clouds and atmospheric dust. Most of the remainder penetrates to the earth's surface because it is at high frequencies that are not directly absorbed by atmospheric gases. Absorption of this radiation heats the land and oceans, and because the earth's surface temperature (approximately 20°C) is lower than the sun's, the earth then radiates this heat back into the atmosphere at lower frequencies than the incoming radiation. This outgoing radiation occupies a range of the

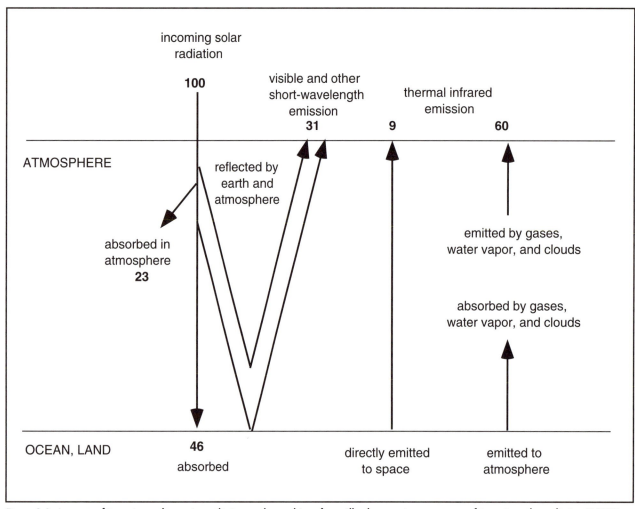

Figure 8.6 Amounts of incoming and outgoing radiation on the earth's surface. All values are in percentages of incoming solar radiation (100%). Adapted from Mitchell (1988).

spectrum referred to as "thermal infrared" (thermal IR) and contains a number of frequencies that are specifically absorbed by H_2O, CO_2, CH_4, and other gases in the atmosphere (Fig. 8.5). Absorption results in higher atmospheric temperatures by a process analogous to the heating of a greenhouse, which is caused by the fact that incoming solar radiation is at high frequencies that can penetrate the glass but outgoing radiation in the infrared range cannot get back through the glass. Thus, the greenhouse warms up, and the comparable process in the atmosphere is known as the "greenhouse effect."

THE OCEANIC CONVEYOR BELT Water that is colder and/or saltier than normal ocean water has a higher density and sinks. In the present earth, but not at all times in the past, this density difference results in a "conveyor belt" that moves water from the North Atlantic Ocean into the north Pacific (Fig. 8.7). It is initiated because

trade winds carry moisture from the air above the Atlantic Ocean across the narrow isthmus of Central America to the Pacific Ocean. Consequently, an excess of evaporation over precipitation makes the water of the Atlantic Ocean saltier and denser than Pacific water, and in the North Atlantic the cold temperatures and melting ice make the water colder. This cold, salty water sinks to the floor of the Atlantic and moves southward as a mass referred to as North Atlantic Deep Water (NADW). At about the equator it begins to mingle with similar water moving northward along the ocean floor from the Antarctic (Antarctic Bottom Water – AABW). The NADW then moves around Africa, rises to slightly higher levels in the Indian Ocean, filters through the Indonesian area, finally arrives, much diluted by other water, in the Pacific, and rises to the surface. Some of it returns to the North Atlantic along the ocean's surface,

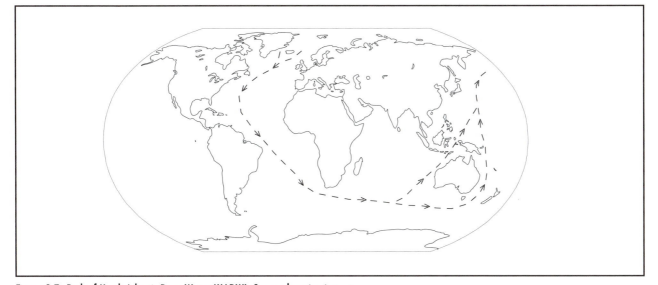

Figure 8.7 Path of North Atlantic Deep Water (NADW). See explanation in text.

following essentially the reverse route that it took on the way out.

The flow of NADW brings oxygen and cold temperatures to the lower part of the Atlantic Ocean, permitting some organisms to live at those depths. This cold water of the NADW also dissolves CO_2, forming a reservoir of CO_2 in the deep ocean that contains about 90% of all of the CO_2 that is mobile on the earth's surface (Fig. 8.4). The flow of NADW also affects climates because it transfers cold water southward at depth and, thereby, transfers heat northward on the surface. This northward transfer of heat combines with eastward-blowing winds (westerlies) and oceanic circulation patterns to keep Europe warmer than other land areas at comparable latitudes. Some investigators have speculated that the NADW conveyor belt partly or completely shuts down when global temperatures rise only a few degrees C, and we discuss the effects of its absence in Section 8.2.

FURTHER READING: Speidel (1988); Millero and Sohn (1992).

8.2 HUMAN ACTIVITY AND THE RECENT HISTORY OF ATMOSPHERE AND CLIMATE

Once we know for sure, it'll be too late. – Swiss Re, an insurance underwriting company, commenting on mounting insurance losses from natural disasters and their possible relationship to emissions of greenhouse gases, cited by the Canadian Globe and Mail, *July 27, 1996, p. A6*

Armed with the theoretical information provided in Section 8.1, we now investigate the recent history of the atmosphere and climate, the degree to which they have been influenced by human activity, and finally the possible consequences of global warming.

Recent history of the atmosphere and climate

Atmospheric data for more than 200,000 years can be obtained from cores drilled into the Antarctic and Greenland ice caps and smaller ice fields in mountainous areas. Ice caps undergo slow, steady accumulation of ice, and the ages at various depths can be dated using methods that we will not discuss here. The ice in these cores trapped small inclusions of the atmosphere as it crystallized and thus provides a record of changes in atmospheric composition through time. The deepest core is in Greenland, but the rate of accumulation of ice in Greenland is much higher than in the Antarctic, and the base of the Greenland core is "only" 110,000 years old. By contrast, the deepest core in the Antarctic ice cap is 2,546 m long and provides a record going back more than 250,000 years at the Russian research station Vostok. Abundances of CO_2 and CH_4 in the Vostok core are plotted against age back to 220,000 years ago in Figure 8.8 and correlated with estimated deviations from present average temperatures at Vostok. The CO_2 concentrations were at a high of approximately 275 ppm (0.0275%) in the last interglacial (warm) period, gradually declined to the peak of glaciation approximately

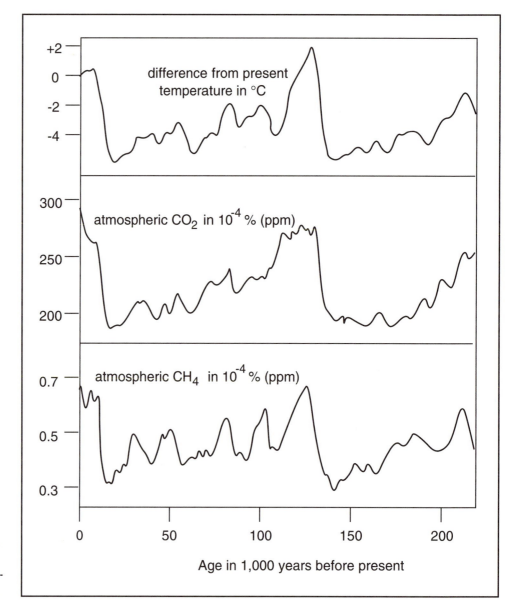

Figure 8.8 Temperatures, CO_2 and CH_4 concentrations in ice core at Vostok, Antarctica. Generalized from Mintzer (1992).

15,000 years ago, and then began to rise again. The CH_4 concentrations also fluctuated, with high levels (maximum of 0.7 ppm) during interglacial periods because of greater rainfall and more extensive marsh and tundra when the earth is warmer (discussion of CH_4 release in Section 8.1).

The CH_4 concentrations were too low to have any significant effect on atmospheric temperature at any time before the past 100 years, but the correlation between temperature and CO_2 raises the issue, which we have encountered before, of what is cause and what is effect. Many investigators have used the correlation to support the concept, discussed in Section 8.1, that increase in atmospheric CO_2 causes an increase in global tempera-

tures. Other investigators, however, have proposed that the high CO_2 concentrations in interglacial periods may simply be the result of the smaller volume of polar water during those times. Because this cold water contains higher concentrations of plankton than warm water, when the earth becomes warmer the decrease in the volume of plankton releases more CO_2 into the atmosphere.

Information on the atmosphere and climate in the past 1,000 years is obtained from various sources and provides more detail than is possible for the 220,000-year period just described. Atmospheric composition prior to about 100 years ago can be measured only from ice cores, but since the late 1800s we have been able to sample the air and measure its composition directly; some recent measurements have been made using remote-sensing methods from satellites. Global average temperatures are measured in various ways. For the past 100 to 200 years, temperature is obtained from year-round measurements at recording stations all over the earth. The data for each station show diurnal (daily) and seasonal fluctuations that can be combined to yield global average temperatures for each year. Before about 1800, temperatures must be estimated largely from unquantified written records, and they are obviously less accurate than recent data. For example, the "Little Ice Age" is well recorded in European records but only poorly, if at all, in most of the rest of the world. Temperatures can also be determined by studies of tree rings, which are wider because of faster growth when temperatures are higher, and worldwide sampling of tree rings provides a well-established temperature record for the past several thousand years.

Temperature and atmospheric compositions obtained by methods described here are plotted for the past 1,000 years in Figure 8.9a. The total temperature variation is only 1°C, but within that small range we can discern periods of relatively low and relatively high temperatures. The Little Ice Age, the most recent period of general coolness, lasted from approximately 1500 to 1800, although it was punctuated by several periods of warmth lasting several tens of years. This period of coolness followed the Little Climatic Optimum, a time of comparative warmth lasting from approximately 1200 to 1400, roughly the latter part of the Middle Ages in Europe. During this warm period, farms were established as far north as Greenland, but they had to be abandoned when the Little Ice Age arrived. The cause of the variation is highly controversial, but well-studied astronomical records for the past 500 years show that periods of coolness coincide with times when the sun had a radius slightly larger than normal and was rotating more slowly than in periods of warm climates. The principal variations in atmospheric composition, and possibly also in temperature, shown in Figure 8.9a are in the past century, and because they are related to human activity we discuss them here.

Because of their short duration, Figure 8.9a does not show some cool periods that lasted only a few years and are readily explained by massive volcanic eruptions that blew so much dust in the atmosphere that incoming solar radiation was reflected directly back into space. Five of these eruptions in the past 150 years have resulted in estimated global cooling of a few tenths of a degree centigrade. They include Krakatau, a small island west of Java, Indonesia, which exploded in 1883; Santa Maria, Guate-

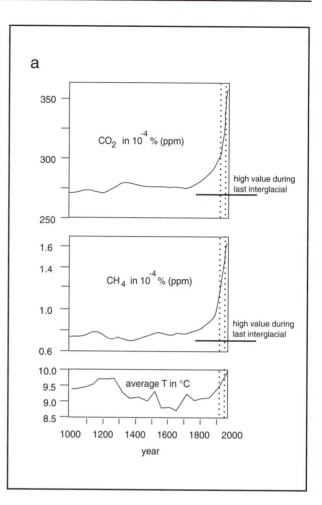

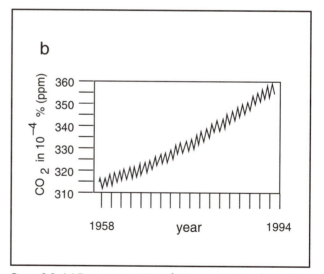

Figure 8.9 (a) Temperatures, CO_2 and CH_4 concentrations during the past 1,000 years, with the past 100 years shaded. High values for CO_2 and CH_4 during last interglacial period are from Figure 8.8. CO_2 and CH_4 data from Mintzer (1992) and Houghton et al. (1995, 1996). Temperature data for England from H. H. Lamb (1995). (b) CO_2 concentrations measured at Mauna Loa, Hawaii, for 36 years.

BOX 8.2 THE YEAR WITHOUT A SUMMER

During early April 1815, at distances up to 500 miles around the island of Soembawa, to the east of Java, residents of the Indonesian islands (then Dutch East Indies) began to hear strange sounds. By April 5, soldiers in Java were searching for an invading army, and the crew of a vessel about 200 miles from Soembawa thought that a naval battle was in progress somewhere near them. People on and near Soembawa found themselves under a darkening sky and a rain of fine ash. These preliminaries came to a head in the night of April 10–11 when Tambora volcano on a peninsula of Soembawa finally blew up. The explosion covered the surrounding land with ash and pyroclastic flow deposits (nuées ardentes; see Section 3.4), rained fragments into the adjacent ocean, and even produced floating "rafts" of pumice up to 3 miles long. The total eruption produced somewhere between 50 and 125 cubic kilometers of ash and pyroclastic flow deposits (Section 3.5), leaving the survivors on Soembawa and adjacent islands in a pitiable condition.

The eruption of Tambora affected not only the local area but the history of much of the earth. During the peak of the eruption, some of the ash rose to a height of nearly 10 miles and entered the worldwide atmospheric circulation pattern. This ash fell out of the atmosphere slowly enough that it reduced global temperatures by an estimated 0.5° to 1°C in 1816 and to a lesser extent in the succeeding 2 years. The climatic consequences of this cooling gave 1816 the name of "Year without a Summer." Summer temperatures in North America and Europe appear to have been 3° to 4°C colder than normal in 1816 and somewhat less extreme in 1817.

Much of North America was affected by both low temperatures and drought (a cold one, not a hot one!) from the spring into the fall of 1816. Crops as far south as New Jersey and Maryland were destroyed by killing frosts in June. The lack of rainfall is recorded by vegetation in much of the midwestern United States and mountainous regions farther west. The news

was not all bad, however – although part of Hudson Bay remained clogged by ice during the entire summer, the surrounding area actually benefited agriculturally. The cold weather increased the potato crop, and low lake and river levels from the drought permitted a bountiful harvest of wild rice. These harvests offset the scarcity of animals that normally constituted a large part of the food supply.

Although most of Europe was cold, conditions there differed in one important aspect from those in North America. The continent was not affected by drought but by excessive rain as cold, moist, air from the Atlantic Ocean was drawn eastward. The combination of cold temperatures and flooding, except in Scandinavia, destroyed many grain crops, and snow in the Swiss Alps was so thick and at such low elevations that farmers were never able to take their cattle to summer pastures in the mountains. These effects brought food prices to very high levels in most of Europe just as Europeans thought they were ready to recover from the economic misery brought about by the Napoleonic wars.

Records outside of Europe are commonly too incomplete or inaccurate to document the effects of 1816 elsewhere in the world. One notable exception is China, where snow fell in many areas of the country normally regarded as hot and rainy. Snow was even recorded in Lhasa, Tibet, in the summer (snow is hardly a surprise in Lhasa, but it is in June!). And one more bizarre note – the maritime, rainy, island of Taiwan had an ice storm in the winter of 1816.

The eruption of Tambora was extreme, the largest one that was recent enough to be recorded adequately in climate records. It is not, however, the only one, and in the text we briefly note five more large eruptions that have affected world climate for 1 or 2 years since 1883.

FURTHER READING: Harrington (1992).

---

mala, in 1902; Gunung Agung, in Bali, Indonesia, in 1963; El Chichon, Mexico, in 1982; and Pinatubo, Philippines, in 1991. The eruption of Pinatubo had a particularly interesting effect on world climate. It occurred during a period in which worldwide temperatures were rising to the levels at which scientists felt that they could convince nonscientists of the reality of human-induced global warming. The evidence was that 8 of the

10 hottest years of the 20th century were from 1985 to 1995. The two years that did not set a record were 1991 and 1992, which not incidentally followed the eruption of Pinatubo in 1991. None of these volcanoes reached the estimated 0.5°C cooling caused by the eruption of Tambora on the island of Soembawa, Indonesia, in 1815, which produced the "Year without a Summer" (see Box 8.2 for further discussion).

Possible effects of human activity

The first indication that people were having a significant effect on the atmosphere was shown by measurements of CO_2 concentration in the atmosphere from 1958 to the present at a monitoring station on the top of Mauna Loa, a 13,675-foot volcanic peak in Hawaii (Fig. 8.9b). The data from Mauna Loa show the standard increase in CO_2 in the winter and decrease in the summer, when plant photosynthesis is more active than in the winter. The particular significance of measurements in Hawaii is that the islands do not have an industrial base of their own and are not closely downwind from industries in Asia. Therefore, atmospheric compositions measured in Hawaii represent the atmosphere of most of the earth rather than local "pollution," and similar results have now been obtained from numerous other locations.

The shaded part of Figure 8.9a shows temperature and concentrations of CO_2 and CH_4 for the past century, a time in which the average global temperature has apparently increased by approximately 0.5°, although the error in the estimate is probably large. To place this apparently small temperature increase in perspective, we call attention to the estimated increase in temperature of less than 5°C on the Antarctic ice cap since the peak of the last glacial period about 15,000 years ago (Fig. 8.8). Although good data are not readily available for global temperatures, the worldwide average temperature increase during this 15,000-year period is almost certainly less than the change in the Antarctic. Thus, natural warming since the last glaciation has increased global temperatures by an average of 0.025°C or less per century (5/20,000 or less). In this context, the increase of 0.5°C in the past century is at least 20 times the natural rate of change.

With the information in Figure 8.9a and b, we look for a cause for the rapid temperature increase of the past century. The possibility of an increase in solar radiation is eliminated because the past century has been a time when sunspot activity (and intensity of solar radiation) has not been particularly high and cannot be responsible for the warming. The principal explanation given by many atmospheric scientists is an increase in absorption of thermal IR, which could be caused by buildup of any combination of greenhouse gases, principally CO_2, CH_4, and CFCs (an explanation of absorption appears in Section 8.1; a discussion of CFCs, in Section 8.3). The abundances of all greenhouse gases increased during the past century, and although CH_4 and CFCs do not currently account for a large proportion of thermal absorption, the 25% increase in concentration of atmospheric

Table 8.2 *Percentages of thermal infrared radiation absorbed by various gases*

Gas	% absorption
H_2O	65
CO_2	32
CH_4	1
N_2O	1
O_3	1
Total CFC	0.1

Notes: Total CFC is the sum of the percentages of all chlorofluorocarbons.
Sources: Data adapted from Mitchell (1988).

CO_2 causes it now to absorb 32% of the thermal IR from the earth's surface (Tables 8.1 and 8.2). The increase in concentrations of all three greenhouse gases is clearly the result of human activities (Section 8.1).

Based on the controversial premise that emission of greenhouse gases by human activity is responsible for recent temperature increase, we attempt to extrapolate the experience of the 20th century into the 21st in order to predict the amount of temperature rise and, consequently, its effects. We make two assumptions. First, we assume that solar intensity will fluctuate no more widely during the next century than it has in the past 1,000 years, during which it apparently has generated temperature changes smaller than the ones that we predict here. Our second assumption is that the temperature increase will be at least approximately proportional to the increase in concentration of greenhouse gases. Increase in CO_2 during the 20th century has not caused an exactly proportional increase in absorption because some (not all) of the CO_2 absorption wavelengths are already "saturated" by thermal IR; that is, they cannot absorb any more. Other gases, particularly CH_4, however, are increasing very rapidly and are still below their limits of absorption saturation. Thus, the assumption of approximate proportionality between absorption and temperature is reasonable.

Using these and other assumptions, at the beginning of 1996 the Intergovernment Panel on Climate Change (IPCC) predicted that the average temperature of the earth's surface will increase between 2°C and 6°C by the end of the 21st century. The prediction was obtained from computer models of climate that use such input variables as intensity of solar radiation, amount of reflection of radiation from clouds and ice, amount of heat absorbed in the atmosphere, distribution of heat by ocean currents and wind, amount of heat absorbed by the oceans, and many factors of less significance. The computer output is the average temperature of the earth, the distribution of temperature at all places on the earth during various seasons, and the average rainfall and its distribution around the earth in different seasons. The IPCC estimate was published at the same time that global temperature data showed that 1995 was the hottest year ever recorded on the earth. Between 1994 and 1995, global average temperatures increased 0.07°C. This increase may not appear to be very much, and temperatures from year to year are highly variable, but we should point out that 0.07° per year for 100 years is 7°, consistent with the IPCC prediction.

How good are the estimates made by IPCC and other organizations? The people who do climate modeling have become fairly adroit at matching their outputs with actual observations of worldwide temperature and precipitation, but can they predict future climates? The way to check their ability is to determine how good they are at predicting past climates, and we illustrate one very fine example of this ability in Box 8.3. Because of the speed of the increase in greenhouse gas concentrations, however, we are not sure whether we (or IPCC or anyone else) can make such assumptions as an approximate proportionality between absorption and temperature. We now discuss six processes, of possibly many more, that could affect simple models of climate change.

1   The oceans could absorb more CO_2, thus reducing the rate of buildup in the atmosphere.
2   Increasing cloud cover might keep temperature down because a warmer atmosphere can absorb more water and form more clouds that reflect some of the sunlight back into space before it can be absorbed by the atmosphere. Paradoxically, the same increase in reflectance can be obtained by pouring industrial dust and aerosols into the atmosphere, possibly at the same rate as CO_2 and CH_4 (raising the interesting possibility that we should demand more, not less, pollution from our heavy industries).
3   An increase in CO_2 may speed up plant growth, creating a larger mass of plants and, indirectly, a larger mass of animals that live on plants and of carnivores that live on other animals. The extra carbon extracted from the atmosphere as CO_2 need not be stored solely in living organisms. If carbon from dead organisms is buried in soil or sediments, it is effectively removed from recycling into the atmosphere.

4   Increasing atmospheric CO_2 may warm the oceans so that more CO_2 is driven out, increasing atmospheric concentrations further and causing more warming (discussion in Section 8.1). This mechanism is significant only if temperatures remain high for 1,000 years or more so that the deep oceans warm up as well as the surface layer (Section 8.1).
5   Melting of snow in polar regions could increase absorption of heat by the earth because snow reflects sunlight back through the atmosphere without converting it to thermal infrared radiation, and thus reduces the amount of atmospheric absorption.
6   Reduction or elimination of the conveyor belt generated by cold North Atlantic water (NADW; see Section 8.1) could occur if, as models predict, warming in high latitudes is above, and in low latitudes below, the average global temperature increase, thus decreasing the present temperature gradient between the equator and the poles. Under these circumstances, polar temperatures would not be low enough to produce the cold, salty water, and there would be no dense water mass that could set the conveyor belt in motion. Shutdown of the belt has been proposed for particularly warm intervals in the past several tens of thousands of years, and when that has happened global temperatures have risen by more than 5° in less than a century.

Possible consequences of global warming

The upshot of all of these predictions, models, and complications is that the world may become warmer, but we don't know exactly how much and when. So what if it does? Maybe that would make the world a better place. Some of us are tired of winter. On the other hand, let us consider the effects of an increase of 2°C in average global temperature, which is at the low end of IPCC estimates but would make the earth warmer than it was in the last interglacial period (Section 8.1; Fig. 8.8). Temperatures should increase more in the Northern Hemisphere, where most continents are located, than in the south, but both hemispheres would show a movement of climate zones away from the equator and toward the poles. This movement of climate zones would have several effects on both climate and resulting agricultural productivity:

• Equatorial regions with high precipitation and trade winds should become hotter and would expand both north and south. Agricultural yields, however, would

BOX 8.3 A LAKE AND ITS CLIMATE

Have you heard the story about the Swedish crocodile? Probably not. That is because there are no Swedish crocodiles. Crocodiles and alligators, their American relatives, live only in tropical and semitropical areas. Presumably their predecessors did too, and their presence at latitudes as far north as 45° north in North America about 40 million years ago (the Eocene) attracted attention. The crocodiles occur in Eocene river and lake sediments (Green River Formation) that occupy parts of Utah, Colorado, Wyoming, and even as far north as Montana. In addition to the crocodiles, these sediments contain leaves and other plant debris that also clearly indicate deposition in a near-tropical climate. The question is, Why was the area that hot?

The Eocene was known to be a time of general warmth on the earth, with a relatively low temperature difference between the equator and the poles. The isolation of the Antarctic and development of its ice cap had not yet started (see text), and the temperature difference between land and oceans must have been smaller than it is today. The latitudinal distribution of land and ocean, however, was approximately the same as the present, and the solar radiation was virtually identical. Putting all of these known variables (land distribution, ocean temperature, etc.) into a computer model for North America generated the temperature distribution shown by the dashed line in Figure 8.10 for an area in central North America at a latitude of 40° north. The model for the continent without a lake shows the expected alternation of winter and summer temperatures, with winter temperatures of somewhat below 0°C.

The model in Figure 8.10 fits expected variations extremely well except for one small detail – those pesky crocodiles and the plants that they lived with. Being cold blooded and not possessing fur coats, crocodiles could not possibly have survived low winter temperatures. Therefore, the climate model needed adjusting. A modification tried by Dr. L. C. Sloan, of the University of California at Santa Cruz, was based on the environment occupied by the crocodiles. Stratigraphic evidence showed that they lived mostly in swampy areas around a lake with an area of several thousand square miles. This lake modified the local meteorological conditions in two ways. It provided moisture to

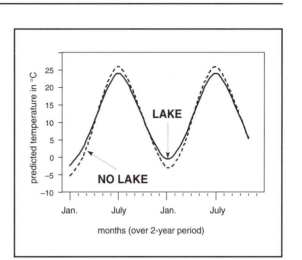

Figure 8.10 Predicted temperatures for a 2-year period in the Green River area of North America during the Eocene (40 million years ago) without assuming any effect of a lake (dashed line) and with the effects of a large lake (solid line). Diagram from Sloan (1992a, 1992b), with further explanation in text.

the atmosphere and, in the same way that oceans modify the temperatures of adjoining lands (Section 8.1), its presence reduced summer temperatures and increased winter temperatures.

The model including the effect of a large lake yielded the prediction shown as a solid line in Figure 8.10, with winter temperatures that would not kill the vegetation and would keep crocodiles warm and comfy through the winter, but their significance did not end there. The ability to produce a model of ancient temperatures that fit the pattern required by paleontological evidence demonstrated the validity of the modeling process. Given the proper data, it now seems possible that we can predict future climate variations. The predictions will not be perfect, but they cannot be ignored, and we feel comfortable in using them for this book.

FURTHER READING: Sloan (1992a, 1992b).

---

probably decrease because any increase in precipitation should be offset by increased evaporation, leading to reduction in soil moisture that would require more irrigation and a resulting increase in the cost of crop

production. Also, higher CO_2 levels should have little "fertilizer effect" because C4 plants, such as corn (Section 2.2), are relatively unaffected by increase in CO_2 above present concentrations.

- As the tropics expand, present desert regions that are now centered at approximately 20° north and south latitude would also move north and south. Depending on the extent of movement, arid areas of northern Mexico and the western United States could expand northward, and the southeastern United States might become drier and hotter. Arid conditions also could expand northward from Africa into southern Europe, push the southern border of the Sahara Desert southward, and encroach southward from the desert regions of central Australia.

- Both the northern and southern temperate zones would shift toward the poles by several degrees of latitude. Expansion of crop production northward in North America and Siberia could compensate for loss of temperate, productive climates in low latitudes. If the flow of NADW is reduced or eliminated, however, the British Isles could become colder while mainland Europe warms up.

The movement of climate zones may be less disruptive than two other effects of global warming. One is more extreme weather such as hurricanes, droughts, and, paradoxically, cold spells in the winter. The year 1995, the hottest in recorded history, had more hurricanes in the Atlantic Ocean than ever before, and the entire world experienced violent storms and temperature extremes. Then in the Northern Hemisphere winter of 1996, North America set records for cold weather and blizzards, and parts of South America had their hottest summer ever. Some climate scientists have attributed these abnormalities to overall global warming.

The second, and most drastic, effect on a worldwide scale may be rising sea level. A 2°C increase in temperature is estimated to cause a sea level rise of 1 to 2 meters, about half by expansion of the ocean as it heats up and half by melting of the Greenland ice cap and smaller ice fields in Iceland and mountainous regions. This rise is potentially serious for the ~5% of the world's population that live along coastlines, in river deltas, on gently sloping coastal plains, and on small islands, all within 2 m of present mean sea level. A sea level rise of just 2 m could push shorelines up to 5 miles inland in these areas, and hurricanes and other storms would push water many miles farther than it now reaches. Most of the disruption would be borne by the world's poor people, who live as subsistence farmers in deltas of the world's major river systems (Section 3.7). The richer part of the world would also be affected, largely by flooding of port cities; a potpourri of these ports includes Rotterdam (and adjoining areas of the Netherlands that are below sea level now),

Copenhagen, Tokyo/Yokohama, Bangkok, and Miami/Port Everglades.

None of the potential disruptions of people that would be caused by global warming lack solutions. Grain production can be shifted to higher latitudes, and we might enjoy the bounty of the Siberian wheat fields and satisfy our sweet tooth from the sugarcane harvest of South Dakota. The stress placed on the world's food supply by reduced production in tropical regions might be alleviated by more intense irrigation to replace soil moisture lost if rainfall is inadequate. Coastal plains and low-lying islands can be evacuated. Ports can be moved inland and rebuilt. Only three things are needed:

1 Time. Can these dislocations be accomplished in 1 century? Yes, but see points 2 and 3.
2 Will. Some of the dislocations of food and people probably cannot be accommodated within present countries or within present states and provinces. Where will people displaced from the Nile Delta go? What northern country will send more food to the Middle East and southern Asia? Where will poor native people of the far north go when rich people from farther south arrive with money to buy their land for dairy farms? Are the world's people ready to work together to solve global problems?
3 Money. Rebuilding thousands of miles of the world's coastline and moving 5% of the world's population is hideously expensive even if it is politically possible. In 1995, the world's major insurance companies began to ask for international political and industrial action to reduce the threat of global warming because even normal expectations of loss would bankrupt the insurance system (Section 8.5). Thus, it may be necessary for the world economy to reallocate resources away from other human activities in order to do so.

Policy questions

Should we attempt to reduce emissions of carbon dioxide by reducing the use of fossil fuels for industry, transportation, and other purposes? If so, should this be done by: (1) switching to energy sources such as nuclear power (but see Sections 5.4 and 7.4); (2) placing such a heavy tax on emission of carbon dioxide that people and organizations will consume less energy; or (3) some other method of your devising?

FURTHER READING: Flohn and Fantecci (1984); Mitchell (1988); Lamb (1991); Mintzer (1992); Kaiser and Dennen (1993); Houghton et al. (1995, 1996); Lamb (1995).

8.3 EXTINCTIONS AND SPECIES DIVERSITY

A tree is a tree.... Seen one, you've seen them all. — Unpublished comments by Ronald Reagan as a candidate for governor of California

To people, a pack of wolves hunting its prey or a shark cruising through the ocean is a frightening sight. To a mouse, your cat is even more terrifying. These hunters are carnivores that live by predation, the hunting of other animals for food (although your cat, in lieu of a mouse, can simply wait by the supper dish until you open a can of cat food). Some animals are omnivores, eating both animals and plants; an example is a bear that feasts on both fish and berries. Occasionally, predatory animals kill other predators for protection, although this is not a common practice. With one exception, animals do not kill for the enjoyment of killing, and they rarely waste food that has been obtained with such effort. The exception, of course, is humans.

To the extent that paleontologists can determine the geologic record, people appear to be the most fearsome and wasteful predators in the entire 3.5-billion-year history of life on earth. Partly the carnage results from the fact that we kill for sport – the bearskin rug on the floor of the den; the moose antlers on the wall of the ski lodge; the mounted swordfish in the seafood restaurant. More important, however, is the extraordinary efficiency with which we kill. We have already addressed (in Section 2.4) our ability to sweep the oceans of fish, perhaps to the point that breeding stocks have been so depleted that some species will die out. Also, some of our efficiency results from our ability to modify the environment by non-hunting activities, such as clearing swamps to reduce malaria, building dams to control floods, clearing forest for agriculture, and burning fossil fuels.

Before we discuss human effects on the ecosystem in this section, we should warn readers who are ready to rush to the next town meeting to protest cutting down a tree in a park or trapping beavers that have been polluting a local lake. More people now live better than at any other time in human history, and more of some types of animals, principally domesticated ones such as dogs and sheep, are now alive than at any other time in earth history. Furthermore, industrial activity is, in many ways, no more responsible for modification of the ecosystem than the life-styles of people who have a nonindustrial lifestyle. As we saw in Chapter 7, many industrialized countries are environmentally cleaner than poorer ones. In fact, enormous carnage was caused by people during and following the last ice age, approximately the past 50,000 years. It is referred to as the "great megafauna

extinction," and we describe it before considering the present.

Extinction of the megafauna

Numerous types of large land animals (megafauna) became extinct in the past 50,000 years. They include the dodo, a large flightless bird on the Indian Ocean island of Mauritius that died out before European settlers arrived and bequeathed to us the phrase "dead as a dodo." Other extinct megafauna include such diverse varieties as mammoths and saber-tooth tigers in northern Eurasia and North America, large marsupials (diprotodon) in Australia, and the moa, a New Zealand flightless bird considerably larger than the average human and probably related to the dodo. This die-off occurred as the last ice age was slowly reaching its peak about 20,000 to 15,000 years ago, and following the peak as the world warmed up rapidly (see the discussion in Section 8.1). The last major casualty was probably the moa, which was still thriving about 1,000 years ago. Because of these extinctions, modern humans live in a world with less species diversity than our ancestors.

Geologists have puzzled about the megafauna extinction since it was first recognized during the early part of this century. A leading theory has been that the encroaching ice sheets in the north and the decreasing temperatures worldwide caused environmental stress that the animals could not adapt to, and they died. That is, in the game of "survival of the fittest," these animals lost. This explanation is identical to those proposed for extinction and evolution throughout geologic history. The problem, however, is that the megafauna extinction event is both too short and too long and provides too much detail for conventional geologic investigation. It seems instantaneous to geologists accustomed to dealing in millions and billions of years, within which a period of 50,000 years could not be discerned by any available dating methods. Thus, if this extinction had occurred 50 million years ago, we would have thought that all of the species had died at the same instant. Because of the detail, however, we know that animals related only by the fact that they were large and lived on land died out sequentially in different parts of the world over a period of a few tens of thousands of years.

Why? What was special about the last ice age? For two reasons, the diversity of animals that became extinct and the range of times at which they did so seem to require an alternative explanation to the theory of climate stress imposed by the ice age. First, any satisfactory explanation must account for the fact that no megafauna extinction appears to have occurred during any of the 15 to 20 pre-vious ice ages (glacial advances) in the past 2 million years (Section 8.1). Second, climatic effects alone could have caused animals to migrate but probably not to have become extinct. Moving southward during glacial ad-vance in the Northern Hemisphere or to lower eleva-tions anywhere on the earth would have permitted ani-mals to have remained in a climate to which they were adjusted. Also, the retreat of seas from continental shelves caused by the approximately 500-foot (300-m) lowering of sea level at the height of glaciation would have opened large areas at low elevation to colonization by migratory species.

Perhaps we can find an explanation for the megafauna extinction in the fact that humans began to stir from their earliest homelands about 50,000 years ago, an "evo-lution" in human ability that is still a mystery to students of prehistory. When these migrating people arrived in virgin lands where humans had never been before, the local animals were brought into contact with the most ridiculous animal they had ever seen. After all, we are not covered in fur that keeps us warm or scales that pro-tect us, we do not have humps to store water for a desert journey, our teeth and fingernails are useless as weapons, and our sense of smell is too limited to warn us of ap-proaching danger. In short, we do not look dangerous, particularly to animals that have never before encoun-tered humans. By the time of the early migrations, how-ever, people had progressed to the point that they could use fire, engaged in primitive agriculture, could use clubs or other primitive weapons, and were learning to venture out to sea in crude vessels capable of carrying them at least short distances over the water. These abilities made humans the dominant species on the earth.

The possibility that people were responsible for the megafauna extinction rests on the dates at which various species died out. To the extent that the dating is accurate (it is only approximate for many animals), the extinction of many species seems to have occurred shortly after the area was first colonized by migrating humans (Fig. 8.11). Most of the species that died fell into one of two cate-gories. One included relatively harmless animals, such as mammoths, and flightless birds, such as moa, both of which were clearly a valuable food source to any hunter capable of killing them. The other species that became extinct were predators that posed a threat to humans, such as saber-tooth tigers. These relationships strongly suggest that people were the cause of an extermination that has been referred to as a "blitzkrieg" by P. S. Martin of the University of Arizona.

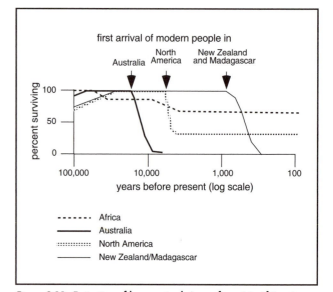

Figure 8.11 Extinction of large animals (megafauna) in Africa, Aus-tralia, North America, and New Zealand/Madagascar during the past 100,000 years. Percent surviving is the percentage of the number in-ferred to have been present 100,000 years ago. The diagram shows times at which rapid increase in extinction rates coincided with first ar-rival of humans in the areas plotted. Adapted from Martin and Klein (1984).

The preceding reasoning is supported by relics of a prehuman past still present in a few places when the first people capable of making written records arrived. Much of this history is in the southwestern Pacific, on Australia and the islands that stretch east and south from the Southeast Asia mainland. Possibly the first to arrive were the Australian aborigines, and successive waves of people finally occupied almost all of the islands and other land areas by a few hundred years ago. The trash piles of these early people show massive kills of native species, both animals and birds, as the first settlers enjoyed the luxury of seemingly boundless food resources. One island, Lord Howe Island, east of Australia, apparently was not lo-cated until 1788, when a European exploration ship was attracted to it by the hordes of seabirds that occupied it. One European visitor wrote: "When I was in the woods amongst the birds, I cd [could] not help picturing to my-self the Golden Age. . . . we had nothing more to do than stand still a minute or two and knock down as many [birds] as we pleased wt [with] a short stick." (A. B. Smyth, cited by Flannery, 1994, p. 177).

The social consequences of rapid overhunting and ex-tinction were most readily displayed in islands or other restricted areas, and we use the South Pacific as examples. The Maori, who arrived in New Zealand about 1,000 years ago, originally used the bounty of moas and other game to create a prolific civilization that built cities, farmed agricultural areas, raised domestic animals, and engaged in trade. By the time that the first Europeans arrived in the mid 1800s, however, the food supply had been so destroyed that the remaining Maori had retreated into almost impregnable fortresses and survived partly by cannibalism, using their vanquished enemies as food. Similarly on Easter Island, one of the world's most isolated places, early settlers found enough food and other necessities that they could, with great effort, erect the great stone statues that are such a hallmark of the island. When the first European exploration vessels arrived in the 18th century, however, the early forests had completely disappeared, the population had been reduced to not more than 25% of its former size, the remaining people lived largely by scavenging remaining resources, and warfare and cannibalism between the survivors were endemic. We emphasize, as elsewhere in this book, that an absence of resources brings out people's most vicious instincts.

Now we switch to the issue of the degree to which modern humans are affecting the world's ecosystem. We do so in three categories: (1) modern species diversity; (2) extinction and endangered species; and (3) deforestation.

Modern species diversity

We all agree that biological diversity is good. We like to see different plants and animals, and we make use of this diversity for agricultural purposes (Chapter 2). Similarly, virtually no one wants species to become extinct, and worldwide lists of "endangered" species have been prepared to aid efforts to prevent extinction. Although diversity and extinction are obviously closely related, we will discuss them separately, but before doing so, we must consider the term "species."

The definition of species is not simple. For animals, "species" mostly means a group of organisms that can breed with each other to produce fertile offspring. Thus all humans, regardless of their differences in such characteristics as skin color, are all one species. The concept encounters subtle problems, however. For example, ordinary domestic dogs are capable of mating with wolves and coyotes, both of which would be considered separate species by any normal definition. Thus, animal species are commonly defined on the basis of a set of broad characteristics, with minor differences identifying "varieties"

that are capable of interbreeding. The problem of species definition becomes somewhat more acute with plants, many of which can be cross-pollinated or grafted onto each other. Consequently, much of the worldwide discussion about plant preservation is really concerned with varieties rather than with wholly independent species.

Once we have decided, on somewhat arbitrary grounds, what a species is, we can tabulate the world's species and their abundances. Because of the problems mentioned earlier and because of the disproportionate amount of effort given to different kinds of organisms, we obtain highly variable results. At present, depending on the classification used, about 2 million different species have been officially described and named. Roughly 4,000 of them are mammals, 9,000 birds, 20,000 fish, and more than 1 million are arthropods, mostly insects (beetles account for more than 20% of the world's named species). Among all organisms as a whole and within individual groups of organisms, the number of species decreases from the equator to the poles. The largest number of species is in tropical rainforests, primarily insects but also other animals and numerous varieties of plants.

DIVERSITY   What do we mean by "diversity"? Qualitatively, it may mean the number of species within a local area, the rate of change of species from one habitat to another, the number of different usable resources within a local area, the variety of habitats within a region, or the total variability of species and habitats over a broad region. Quantitatively, various methods are available to measure diversity, the simplest measure being the number of species living in some designated area. This measure encounters the problem of dealing with species of altogether different kinds of organisms (moths, snakes, cacti, etc.) and also the problem of accounting for vast differences in the abundances of species. In order to overcome (at least partially) this problem, various quantitative indexes have been designed. One of many measures of this type is the Simpson Index (SI), calculated from the equation

$$SI = 1 - \sum_{i=1}^{\text{no. species}} (p_i)^2$$

where p_i is the proportion of species i. The Simpson Index varies from 1.0, if only one species is present, to very low numbers if numerous species are present. It is useful only within some identifiable group of organisms, such as birds or flowering plants and not for a mixture of species from different groups. Assuming that we can make accurate counts of the numbers and abundances of species present, which is not easy with migratory species such as birds, we can calculate an index for the area surveyed. We can make similar calculations for other habitats and then combine them in some way to determine species diversity over a region.

When we make these calculations, we establish an important relationship between diversity and area. Counting species and their abundances in areas of different sizes yields an exponential relationship with the general equation

diversity = (area)^a

where a = coefficient of diversity. The coefficient a varies with the type of organism surveyed but is commonly between 0.2 and 0.4. This relationship means, for example, that the total number of species of one type of organism in two forests with areas of 1 square mile each is generally less than the number of species in one forest with an area of 2 square miles. Consequently, people and organizations that try to maximize the number of species of some organism (birds, ferns, etc.) want to set aside areas of maximum available size.

Quantitative data on species' abundance, calculated indexes of diversity, and the recognition of the importance of the size of area involved all provide important information for making decisions on ecological-environmental issues. They do not, however, lead to simple solutions. Consider the following questions:

- Are we concerned about the number of species or also about their abundance?
- Are we concerned about the diversity within a group, such as fish in a river, or the total number of different groups in the river, such as fish, water fowl, and insects?
- Are we concerned about the diversity within a broad region rather than in a local area? For example, if each of five ponds holds only 2 species of fish that are not related to fish in other ponds, making 10 different species overall, is that better than having 6 species of fish distributed in all five of the ponds? That is, do we want to maximize diversity within each pond or among all of them?
- Are we more interested in a diversity of habitats than a diversity of species? For example, is a large forest that contains 10 species of birds more or less valuable than a small forest with 6 species of birds, with the other part of the forest replaced by a dammed lake that contains 6 species of fish?

Clearly, none of these questions has an objective answer. All of them must be answered by groups of people involved, commonly by people elected or appointed to carry out the will of the local or national population. We illustrate some of the problems by discussing the plight of the Florida panther in Box 8.4.

Finally we raise the issue, why should we care about species diversity anyway? A high proportion of the world's food (exact figures uncertain) comes from a few species of animals and plants, many of them with numerous interbreeding variants – one major species of cow, one of sheep, one of corn, one of wheat, and so forth (see further discussion in Chapter 2). This standardization of food products has enabled us to bring more and better food to more people than ever before in earth history. Considering the importance of this achievement, is it worthwhile to worry about the number of species of songbirds in our garden? Should we be concerned if a marsh is drained for agricultural development, thus eliminating a habitat and its inhabitants? Are we concerned only about aesthetics or justice, or is there some practical concern about loss of species diversity?

Yes, there is a practical concern. In Section 2.5 we discussed the economic effects of monocultures, areas whose agriculture is oriented almost wholly to one type of food. One danger with monocultures is the possibility of disease, such as the Irish potato blight that destroyed so much of the crop that it caused widespread hunger. Our exposure to this type of disease and its consequences for world food supply increase as the number of species and number of variants decrease, that is, as genetic diversity decreases. With this fact in mind, consider that all of the coffee bushes in Central and South America are descended from one bush cultivated in a greenhouse in the Netherlands in the early 1800s. Also, all navel oranges in the United States are clones from one original orange tree, and all of the sheep in New Zealand are descended from a few breeding pairs brought to the islands by the first British settlers. It is hardly necessary to point out the risk that a disease could run rampant through an entire population of any of these plants or animals and possibly render the group extinct or at least so small that it is no longer of economic use. Diversity is not only aesthetically pleasing, it is also an insurance policy against future environmental disaster.

Extinctions and endangered species

The extinctions caused by the first human occupation of virgin lands in the past 50,000 years – the "blitzkrieg" discussed earlier – are historically documented. Now we

BOX 8.4 *FELIS CONCOLOR CORYI*

Felis concolor coryi is the scientific name for a cat (*Felis*) of essentially one color (*concolor*) first precisely described by the 19th-century naturalist Charles B. Cory (*coryi*). This cat is also popularly known as the Florida panther. It is a variety (subspecies) of a group of large (approximately 6 feet long) tawny-colored cats known in different places by names such as puma, cougar, and mountain lion (including the Nittany lion). These cats used to occupy most of the United States and Mexico but now are almost completely restricted to the mountainous west. Characteristics of the survivors vary from place to place, and the Florida panther is distinguished by a color that is slightly more reddish brown than other panthers and a few other minor features.

Until approximately 1900, the Florida panther was more fortunate than its relatives in other parts of the eastern United States. As population increased in most of the United States and agricultural lands were expanded at the expense of forests, large predator cats were either driven into smaller stands or simply hunted down. The interior of south Florida, however, is a land of inhospitable palmetto and large swamps (including the Everglades) that resisted human intrusion. Thus, the panther was relatively unharmed by hunting and ranged wild through an area large enough to support the deer population that was its primary food source.

By about 1950, the good fortunes of the panther had been reversed and the population had declined to very low numbers – good data are nonexistent – because of human incursion. People discovered that at least part of the area could be used for agriculture, particularly cattle ranches (also see Box 4.7 for a discussion of water problems). These ranches soon decimated the deer population not only because of reduction in the area available to wild animals but also because efforts to control tick-borne diseases in the cattle led to eradication efforts that caught a large proportion of the deer population. Efforts by hunters to prevent the eradication program were unsuccessful, and in a short period of time the panther was left with little food except for the small population of wild pigs.

In the 1950s, a variety of factors led to a rapid increase in the deer population, but the numbers of panthers apparently continued to decline. They did so because, by this time, the number of panthers was so small that effective breeding was becoming impossible. By 1980, only 20 panthers were believed to be still alive, and their numbers were being reduced at a rate of approximately 1 per year by automobile accidents on newly developed high-speed roads through the swamps and adjacent areas (that 1 panther represents 5% of the total population). At this point, new types of vehicles made travel through the swamps easier, and hunters – whose earlier policies would, if followed, have protected the panther – now wanted greater access to the few wild areas left.

What to do? The options can be summarized easily, but they all have problems.

- Nothing. The Florida panther is so similar to other varieties of panthers/pumas/cougars in North America that, if it does not survive on its own, it will not be seriously missed. The problem with this solution is that some people would miss it.
- Breed it artificially. Three panthers in captivity have produced kittens that could be reintroduced in the wild. The problem with this breeding is that it is starting from a very small genetic stock, and the likelihood of defects caused by inbreeding is high. There is the further issue that not all of the captive panthers are native to Florida, in which case the kittens would not really be Florida panthers, which may or may not be a problem depending on your point of view.
- Introduce panthers from other areas into the wild in the hope that they will breed with the surviving Florida panthers. But – again – the new panthers would not really be Florida panthers.

Regardless of the specific solution (if any) chosen, the panther is like other wild animals in requiring a large connected area to roam in if it is to survive (see text for discussion of species diversity and area). Unless that type of wilderness area can be established in South Florida, both the panther and probably several other animals cannot survive.

FURTHER READING: Alvarez (1993).

ask whether human activity continues to cause extinction. Table 8.3 shows estimates of the number of animal species that have become extinct since approximately A.D. 1600. It is based on written records where they are available, paleontological/archaeological records in areas where written records were not kept until recently, and on present searches for the animals involved. We can hardly list all of the 464 species believed to have become

Table 8.3 *Selected data on animal extinctions*

Type of animal	Number lost			
	On islands	On continents	Total	% lost on islands
Molluscs	151	40	191	79
Birds	104	11	115	90
Mammals	34	24	58	59
Other	74	46	120	62
Total	363	121	464	75

Notes: Other includes reptiles, amphibians, fish, etc. but is mostly insects.
Sources: Information from World Conservation Monitoring Centre (1992).

extinct in the last 400 years, but here are a few examples: the St. Helena earwig (an insect on St. Helena Island – was it still alive when Napoleon was exiled to St. Helena?); the Jamaica tree snake; the great elephantbird of Madagascar (similar to the dodo and the moa); the great auk, which formerly ranged from Canada to Iceland and the Faeroe Islands; the St. Lucia rice rat (on St. Lucia Island in the Caribbean); the New Zealand laughing owl; several varieties of rocksnails, the Utah sculpin (a fish), and the Three-Tooth caddis fly, all in the United States.

Table 8.3 suffers from at least four problems. One is defining a species, which is done differently by different investigators. A second problem consists of proving that a species is extinct. Does the mere fact that no one can find a living specimen mean that the species is extinct, or might a few individuals have survived somewhere in sufficient numbers that they can become a reasonable breeding population? The size of a population that can successfully breed in the wild varies with the species but is estimated to range from a few hundred to many thousand. Ecologists refer to species that reappear after they were thought to have become extinct as "Lazarus species."

Two other problems with Table 8.3 center around the availability of information. For example, we do not know how many species have not been described and cataloged and, obviously, what their recent fate has been. Animals such as snails, mammals, and birds have been much better studied than reptiles, which probably are underrepresented in the data. Insects, which are thought to constitute approximately 60% of all animal species on earth, probably have not been adequately tabulated. Finally, the data clearly vary enormously from one geographic area to another. For example, the Soviet Union claims

that it has lost only two species (one bird and one mammal) in the past 400 years, and U.S. data show that it has lost animals more rapidly than lesser-developed parts of the world. Are the Soviets correct, or have they simply not had the resources to conduct adequate surveys or the political will to admit ecological problems? (We suspect the latter.) Is the United States really that dangerous, or has it simply been studied more carefully? (Again we suspect the latter.)

The problems discussed here make us use Table 8.3 with great caution, but regardless of the accuracy of the data, we find one very important conclusion – most extinctions have occurred in islands or other geographically restricted areas. Clearly species that live in small areas and cannot migrate from them are particularly vulnerable to any change that threatens their existence. This conclusion is consistent with estimates – they are only estimates – that about 30% of the extinctions in the past 400 years have resulted from introduction of new animals to habitats unprepared for them or from accidental habitat destruction. A prime example is Hawaii, where European and U.S. visitors and settlers deliberately introduced pigs and goats in order to produce meat and milk for colonists and passing ships. The goats cropped the vegetation to the roots (Section 2.3), and the pigs gnawed through the forests, both causing such ecological destruction that dozens of plant and animal species are apparently lost forever. (Also see Box 2.4 for a related example of the introduction of nonendemic organisms that has resulted in destruction but, so far as is known, no extinction.)

Now we ask whether the extinctions shown in Table 8.3 are unusual in the geologic record. Organisms have evolved and become extinct for more than 3 billion years, but has it become faster during human time? We can estimate the change in rate from the following data: the earth now contains about 2 million known animal species (depending on the classification used). Paleontological records suggest that most species survive about 4 million years before becoming extinct. Because this value is derived from paleontological records that predate the evolution of humans, it provides a "background" rate of extinction unaffected by people, although the value is probably too long because the measurements are biased against finding very short-lived species. Using the same relationship between life expectancy and birth–death rate that we used in Section 1.3, we calculate that approximately 0.5 species should become extinct each year by normal processes ($2 \times 10^6 / 4 \times 10^6 = 0.5$). An extinction rate of 0.5 species per year should result in 200 extinctions in the 400-year period covered by Table 8.3, which is smaller than the 464 shown. A similar calculation applied to the 3,000 mammal species indicates that 0.3 (roughly one-half) mammal species should have become extinct in the past 400 years, in contrast to the 58 actually shown. Therefore, regardless of the inaccuracies in Table 8.3 it is clear that extinctions have been more rapid in the recent past than during any part of geologic history before the arrival of humans, and we can reasonably blame ourselves for them.

Should we do anything about the increased extinction rate? Does it matter that the three-tooth caddis fly and the Jamaica tree snake are extinct? Possibly we are better off without the St. Lucia rice rat, but the laughing owl might have been fun to have around. Actually, several programs are now preserving species and varieties that might otherwise become extinct. Nonwild preservation occurs in zoos, botanical gardens, sperm and egg banks, seed banks, and private organizations. Some varieties of plants are apparently extinct in the wild and are available only through horticulturists, and some organizations such as Heirloom Seeds and the Carnivore Preservation Trust maintain as much genetic diversity as possible in the groups of organisms that they serve.

Efforts to preserve organisms in the wild center around the concept of "endangered species." An endangered species is one that has been determined by one or more organizations as in danger of becoming extinct. Most countries have national parks that attempt to preserve wilderness and prohibit picking of flowers or killing and trapping of animals. This concept works well except where the preservation effort is overwhelmed either by economic desperation or rapacity. In many poor countries, poaching of animals either provides necessary food

for starving people or a source of economic benefit if the animal or its parts can be traded. In industrial countries, parks are sometimes viewed as places that are unnecessarily shielded from development, either for minerals, fossil fuel, or simply tourism and other commercial development. Each country, and each reader, must decide what attitude to take toward efforts to develop nature preserves.

In addition to local preservation, 120 countries have banded together to preserve endangered animals and plants by establishing the Convention on International Trade in Endangered Species (CITES). In an effort to curb the illegal trading and preserve some species that are nearing extinction, CITES has established a list of approximately 700 species or varieties of organisms on a critical list (appendix I of the CITES treaty). They include all rhinoceros and tigers; most elephants; and several (although not all) varieties of whales, monkeys, boa constrictors, ostriches, and cacti (Fig. 8.12). Well-known animals on the list include the giant panda, the cheetah, Californian and Andean condors (birds), and a number of birds from the Amazonian rainforest. The treaty organization has virtually no power to enforce its edicts, however, and illegal trade continues to flourish.

Much of the illegal trade passes through China and South Korea, which are technically members of CITES, and Taiwan, which is not. A glaring example is rhinoceros horn. The population of African rhinoceros (several species) in the 1990s is approximately 10% of its size in 1970. The horn is used for ceremonial daggers in Yemen, and powdered horn is regarded as a remedy for fever in eastern Asia. Most, but not all, of the horn is from Zimbabwe, partly poached by people crossing the borders from other countries. Zimbabwe has recently imposed stringent penalties, including shooting of poachers, in an effort to stop the trade and preserve its rhinoceros population, but a price of more than $1,000 (U.S.) per pound in eastern Asia lures hunters in search of easy money. Another endangered species is the tiger. Fewer than 5,000 tigers, all in Asia, remain in the wild. They are hunted to provide skins and heads for souvenirs, their bones are used in Asiatic remedies for skeletal and muscular problems, and tiger brains rubbed on the body are thought to cure acne. A tiger skin may sell for $10,000 (U.S.) in eastern Asian shops.

Paradoxically, some of the CITES efforts may actually harm animals that they want to preserve. The principal example is the prohibition, agreed to by virtually all countries, against the sale of ivory. The ban was installed largely to protect African elephants, which were being il-

(a)

(b)

Figure 8.12 Organisms endangered by human activity: (a) rhinoceros in East Africa, endangered by poaching for horns and also elimination of their natural habitat – Courtesy Kathleen Henderson; (b) welwitchia in Namibia, endangered by mineral exploration; the plant lives on dew in this very arid region, and the odd shape of the leaves permits dew to trickle down to the plant's roots.

legally killed for their tusks, but most of the killing occurred in Kenya, which was unable or unwilling to control the poaching. The worldwide ban, however, also affected countries that had elephant populations so large that they had to be controlled by selective hunting (a process known as "culling"). Zimbabwe and South Africa, for example, had healthy elephant populations because they sold the ivory from culled animals and used the profits to support and manage their wildlife. When this source of income disappeared, the two countries either had to allocate money from other sources to wildlife management or decrease their preservation programs.

The United States has made a similar effort to preserve animals and plants that are in danger of becoming extinct. The principal mechanism is the Endangered Species Act of 1973, administered by the National Biological Service, which attempts to prevent destruction of organisms whose numbers are so limited that they might become extinct without protection. At the time of listing as endangered species, most animal groups have an estimated fewer than 1,000 individuals and plant groups fewer than 100. Restrictions apply to less than 1% of known U.S. species. The act applies not only to species but also to specific varieties of species, some of them recognized by their limited geographic range. The Florida panther, which we discussed in Box 8.4, is one example of this restricted listing.

Because of its impact on commercial activity, the Endangered Species Act receives continual criticism and numerous challenges to the various ways in which it attempts to provide protection. The least-controversial method consists of the establishment of wilderness areas, in some cases purchasing new wilderness areas with the proceeds from the sale or exchange of noncritical public land elsewhere. A second and far more contentious method is to take governmental or legal action to prevent commercial or other developmental activity that might destroy a habitat that a protected species needs. A prime example is the prevention of logging in "old-growth" forest in the Pacific Northwest, and because of the complexity of this issue we have reserved Box 8.5 for its discussion. A third method of preserving animals in local areas consists of deliberately reintroducing them to places that they had occupied before. Unfortunately, many of these animals are predators that had been wiped out in order to protect domesticated animals. Thus, when the federal government began to bring gray wolves from Canada to Yellowstone Park, local ranchers protested that the wolves would attack their cattle. (We mention, parenthetically, that farmers in Nepal made the

same complaints to their government about protection of rhinoceros in an adjacent park.)

Deforestation

"Leave a Tree in Iceland" reads the sign over the box in the departure lounge of the Reykjavik airport. Passengers are urged to deposit their excess Icelandic money, or any other money for that matter, as they leave the country. The money collected is used in reforestation projects at numerous places around the island. The reforestation is necessary because the first Viking settlers inherited an island that was approximately one-third covered by trees and proceeded to chop so many down that now the country is virtually treeless. In the past 10 years, the program has worked well, and although tree growth is slow in Iceland's Arctic climate, presumably at some time in the next century the island will be restored to its former state.

Iceland is not alone among countries that have been virtually deforested (Fig. 8.13). In the first millennium B.C. the famed cedars of Lebanon provided timber for a Phoenician fleet that was so large that now the cedars are restricted to a minuscule area relative to their former widespread extent. Another example is the ecologic holocaust on Easter Island, where early settlers not only destroyed much of the animal life but also cut down all of the trees. Some of the cutting was done so that the large stones that were erected to make the famous statues could be rolled into place from their locations higher in the mountains. By the time European visitors arrived, not only were the natives destitute, but they also did not even have enough wood to build canoes to escape from the island.

The most extensive deforestation was caused not by cutting but by fire. Early European voyagers to the South Pacific saw fires on most of the land areas, and Captain James Cook, in the 1700s, referred to Australia as "this continent of smoke." The fires were caused accidentally, but mostly deliberately, by early human colonists. Called "firestick farming," the technique was to burn off forest and scrubland to open it to the agriculture and hunting practiced by Australian aborigines. This type of farming was also practiced by early inhabitants in grasslands and plains of midwestern North America and is still used today in parts of Africa and a few other relatively poor areas. As we discussed in Section 2.2, burning stubble returns nonvolatile nutrients (phosphorus and potassium) to the soil but drives off virtually all of the nitrogen. In areas that are already semiarid, this burning places so much stress on the vegetation and the animals that use it

BOX 8.5 IS THE SPOTTED OWL A RED HERRING?

A traditional method of finding fugitives from justice, and even people who are simply missing, is to track them with bloodhounds. The dogs' sense of smell enables them to follow trails through tangles of different odors and arrow in to their quarry. Naturally, many of the quarries do not want to be found, and an also-traditional device for evading detection is to provide a smell that distracts the dogs. Red herring – that is, herring that are rotten – have a strong odor that serves this purpose admirably, and a fugitive who can draw a red herring across his track can sometimes send the dogs off into the wrong direction. This ruse has given us the term "red herring" as something that deflects attention from the real issue.

As in other parts of the world, the forests of the Pacific Northwest of the United States can be divided between "old-growth" forests, a term that generally means an abundance of trees more than 50 to 100 years old, and "new-growth" forests, which contain trees generally less than 50 years old. The old-growth forests are mostly on public land, but the new-growth forests occur on a mixture of public and private land. They occupy sites burned over by forest fires or, more commonly, areas that were extensively logged by lumbering interests or cleared for farms that have now been abandoned. Many of the new-growth forests are tree farms planted by large companies and owned as private land, but some are public land.

The difference between public and private land is important. Private land is managed by the owners, mostly large corporations, and is lightly regulated by a combination of federal, state, and local agencies. Public land, however, is controlled largely by the U.S. National Forest Service, which sharply regulates the amount of logging and other economic use (grazing, recreation, etc.). The concentration of old-growth forests on this public land has caused major conflicts about the amount of logging that can be permitted. Small companies or individual operations require a steady supply of uncut public forest in order to stay in business, and many of them need to use the old-growth forests for that supply. Large corporations with private land also need to in-

crease their logging on public lands because they have slightly exceeded the sustained-yield capacity of their own holdings.

The pressures to increase logging yields on public lands raise two questions: How much logging is possible without long-term damage, and who makes this decision? This leads to a now-classic confrontation. The independent loggers must cut trees in order to survive. Large companies must increase their acreage in order to regain a sustained-yield operation. The Forest Service must preserve trees in order to prevent complete deforestation. The fishing industry must have pristine streams in un-logged forests so that the fish they want to catch can spawn and begin their growth in fresh water. The tourist industry needs forests that attract people who want to see "real" nature.

Enter the spotted owl, which unfortunately thrives only in the old-growth forests of the Pacific Northwest. Although not a species distinct from several other types of owls, the spotted owl is a distinct variety, and logging in the old-growth forests threatens its habitat and possibly its existence. "Environmentalists" have decided to use the owl as the principal tactic to prevent, or at least delay, cutting, and for this purpose they have invoked the Endangered Species Act (see text). This effort has led to court suits to prevent any logging at all and other suits to open all of the old-growth forest to immediate timbering. The argument has basically developed into a shouting match between people who want to save the spotted owl and those who want to save the jobs of lumbermen. Other voices have been drowned out, including those of people who know that excessive clearcutting would weaken or destroy the fishing industry and also jeopardize the jobs of future generations of lumbermen.

In short, we should be arguing about the relative economic interests of fishing and lumbering and the relative importance of jobs for present lumbermen and those of their offspring. Instead, we are arguing about the spotted owl. It is undeniably an owl, but it is also a red herring.

FURTHER READING: Carroll (1995).

as the base of the food chain that the ecology of the area may be almost permanently damaged. Some investigators speculate that the arid "red center" of Australia would be much less harsh today if the original inhabitants had not brought firestick farming with them.

We can do nothing about the depredations of 1,000 or more years ago, so we proceed to the question of what we are doing today to the world's forests. First, the good news! In the mid 1990s, most western European countries have an annual growth of new wood that is 25%

Figure 8.13 Areas that have been deforested: throughout much of human habitation in Scotland, for example, where this river valley used to be lined with trees.

greater than net loss by cutting and natural causes and are experiencing an increase in forest-covered land at rates ranging from very low up to 1% per year. Much of this forest growth is caused by efforts by some European countries to remove land from agricultural use, partly to reduce overproduction of food and partly for the purpose of environmental restoration (see Section 2.2 for a discussion of European agriculture). The forest expansion has occurred at a time when wood products are a high percentage of the export trade in several countries, mostly in Scandinavia, with vigorous tree cutting for lumber, paper, and various types of processed wood (such as plywood).

On a slightly longer time scale than the present, forests in the United States have expanded in the past century. Through the 1800s, most of the population was engaged in farming, largely on small plots owned by individual families. As industrial activity became a more important part of the economy and the efficiency of farm production increased, the small family farms were no longer economically viable, and many have reverted to forest, particularly in the eastern United States. Approximately two-thirds of the land area of the state of North Carolina, for example, was devoted to farms in the 1700s and 1800s, but now only one-third is farms, and much of the state is covered by forests that have grown in the past 50 to 100 years. This expansion in the east compensates for the loss of forests in the western United States, where

timber production and clearance for other development purposes has somewhat outstripped growth. Even in these areas, major timber organizations have placed much of their (private) holding on a "sustained yield" basis, logging only at the replacement rate.

Now for the bad news! Some European countries, principally the semiarid Spain and Greece, are losing forest. Pressures to log previously uncut ("old-growth") forest in the northwestern United States are leading to rapid depletion of woodland in this region (see our discussion of the spotted owl in Box 8.5). The worst news, however, comes from the tropics and lesser-developed countries in general. Virtually every forested country in South America, Africa, and southern Asia is losing forest at ~0.5% to 1.0% per year, although some are notably higher, with those being deforested at significantly more than 1% per year shown in Table 8.4. Of these countries, some have almost no forest left – they include Bangladesh, Pakistan, the Philippines, and Grenada. The highest reported loss rate is 5.3% in Jamaica, with Haiti close behind at 3.9% (Box 8.6). Some of the stripped forest is undergoing a secondary growth, but the new trees will need a century or more to regenerate the tropical ecosystem that has been destroyed.

Tropical deforestation occurs for three reasons. First, some forests are clearcut in the expectation that the cleared land will be agriculturally productive. A second reason is that some wood is used for fuel, particularly in

Table 8.4 *Selected data on the world's forests*

Region or country	% gain (+) or loss (−)
Changes in woodlands in the 1980s	
Canada	+5.8
New Zealand	+4.3
Japan	+0.4
United States	−1.1
Australia	−0.9
Ecuador	−21.0
Nicaragua	−23.5
The Gambia	−26.2
Paraguay	−27.7
Viet Nam	−28.8
Haiti	−30.0
El Salvador	−31.6
Estimated total loss of tropical forests by 2000	
Indonesia	−10
Malaysia	−25
Brazil	−33
Colombia	−33
Guatemala	−33
Madagascar	−33
Mexico	−33
Ecuador	−50
Honduras	−50
Nicaragua	−50
Thailand	−67
Costa Rica	−80
Ivory Coast	−100
Nigeria	−100

Sources: Information on industrial regions and countries and information on loss rates during the 1980s from World Resources (1994). Estimated losses of tropical forests from Park (1992) and references cited therein.

semiarid regions where population growth has outstripped natural resources. In the Sahel region just south of the Sahara Desert, for example, people commonly walk many miles per day to scavenge enough firewood (mostly from scrub) to prepare an evening meal (Section 2.3). Most of the clearing of tropical forests, however, is for wood to be used either as lumber or as pulp for paper, plywood, and similar products, and it is very wasteful. Trees usable for lumber constitute only a small percentage of the typical tropical forest, but the efficient way to

log them is to clearcut everything. For example, mahogany trees contain an ornamental wood that is extremely valuable for expensive furniture and other uses, but the trees are so rare that it may be necessary to clearcut several acres of forest to find one. Rough estimates are that only about 5% of the trees cut down in typical tropical logging operations are usable. The others are simply left to decay on the ground or are removed in bulk for conversion to sawdust, paper, or other wood composition products.

The effects of logging or burning tropical forests commonly do not become evident for several years. Nutrients in the soil can yield food crops or grass for animals for a few years, after which the natural low productivity of the soil either terminates use of the area or requires massive import of fertilizer (Section 2.2). For this reason, clearing of the Amazon and similar rainforests has not produced land for individual peasant farmers but only for large landowners who can afford modern agricultural techniques. Other adverse effects include the destruction of the fauna and flora that are uniquely adjusted to life in tropical forests and the enhanced soil erosion and reduced groundwater quality that accompany tree removal. Reduction in water quality results from the fact that the shallow roots of forest vegetation soak up nutrients (such as calcium, nitrogen, and phosphorus) that enter streams and cause eutrophication (Section 7.5). On a broader scale, approximately 10% of the oxygen returned to the atmosphere, and carbon dioxide removed from the atmosphere, by photosynthesis each year comes from tropical forests.

The message from tropical forests is fairly clear – leave them alone! This is not easy to do. Some wealthy countries need wood that they are not able to produce for themselves. Japan and Taiwan import much of their wood as logs and cut it themselves rather than import lumber that has been milled in the country of origin. Other countries use wood from tropical forests for the manufacture of high-priced merchandise, largely furniture (e.g., mahogany, already mentioned, and much larger quantities of teak wood). For many tropical countries, the income from these lumbering operations provides a much-needed source of income, in some countries almost the only one. A major problem, however, is that this income commonly goes to a small group of already wealthy people rather then being spread out among the populace as a whole.

Policy questions

The topics of this section are of direct interest to most people, and we ask two general questions.

BOX 8.6 THE TREES OF HAITI _____

The folk artists of Haiti have produced some of the world's most charming paintings. Many of them show a tranquil rural landscape of mingled forests, grasslands, and croplands. The productivity, and possibly the beauty of the country, enabled the Haitian people to withstand the generations of dictators or other unelected rulers that have been their lot since the country gained its independence in a revolution in 1804.

In the late 1900s, the tranquillity and much of the beauty vanished at an accelerating pace. Both disappearances are probably related. Politically, the last round of despotism overthrew the only freely elected president in Haiti's history, ultimately leading to U.S. intervention to restore him in 1994. The loss of beauty was the result of deforestation and overuse of land as Haiti's population doubled and redoubled. By 1990, the population density was 700 per square mile, and Haitian women had a fertility rate of 6 children apiece.

Deforestation in Haiti resulted partly from the need to clear more land for marginal agriculture but was mostly the result of cutting trees for charcoal. The only fuel, other than raw wood, available to most Haitians is charcoal. In order to supply the needs of 6 million Haitians, the pace of logging far outstripped the rate of new growth. The cutting encompassed not only mature trees but, in the need for fuel, new shoots that would have been the forests of the future. In the late 1900s, Haiti's forests were disappearing at a rate of more than 3% per year, leading to predictions that virtually no trees would be left in the country by the early 21st century.

Overcutting affected not only forests, a small and dwindling part of Haiti, but also agricultural land. Scattered trees outside of forests are also cut. Removal of tree and leaf cover exposed grasslands and tilled fields to erosion by rain and rills and streams. Whole regions had their topsoil picked up by rivers and dumped into the ocean. The result is that the once-green color of much of Haiti has now turned to red, the color of the underlying tropical (lateritic) soil laid bare by removal of the organic- and nutrient-rich topsoil.

The stripped red soils still can be coaxed into production of food, and by the 1990s, the total crop was little changed from 10 to 20 years earlier. But a population growth rate of 2% to 3% a year needed a comparable increase in production that was not forthcoming. By 1990, the average Haitian could count on less than 90% of the nutrition (calories) needed for a reasonable life. The agricultural and ecologic chaos enhanced the already deep divisions between haves and have nots — the haves desperate to maintain a good life and the have-nots desperate to maintain life at all.

Political reorganization is obviously important to Haiti. It cannot be successful, however, without reforestation and other ecologic repair of the country.

FURTHER READING: Repetto and Gillis (1988).

- Should we preserve wildlife even if that preservation causes economic hardship to some people? Can arrangements be made for preservation and economic development at the same time?

- Do countries or organizations that buy wood from tropical forests or buy food produced in them have any right to dictate the way in which the forests are used? For example, should wealthy countries refuse to buy forest products from land that is being environmentally degraded? Should the price paid for forest use be higher than the world market price in order to compensate for the environmental loss?

FURTHER READING: on species diversity – Martin and Klein (1984); Tobin (1990); Spellerberg (1991); World Conservation Monitoring Centre (1992); Bonner (1993); Flannery (1994); Hemley (1994); Pyne (1995); Rosenzweig (1995). on deforestation – Repetto and Gillis (1988); Park (1992); Kuusela (1994).

8.4 CHEMICAL MODIFICATION OF THE EARTH'S SURFACE

If rich countries are so concerned about depletion of the ozone layer they are going to have to pay [poor] countries not to repeat their mistakes. – Former minister of the environment of India, cited by R. Lamb, "Public information"

Much of the effluent of our lives is a local problem. We produce solid waste, biological waste and excess fertilizers, and a variety of chemically active materials that must be disposed of safely. We discuss these issues in Chapter 7 and here consider only problems posed by chemicals that have a worldwide distribution and have the capacity to modify the global environment. We do so in two categories – chlorofluorocarbons (CFCs) and their effect on the ozone layer, and lead.

Chlorofluorocarbons (CFCs) and the ozone layer

Scientists "wintering over" in British and U.S. research bases in the Antarctic face a long period of cold and dark. Some of them pass part of the time with balloons, periodically sending up a panoply of devices to measure the physical properties and chemical composition of the atmosphere. In September 1985, as the winter waned and the Austral spring began to arrive, ground temperatures shot up to between −10°C and −20°C, and the atmospheric composition showed an unexpected anomaly. In the stratosphere, at heights of 12 to 50 km, the abundance of ozone (O_3) was lower than they had recorded at other times and also lower than the worldwide average. The low O_3 concentrations persisted for about 2 months, rising to normal as the Antarctic summer arrived (and ground temperatures during local heat waves were occasionally above freezing).

The initial observations of O_3 depletion have been exhaustively followed by further balloon measurements and bolstered by surveys from satellites equipped with scanning instruments that obtain comprehensive analyses of most atmospheric gases. We now know that during the Austral spring, an "ozone hole" appears over the Antarctic and extends as far north as southern Australia, New Zealand, and South America (Fig. 8.14). Ozone concentrations in the hole fall below 50% of the atmospheric norm of 0.3 ppm ($0.3 \times 10^{-4}\%$, or 300 Dobson units) by volume. At first the ozone depletion was confined to the Southern Hemisphere, but in the early 1990s widespread ozone depletion of 5% to 15% was detected throughout much of the Northern Hemisphere (although generally not as a well-defined hole). Because both the Southern and Northern Hemispheres had been surveyed for atmospheric ozone concentrations for many years before the reduction in concentration was discovered, it is clear that ozone depletion is a new phenomenon, not simply one that had already been present and not observed. At the present rate of worldwide decrease in O_3 concentrations, the concentration will be less than half of normal by the middle of the 21st century.

We worry about reduction in O_3 concentration because O_3 absorbs solar radiation in the ultraviolet (UV) part of the spectrum (see Section 8.1 for a discussion of radiation spectra and selective absorption). In particular,

O_3 absorbs in the high-energy (UVB) part of the spectrum at a frequency of ~300 nm (Fig. 8.5). This absorption shields the earth from UVB bombardment, and reductions in O_3 concentrations permit the UVB reaching the earth's surface to increase by some 10% to 20% worldwide, with almost no protection left within the ozone holes. Because UVB is in a high-frequency, consequently high-energy, part of the spectrum, it penetrates biological tissue and can cause:

- Sunburn. You can get as bad a sunburn on a cloudy day as a sunny one simply because the radiation that causes the burn is UVB, which is at too high an energy to be absorbed by clouds.
- Skin cancer. Penetrating radiation disrupts the metabolic process of cells and may lead to cancer in the same way as other damaging carcinogens. Present rates of reduction in O_3 concentrations should cause several million more skin cancers worldwide per year within a century.
- Cataracts. Ultraviolet radiation can damage tissues in the eye, which is why many modern sunglasses are specially designed to absorb UVB. The risk to unprotected eyes increases in proportion to the decrease in O_3 abundance.
- Birth defects, possibly even mutation. If penetrating radiation strikes reproductive organs, the resulting eggs or sperm may be damaged, leading to defective offspring.
- Crop reduction. UVB damages both the metabolic and reproductive processes of plants, and its increase could have a significant effect on world food supply.

Assuming that these facts are worrisome, which they are, we need to discover the reason for the recent appearance of ozone holes. Some pundits, generally not scientists, do not believe that the ozone holes exist, or that they are harmful if they do exist, or that they are caused in any way by human activity (we discussed the concept of "denial" at the start of this chapter). Continued atmospheric measurements, however, show that low concentrations of ozone are matched by increased concentrations of chemicals known as chlorofluorocarbons (CFCs). The leading hypothesis for the existence of ozone holes is now the effect that returning spring sunlight has on the reaction of CFCs with ozone and need to explain the reaction.

Chlorofluorocarbons (CFCs) are a set of chemicals that do not occur in nature and are in the atmosphere solely because they are manufactured for a variety of in-

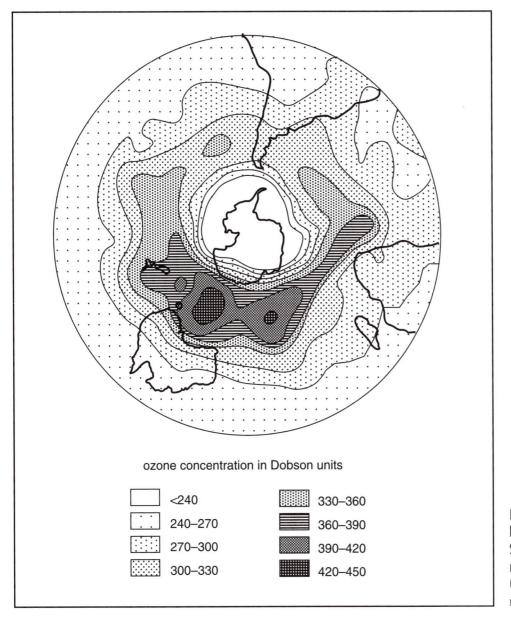

ozone concentration in Dobson units

☐	<240	▦	330–360
☐	240–270	▬	360–390
☐	270–300	▨	390–420
☐	300–330	▦	420–450

Figure 8.14 Antarctic ozone hole during Austral spring on September 11, 1996. Based on mapping by U.S. National Oceanic and Atmospheric Administration.

dustrial and commercial applications, principally as refrigerants. Air conditioners and refrigerators work by using gases in a linked system of compressors outside the area to be cooled and expansion devices within the area (Fig. 8.15). If the compression and expansion take place rapidly enough that there is no exchange of heat with the surroundings (known as "adiabatic" processes), the compressed gas heats up and the expanded gas cools down (see the discussion of heat and temperature in Section 5.1). As examples, when you pump up a bicycle tire the pump heats up because you have compressed the air too rapidly for it to lose heat, and shaving cream cools down when it expands abruptly out of an aerosol can. Because

refrigeration requires energy for compression and for transferring the gases into and out of the refrigerator, in order to keep the cost of operation low the gas used must be efficient at heat transfer. It must also be nontoxic, noncorrosive, and stable. CFCs have all of these properties, and when they were discovered in the mid 1900s, they became the gases of choice for refrigeration systems (the most widely used is freon, CF_2Cl_2). They would still be the gases of choice if it were not for one little problem – their effect on atmospheric ozone.

Ozone is manufactured in the upper atmosphere (stratosphere) by the effect of solar radiation on normal oxygen (O_2). The "photolytic" (light-mediated) reaction

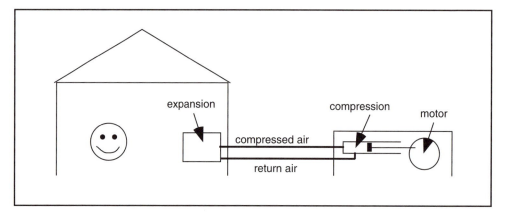

Figure 8.15 Diagram of air conditioner. Energy is used primarily to compress gas on the outside of the house. Expansion of gas on the inside causes cooling and lets gas return to compressor unit on outside of house. See further discussion in text.

results from the breakdown of O_2 into two monatomic O atoms, which then recombine with more O_2 to form O_3 at a normal concentration O_3 of 0.3 ppm. Chlorofluorocarbons reduce the concentration of O_3 by releasing monatomic chlorine (Cl) photolytically. The Cl then combines with O_3 through a series of reactions that result in formation of O_2 and the release of the Cl.

$$Cl + O_3 \rightarrow ClO + O_2$$

$$ClO + O \rightarrow Cl + O_2$$

for a net reaction of

$$O_3 + O + Cl \rightarrow 2O_2 + Cl$$

Regeneration of the Cl after the breakdown reaction makes it available for use in breaking down other O_2 molecules. Thus, the Cl acts as a "catalyst" that merely promotes the reaction rather than as a reactant that is used up in the reaction, and an estimated 70,000 O_3 molecules can be decomposed by a single Cl atom from a single CFC molecule.

The catalytic decomposition of O_3 is most effective on the surface of some solid particle. That is why ozone holes appear over the Antarctic and Arctic during their respective springs. The returning sunlight interacts with ice particles in the stratosphere to promote reaction of Cl and O_3, causing destruction of most of the ozone in the atmosphere until the stratosphere warms up enough to reduce the amount of stratospheric ice. Other particles than ice, however, can be equally effective. The best demonstration of that effectiveness was provided by the eruption of Mt. Pinatubo, in the Philippines in 1991, which placed enough dust in the atmosphere to lower worldwide temperature (Section 8.1) and give sunsets an

orange color for several years. Some of the dust contained sulfur, largely in the form of SO_2 compounds that provided enough surfaces for catalytic decomposition of O_3. About 1 year after the eruption of Mt. Pinatubo, worldwide O_3 concentrations fell almost 25% below their normal values. Experts on atmospheric chemistry now predict that another (or larger) volcanic eruption than Pinatubo would act on an atmosphere that has even less O_3 and could cause almost complete collapse of the earth's ozone shield.

Because of these concerns, the principal CFC-producing countries met in Montreal, Canada, in 1987 and signed the Montreal Protocol, which planned the elimination of CFC production by the year 2000. The intent was to replace CFCs with chemicals that would not damage stratospheric ozone. One of several proposed types is hydrogen-bearing chlorofluorocarbons, known as HCFCs, which have properties very similar to the non-hydrogenated varieties, but are readily decomposed by reactions in the lower atmosphere and do not rise to the levels at which they become a threat to stratospheric ozone. The switch to HCFCs could be accomplished by refitting present equipment, which costs money but generally can be done quickly, and by requiring that all new equipment be designed for HCFCs. Even the complete elimination of CFC production, however, will not cause immediate restoration of the ozone layer. Not only are some recently produced CFCs still rising through the atmosphere, but those already in the stratosphere would not completely disappear for 50 to 100 years (remember the ability of one Cl atom to be recycled through the catalytic reaction that destroys ozone perhaps 70,000 times). The intent of the protocol and similar arrangements, therefore, is to begin the process of restoring ozone, with understanding that complete recovery may require a century.

Regardless of the seeming urgency, the Montreal Protocol and related efforts have been criticized for both economic and scientific (or pseudoscientific) reasons. The economic arguments have some validity. They center around the cost of conversion of old equipment and the inability of some small operations either to convert or to purchase new equipment designed for HCFCs or other chemicals. This problem is proving to be relatively minor for large industries, where both conversion of old and purchase of new equipment is being done without undue delay or cost. Conversion of small operations, however, may cost more money than the user has available, particularly for industries in lesser-developed countries and also for laundries that use CFCs for cleaning and other small enterprises in industrial countries. These users may require more time to phase out CFC use, and instead of refitting old equipment to HCFCs, they may need to use CFCs until present equipment wears out and then buy new equipment designed for HCFCs. This problem of refitting is particularly acute for owners of cars with air conditioners built before the switch to HCFCs. The conversion costs approximately $300 (U.S.) and has helped to spawn a "black market" in CFCs (we are not joking – in the middle 1990s CFCs are being smuggled into the United States for use in air conditioners, laundries, etc.).

The scientific (so-called) arguments against the Montreal Protocol are based on two alleged lines of reasoning, both of which are best regarded as bizarre, although other adjectives might also be appropriate. One is that the ozone hole does not exist, either because atmospheric scientists are incompetent or because they want money to support research and are trying to scare the public into giving it to them (remember our quotation at the start of the chapter). The second argument against doing anything to stop CFC production is based on the idea that volcanoes discharge enormously greater quantities of chlorine into the atmosphere each year than is delivered by CFC leakage. This argument is true if we only count chlorine atoms, but in actuality the Cl emitted from volcanoes is almost entirely in the form of hydrogen chloride (HCl), which forms hydrochloric acid when dissolved in water. Because the HCl dissolves in water, however, most of it is swept out of the lower atmosphere by rain, and the tiny amount that rises to the stratosphere dissociates to so little atomic Cl that it has a much smaller effect on ozone concentrations than the CFCs.

We finish the discussion of ozone depletion with some good news. By 1995, limited evidence suggested that worldwide ozone levels were slightly higher than they had been in the previous 10 years. Apparently the efforts to reduce CFC emission are having some effect, but there is a long way to go, and two central questions remain unanswered, such as who pays for the halt in CFC production, and is the cost too high? We return to these issues in our policy questions at the end of the section.

Lead

It is a relief to consider an environmental issue that has a (mostly) happy ending – reduction in lead pollution. The reason for the success is that no one disputes the danger of lead poisoning. Lead (Pb) has been known for many years as a toxin that works largely by impairing the neural system and brain functions, and even relatively small doses cause mental retardation. Lead poisoning commonly results from increments of small doses rather than one large one. This cumulative poisoning is caused by the difficulty with which lead passes through both the liver and the kidney, thus preventing effective excretion and leading to a continual buildup of ingested lead until it reaches some critical level. When the lead problem was fully realized about 1980, most of the world embarked on a cleanup that eliminated its use in many processes. In this section, we tell some of the story of how we came to understand the problem, what has been done about it, and what should be done further.

We have already made use of the atmospheric record provided by the Greenland and Antarctic ice caps to document the effect of CO_2 on climate (Section 8.2) and even to ponder the fate of the "lost continent" of Atlantis (Box 3.2). Now we use the ice caps to follow the world's use of lead (Fig. 8.16). The Greenland ice contains low lead values of approximately 1 ppb (part per billion; 10^{-7}%) from several thousand years ago until the latter part of the 1700s, when concentrations began to increase rapidly. The increase was caused by the industrial revolution, which brought the use of lead in storage batteries (Fig. 5.10; about 50% of modern use), in solder for water and sewer pipes, as a base that made paint shiny, and in the production of high-luster glass. When the batteries leaked, water etched the pipes, and paint weathered, the lead was released into water and air and ultimately made its way into the atmospheric circulation pattern. Greater industrialization of the Northern Hemisphere relative to the Southern Hemisphere is shown by the small amount of Pb pollution in Antarctic ice cores.

By far the largest increase in lead in the Greenland ice cores occurred in the 1930s, when the producers of automobile engines and the gasoline that drove them discov-

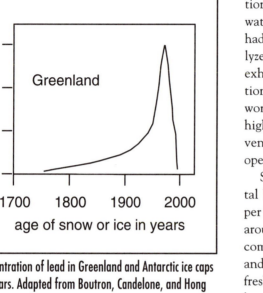

Figure 8.16 Concentration of lead in Greenland and Antarctic ice caps during past 250 years. Adapted from Boutron, Candelone, and Hong (1994).

ered that the compound lead tetraethyl (Pb(C$_2$H$_5$)$_4$) made engines run more efficiently. "Ethyl" was referred to as an "antiknock" additive to gasoline because it prevented engines from "pinging" as unburned fuel and air went through into the exhaust. Emission from automobile exhausts placed lead into the atmosphere far more effectively than all of the industrial processes combined and distributed lead-filled air throughout the Northern Hemisphere from the 1930s to the present.

Confirmation of widespread lead pollution came, paradoxically, from the lunar program. In the 1960s, when laboratories in numerous countries were preparing to analyze rocks returned to earth from the moon, one of the principal tasks was to construct mass spectrometers capable of measuring very low isotopic abundances. Because isotopes of lead are produced by decay of uranium and thorium (see the discussion of radioactivity in Section 5.4), the isotopic abundances can be used to date rocks and also to interpret various aspects of lunar history. Remote sensing of the lunar surface, however, had strongly indicated that rocks on the moon would contain lead and other isotopes produced by radioactive decay in

far lower concentrations than in terrestrial rocks. Thus, it became necessary to build better and better spectrometers, and, once built, the spectrometers were also used for better analyses of terrestrial materials.

When good measurements of lead abundances became available, it was apparent that many earlier analyses were incorrect. Partly they were wrong because of inadequate instruments, and partly they suffered from contamination during the process of collection and analysis. Ocean water, for example, had been collected in containers that had not been stringently cleaned of lead and then analyzed in laboratories that contained air polluted with the exhaust from leaded gasoline. The problem of air pollution is particularly severe, and all modern lead-isotopic work is done in rooms that contain purified air at a higher pressure than the outside atmosphere, thus preventing outside air from flowing in when the door is opened.

Some of the first good measurements of environmental lead concentrations were reported in a landmark paper by D. M. Settle and C. C. Patterson in 1980, centered around the lead concentration of canned albacore (a common variety of tuna). To summarize briefly, Settle and Patterson, and others following them, reported that fresh albacore meat contained 0.3 ppb lead, whereas albacore from a lead-soldered can contained 140 ppb, an increase of approximately 500 times caused by the can. Their investigation continued by comparison of the bones of modern people with those of people who lived before lead was used or in areas where it was not used. The lead reaches the bones by being ingested and then working its way through the blood to the sites where bone is synthesized. Modern Europeans and Americans contain an average of 150 ppb lead in their bones. By contrast, Incas who lived in the high Andes well before the arrival of Europeans contained only 0.2 ppb lead in their bones. The only metal mined by the Incas was gold, which did not produce lead as a byproduct, and thus the Incas lived in an environment as lead free as possible, and they had a skeletal lead concentration 750 times less than the concentration in modern people from industrial societies.

The effects of the industrial revolution on lead distribution receive an interesting verification from study of the Roman Empire. The only anomaly in the lead of the Greenland ice cores before 1750 was in the first century A.D., when the normal concentration of 1 ppb increased briefly to about 3 ppb. Further investigation showed that aristocratic Romans had high lead concentrations in their skeletons, whereas the common folk did not. The difference was that aristocrats could afford pottery with nice shiny glazes as storage casks for wine and various foods, but poor people had to make do with unglazed pots. Because wine is acidic, it leaches various elements from the containers in which it is stored. We leave the reader to figure out what the Romans used to make glazes shine.

The remedies to the problem of lead contamination are fairly simple and have been introduced in most industrial countries. They include: (1) elimination of lead from gasoline, with antiknock qualities provided by organic compounds that burn like gasoline; (2) substitution of lead-free solder, both in cans and water pipes; and (3) elimination of lead from paint (also see Section 4.5 for a discussion of lead in drinking water). Lead is still used in batteries because alternatives are considerably more expensive. Unfortunately, many of the safeguards have not been implemented in lesser-developed countries, where leaded gasoline and paint are still freely available. Because most gasoline is burned in industrial countries (Chapter 5), however, worldwide atmospheric cleanup has already begun, with a precipitous decline in lead in the Greenland ice cap beginning in the 1970s.

As a target for the cleanup, the U.S. Food and Drug Administration (FDA) selected a maximum lead concentration of 20 parts per billion (20 ppb = $20 \times 10^{-7}\%$) in blood. Above that level, a person would be considered as having some degree of lead poisoning. Recently, the FDA lowered the maximum concentration to 10 ppb, thus automatically reclassifying some people hitherto considered healthy to a possibly contaminated status. Reduction to 10 ppb sparked controversy. The issue is whether there is a threshold level below which blood Pb has no effect and, if so, whether the threshold is above 20 ppb (we discussed the threshold problem more extensively in Section 7.4).

In addition to setting standards, federal, state, and local governments have ordered extensive efforts to remove lead from public buildings, particularly schools. This has led to expenditures of some billions of dollars to scrape paint off of walls or to cover it with new paint. As with the great asbestos panic (Box 7.1), much of this money is wasted. We already know who is at greatest risk of lead poisoning. They are children who are poor and live in old houses with peeling lead-based paint. These paint chips have a sweet taste, and children sometimes chew them as a substitute for candy. Experts on public health generally feel removal of this lead and intensive medical testing of these children's blood would accomplish more with a far smaller expenditure of time and

money. Before you say "How can we ration money when children might be at risk?" remember that the money and effort spent on unnecessary lead removal might have been expended on other, more-effective, efforts to improve health.

Policy questions

Because the necessity of removing lead from the human environment is not controversial, we confine our questions to the issue of ozone depletion. They include

- Should large industries in wealthy countries be required to convert now to equipment that does not use CFCs, and should they pay the cost of the conversion?
- Should industries in lesser-developed countries be required to convert now? If so, should they pay the cost or should some unspecified international fund help them if the cost would cause the industries to go out of business?
- Should industrial countries require immediate conversion of equipment in laundries, automobile air conditioners, and similar machinery that is individually owned? If so, do the individual owners have to pay the cost, or should the society that demands this transformation bear some of the cost?

FURTHER READING: on ozone – Litfin (1994); on lead – Settle and Patterson (1980); Whitten and Prasad (1985); Nisbet (1991).

8.5 SUMMARY AND CONCLUSIONS

We close this chapter by recognizing that people have changed the earth's surface environment and asking three questions about the modification – how much, is it dangerous, and is it reversible?

"How much?" varies with the type of change, and for some types we do not have sufficient evidence to reach a firm conclusion. It is clear, however, that we have changed the chemical composition of water, air, and soil. Industrial activity has increased the amount of carbon dioxide in the atmosphere by about 25%, increased the methane content by apparently larger factors, and placed CFCs in an atmosphere that contained none before modern industries put them there. And we have distributed lead, primarily by burning leaded gasoline, to such an extent that it appears in the ice of Greenland and the oceans now have more than 1,000 times as much lead as they contained before people began extracting metal from ores.

How much change there has been in animal and plant life is more difficult to assess than chemical change. Early, prehistoric people certainly exterminated several types of animals (megafauna) as they expanded into new areas, and destruction of forests has left many areas virtually treeless. The clearcutting is continuing in tropical forests, and at least local elimination of animal and plant species is occurring in islands and other areas where the ecosystem is particularly fragile. Thus, we can conclude that we have reduced, and continue to reduce, biodiversity, but quantitative information on the rate is nearly impossible to obtain.

"Is it dangerous?" is more difficult to answer because the long-term effects of our human-induced changes are unclear. The effects of our chemical changes are somewhat more certain than those of biodiversity changes, but even the chemical modifications are controversial. Although many people find evidence that the earth's atmosphere and surface are becoming warmer, that conclusion has been challenged on the basis of statistically inadequate data. Also, even if the data are correct, we have not yet proved that emission of greenhouse gases (CO_2 and CH_4) is the cause of the warming. Evidence that emission of CFCs has reduced the ozone content of the atmosphere and increased exposure to ultraviolet radiation is clearer, but the medical and agricultural effects of this radiation have not been firmly established. And although everybody agrees that lead pollution is dangerous, we have not demonstrated that global increase in lead exposure has led to global medical problems.

The dangers of reduction in biodiversity and natural habitats are impossible to quantify. Many of the concerns are aesthetic, which is of little interest to many people, but some real dangers may result both from the extinction of organisms and also from the worldwide propagation of a few species of domestic plants and animals at the expense of a more diverse population. By reducing the genetic variability of some of our most vital food sources, we may be exposing ourselves to the type of global reduction in food supplies that has occurred locally at several times in our history. On the other hand, no global catastrophe has occurred yet, so there is no good evidence that it will.

"Is it reversible?" is probably the easiest question to answer. The carbon dioxide that we have added to the atmosphere will be there for a long time. Some reduction will occur by absorption into the oceans, but unless we completely abandon our use of fossil fuels and manufacture of concrete (both of which are obviously impossible), we will continue adding CO_2 to the atmosphere. We will make similar additions of methane as long as we use fossil fuel and raise animals. With these two chemicals, the best that we can hope for is to reduce their rate of increase. We can, however, reduce CFC concentrations, and some efforts to do this seem to have moderate success. Our best success, however, has been reduction of lead contamination, and although there is no evidence that our efforts have had an effect on oceanic lead concentrations, we may hope that that will occur within the foreseeable future. Reducing the loss of biodiversity is probably more within our grasp than reducing lead, but because it is less obviously dangerous, we are making little effort to do so.

To sum up. We have made some changes in the earth's surface environment. Some are probably dangerous. Some may not be. Some may be reversible. Some may not be. Anything else you would like to know?

PROBLEMS

1 If the earth's average surface temperature increases in the future in proportion to the concentration of CO_2 in the atmosphere, what would the temperature be if the CO_2 concentration doubled? Use data from Section 8.2.

2 An estimated 40 million automobiles in the United States have old air conditioners that use freon. The cost of converting to coolants that are not hazardous to stratospheric ozone is about $300 per car. If the cost were paid by the federal government, how much would each taxpayer have to pay?

3 How long can a country lose forest at a rate of 5% of the existing forest each year before only one-half of the original forest is left? (Note – this question can be answered by precise calculation or by approximation.)

4 A biological survey finds seven species of birds, with the percentages of the various species being 34%, 19%, 15%, 12%, 9%, 7%, and 4%. Calculate the Simpson Index for this population.

5 If the coefficient *a* in the diversity–area relationship on p. 307 is 0.3, how much greater would be the species diversity in one area of 10 square miles than in five areas of 2 square miles each?

6 Use the relationships in Figure 8.4 to calculate the residence times of CO_2 in both the shallow and deep oceans.

7 Use Figure 8.16 to estimate how much of the lead in the atmosphere was caused by burning leaded gasoline in the 1970s. Assume that the concentrations of lead in Greenland ice are proportional to the amounts of lead released by industrial activity.

8 Could houses be air-conditioned by bringing cylinders of compressed air into them and releasing the air rapidly through a nozzle? Is this a practical way to cool houses?

9 Nitrogen (N_2) constitutes approximately 75% of the atmosphere and CO_2 approximately 0.0003%. In plants, the ratio of C : N is about 7 : 1. Compare the residence time of N_2 in the atmosphere to that of CO_2.

REFERENCES

Alvarez, K. (1993). *Twilight of the Panther: Biology, Bureaucracy and Failure in an Endangered Species Program*. Sarasota, Fl.: Myakka River.

Bonner, R. (1993). *At the Hand of Man: Peril and Hope for Africa's Wildlife*. New York: Knopf.

Boutron, C. F., Candelone, J. P., and Hong, S. (1994). Past and recent changes in the large-scale tropospheric cycles of lead and other heavy metals as documented in Antarctic and Greenland snow and ice: A review. *Geochimica et Cosmochimica Acta* 58: 3217–25.

Carroll, M. S. (1995). *Community and the Northwestern Logger: Continuities and Changes in the Era of the Spotted Owl*. Boulder, Colo.: Westview Press.

Dlugokencky, E. J., Masaire, K. A., Lang, P. M., Tans, P. P., Steele, Ò. P., and Nisbet (1994). A dramatic decrease in the growth rate of atmospheric methane in the northern hemisphere during 1992. *Geophysical Research Letters* 21: 45–8.

Flannery, T. F. (1994). *The Future Eaters: An Ecological History of the Australasian Lands and Peoples*. Chetwood, New South Wales: Reed Books.

Flohn, H., and Fantecchi, R. (1984). *The Climate of Europe: Past, Present and Future*. Dordrecht: D. Reidel.

Gould, L. M. (1931). *Cold: The Record of an Antarctic Sledge Journey*. New York: Brewer, Warren & Putnam.

Harrington, C. R., ed. (1992). *The Year without a Summer?: World Climate in 1816*. Ottawa: Canadian Museum of Nature.

Hemley, G., ed. (1994). *International Wildlife Trade: A CITES Sourcebook*. Washington, D.C.: Island Press.

Houghton, J. T., Meira Filho, L. G., Bruce, J., Lee, H., Callander, B. A., Haites, E., Harris, N., and Maskell, K. (1995). *Climate Change, 1994*. Cambridge: Cambridge University Press.

Houghton, J. T., Meira Filho, L. G., Callander, B. A., Harris, N., Kattenberg, A., and Maskell, K., eds. (1996). *Climate Change 1995 – The Science of Climate Change*. Cambridge: Cambridge University Press.

Kaiser, H. M., and Dennen, T. E., eds. (1993). *Agricultural Dimensions of Global Climate Change*. Delray Beach, Fl.: St. Lucie Press.

Kuusela, K. (1994). *Forest Resources in Europe*. Cambridge: Cambridge University Press.

Lamb, H. H. (1995). *Climate, History, and the Modern World*. London: Routledge.

Lamb, R. (1991). Public information and attitudes. In J. Jaeger and H. L. Ferguson, eds., *Climate Change: Science, Impacts and Policy – Proceedings Second World Climate Conference*, pp. 367–9. Cambridge: Cambridge University Press.

Litfin, K. T. (1994). *Ozone Discourses*. New York: Columbia University Press.

Martin, P. S., and Klein, R. G., eds. (1984). *Quaternary Extinctions: A Prehistoric Revolution*. Tucson: University of Arizona Press.

Millero, F. J., and Sohn, M. L. (1992). *Chemical Oceanography*. Boca Raton, Fl.: CRC Press.

Mintzer, I. M. (1992). *Confronting Climate Change: Risks, Implications and Responses*. Cambridge: Cambridge University Press.

Mitchell, J. F. B. (1988). The "greenhouse effect" and climate change. *Reviews of Geophysics* 27: 115–39.

Nisbet, E. G. (1991). *Leaving Eden*. Cambridge: Cambridge University Press.

Park, C. C. (1992). *Tropical Rainforests*. London: Routledge.

Pyne, S. J. (1995). *World Fire*. New York, Henry Holt.

Repetto, R., and Gillis, M. (1988). *Public Policies and Misuse of Forest Resources*. Cambridge: Cambridge University Press.

Rogers, J. J. W. (1993). *A History of the Earth*. Cambridge: Cambridge University Press.

Rosenzweig, M. L. (1995). *Species Diversity in Space and Time*. Cambridge: Cambridge University Press.

Settle, D. M., and Patterson, C. C. (1980). Lead in albacore: Guide to lead pollution in Americans. *Science* 207: 1167–76.

Sloan, L. C. (1992a). A comparison of Eocene climate model results to quantified paleoclimatic interpretations, *Palaeogeography, Paleoclimatology, and Palaeoecology* 93: 183–202.

Sloan, L. C. (1992b). Equable climates during the earliest Eocene: Significance of regional paleogeography for North American climate. *Geology* 22: 881–4.

Speidel, D. H. (1988). *Perspectives on Water: Uses and Abuses*. New York: Oxford University Press.

Spellerberg, I. F. (1991). *Monitoring Ecological Change*. Cambridge: Cambridge University Press.

Tobin, R. (1990). *The Expendable Future: U.S. Politics and the Protection of Biological Diversity*. Durham, N.C.: Duke University Press.

Whitten, R. C., and Prasad, S. S. (1985). *Ozone in the Free Atmosphere*. New York: van Nostrand Reinhold.

World Conservation Monitoring Centre (1992). *Global Biodiversity: Status of the Earth's Living Resources*. London: Chapman and Hall.

World Resources Institute (1992, 1994). *World Resources*. Oxford: Oxford University Press.

9 | A FINAL WORD

We have reached the end. How are people and the earth getting along? Let's give a brief summary chapter by chapter.

PEOPLE AND LAND

The population is growing rapidly in most of the world, but growth in industrial countries has slowed almost to zero. Except in oil-rich countries, where export income provides a general national wealth regardless of growth rate, high rates of population increase are almost invariably associated with national poverty. One of the world's major controversies is whether high growth rates cause poverty or are the result of poverty. Regardless of the answer to this question, it is clear that high population densities have virtually no relationship to poverty, with some very crowded countries being able to utilize their resources in ways that make them wealthy.

Patterns of land use have clearly changed throughout human history as people occupied more of the earth and developed an agricultural society. These changes, however, seem to have slowed down. Current patterns of land use are mostly determined by the basic climatic and geographic factors that existed when the human race developed. Most cropland has naturally fertile soil and suitable climate, and land used as pasture ranges from natural grassland to brushy areas. Although local deforestation has occurred, some areas have recently been returned to forest, and the total area of forest and woodland is not much smaller than it was thousands of years ago. Similarly, areas designated as wasteland because they are covered by ice or are deserts or rocky hillslopes are still little utilized by people, and the conversion of pasture and cropland to human "wasteland" by building cities has affected only a very small part of the total land surface of the earth.

We conclude that the human race has not yet exceeded the capacity of the land to hold it. The problem is not numbers of people but the rate of growth in some parts of the world.

FOOD

The general increase in the human population during the past thousands of years, and the very rapid increase of the past 50 years, is not responsible for hunger. Total world food production has risen more rapidly than population and now could provide a nutritious diet for everyone. Although famine is endemic in some areas and occurs sporadically in others, it is not caused by a lack of food. Hunger basically results from the ability of some people, largely in industrial countries, to pay prices for food that are beyond the reach of poor people. On a short term, the problem of hunger could be solved if: (1) poor countries devoted their land to subsistence agriculture rather than export crops; (2) rich people did not use so much grain as cattle feed; and (3) hunger was not used as a weapon of warfare in civil and international strife, mostly within the lesser-developed world.

On a long term, the present high productivity of the world may not be sustainable. It is currently maintained by application of large doses of fertilizer, pesticides, and irrigation water. Both the fertilizer and pesticides have serious environmental effects, and irrigation water is being used at a rate greater than its natural recharge rate (for more discussion, see Chapters 4, 7, and 8). Consequently, we do not know whether there will be famines in the future, and if there are, whether they will be caused by natural factors that reduce food production or by human practices that could be avoided.

NATURAL HAZARDS

The risk that is an inherent part of being alive is accentuated in some areas by geologic and climatic hazards. Although we cannot predict extreme disasters such as large earthquakes and excessively high floods, we can outline general areas of risk and the consequences of living in them. For example, geologists know the location of the earth's major seismic zones and most dangerous volcanoes, we can foresee the height of normal floodwaters, and we can designate areas at greatest risk from shoreline erosion and land movements. Geologists can also help design and locate structures to minimize damage. What geologists cannot do is make people pay attention to the information that they provide.

The problem with natural hazards is largely political and social. Many people either do not believe that they will be affected by natural disasters, or they believe that someone else will pay for their losses if they sustain any. For this reason, we continue to build on floodplains, unstable slopes, and eroding beaches, and when we build in areas of inescapable hazards such as earthquakes, we commonly avoid the cost of safe structures and of insurance to pay for loss. Until societies place the responsibility for a loss directly on people who do not pay attention to geological information, these unsafe and costly practices will probably continue.

WATER

Fresh water is very unevenly distributed over the earth. Areas with the least water are deserts centered around 20° north and south latitudes, with more water available in the tropics and in temperate regions from about 30° to 50° latitudes. The combination of water availability and climate makes temperate regions the most agriculturally productive and has led to the establishment of industrial societies that would have been virtually impossible to develop in deserts and difficult to develop in tropical areas where high rainfall is combined with perpetually high temperatures.

Water problems largely arise because people attempt to expand water use beyond the limits imposed by cli-

mate. These efforts include pumping of groundwater that has accumulated over thousands of years (water mining) and running pipelines from available surface water into dry areas. In many places these efforts have led to agricultural production and development of large cities that use water at rates well beyond those that could be supported by local rainfall. As world population increases, access to water will become more contentious, leading either to increasing strife or, hopefully, local and international agreements to share it.

ENERGY

Although wood is the only fuel available to most of the world's poor people, a large variety of industrial energy sources are either in use now (fossil fuel, nuclear fission, hydropower) or potentially usable in the near future (nuclear fusion, solar). The world's industrial economies consume a far higher share of this fuel than lesser-developed countries. The proportion of energy produced by coal, oil, gas, and nuclear reactors varies from country to country, but presently the most indispensable fuel is oil, which almost single-handedly enables the world's transportation system to operate.

Despite their dependence on oil, industrial nations produce only a small fraction of the amount that they consume. Most of the production is centered around the Arabian Gulf, and because the area also contains by far the largest reserves, dependence on this region will almost certainly increase in the near future. Consequently, if industrial nations do not want a reduction in their consumption of energy, presumably resulting in a decline in their standards of living, these countries must either accept the diplomatic consequences of dependence or must develop alternative energy sources and the technology to use them for transportation.

MINERAL RESOURCES

Most people are unaware of the enormous variety of materials that we obtain from the earth and use either directly or after processing. The greatest bulk is building material, including quarried stone, clay to be converted to bricks and ceramics, and concrete, which is made by processing a mixture of sand, gravel, and limestone. Other major nonmetallic products include phosphate and potassium salts for fertilizer and quartz sand for glass. The principal metals are iron, converted to numerous steel alloys, and the "base" metals copper, lead, and zinc.

Economic deposits of most minerals are not broadly distributed around the earth. Although nonmetallic materials are more widely available than metals, differences in quality make some deposits more valuable than others, and most metals can only be mined profitably from a few places in the world. This restriction means that operation of their industrial economies requires the United States, western Europe, and Japan to import large quantities of metals. A few countries, particularly Canada, South Africa, and Australia, export several commodities, but some countries have only one exportable resource, including aluminum from tropical countries, copper from Peru and Chile, and tin from Bolivia.

Because countries that use minerals are generally not the same as those in which the resources can be mined profitably, the international trade in minerals will continue into the near future. Some reduction in trade could result from additional efforts by mineral-using countries to recycle, to subsidize production from deposits not now profitable, and/or to find substitutes for metals that they currently import. Until one or more of these changes are made, we have to accept the interdependence of countries and the international nature of the world's economy.

WASTE AND POLLUTION

If we define pollution as harmful modification of the earth's surface environment by human activity, then virtually all of the earth's surface is polluted. The types and amounts of contaminating materials, however, vary from place to place and from time to time. Because of these differences, the magnitude of the pollution problem is also highly variable.

Industrial countries, which produce much of the world's waste, are managing to reduce the levels of contamination for most pollutants. They have achieved this result by a series of measures that include reducing the amount of bulk waste, banning particularly harmful chemicals or shipping them abroad, improving sewage treatment, and reducing emissions into the atmosphere from automobiles and factories. The major problem that they have not solved is the disposal of radioactive waste, and this is more for political than technical reasons. Unfortunately, lesser-developed countries have either not been able to afford, or have not wanted, to take similar steps to reduce pollution. This inaction, coupled with

the shipment of hazardous waste abroad, means that environmental damage continues to increase in much of the world while it decreases in rich areas.

GLOBAL CHANGE

People have caused some aspects of the earth's surface environment to change more rapidly than they would have through normal geologic processes. They include both additions to, and subtractions from, the environment. The principal additions are carbon dioxide and methane to the atmosphere, lead to water and soil, and CFCs (chlorofluorocarbons) and other industrially produced chemicals throughout the earth's surface. Subtractions include large animals (megafauna) apparently killed during early stages of human radiation into new lands, and forests destroyed throughout much of human history.

Despite these very clear modifications, we are not sure how much change the earth has undergone. Addition of CFCs has certainly damaged the ozone layer, and some animals are gone forever, but whether the emission of greenhouse gases makes people responsible for global warming, or whether forests are being restored at roughly the same rate as they are being cut down, is not clear. Furthermore, we have recognized some of our mistakes and are beginning to eliminate lead pollution, have made a start on reducing CFCs, and are actively replanting forests in many industrial countries.

A FINAL WORD

People are people. We do some things well and some things poorly, and commonly we do well at one time and poorly at another. Through our successes and failures, we have to go about our business of living on the earth, and this requires us to make decisions about what we take from the earth (its resources) and what we put back into it (our effect on its surface environment).

We chose the topics of this book to underscore our main philosophy about the relationship between the earth's resources and its environment. Unfortunately, these topics have become entangled in a myth – that a person cannot be interested in both resources and environment, that people are divided between those who want to use the earth and those who want to preserve it. The first are sometimes called developers, exploiters, plunderers, and other terms too nasty to mention here.

The second are sometimes called idealists, greenies, tree-huggers, and other words that again won't be repeated.

We find no conflict between using the earth and preserving it. Our generation needs coal to power its industries, but the next generation needs us to stop pumping so much carbon dioxide into the atmosphere. We cut down trees to build buildings and save trees so that our grandchildren will also be able to build. The cost of depleting a source of groundwater may not be borne by us but by our offspring. In all of these cases, the conflict is between competing economic interests, not environmentalists and developers: lumber interests that want to cut trees along salmon-spawning streams and fishing companies that want large stocks of fish; people who want cheap hamburgers now and those who are concerned that our agricultural policies are so shortsighted that the soil will not be productive in 50 years. These are real problems that do not have simple solutions but demand immediate answers.

In short, most of the earth is a commons. Preservation of that commons for ourselves, our descendants, and all of humanity requires us to care about its health. Therefore:

- We cannot use the earth in any way that we wish. It does not provide infinite resources and will not accept infinite amounts of our waste products without degradation of the environment that we live in.
- We must understand how the earth works, the science of our planet, in order to coax the maximum long-term use from it. If we do not understand the science, we may bequeath a badly damaged planet to future generations.

Writing this book (though somewhat alarming at times) has been fun. We hope that you have enjoyed it and will make good use of it.

AUTHOR INDEX

Abbott, P. C., 69, 70
Ahmed, A. U., 32
Ahmed, M., 32
Aird, J. W., 29, 32
Alexander, D., 121, 123
Alvarez, K., 309, 325
American Petroleum
        Institute, 200, 212
Amin, R., 30, 32
Anderson, B., 64, 70
Andreas, P., 24, 32
Appleman, P., 11, 12, 32
Aschenbrenner, S. E., 123

Baldwin, J. E., II, 82, 123
Balter, M., 204, 213
Bannister, A., 32
Ben-Menahem, A., 77, 123
Beven, K., 104, 123
Bignon, J., 286
Bird, E. C. F., 113, 123
Blackman, W. C., Jr., 249,
        265, 286
Blanken, J., 32
Blunden, J., 17, 32
Bolt, B. A., 82, 89, 123
Bonner, R., 317, 325
Boubel, R. W., 284, 286
Boutron, C. F., 322, 325
Bowes, A. de P., 41, 70
Bridges, E. M., 78, 123
Bridgman, H. A., 284, 286
Broadus, J. M., 265, 286
Browning, J. M., 116, 123
Bruce, J., 325
Burt, A. T., 70

Callander, B. A., 325
Cameron, E. M., 245, 246

Candelone, J. P., 322, 325
Carling, P., 104, 123
Carroll, M. S., 314, 325
Carson, R., 265, 286
Chadwick, D. J., 69, 70
Chang, F. K., 190, 212
Cheremisinoff, P. N., 155,
        160, 167
Chopin, K., 282, 286
Chowdhury, J., 32
Chrispeels, M. J., 52, 70
Chyba, C. F., 77, 123
Coates, D. R., 113, 123
Coch, N. K., 81, 119, 123
Cohn, J. P., 159, 167
Corn, M., 286
Craig, J. R., 223, 246
Crowson, P., 235, 246
Cushing, D. H., 63, 70

Davies, H. R. J., 124
de Haen, H., 32
Degen, C., 106, 123
Degen, P., 106, 123
Dennen, T. E., 304, 325
Dlugokencky, E. J., 294, 325
Dold, C., 6, 286
Doornkamp, J. C., 119, 123
Druitt, T. H., 123
Dudley, W. C., 98, 123

Eden, G. E., 279, 286
Ehrlich, P., 21, 26, 31, 32
Eiby, G. A., 89, 123
Environment, 283, 286
Eumorphopolous, L., 76, 123
Evans, A. M., 232, 246

Fainberg, A., 179, 212

Fantecchi, R., 304, 325
Farzin, Y. H., 50, 70
Finkl, C. W., 159, 167
Flannery, T. F., 317, 325
Flawn, P. T., 235, 246
Flemming, N. C., 123
Flohn, H., 304, 325
Fradkin, P. L., 145, 167
Francis, B. M., 247, 265, 286
Francis, P., 96, 123

Galanopolous, V. P., 123
Gambini, D.-J., 274, 269,
        286
Gardner, G., 136, 167
Garrett, F. E., 279, 286
Gee, J. B. L., 286
Gershoff, S. M., 40, 70
Gilchrist, J. A., 121, 123
Gillis, M., 317, 317, 325
Glasscock, C. B., 243, 246
Gleick, P. H., 145, 167
Gordon, R. L., 200, 212
Gould, L. M., 288, 325
Granier, R., 274, 269, 286
Griggs, G. B., 121, 123
Guilbert, J. M., 223, 232, 246
Guthrie, G. D., 251, 286

Haigh, M. D. F., 279, 286
Haites, E., 325
Hall, C. W., 52, 70
Hammer, M. J., 277, 279, 286
Hammer, M. J., Jr., 277, 279,
        286
Harben, P. W., 246
Hardin, G. J., 31, 32, 63, 70
Hardy, D. A., 80, 123
Harrington, C. R., 294, 325

Harris, J. H., 240, 246
Harris, N., 325
Harrison, R. M., 284, 286
Heath, R., 134, 167
Hemley, G., 317, 325
Hill, R., 101, 124
Hong, S., 322, 325
Houghton, J. T., 291, 299,
        304, 325
Howes, R., 179, 212
Hudson, N., 1, 5, 32
Hughes, J. S., 213
Hunt, J. M., 200, 212

Iacopi, R., 78, 123

James, C. I. D., 167
Johansson, T. B., 179, 212
Johnson, P., 32
Jones, H. R., 15, 32

Kaiser, H. M., 304, 325
Kattenberg, A., 325
Kaufman, W., 113, 123
Kay, D., 130, 135, 138, 165,
        167
Keller, J., 123
Keller, W., 108, 123
Kelley, H., 212
Kesler, S. E., 215, 240, 246
Klein, R. G., 306, 317, 325
Knott, D., 192–193, 212, 264,
        286
Kolars, J. E., 102, 123
Kovacs, M. G., 108, 123
Kraft, J. C., 76, 123
Krauskopf, K. B., 267, 274,
        286
Kromm, D. E., 155, 167

Kuusela, K., 317, 325

Lamb, H. H., 299, 304, 325
Lamb, R., 304, 317 325
Lang, P. M., 325
Lappe, F. M., 34, 55, 70
Lee, H., 325
Leighton, F. B., 113, 123
Leopold, L. B., 98, 123
Linthicum, R., 2, 123
Litfin, K. T., 287, 323, 325
Livi-Bacci, M., 11, 32
Lowi, M. R., 160, 162, 165, 167
Lutz, W., 26, 32

MacKenzie, D., 265, 286
Malle, K.-G., 266, 286
Malthus, T. R., 15, 32
Manning, J. C., 145, 167
Marples, D. R., 204, 212
Marsh, J., 69, 70
Martin, P. S., 317, 306, 325
Masaire, K. A., 325
Maskell, K., 325
Mason, B. J., 284, 286
Mather, J. R., 131, 138, 155, 167
Mather, R., 52, 55, 70
Matzke, R. H., 193, 212
McDonald, A., 130, 135, 138, 165, 167
McGregor, D. F. M., 119, 121, 123
Medvedev, Z. A., 204, 212
Meira Filho, L. G., 325
Meyers, S., 171, 213
Micklin, P., 156, 167
Miller, E. W., 211, 212
Miller, R. M., 211, 212
Millero, F. J., 297, 325
Mining Magazine, 223, 246
Mintzer, I. M., 298, 299, 304, 325
Mitchell, J. F. B., 296, 301, 304, 325
Mitchell, W. A., 102, 123
Moore, J. M., 128, 167, 168
Morehouse, W., 262, 286
Mossman, B. T., 251, 286
Mounfield, P. R., 206, 212
Murphy, C., 252, 254, 286
Musa, S. B., 124

Myles, D., 97, 98, 123

Newbury, C., 19, 32
Nisbet, E. G., 323, 325
Nisbet, J., 325

O'Brien, T. F., 234, 246
O'Conner, R., 106, 124
O'Hara, S. L., 66, 70
Orange County Landfill Authority, 286
Organization for Economic Cooperation and Development, 249, 286
Ornstein, R., 21, 26, 31, 32

Palm, R., 121, 124
Park, C. C., 316, 317, 325
Park, C. F., 223, 232, 246
Patterson, C. C., 322, 323, 325
Pearce, F., 177, 212
Pedersen, G., 10, 32
Perlman, H. A., 167
Peters, W. C., 239, 246
Pierce, R. R., 167
Pilkey, O. H., 113, 123
Pimentel, D., 52, 70
Pinneker, E. V., 145, 167
Population Reference Bureau, 1, 3, 13, 32
Prasad, S. S., 323, 326
Price, M., 155, 167
Pyne, S. J., 317, 325

Rabbinge, R., 69, 70
Radojevic, M., 284, 286
Rapp, G., Jr., 123
Rathje, W., 252, 254, 286
Reddy, A. K. N., 212
Reich, R. B., 21, 26, 32
Reisner, M., 145, 167
Remnick, D., 168, 212
Repetto, R., 317, 317, 325
Reyburn, W., 279, 286
Rhodes, R., 200, 206, 212
Rifkin, J., 55, 70
Rodale's Organic Gardening, 279, 286
Rodda, J. C., 167
Rogers, J. J. W., 78, 124, 289, 325
Rosenzweig, M. L., 317, 325

Ross, M., 251, 286
Roxburgh, I. S., 274, 286

Sadava, D. E., 52, 70
Samatar, I. A., 50, 70
Sarre, P., 17, 31, 32
Schipper, L., 171, 213
Scientific American, 179, 213
Seaton, A., 286
Selley, W. B., 135, 137, 167
Settle, D. M., 322, 323, 325
Shaw, E. M., 145, 150, 155, 167
Sheldon, I. M., 69, 70
Shell Oil Co., 200, 213
Shipes, H. R., 223, 246
Sitar, N., 82, 123
Skinner, B. J., 246
Sloan, L. C., 303, 325
Smith, K., 78, 124
Smith, Z. A., 155, 165, 167
Sohn, M. L., 297, 325
Speidel, D. H., 138, 167, 297, 326
Spellerberg, I. F., 317, 326
Steele, O. P., 325
Stegner, W., 145, 167
Stone, R., 204, 213
Street-Perrott, F. A., 70
Subramaniam, M. A., 262, 286
Sullivan, L. R., 177, 213

Tans, P. P., 325
Thomas, G., 94, 124
Thomas, P. J., 123
Thompson, D. A., 119, 121, 123
Tobin, R., 317, 326
Todd, D. K., 126, 167
Toniolo, E., 101, 124
Toussaint-Samat, M., 34, 70
Troise, F. L., 126, 167
Tyler, P. E., 177

United Nations *Demographic Yearbook*, 1, 15, 32
United Nations Development Programme, 16, 32
United Nations Environmental Programme, 16, 32, 274, 286

United Nations Food and Agriculture Organization, 34, 37, 47, 48, 49, 57, 60, 62, 70
United Nations *Statistical Yearbook*, 1, 20, 23, 26, 32
United States Energy Information Administration, 189, 208, 211, 213
United States Environmental Protection Agency, 265, 286
United States *Statistical Abstract*, 1, 23, 26, 32

Van Andel, T. H., 73, 78, 124
van der Leeden, F., 126, 160, 167
Vartanov, R. V., 265, 286
Vaughan, D., 246
Vitek, J. D., 113, 123
Viviani, M., 237, 246
von Braun, J., 19, 32

Walsh, R. P. D., 101, 124
Walton, J., 146, 167
Wannenburgh, A., 12, 32
Waterbury, J., 125, 142, 167
Wells, P. G., 195, 213
White, S. E., 155, 167
Whitten, R. C., 323, 326
Wild, A., 52, 70
Williams, N., 204. 213
Williams, R., 212
Witts, M. M., 94, 124
Wolfe, J. A., 223, 246
Woodside, G., 265, 286
World Conservation Monitoring Centre, 310, 317, 326
World Energy Council, 211, 213
World Resources Institute, 1, 3, 13, 32, 208, 213, 316, 326

Yergin, D., 180, 187, 213
Young, G. J., 165, 167
Youngquist, W., 245, 246

Zahnle, K. J., 123

SUBJECT INDEX

Aborigines (in Australia), 306
accessory rights to ores, 240
acid rain, 280
Afghan nomads, 9–10, 9
age distribution, 14–5
ages of people (prehistory to present), 215
agriculture
  environmental issues, 63–5
  money, 65–8
air conditioner, 319–20
air pollution, 279–84
  cities with problems, 282–3
aldicarb, 260
alpha particle, 200
amino acids, 37
angle of repose, 116, 118
animal production, 52–5, 57
  regional preferences, 53–5
animal raising, 53–5
  in pastures, 6–7, 53–6
  in pens, 53
anoxic decay, 180–2
Antarctic, 320–2
Antarctic bottom water (AABW), 296
anticline, 182–3
aquiclude, 147
aquifer, 145
Arabian Gulf, 184–5
Aral Sea, 156
artesian system, 150–2
asbestos, 251
Asbestos Hazard Emergency Response
  Act (AHERA), U.S., 251
Asian countries, 16–7
  not Islamic, 16–7
Aswan High Dam, Egypt, 142
atmosphere
  composition, 288, 291–3

controls, 288–97
history, 288–97
Australia, 313

banded iron formation (BIF), 232
Bangladesh family planning, 30
barrier island, 109–10
Basel Convention on hazardous waste, 265
bauxite, 231
bay, 109
beach, 109–10
beach replenishment, 113
beneficiation of ores, 236–8
Bessemer process, 237–9
beta rays, 200
Bhopal, India, 262
biodegradable pesticides and
  herbicides, 259
biologically active waste, 274–5
biologically mediated mineral deposits, 231–2
birthrate, 11–16
Black triangle, eastern Europe, 281–2
black-body radiation, 295
blitzkrieg and megafauna extinction, 306
blue clay, 118
Bolivia, 234
breakwater, 113
Brent Spar, 264
British thermal unit (Btu), xxi–xxii, 170–1
Bronze Age, 215
building stone, 224–5, 228
bulk waste, 249–54
  burning, 250
  history, 249–50

landfill (see landfill)
recycling, 250–3
Bushmen, 12
Butte, Montana, 242–3

California–Mexico border, 24
calorie, xxi–xxii, 41, 170–1
Canada
  energy consumption, 208
  energy production, 208
  gas production, 189
  gas reserves, 189
  land use, 3
  oil production, 189
  oil reserves, 189
  population density, 3
  surface-water supply, 135
carbamates, 259–60
carbaryls, 259–61
carbohydrates, 34
carbon dioxide, 291–3
  in atmosphere, 291–3
  in oceans, 291–3
  reservoirs, 291–3
  transfer rates, 291–3
carbonate solubility in oceans, 292
carboxyl, 37
carrying capacity, 26–7
cash crops, 68
Caspian Sea, 192–3
catalysis, 220
cellulose, 34
centigrade scale, 169
Central American and Caribbean
  countries, 16–17
Chernobyl, Ukraine, 204
  Belarus contamination, 204
  nuclear accident, 204
Chile, 234

China
    birth control, 29
    Three Gorges dam (*see* Three
        Gorges dam, China)
chlordane, 259
chlorofluorocarbons (CFCs), 318–21
    effect on ozone, 319–20
chlorophyll, 33
cholesterol, 38
clay, 225
climate
    controls, 293–7
    effect of human activity, 301–2
    future changes, 302–4
    history of past 1000 years, 297–300
    history throughout geologic time,
        288–9
Coal Age, 215
coal, 197–9
    consumption, 198–9
    geology, 197–9
    production, 198–9
    reserves, 198–9
    smoke, 280
    transportation, 198–9
coastal erosion, 104–13
coastal flooding, 104–13
coastline, 104–13
    effects of human activity, 111–13
    features, 109–10
    risk of living along, 106–12
Colorado River, U.S., 140–5
combustion of organic matter, 171–3,
        180–2
commons, 52–3
condensation of water, 128
cone of depression (*see* drawdown
        around water wells)
conservation of energy, 74
continental margins, 75–6, 109
        active, 75–6, 109
        passive, 75–6, 109
continental-basin mineral deposits,
        229–31
convection in earth's mantle, 74
Convention on International Trade in
        Endangered Species (CITES),
        311–13
Copper Age, 215
core of earth, 74
corn (maize), 45
    chemical composition, 45
cosmic ray, 268
creep, 113–14
Creutzfeldt-Jakob disease, 56
critical mass, 202
crop production, 40–52
crop requirements, 42–5
        light, 42

nitrogen, 45
nutrients, 43–5
phosphorus, 45
potassium, 45
water, 42–3
cropland, 4, 6–7
crushed rock, 225
crust of earth, 74
crystal segregates as mineral deposits,
        226
curie, 267
cyclodienes, 259, 259

dams, 141–3, 175–7
DDT, 257–8
death rate, 11–14
debris avalanche, 113
deforestation, 313–17
delta, 109
demographic transition, 14–15
desalination, 132, 313–17
desert regions, 131, 133
dioxins, 257
direct energy, 171–3
disasters, 78
        worldwide cost, 119–21
doubling time, 5–8
drainage basin, 139–41
drainage divide, 139–41
drawdown around water wells, 151–3
drinking water, 155–9
        bacteria, 155–7
        chlorinated hydrocarbons, 157
        hardness, 159
        iron, 159
        lead, 157–9
        parasites, 155–7
        salt content, 159
        volatile hydrocarbons, 157
dunes, 109–10

earthquake damage, 83–7
        disease, 87
        disruption of services, 87
        fire, 86–7
        ground displacement, 86
        landslides, 87
        liquefaction, 87
        shaking and collapse, 86
        subsidence, 86
        tsunamis (*see* tsunami)
earthquakes, 82–9
        causes, 83
        faults, 83, 87
        frequency–magnitude relationship,
                83–5
        kinds, 83
        Richter magnitude, 83
        risk analysis, 86–9

Easter Island, 307
Eastern Europe, 256, 281–2
education in lesser-developed
        countries, 17–20
Einstein equation for equivalence of
        mass and energy (*see* mass-energy
        conversion)
elastic limit, 83
elastic strain, 83
electricity, 173–7
        generator, 173–4
        production, 209–10
electromagnetic radiation, 294–6
        spectrum, 295
employment in lesser-developed
        countries, 17–20
endangered species, 312–3
        Endangered Species Act (U.S.), 313
endothermic process, 171
energy, 40–2, 169–71
        in human nutrition, 40–2, 62
        relationship to population density,
                11
        non-food sources, 171–9
        value of foods, 40–2
energy consumption, 206–11
        future changes, 209–10
        present use, 207
enzymes, 38–40
estrogen, 38–9, 275–6
estrogen-mimicking compounds, 275–6
Euphrates River, Iraq, 102, 107–8
eutrophication, 277–9
evaporation of water, 128, 132, 230
Everglades, Florida, 158–9
Exclusive Economic Zone (EEZ),
        58–9
exothermic process, 172
exponential curve, 5–11
external heat engine of the earth, 73
extinction, 305–17
        rate, 310–11
Exxon Valdez, 194–5

Fahrenheit scale, 169
fats, 34–7
faults (*see* earthquake faults)
ferrous metals, 224
fertility rate, 11–14
fertilizer, 45, 224
fibers, 51–2
firestick farming, 313
fish, 55–63
        food chain, 55–9
        nurture, 55–9
        production, 59–61
        reproduction and growth, 55–9
fishing, 59–61
        industrial, 59–61

fission products, 269–70
    mass distribution, 269–71
flood insurance (*see* National Flood
    Insurance Program)
floodplain, 98–9
Florida panther, 309
food pyramid, 40
forests and woodlands, 5–7, 313–17
    Europe, 315–16
    new growth, 313
    North America, 313, 316
    old growth, 315
    tropical, 315–16
fossil fuel, 180–200
Freedom Food, 56
Fresh Kills, New York, 252
freshwater, 127–30
    annual water flux, 129
    budget, 127–30
    consumption, 135–8
    diurnal cycle, 128–9
    human use, 131–8
    latitudinal variation, 130–11
    regional availability, 130–11, 137
    sources, 132–5
fruits and berries, 51

gaining stream, 150–1
gamma ray, 201
Ganges Delta, Bangladesh, 106
gas (*see* natural gas)
geothermal energy, 173
geothermal water, 173, 227
Germany in Second World War, 186–7
Gilgamesh, 107–8
glacial periods, 288–9
glaciation, 289–90
global warming, 301–4
    evidence, 301–2
    possible consequences, 302–4
glowing clouds, 90–1
glucose, 34, 36
glue, 256–7
Golden Apple Snail, 64
Golden fleece (*see* Jason and the
    Argonauts)
goodies (herbs, spices, and other
    flavorings), 51
grain production, 45–9
    animal feed, 48, 53
    European history, 46
    trade and aid, 48–9
    world history, 46–9
Grand Banks, Newfoundland, 58–9
Great Flood (biblical), 107–8
Great Man-Made River, Libya, 136
green revolution, 65
Green River lake, western U.S., 303
greenhouse effect, 296

greenhouses in climate history, 288–90
Greenland ice cores, 297, 321–2
ground nuts, 138
groundwater, 145–55
    extraction, 150–1
    flow velocity, 149–9
growth rate of populations, 11–14
Gulf of Mexico, 184
Gulf of Thermakos, Greece, 76

Hadley cells, 130–1
Haiti, 317
hazardous chemicals, 254–61
heat, 169–71
    of combustion, 41, 180–2
herbicides, 257–61
Hidrovia project, South America, 134
High Plains aquifer, 151–5
Hiroshima, Japan, 268
hormones, 38–40
horsepower, 171
Huascaran, Peru, 116
human metabolism, 34–40
human nutrition, 34–40
    animals and fish, 61–3
    calcium, 37–8
    energy (*see* energy in human
      nutrition)
    fiber, 34
    iodine, 38
    iron, 38
    minerals, 37–8
    phosphorus, 37–8
    vitamins, 38–40
humidity, 128–9
hunting, 53
hurricanes, 109–12
hydraulics, 149
    conductivity, 149–50
    gradient, 149
    head, 147–9
hydrocarbons (*see* oil and gas)
hydrochlorofluorocarbons (HCFCs),
    321
hydrogen
    atmospheric, 288
    energy storage system, 179
hydrologic cycle, 127–8
hydropower, 174–5
hydrothermal mineral deposits, 227–8

icehouses in climate history, 288–90
Iceland, 313
immiscible-sulfide mineral deposits,
    226
income in lesser-developed countries,
    20
Indonesia, 184, 186
industrial chemicals, 255–7

industrial countries (*see* Organization
    for Economic Cooperation and
    Development)
industrial minerals, 225
Industrial Revolution, 11, 322
industrial rocks, 225
infrared (IR), 295–6, 301
Intergovernment Panel on Climate
    Change (IPCC), 301–2
internal combustion engine, 172
internal heat engine of the earth, 73
ionization by radiation, 267
Irian Jaya, Indonesia, 222–3
Iron Age, 215
irrigation, 135
Islamic countries, 17
islands and extinctions, 310
Israel, 162

Japan
    oil in the Second World War,
      186–7
    population, 3, 4
Jason and the Argonauts, 216
Johnstown flood, Pennsylvania, 105–6
Jordan, 162
Jordan River, 162
joule, xxi–xxii, 171

Kalahari desert, 12
Kazakhstan, 156, 192–3
Khartoum, Sudan, 101
Kilauea, Hawaii, 95
kilocalorie, xxi–xxii, 41, 171
kilowatt, xxi–xxii, 171
kilowatt-hour, xxi–xxii, 171
kinetic energy, 73–4, 147–8
Kobe, Japan, 83–5

lagoon, 109
Lake Nyos, Cameroon, 91–3
Lake Okeechobee, Florida, 158–9
Lake Patzcuaro, Mexico, 66–7
lake storage of energy, 179
land use, 3–4
landfill, 250, 253
landslide, 113–19
    causes, 116–18
    human influences, 118–19
lava, 90–2
Law of the Sea, 232
lead
    blood level, 157
    concentration in water, 157–9
    environmental concentration,
      321–3
    industrial use, 321
    medical effects, 157
    Roman use, 322–3

lead (*cont.*)
    tetraethyl in gasoline, 322
leaf vegetables, 51
legumes, 49–51
lesser-developed countries, 17–21
life expectancy, 13–15
light metals, 224
lipids (*see* fats)
lithosphere, 74–6
Little Climatic Optimum, 299
Little Ice Age, 299
Loma Prieta earthquake, San
        Francisco, 81–2
longshore drift, 112–3
Lord Howe Island, South Pacific, 306
losing stream, 150–1

Mad Cow Disease (bovine spongiform
        encephalopathy; BSE), 56
magma, 89–90
magmatic mineral deposits, 225–7
Mammoth Mt., California, 95–6
mammoths, 306
manganese nodules, 232
mantle of the earth, 74
    plumes, 74
Maori, 307
mass–energy conversion, 201
maturation of organic matter, 180
Mauna Loa, Hawaii, 299
mechanized farming, 41
megafauna, 305–12
metals
    crustal abundance, 217–18
    economic cut-off values, 217–18
methane (*see* natural gas)
methane in atmosphere, 293–4
methyl isocyanate, 259, 262
microwave, 295
mid-ocean ridge, 75
mineral deposits
    classification, 223–5
    location, 233–5
    modes of formation, 225–32
    types, 225–32
mineral reserves, 218–20
    inventories, 218–20
    possible, 219–20
    proven, 219–20
    scarcity, 217–18
    undiscovered, recoverable, 220
mineral resources
    discovery risk, 221–3
    future prospects, 245
    inequality of distribution, 233
    legal issues, 240–5
    location, 233–5
    U.S. public lands, 241–5
Mining Law of 1872 (U.S.), 244

mining methods, 235–9
    history, 239
Mississippi River flood, 99
Missouri-Valley-type (MVT) mineral
        deposits, 229–31
moa, 305
monocultures
    food, 68
    mineral production, 234–5
Montreal Protocol on ozone, 320–1
mother lode, 229
Mt. Pelée, Martinique, 93–5
Mt. Pinatubo, Philippines, 72–3, 300
Mt. St. Helens, 71–2, 93

Nagasaki, Japan, 268
National Flood Insurance Program
        (NFIP) in the U.S., 120–1
natural disasters, 119–21
    costs, 119–21
    government action, 119–21
    insurance, 119–21
    responsibility for cost, 119–21
natural gas, 180
natural hazards and risk, 76–8
natural levee, 98
Nauru Island, South Pacific, 237
nekton, 57
Neolithic Age, 215
Nevado del Ruiz, Colombia, 95
New Orleans, Louisiana, 106
Nile River, 2, 101
nitrate, 234
nitrogen in atmosphere, 288
nonferrous (base) metals, 224
nongrain plants, 49–52
nonsustainable resources, 215–17
North Atlantic deep water (NADW),
        296–7
North Sea, 184–6
North Slope, Alaska, 184, 191
Nuclear Age, 215
nuclear fission, 201–2
nuclear fusion, 202
nuclear power, 200–6
nuclear reactors, 202–5
    fission (breeder), 205
    fission (burner), 202–3
    fuel, 205–6
    fusion, 202
    safety, 203
nuclear weapons, 205
nuées ardentes (*see* glowing clouds)

ocean disposal, 264
oceanic conveyor belt, 296–7
offshore bar (*see* barrier bar)
oil and gas, 180–97
    consumption, 189–93

exporting countries, 189–93
geology, 182–5
hydrocarbon varieties, 181
importing countries, 189–93
oil and gas basins, 183–8
oil and gas fields, 184–6
production, 189–93
regional distribution, 187–9
reserves, 187–9
reservoirs, 182–3
transportation, 189–93
traps, 182–3
world abundance, 187
oil shale, 193–7
oils as food from plants, 51
Ok Tedi, New Guinea, 222–3
open-pit mining, 235, 239
ore (*see* mineral resources)
organic compounds, 34
    benzene rings, 35
    chains, 35
    rings, 35
organic formulas, 35
organic gardening, 45
organic solvents, 255
Organization for Economic
        Cooperation and Development
        (OECD), 17, 100, 264
organophosphates, 258–9
overgrazing, 63
oversteepening of slopes, 118–19
Owens Valley, California, 146
oxygen in atmosphere, 288
oxygen isotopes, 289
ozone, 282, 318–21
    hole, 318–20
    pollution near ground, 282
    ultraviolet shield in stratosphere,
        318

paint, 256–7
paleoclimate models, 303
Paleolithic Age, 215
Papua New Guinea, 222–3
parabolic mirror, 174, 177
parathion, 259
pastoralism, 53
pasture, 5–7
perched water table, 151–2
permeability, 145, 148
Peru, 234
pesticides, 257–61
    alternatives, 260–1
Petroleum Age, 215
pH, 292
phosphate minerals, 231–2, 237
photoelectric cell (*see* photovoltaic
        cell)
photolysis, 319

photosynthesis, 33, 291
photovoltaic cell, 174–8
placer mineral deposits, 228–9
plankton, 55
plants
    C3, 42
    C4, 42
    metabolism, 41
    nutrition, 41
plastics
    types used in households, 252–4
plate tectonics, 74–6
plumes (*see* mantle plumes)
plutonium–239, 202, 205, 267–9
polar front, 131
pollutant dispersal, 248–9
polychlorinated biphenyls (PCBs),
    255–7
polymers, 34
polysaccharides, 34
population
    control, 28–9
    density, 3–4
    growth, 5–11
    history, 5–11
porosity, 145
potential energy, 73–4, 147–8
power, 169–71
precious metals, 224
precipitation of water, 129, 132
preservatives, 256–7
Prince William Sound, Alaska, 194–5
proteins, 37
pumice, 226–7
pyrethrins, 259–60

quad (quadrillion Btu), 170–1
quality of life, 27–8
quarrying, 235, 238
quotas on food imports, 67–8

rad (radiation absorbed dose), 267
radiation, 267–9, 296
    absorption in atmosphere, 294–6,
        301
    balance of earth, 295–6
    dose–response curve, 267–8
    natural background, 268–9
    threshold, 268–70
radioactive decay, 200–1
    daughter element, 201
    half life, 201, 271
    parent element, 201
radioactive waste, 267–74
    danger, 267–9
    high level, 269–73
    low level, 273–4
    permanent storage, 272
    retrievable storage, 272

underground storage, 272–3
    ocean disposal, 273
rainfall, 42–3
    world distribution, 43
rapidly industrializing countries of east
        Asia, 16–17
reclamation of fresh water, 133
recycling of minerals, 217
regalian rights to ores, 240
rem (roentgen equivalent man), 267
reprocessing of nuclear waste, 270–1
residence times of water, 129–30
residue from metal industry, 257
retorting of oil shale and tar sand,
        195–6
Rhine River, 266
risk, 78–81
    cyclicity, 79
    individual, 78
    predictability, 79
    societal, 78–9
river floods, 98–104
    cyclical, 98–9
    dam failure, 104–6
    discharge, 100–2
    human influences, 104
    random (unpredictable), 99–104
    recurrence interval, 102–3
    risk, 103
    seasonal, 98–9
    urbanization effects, 104
river terraces, 98–100
rock phosphate (*see* phosphate
        minerals)
roentgen, 267
roll-front uranium deposits, 231
rotation of animals and crops, 53
ruminants, 34
Rwanda, 19

salt dome, 183
saltwater intrusion of aquifers, 151–3
San Francisco earthquakes, 81–2, 85
Santorini, Aegean Sea, 80
sarin, 258–9
saturated fats (*see* fats)
sea level and glaciation, 289–90
seawall, 113
sedimentary basins, 183–6
sedimentary mineral deposits, 226–9
seismicity, 86–9
seismograph, 83
sewage, 274–7
    biochemical oxygen demand, 276
    composition, 277
    nitrogen content, 277
    phosphorus content, 277
    treatment plants, 276–7
Simpson Index, 307

slope stability, 116–18
slump, 115
smog, 280–3
soil, 43–5
    erosion, 63–5
    formation, 43–5
    profiles, 44
solar energy
    active, 174–7
    passive, 172–3
Somalia, 50
South American countries, 17
South Korea, 31
Soviet pipelines, 294
specialty metals, 215, 224
species, 307–13
    definition, 307
    numbers of, 307
species diversity, 307–13
    dangers of restriction, 308
    indices, 307
spectral absorption of radiation, 294–6
Spotted Owl, 314
Spratly Islands, South China Sea, 190
starch, 34
steel, 237–9
steroids, 38–9
storage battery, 177–8
stored energy, 177–9
Straits of Hormuz, Arabian Gulf, 191
stream gauge, 100
strip mining, 235, 238
strontium–90, 267
subduction, 75
sub-Saharan Africa, 16–17
subsidies of food production, 65–7
subsistence agriculture, 41
subtropical convergence, 130
sugar, 34, 51
superphosphate, 45
surface water, 138–45
    extraction, 141–3
    runoff, 139–41
surficial mineral deposits, 226, 231
sustainable resources, 215–17
sustained yield of forests, 315
Syria, 162

Taiwan, 311
Tambora, Indonesia, 91, 300
tar sand, 193–7
tariffs on food imports, 67–8
temperature, 169–71
tephra, 90
testosterone, 38–9, 275–6
Thera (*see* Santorini)
thermal infrared, 296
    absorption by gases, 296
Three Gorges dam, China, 176–7

Tigris River, Iraq, 107–8
toxic chemicals (*see* hazardous
    chemicals)
trade winds, 131, 293
transuranic elements, 269–70
transform plate margin, 75
transpiration, 129
Troy, 249
tsunami, 87, 96–7
tubers, 49
Tunguska, Siberia, 77

ultraviolet (UV), 295, 318
    dangers, 318
underground mining, 235–9
United Kingdom
    animal production, 57
    birth rate, 13
    fertility rate, 13
    fertilizer use, 47
    gas production, 189
    gas reserves, 189
    grain production, 47, 48
    grain trade and aid, 49
    land use, 3, 5
    life expectancy, 13
    nutrition energy, 62
    oil production, 189
    oil reserves, 189
    population density, 1, 3, 48
United States,
    animal production, 57
    birthrate, 13, 23
    education, 23
    electricity production, 210
    energy consumption, 208
    energy production, 208
    fertility rate, 13, 23, 25–6
    fertilizer use, 47
    gas production, 189
    gas reserves, 189
    grain production, 47–8
    grain trade and aid, 49
    immigration, 26
    income, 25–6
    land use, 3

life expectancy, 13, 23
life-styles, 21–5
nutrition energy, 62
occupations and employment,
    21–5
oil consumption history, 195
oil production, 189, 195
oil reserves, 189
population characteristics, 20–6
population density, 3
population history, 21–2
races and ethnic groups, 25–6
water consumption, 137
wealth, 23–6
units (definitions and equivalents),
    xxi–xxii, 170, 171
unsaturated fats (*see* fats)
uranium–235, 200, 202
uranium–238, 200, 202
Uzbekistan, 156

Vietnam, 64
Vistula River, Poland, 281–2
vitamins, 38–40
    vitamin D, 38–40
volcanic eruption, 89–96
    ash, 90
    blocks, 90
    gas, 90
    hazards, 91–3
    prediction, 93–6
    risk, 93–6
    tephra, 90–3
volcanogenic massive sulfide mineral
    deposits, 228
Vostok ice core, 297–8

waste disposal methods, 261–5
wasteland, 5–7
waste oil, 255
waste treatment plants (*see* sewage)
water, 126–31
    control and allocation, 160–5
    global abundance, 126–7
    global distribution, 126–7
    pressure, 148

saturation in air, 128–9
water ownership, 161–5
    absolute territorial integrity, 164
    absolute territorial sovereignty,
        163
    beneficial use, 164–5
    common jurisdiction, 164
    equitable utilization, 164
    groundwater, 164–5
    reasonable use, 164–5
    riparian doctrine, 163
    surface water, 161–4
water quality, 155–60
    animal use, 159–60
    boilers for electricity production,
        159–60
    cooling water, 159–60
    drinking water (*see* drinking water)
    irrigation, 159–60
water table, 145, 151–3
water wells, 150–1
watershed (*see* drainage basin)
watt, xxi–xxii, 171
wealth in lesser-developed countries,
    relationship to birthrate, 15–17
    relationship to energy consumption,
        206–7
    relationship to population, 15–17
West Siberia, 184
westerlies, 293, 131
Western Europe
    energy consumption, 208
    energy production, 208
Williston basin, 184–6
wind and earth's rotation, 131, 293
women in lesser-developed countries,
    20–1
wood as energy source, 172

x-rays, 295

Yangtze River, China (*see* Three
    Gorges Dam)
Year Without a Summer, 300

Zimbabwe, 313